Angiogenesis in Cancers

Angiogenesis in Cancers

Editor

Domenico Ribatti

MDPI • Basel • Beijing • Wuhan • Barcelona • Belgrade • Manchester • Tokyo • Cluj • Tianjin

Editor
Domenico Ribatti
Università degli Studi di Bari
Italia

Editorial Office
MDPI
St. Alban-Anlage 66
4052 Basel, Switzerland

This is a reprint of articles from the Special Issue published online in the open access journal *Cancers* (ISSN 2072-6694) (available at: https://www.mdpi.com/journal/cancers/special_issues/Angiogenesis_Cancers).

For citation purposes, cite each article independently as indicated on the article page online and as indicated below:

LastName, A.A.; LastName, B.B.; LastName, C.C. Article Title. *Journal Name* **Year**, *Volume Number*, Page Range.

ISBN 978-3-0365-6916-1 (Hbk)
ISBN 978-3-0365-6917-8 (PDF)

© 2023 by the authors. Articles in this book are Open Access and distributed under the Creative Commons Attribution (CC BY) license, which allows users to download, copy and build upon published articles, as long as the author and publisher are properly credited, which ensures maximum dissemination and a wider impact of our publications.

The book as a whole is distributed by MDPI under the terms and conditions of the Creative Commons license CC BY-NC-ND.

Contents

About the Editor vii

Preface to "Angiogenesis in Cancers" ix

Roberto Ronca, Sara Taranto, Michela Corsini, Chiara Tobia, Cosetta Ravelli, Sara Rezzola, et al.
Pentraxin 3 Inhibits the Angiogenic Potential of Multiple Myeloma Cells
Reprinted from: *Cancers* **2021**, *13*, 2255, doi:10.3390/cancers13092255 1

Maryam Nakhjavani, Eric Smith, Kenny Yeo, Helen M. Palethorpe, Yoko Tomita, Tim J. Price, et al.
Anti-Angiogenic Properties of Ginsenoside Rg3 Epimers: In Vitro Assessment of Single and Combination Treatments
Reprinted from: *Cancers* **2021**, *13*, 2223, doi:10.3390/cancers13092223 13

Sarah Sabatasso, Cristian Fernandez-Palomo, Ruslan Hlushchuk, Jennifer Fazzari, Stefan Tschanz, Paolo Pellicioli, et al.
Transient and Efficient Vascular Permeability Window for Adjuvant Drug Delivery Triggered by Microbeam Radiation
Reprinted from: *Cancers* **2021**, *13*, 2103, doi:10.3390/cancers13092103 33

Catarina Nascimento, Andreia Gameiro, João Ferreira, Jorge Correia and Fernando Ferreira
Diagnostic Value of VEGF-A, VEGFR-1 and VEGFR-2 in Feline Mammary Carcinoma
Reprinted from: *Cancers* **2021**, *13*, 117, doi:10.3390/cancers13010117 49

San-Hai Qin, Andy T. Y. Lau, Zhan-Ling Liang, Heng Wee Tan, Yan-Chen Ji, Qiu-Hua Zhong, et al.
Resveratrol Promotes Tumor Microvessel Growth via Endoglin and Extracellular Signal-Regulated Kinase Signaling Pathway and Enhances the Anticancer Efficacy of Gemcitabine against Lung Cancer
Reprinted from: *Cancers* **2020**, *12*, 974, doi:10.3390/cancers12040974 63

Diana Papiernik, Anna Urbaniak, Dagmara Kłopotowska, Anna Nasulewicz-Goldeman, Marcin Ekiert, Marcin Nowak, et al.
Retinol-Binding Protein 4 Accelerates Metastatic Spread and Increases Impairment of Blood Flow in Mouse Mammary Gland Tumors
Reprinted from: *Cancers* **2020**, *12*, 623, doi:10.3390/cancers12030623 85

Fatema Tuz Zahra, Md. Sanaullah Sajib and Constantinos M. Mikelis
Role of bFGF in Acquired Resistance upon Anti-VEGF Therapy in Cancer
Reprinted from: *Cancers* **2021**, *13*, 1422, doi:10.3390/cancers13061422 107

Sofia S. Pereira, Sofia Oliveira, Mariana P. Monteiro and Duarte Pignatelli
Angiogenesis in the Normal Adrenal Fetal Cortex and Adrenocortical Tumors
Reprinted from: *Cancers* **2021**, *13*, 1030, doi:10.3390/cancers13051030 125

Barbara Muz, Anas Abdelghafer, Matea Markovic, Jessica Yavner, Anupama Melam, Noha Nabil Salama, et al.
Targeting E-selectin to Tackle Cancer Using Uproleselan
Reprinted from: *Cancers* **2021**, *13*, 335, doi:10.3390/cancers13020335 139

Nils Ludwig, Dominique S. Rubenich, Łukasz Zareba, Jacek Siewiera, Josquin Pieper, Elizandra Braganhol, et al.
Potential Roles of Tumor Cell- and Stroma Cell-Derived Small Extracellular Vesicles in Promoting a Pro-Angiogenic Tumor Microenvironment
Reprinted from: *Cancers* **2020**, *12*, 3599, doi:10.3390/cancers12123599 155

Antonio Giovanni Solimando, Simona De Summa, Angelo Vacca and Domenico Ribatti
Cancer-Associated Angiogenesis: The Endothelial Cell as a Checkpoint for Immunological Patrolling
Reprinted from: *Cancers* **2020**, *12*, 3380, doi:10.3390/cancers12113380 171

Panagiotis Ntellas, Leonidas Mavroeidis, Stefania Gkoura, Ioanna Gazouli, Anna-Lea Amylidi, Alexandra Papadaki, et al.
Old Player-New Tricks: Non Angiogenic Effects of the VEGF/VEGFR Pathway in Cancer
Reprinted from: *Cancers* **2020**, *12*, 3145, doi:10.3390/cancers12113145 195

Antonio Giovanni Solimando, Tiziana Annese, Roberto Tamma, Giuseppe Ingravallo, Eugenio Maiorano, Angelo Vacca, et al.
New Insights into Diffuse Large B-Cell Lymphoma Pathobiology
Reprinted from: *Cancers* **2020**, *12*, 1869, doi:10.3390/cancers12071869 223

Harman Saman, Syed Shadab Raza, Shahab Uddin and Kakil Rasul
Inducing Angiogenesis, a Key Step in Cancer Vascularization, and Treatment Approaches
Reprinted from: *Cancers* **2020**, *12*, 1172, doi:10.3390/cancers12051172 245

About the Editor

Domenico Ribatti

Domenico Ribatti was awarded his M.D. degree in October 1981, with full marks. In 1983, Dr. Ribatti joined the Medical School as Assistant at the Institute of Human Anatomy, University of Bari. In 1984, he chose his specialization in Allergology. In 1989, he spent one year in Geneva, working at the Department of Morphology. In 2008, he received an honoris causa degree in Medicine and Pharmacy from the University of Timisoara (Romania). His present position is full professor of Human Anatomy at the University of Bari Medical School. Domenico Ribatti is the author of 867 publications, as reported in PubMed (scopus "h" index: 95), and of 15 monographs.

Preface to "Angiogenesis in Cancers"

The interaction between neoplastic cells and blood vessels, both newly formed during angiogenesis or pre-existing normal vessels, is one of the fundamental biological events involved in the development and progression of most solid and hematological tumors and the formation of metastases. Tumor angiogenesis is viewed as the consequence of an angiogenic switch, i.e., a genetic event that endows the tumor with the ability to recruit blood vessels from the neighboring tissue. The newly formed tumor blood vessels have specific characteristics that allow discrimination from resting blood vessels. They are characterized by rapid proliferation, increased permeability, and disorganized architecture. Initially thought to be a must for the growth and progression of tumors, the formation of new vessels was regarded as one of the hallmarks of cancer. However, this has turned out not to be the case, as it was discovered that tumors can also grow without neo-angiogenesis, mainly by co-opting pre-existing vessels but also through vascular mimicry. Since its discovery by Dr. Judah Folkman, tumor angiogenesis has been proposed as a target for novel tumor therapies. However, the success in the clinic of anti-angiogenic compounds has been limited in contrast to many preclinical results obtained in animal models. This is in part due to the fact that tumors can be non-angiogenic and in part due to several newly discovered mechanisms of resistance due to the biology of both the cancer cells and of the endothelium.

Domenico Ribatti
Editor

Article

Pentraxin 3 Inhibits the Angiogenic Potential of Multiple Myeloma Cells

Roberto Ronca [1,*], Sara Taranto [1], Michela Corsini [1], Chiara Tobia [1], Cosetta Ravelli [1], Sara Rezzola [1], Mirella Belleri [1], Floriana De Cillis [2], Annamaria Cattaneo [2,3], Marco Presta [1] and Arianna Giacomini [1,*]

[1] Department of Molecular and Translational Medicine, University of Brescia, 25123 Brescia, Italy; sara.taranto@unibs.it (S.T.); michela.corsini@unibs.it (M.C.); chiara.tobia@unibs.it (C.T.); cosetta.ravelli@unibs.it (C.R.); sara.rezzola@unibs.it (S.R.); mirella.belleri@unibs.it (M.B.); marco.presta@unibs.it (M.P.)

[2] Biological Psychiatry Unit, IRCCS Istituto Centro San Giovanni di Dio Fatebenefratelli, 25125 Brescia, Italy; fdecillis@fatebenefratelli.eu (F.D.C.); acattaneo@fatebenefratelli.eu (A.C.)

[3] Department of Pharmacological and Biomolecular Sciences, University of Milan, 20122 Milan, Italy

* Correspondence: roberto.ronca@unibs.it (R.R.); arianna.giacomini@unibs.it (A.G.)

Simple Summary: Bone marrow (BM) angiogenesis represents a key aspect in the progression of multiple myeloma (MM) and is strictly linked to the balance between pro-angiogenic and anti-angiogenic players produced by both neoplastic and stromal components. It has been shown that Fibroblast Growth Factors (FGFs) play a pivotal role in the angiogenic switch occurring during MM progression. Accordingly, the natural FGF antagonist Long Pentraxin 3 (PTX3) is able to reduce the activation of BM stromal components induced by FGFs. This work explores, for the first time, the anti-angiogenic role of PTX3 produced by MM cells demonstrating that the inducible expression of PTX3 is able to impair MM neovascularization, the onset of a proficient BM vascular niche and, ultimately, to impair tumor growth and dissemination.

Abstract: During multiple myeloma (MM) progression the activation of the angiogenic process represents a key step for the formation of the vascular niche, where different stromal components and neoplastic cells collaborate and foster tumor growth. Among the different pro-angiogenic players, Fibroblast Growth Factor 2 (FGF2) plays a pivotal role in BM vascularization occurring during MM progression. Long Pentraxin 3 (PTX3), a natural FGF antagonist, is able to reduce the activation of stromal components promoted by FGF2 in various in vitro models. An increased FGF/PTX3 ratio has also been found to occur during MM evolution, suggesting that restoring the "physiological" FGF/PTX3 ratio in plasma cells and BM stromal cells (BMSCs) might impact MM. In this work, taking advantage of PTX3-inducible human MM models, we show that PTX3 produced by tumor cells is able to restore a balanced FGF/PTX3 ratio sufficient to prevent the activation of the FGF/FGFR system in endothelial cells and to reduce the angiogenic capacity of MM cells in different in vivo models. As a result of this anti-angiogenic activity, PTX3 overexpression causes a significant reduction of the tumor burden in both subcutaneously grafted and systemic MM models. These data pave the way for the exploitation of PTX3-derived anti-angiogenic approaches in MM.

Keywords: multiple myeloma; long pentraxin 3; FGF/FGFR system; angiogenesis

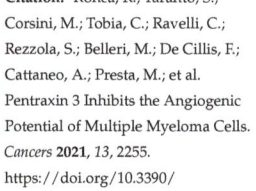

Citation: Ronca, R.; Taranto, S.; Corsini, M.; Tobia, C.; Ravelli, C.; Rezzola, S.; Belleri, M.; De Cillis, F.; Cattaneo, A.; Presta, M.; et al. Pentraxin 3 Inhibits the Angiogenic Potential of Multiple Myeloma Cells. *Cancers* **2021**, *13*, 2255. https://doi.org/10.3390/cancers13092255

Academic Editor: Marcel Spaargaren

Received: 2 April 2021
Accepted: 5 May 2021
Published: 8 May 2021

Publisher's Note: MDPI stays neutral with regard to jurisdictional claims in published maps and institutional affiliations.

Copyright: © 2021 by the authors. Licensee MDPI, Basel, Switzerland. This article is an open access article distributed under the terms and conditions of the Creative Commons Attribution (CC BY) license (https://creativecommons.org/licenses/by/4.0/).

1. Introduction

Multiple myeloma (MM) represents a life-threatening hematological disorder, being the second most common blood cancer diagnosis, with over 12,000 deaths estimated per year and a 5-year survival rate around 54% (www.cancer.net, accessed on 7 May 2021). Notwithstanding the introduction of novel therapeutic approaches that have led to steadily increased survival rates over the last decade, a complete eradication has not been obtained so far, and MM remains an incurable disease [1,2].

In the complex neoplastic MM microenvironment, tumor growth and resistance are fostered not only by plasma cells themselves, but also by bone marrow (BM) stromal cells (BMSCs), including endothelial cells (ECs). Indeed, angiogenesis represents a key feature of MM progression, driving the transition from the avascular state of monoclonal gammopathies of undetermined significance (MGUS) to the widely vascularized condition of active MM [3,4]. Accordingly, BM microvascular density represents a significant prognostic factor for progression free and overall survival in MM patients [5,6].

Among the pro-angiogenic factors, Fibroblast Growth Factor 2 (FGF2) has been shown to play a relevant role in different tumor types, including MM [7–9], its blockade resulting in significant anti-tumor and anti-angiogenic activities [10,11]. Accordingly, the pattern recognition receptor Long Pentraxin 3 (PTX3), a secreted/stromal component of innate immunity able to bind and inactivate various members of the FGF family, including FGF2, has revealed potent anti-angiogenic and anti-tumor properties in different FGF-dependent tumors [12–17]. In a translational perspective, a PTX3-derived FGF trap molecule has been proposed for the treatment of these tumors, including MM [11,18–20].

PTX3 and FGF2 are actively produced by BM plasma cells, ECs, and fibroblasts in normal, MGUS, and MM settings. However, the ratio between PTX3 and FGF2 decreases during the transition from MGUS to MM, leading to more abundant levels of FGF2 in these cells, representing a brake release mechanism to promote an angiogenic switch in MM [21]. Accordingly, in vitro observations have shown that treatment with recombinant PTX3 impaired FGF-mediated viability, chemotaxis, and migration of ECs and fibroblasts isolated from the BM of MM patients, as well as plasma cell adhesion to these cells [21]. These data suggest that restoring physiological PTX3/FGF ratio in plasma cells and BMSCs might impact MM.

To date, no data are available about the effect exerted in vivo by PTX3 of plasma cell origin on the growth and vascularization of MM. Here, taking advantage from PTX3-inducible human MM cell models, we demonstrate that PTX3 upregulation in plasma cells strongly impact MM growth and dissemination, mainly impairing FGF-mediated angiogenesis. These data add new hints regarding the role of PTX3 in MM and pave the way for the exploitation of PTX3-derived anti-angiogenic approaches in MM.

2. Materials and Methods

2.1. Cell Cultures and Reagents

KMS-11 cells were obtained from the Japanese Collection of Research Bioresources (JCRB, Osaka, Japan) cell bank; GFP/Luciferase–expressing MM.1S cells were from Dr. Ghobrial (Dana-Farber Cancer Institute, Boston, MA, USA). All cell lines were maintained at low passage in RPMI1640 medium supplemented with 10% heat-inactivated FBS and 2.0 mM glutamine, tested regularly for *Mycoplasma* negativity, and authenticated by PowerPlex Fusion System (Promega, Madison, WI, USA). KMS-11 PTX3/Mock and MM.1S PTX3/Mock cells were obtained by lentiviral transduction using the pLVX-TetOne-Puro vector (Clontech, Mountain View, CA, USA), either harboring or not harboring the human PTX3 coding sequence. Cells were selected adding 1 µg/mL of puromycin to the cell culture medium. Doxycycline (resuspended at 1.0 mg/mL) was purchased from Merck.

2.2. Western Blot Analysis

Cells were washed in cold PBS and homogenized in NP-40 lysis buffer (1% NP-40, 20 mM Tris–HCl pH 8, 137 mM NaCl, 10% glycerol, 2 mM EDTA, 1 mM sodium orthovanadate, 10 µg/mL aprotinin, 10 µg/mL leupeptin). Protein concentration in the supernatants was determined using the Bradford protein assay (Bio-Rad Laboratories, Hercules, CA, USA). Expression of PTX3 was detected using anti-PTX3 rabbit polyclonal antibody (from B. Bottazzi, Humanitas Clinical Institute, Rozzano, Italy). GAPDH antibody (Santa Cruz Biotechnology, Santa Cruz, CA, USA) and red Ponceau staining were used as loading controls for cell lysates and conditioned medium, respectively. Chemiluminescent signal was acquired by ChemiDoc™ Imaging System (Bio-Rad Laboratories, Hercules, CA, USA).

2.3. RT-qPCR

Total RNA was extracted using TRIzol Reagent (Invitrogen, Carlsbad, CA, USA) according to the manufacturer's instructions. Two μg of total RNA were retro-transcribed with MMLV reverse transcriptase (Invitrogen) using random hexaprimers. Then, cDNA was analyzed by quantitative PCR using the following primers: human *PTX3*, 5′-GTGCTCT CTGGTCTGCAGTG-3′ (forward) and 5′-TCGTCCGTGGCTTGCAGCAG-3′ (reverse); human *CCDN1*, 5′-AATGACCCCGCACGATTTC-3′ (forward) and 5′-CATGGAGGGCGGATT GGAA-3′ (reverse); human *GAPDH*, 5′-TGCCATCACTGCCACCCAGA-3′ (forward) and 5-CGCGGCCATCACGCCACAG-3′ (reverse).

2.4. Gene Expression Profiling (GEP)

GEP was performed on KMS-11 PTX3 cells either treated or untreated with DOXA (200 ng/mL) for 96 h. A cut-off of *p*-value < 0.01 (FDR corrected) and Log_2 fold change ±2 was applied to select differentially expressed genes. Total RNA was extracted using TRIzol Reagent according to the manufacturer's instructions (Invitrogen). RNA integrity and the purity of the treated cells were assessed using a Bioanalyzer (Agilent Technologies, Santa Clara, CA, USA). Hybridization to an Illumina Microarray (Illumina) was performed. Robust spline normalization and L2T were performed in R software, using the Lumi package from Bioconductor open-source software (http://www.bioconductor.org/, accessed on 7 May 2021). Normalized data were imported into Partek Genomic Suite 6.6 software (Partek). After quality controls, an analysis of variance (ANOVA) test was performed to assess the effects of PTX3 on pro-/anti- angiogenic gene expression, comparing KMS-11 PTX3 cells that received DOXA vs KMS-11 PTX3 cells that did not receive DOXA.

2.5. Cytofluorimetric Analyses

Cytofluorimetric analyses were performed using the MACSQuant® Analyzer (Miltenyi Biotec, Bergisch Gladbach, Germany). Propidium iodide staining (Immunostep, Salamanca, Spain) was used to detect PI negative viable cells and viable cell counts were obtained by the counting function of the MACSQuant® Analyzer.

2.6. MM/HUVE Cells Co-Cultures

KMS-11 PTX3 and MM.1S PTX3 cells were co-cultured with HUVE cells at 15:1 MM/HUVEC ratio in presence or absence of Doxycycline (DOXA) 200 μg/mL. After 48 h of co-culture, MM cells were removed without detaching endothelial cells and HUVEC viable cell counting was performed by cytofluorimetric analysis.

2.7. In Vitro Immunofluorescence Analysis

HUVE cells were seeded in 2.5% FBS in Ibidi® μ-Slide 8 wells (Ibidi, Martinsried, Germany) at a density of 30,000 cells/cm^2, and co-cultured or not with KMS-11 PTX3 or MM.1S PTX3 in absence or presence of DOXA 200 ng/mL. After 24 h MM cells were removed and HUVE cells were washed twice in PBS, fixed in cold acetone for 5 min, and permeabilized with 0.2% Triton-X100 in PBS for 2 min at RT. After washing in PBS, cells were blocked for 10 min at RT in 1% BSA and then incubated with rabbit anti-pFGFR1 (Tyr766h, Santa Cruz Biotechnology, Dallas, TX, USA) antibody for 1 h at RT. Cells were then washed in PBS and incubated with AlexaFluor 594-conjugated anti-rabbit antibody (Invitrogen) and DAPI for 30 min at RT. Finally, cells were examined under a Zeiss Fluorescence Axiovert 200M microscope (Carl Zeiss, Milan, Italy).

2.8. Chick Embryo Chorioallantoic Membrane (CAM) Assay

Alginate beads (5 μL) containing vehicle or KMS-11 PTX3 cells (40×10^4 cells/implant) with or without DOXA (200 μg/mL) were placed onto the top of chicken embryo CAMs at day 11 of incubation. After 72 h, newly formed blood vessels were quantified as described [22–24].

2.9. Zebrafish Embryo Model

Zebrafish experiments were performed as approved by the local animal ethics committee (OPBA, Organismo Preposto al Benessere degli Animali, Università degli Studi di Brescia, Italy). Embryos from the Tg (fli1:egfp) strain of the zebrafish, *Danio rerio*, were collected, staged, and raised at 28.5 °C, according to standard experimental conditions. Embryos at 48 h post-fertilization (hpf) were anesthetized using 0.04 mg/mL of tricaine (Sigma-Aldrich, St. Louis, MO, USA), and placed onto an agarose gel plate meld for tumor cell microinjection. MM.1S PTX3 cells cultured for 48 h in the presence or absence of DOXA (200 ng/mL) were washed and transferred using a micro-loader tip (Eppendorf, Milan, Italy) into a borosilicate glass needle (outer diameter/inner diameter: 1.2/0.68 mm) connected to a FemtoJet microinjector and InjectMan NI 2 Micromanipulator (Eppendorf). Finally, tumor cells were injected into the perivitelline space of embryos under a stereo-dissecting microscope (Leica, MZ75). Twenty-four hours after tumor injection, the angiogenic response was analyzed by quantifying the cumulative length of sprouts originating from the subintestinal vein vessels after phosphatase staining.

2.10. Subcutaneous Human Xenografts

Experiments were performed according to the Italian laws (D.L. 116/92 and following additions) that enforce the EU 86/109 Directive and were approved by the local animal ethics committee (OPBA, Organismo Preposto al Benessere degli Animali, Università degli Studi di Brescia, Italy). Six- to eight-week old female NOD/SCID mice (Envigo, Udine, Italy) were injected subcutaneously (s.c.) with KMS-11 PTX3 (5×10^6 cells/mouse) in 200 µL of PBS. The day after tumor implantation, mice were randomly assigned to receive DOXA (1 mg/mL) in the drinking water. Tumor volumes were measured with caliper and calculated according to the formula $V = (D \times d^2)/2$, where D and d are the major and minor perpendicular tumor diameters, respectively. At the end of the experimental procedure, mice were injected intravenously with sulfobiotin as previously described [25] in order to label the whole functional vascular network. Finally, tumor nodules were excised and processed for histological analysis.

2.11. Systemic Human Xenograft

Experiments were performed according to the Italian laws (D.L. 116/92 and following additions) that enforce the EU 86/109 Directive and were approved by the local animal ethics committee (OPBA, Organismo Preposto al Benessere degli Animali, Università degli Studi di Brescia, Italy). Six- to eight-week old female SCID Beige mice (Envigo) were injected intravenously (i.v) with GFP/Luciferase expressing MM.1S PTX3 cells (2×10^6 cells/mouse) in 100 µL of PBS. The day after tumor cell injection, mice were randomly assigned to receive DOXA (1 mg/mL) in the drinking water. Tumor dissemination was assessed by bioluminescence imaging analysis performed at 3, 4, and 5 weeks after tumor cell injection.

2.12. Histological Analyses

Tumor samples and femurs were either embedded in OCT compound and immediately frozen or fixed in formalin and embedded in paraffin, respectively.

For immunofluorescence analysis, tumor cryostat sections (5 µm thick) were air dried and fixed with cold acetone (5 min at 4 °C). After blocking with 1% BSA in PBS for 10 min, samples were incubated for 1 h at room temperature with primary antibodies [rabbit anti-PTX3 (from B. Bottazzi, Humanitas Clinical Institute, Rozzano, Italy), rabbit anti-pFGFR1 (Santa Cruz Biotechnology), rat anti-mouse Ki67 (Dako), or rabbit anti-human phospho Histone H3 (Merck)]. After washing with PBS containing 0.05% Tween 20, samples were incubated for 30 min with the appropriate Alexa Fluor 594-conjugated secondary antibody

(Invitrogen). In vivo biotinylated endothelial cells were detected by incubating tumor sections with 488-conjugated streptavidin (Invitrogen). Finally, after mounting in a drop of anti-bleaching mounting medium containing DAPI (Vectashield, Vector Laboratories, Burlingame, CA, USA), samples were examined under a Zeiss Fluorescence Axiovert 200M (Carl Zeiss) microscope.

Formalin-fixed, paraffin-embedded femur samples were sectioned at a thickness of 3 µm, dewaxed, hydrated, and stained with hematoxylin and eosin (H&E) or processed for immunohistochemistry with mouse anti-human CD38 (Novocastra, Wetzlar, Germany), rabbit anti-human PTX3 (from B. Bottazzi, Humanitas Clinical Institute, Rozzano, Italy), or rabbit anti-mouse KDR (Cell Signaling Technology, Danvers, MA, USA) antibodies. A positive signal was revealed by 3,3′-diaminobenzidine (Roche) or Vector blue substrate (Vector Laboratories, Burlingame, CA, USA) stainings. Sections were finally counterstained with Carazzi's hematoxylin before analysis by light microscopy. Images were acquired with the automatic high-resolution scanner Aperio System (Leica Biosystems, Wetzlar, Germany). Image analysis was carried out using the open-source ImageJ software.

2.13. Two-Photon Microscopy

After euthanasia, mice were transcardially perfused with 0.01 M phosphate-buffered saline (PBS) (Sigma-Aldrich, Milan, Italy) followed by 4% paraformaldehyde (PFA) (VWR). After specimen preparation as previously described [26], femurs were incubated for 48 h at 4 °C with rabbit anti-mouse KDR (Cell Signaling Technology) and mouse anti-human CD38 (Novocastra) followed by 4 h incubation with AlexaFluor 594 and AlexaFluor488-conjugated secondary antibodies. Then, samples were stored in PBS at 4 °C. For imaging, bones were held in 1% low melting agarose. Two-photon imaging was performed on a Zeiss LSM880 equipped with an EC Plan-Neofluar 20×/0.50 controlled by Zen Black 2 (Zeiss GmbH, Oberkochen, Germany).

2.14. Statistical Analyses

Statistical analyses were performed using Prism 8 (GraphPad Software). Student's t-test for unpaired data (2-tailed) was used to test the probability of significant differences between two groups of samples. For more than two groups of samples, data were analyzed with a 1-way analysis of variance and corrected by the Bonferroni multiple comparison test. Tumor volume data were analyzed with a 2-way analysis of variance and corrected by the Bonferroni test. Differences were considered significant when $p < 0.05$.

3. Results

3.1. PTX3 Produced by MM Cells Hampers the Proliferation of Endothelial Cells

To assess the role of PTX3 expressed and released by MM cells, we generated two human MM cell lines with doxycycline-inducible expression of PTX3 (KMS-11 PTX3 and MM.1S PTX3 cells). As shown in Figure 1A, KMS-11 PTX3 and MM.1S PTX3 cells express and secrete high levels of PTX3 already 48 h after treatment with 200 ng/mL of doxycycline (DOXA) when compared to untreated (-DOXA) or control (Mock) cells. In keeping with the capacity of MM cells to stimulate ECs by direct interactions and production of pro-angiogenic factors, including FGF2 [2], human umbilical vein ECs (HUVECs) co-cultured with MM cells showed an increased rate of survival and proliferation as well as elevated levels of FGFR1 phosphorylation, when compared to HUVEC monocultures (Figure 1B,C). Notably, PTX3 released by MM cells upon DOXA induction significantly reduced HUVEC proliferation and FGFR1 phosphorylation to the basal levels observed in HUVEC monocultures (Figure 1B,C).

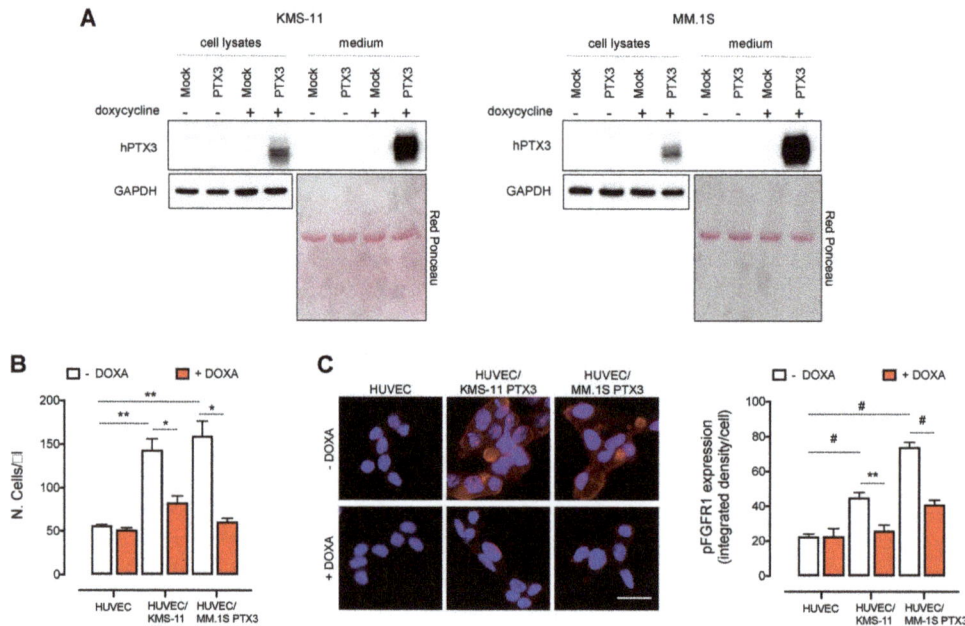

Figure 1. PTX3 released by MM cells impairs HUVEC proliferation by inhibiting FGFR activation. (**A**) Western blot analysis of PTX3 expression (cell lysates) and release (medium) from KMS-11 and MM.1S cells transduced with a doxycycline (DOXA)-inducible PTX3 (PTX3) or a control vector (Mock) and treated or not with DOXA for 48 h. (**B**) Cell count by cytofluorimetric analysis of HUVEC co-cultured or not with KMS-11 PTX3 or MM.1S PTX3 cells for 48 h in the presence or absence of DOXA. (**C**) Left panel: Immunofluorescence analysis of phospho-FGFR1 (red fluorescence) expression in HUVEC co-cultured or not with KMS-11 PTX3 or MM.1S PTX3 cells for 24 h in the presence or absence of DOXA. Scale bar: 50 µm. Right panel: Fluorescence intensity quantification of phospho-FGFR1 by ImageJ software. For each microscopic field, fluorescence intensity values were normalized with the number of nuclei detected by DAPI staining. Data are mean ± SEM of 3 experimental replicates. * $p < 0.05$, ** $p < 0.01$, # $p < 0.001$.

It must be pointed out that, in keeping with previous observations [21], PTX3 upregulation following DOXA induction did not directly affect the survival and proliferation of MM cells (Figure S1), thus ruling out the possibility that the inhibitory effects observed in ECs were due to a reduced survival/proliferation of MM cells. Gene expression profiling performed on KMS-11 PTX3 cells also did not show any significant modulation of the expression of *FGF2* and other pro-/anti-angiogenic genes upon PTX3 induction (Figure S2), indicating that the inhibition of EC proliferation is due to the FGF-trap activity of PTX3 rather than to a modulation of other pro- or anti-angiogenic factors caused by PTX3 overexpression.

Together, these data are in keeping with the capacity of FGF2 to act as a paracrine survival/proliferation factor for ECs and with the potent anti-angiogenic activity of PTX3 consequent to its FGF trap activity that results in the inhibition of the FGFR pathway.

3.2. PTX3 Reduces the Angiogenic Potential of MM Cells

We next assessed the capacity of PTX3 released by MM cells to impair the pro-angiogenic potential of MM in vivo. To this aim, KMS-11 PTX3 cells were grafted onto the top of the chick embryo chorioallantoic membrane (CAM) in the absence or presence of DOXA. As shown in Figure 2A, untreated KMS-11 grafts induced a strong pro-angiogenic response, as shown by the numerous newly formed thin microvessels converging in a spoke-wheel pattern versus the MM cell implant. Notably, this angiogenic response was significantly inhibited by the release of PTX3 from KMS-11 cells following DOXA treatment (Figure 2A).

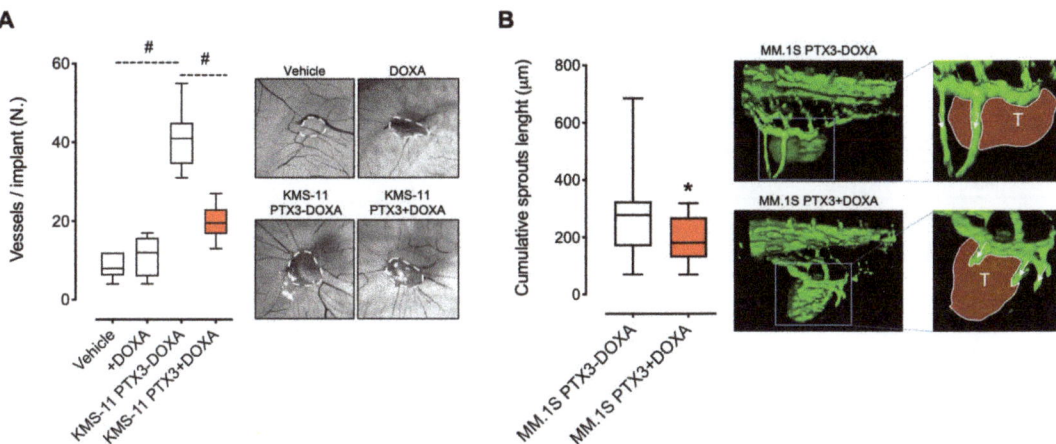

Figure 2. MM cells producing PTX3 are less pro-angiogenic. (**A**) KMS-11 PTX3 cells embedded in alginate pellets containing (KMS-11 PTX3+DOXA) or not (KMS-11 PTX3-DOXA) doxycycline were grafted onto the top of the chick embryo CAM at day 11 of development. PBS (Vehicle) or DOXA were used as control. At day 14, for each embryo, the number of newly formed blood vessels converging towards the implant were quantified. $n = 15$ embryo/group. Representative images of CAMs at day 14 are reported. White dashed lines show the alginate pellet implants. (**B**) GFP-expressing MM.1S PTX3 cells in vitro induced (MM.1S PTX3+DOXA) or not (MM.1S PTX3-DOXA) with DOXA were grafted into the perivitelline space of 48 hpf Tg (fli1:egfp) zebrafish embryos. Twenty-four hours after engraftment, for each embryo, the cumulative length of sprouts deriving from subintestinal vein vessels was quantified. In the magnified images, the tumor mass is highlighted in red and the vessel sprouts are indicated with arrows. $n = 30$ embryo/group. In box and whiskers graphs, boxes extend from the 25th to the 75th percentiles, lines indicate the median values and whiskers indicate the range of values. * $p < 0.05$, # $p < 0.001$.

To confirm these observations, GFP-expressing MM.1S cells were incubated for 48 h in the absence or presence of DOXA and then grafted into the perivitelline space of zebrafish embryos. As shown in Figure 2B, MM.1S cells induced the formation of EC sprouts originating from the subintestinal vein vessels that were significantly reduced in their length by PTX3 production following DOXA pre-treatment. These data confirm the capacity of MM cells to induce strong pro-angiogenic responses and indicate that PTX3 is able to reduce the angiogenic potential of MM cells in vivo.

3.3. MM-Released PTX3 Inhibits Tumor Angiogenesis and Growth In Vivo

To assess the effect of PTX3 released by MM cells on tumor vascularization and growth, KMS-11 PTX3 cells were grafted subcutaneously in immunodeficient mice. In order to detect the whole functional tumor vascular network, endothelial cells were biotinylated in vivo by i.v. injection of sulfobiotin [25]. As shown in Figure 3A, tumors from mice receiving DOXA in the drinking water showed widespread expression of PTX3 that accumulates in the extracellular matrix. Interestingly, PTX3 strongly reduced FGFR1 activation in both sulfobiotin$^+$ endothelial cells and tumor cells as assessed by phospho-FGFR1 immunostaining (Figure 3A). This caused a significant reduction of tumor endothelial cell proliferation as assessed by Ki67/sulfobiotin double immunostatining (Figure 3B). Accordingly, PTX3 expressing xenografts (+DOXA) showed reduced tumor vascularization (Sulfobiotin$^+$ area) and proliferation (pHH3$^+$area) (Figure 3C) that resulted in a significant delay of tumor growth when compared to controls (−DOXA) (Figure 3D). Of note, DOXA did not affect the growth of mock-transfected KMS-11 tumor grafts, thus ruling out any effect exerted by DOXA per se (data not shown).

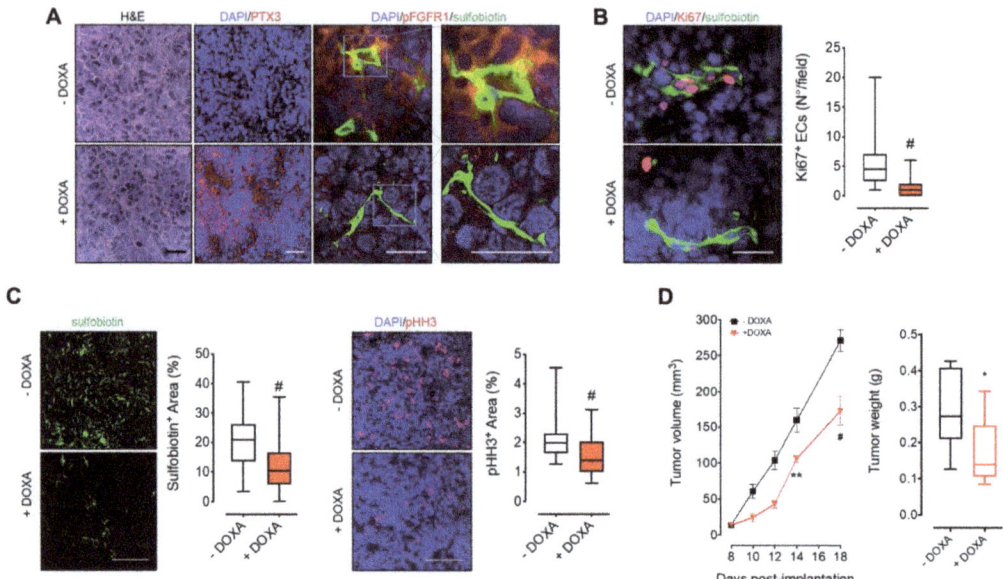

Figure 3. PTX3 released by MM cells reduces tumor vascularization and growth. KMS-11 PTX3 cells were subcutaneously engrafted in NOD/SCID mice receiving (+DOXA) or not (-DOXA) doxycycline in the drinking water. (**A–C**) Histological analyses of tumor sections eighteen days after tumor engraftment. Before sacrifice, mice were injected i.v. with sulfobiotin in order to label the whole functional vascular network. Sulfobiotin and phospho-HH3 positive area were quantified by ImageJ software. Scale bar A, B: 50 µm; scale bar C: 100 µm. (**D**) Left panel: Tumor volumes (mean ± SEM) measured with caliper up to 18 days after tumor implantation. n = 8 mice/group. Right panel: Tumor weights at day 18 post-implantation. In box and whiskers graphs, boxes extend from the 25th to the 75th percentiles, lines indicate the median values and whiskers indicate the range of values. * $p < 0.05$, ** $p < 0.01$, # $p < 0.001$.

3.4. PTX3 Reduces BM Niche Vascularization and Colonization by MM Cells

Since BM angiogenesis plays a pivotal role in MM progression and dissemination [3,4], we investigated the effect of PTX3 on the capacity of MM cells to induce BM angiogenesis while growing in their own microenvironment. To this aim, we took advantage of the MM.1S systemic model of MM by which BM infiltration is detectable 2 weeks after intravenous injection of MM cells, and reaches a peak after 6–8 weeks, when more than 40% of the BM cell population is represented by neoplastic cells [11]. Thus, luciferase-expressing MM.1S cells transduced to express DOXA-inducible PTX3 were systemically injected into SCID beige mice. Notably, the induction of PTX3 expression and release by MM.1S cells (see CD38/PTX3 double immunostaining in Figure 4A) significantly reduced the vascularization of tumor foci growing in the BM, as assessed by CD38/KDR double immunostaining of femurs from mice receiving DOXA in the drinking water (Figure 4A). The reduction of tumor vascularization was confirmed also by 3D two-photon fluorescent imaging of whole femurs (Figure 4B). In keeping with these findings, MM.1S-released PTX3 strongly reduced the systemic spreading and BM colonization of the disease, as detected by in vivo bioluminescence imaging (Figure 4C) and CD38 immunostaining of femur sections (Figure 4D).

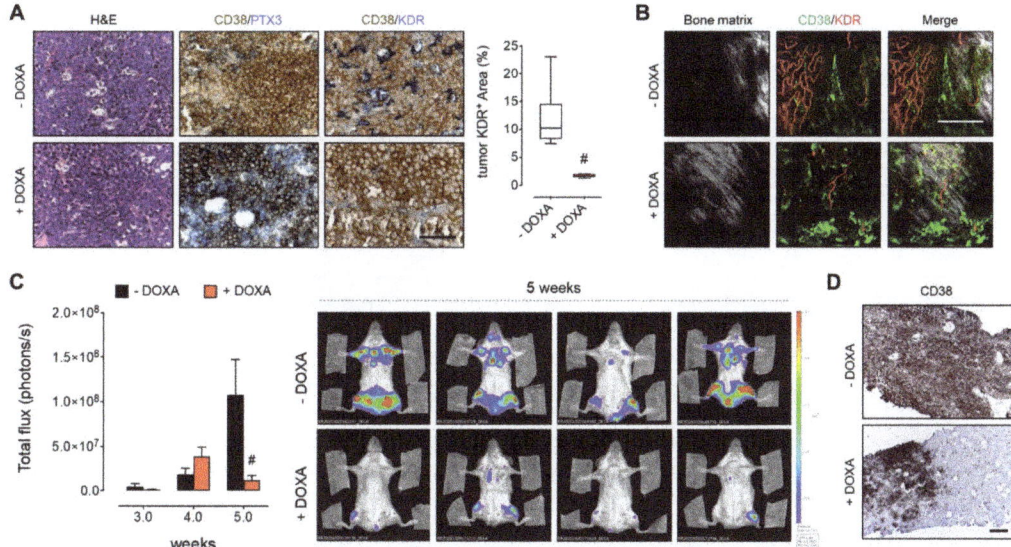

Figure 4. PTX3 reduces MM cell BM colonization. Luciferase-expressing MM.1S PTX3 cells were injected i.v. in SCID Beige mice receiving (+DOXA) or not (-DOXA) doxycycline in the drinking water. (**A**) Histological analysis of femur sections five weeks after tumor cell injection. Tumor cells are detected in brown (CD38) and PTX3 and tumor vessels (KDR) are detected in blue. Scale bar: 100 μm. KDR positive area in tumor spots was quantified by ImageJ software. In box and whiskers graphs, boxes extend from the 25th to the 75th percentiles, lines indicate the median values and whiskers indicate the range of values. # $p < 0.001$. (**B**) Two-photon fluorescence microscopy analysis of femurs five weeks after tumor cell injection. Tumor cells are detected in green (CD38), tumor vessels (KDR) in red, and bone matrix is detected by second harmonic generation in grey. Scale bar: 200 μm. (**C**) Left panel: Quantification of bioluminescent signal of luciferase-expressing MM.1S cells up to 5 weeks after i.v. injection. Data are mean ± SEM, # $p < 0.001$. $n = 8$ mice/group. Right panel: Representative bioluminescence imaging of mice five weeks after MM cell i.v. injection. (**D**) Immunohistochemical analysis of femur sections five weeks after tumor cell injection. Tumor area are detected in brown (CD38). Scale bar: 200 μm.

4. Discussion

BM angiogenesis is a hallmark of MM progression and represents a prognostic factor for MM patients [5,6]. Indeed, like solid tumor cells, highly proliferating plasma cells are able to induce a neovascular response via the release of angiogenic cytokines, giving rise to the so-called "angiogenic switch" [2]. The induction of BM angiogenesis may thus favor the progression from the pre-neoplastic MGUS and non-active MM to active MM, the latter representing the vascular phase of plasma cell tumors. Different mechanisms may take place in order to activate BM angiogenesis and counterbalance the physiological/steady state equilibrium in favor of pro-angiogenic activators. In this context, an analysis performed on BM plasma has revealed an unbalanced ratio between the pro-angiogenic growth factor FGF2 and one of its natural inhibitors, PTX3, in MM patients compared to MGUS patients [21]. This finding suggests that the ratio between PTX3 and FGF2 released from plasma cells and BMSCs decreases during the transition from MGUS to MM, leading to more abundant levels of free FGF2 in the BM. This "stoichiometric" unbalance may actively contribute to pathological BM angiogenesis in MM.

Based on these observations, we hypothesized that the restoration of a balanced ratio between FGF2 and PTX3 in the BM may be able to restrain the avascular–vascular transition in MM. So far, only in vitro data using recombinant PTX3 have been reported showing that addition of exogenous PTX3 reduces the activation of MM-derived ECs and fibroblasts stimulated by FGF2.

Here, we demonstrate and confirm the pivotal role played by the PTX3/FGF2 ratio in MM growth, dissemination, and neovascularization exploiting in vivo MM models characterized by the inducible overexpression of PTX3 by human plasma cells. Our data confirm that MM cells are capable to induce, per se, a strong angiogenic response both in vitro and in vivo by activating FGFR1 in ECs. Interestingly, when PTX3 expression is forced in neoplastic plasma cells, thus leading to an increase of the PTX3/FGF2 ratio, the activation of FGF/FGFR pathway is inhibited and EC activation is reduced both in in vitro co-cultures and in in vivo tumor graft models (i.e., chick embryo CAM, zebrafish and murine xenografts). It must be pointed out that GEP analysis performed on PTX3-expressing MM cells indicates that (i) the increased PTX3/FGF2 ratio appears to be due to PTX3 overexpression and not to PTX3-mediated FGF2 downregulation and (ii) the antiangiogenic response is due to the FGF-trap activity exerted by PTX3 rather than to a modulation of the expression of other pro- or anti-angiogenic factors caused by PTX3 overexpression. As a result of this antiangiogenic effect, MM growth and dissemination is significantly impaired in both subcutaneous and systemic murine models using MM cells overexpressing PTX3.

Beside the antiangiogenic activity exerted by MM-released PTX3, we cannot rule out the possibility that a direct "autocrine" effect exerted by PTX3 overexpression on MM cells may contribute to the observed inhibition of tumor growth and dissemination. Indeed, FGF2 is known to play a pivotal role in the survival and proliferation of MM cells [11]. However, the in vitro data hereby reported, and previous observations from others [21], have shown that endogenous PTX3 overexpression or recombinant PTX3 treatment do not affect the survival and proliferation of MM cells under standard cell suspension culture conditions. On the other hand, we have observed a significant reduction in the expression levels of the proliferation marker *Cyclin D1* when MM cells overexpressing PTX3 were grown embedded in Matrigel (Figure S3). Accordingly, a reduced rate of MM cell proliferation paralleled by a reduced FGFR1 activation occurred in vivo in both endothelial and tumor cells when PTX3 was overexpressed and accumulated in the extracellular matrix of grafted tumors. Together, these data suggest that an interaction with extracellular matrix component(s) may be required to consent and favor the autocrine biological function of PTX3 on MM cells themselves. Further experiments will be required to elucidate this hypothesis.

Altogether the data hereby reported highlight for the first time the role played by endogenous PTX3 released by MM cells on BM-niche components, reinforcing the concept that the PTX3/FGF2 ratio is a key rheostat in MM angiogenesis and progression. In keeping with these findings, we have recently reported a significant reduction of tumor vascularization of subcutaneous KMS-11 xenografts in mice treated with the PTX3-derived small molecule NSC12 [11]. Hence, anti-angiogenic approaches based on PTX3-derived FGF trap molecules may represent a promising future area of research in the field of MM therapy.

5. Conclusions

Our data highlight the role played by endogenous PTX3 released by MM cells on endothelial BM-niche components and demonstrate that the restoration of a balanced ratio between FGF2 and PTX3 affects MM angiogenesis, growth, and progression. Hence, the exploitation of PTX3-derived anti-angiogenic approaches may represent a promising future area of research in the field of MM therapy.

Supplementary Materials: The following are available online at https://www.mdpi.com/article/10.3390/cancers13092255/s1, Figure S1: PTX3 does not affect MM cell proliferation and survival in standard cell culture conditions, Figure S2: PTX3 does not affect the expression of pro-/anti-angiogenic factors, Figure S3: PTX3 affects MM cell proliferation when cells are grown embedded in Matrigel.

Author Contributions: Conceptualization, A.G., M.P., R.R.; methodology, A.G., R.R.; investigation, A.G., R.R., S.T., M.C., C.T., C.R., S.R., M.B., F.D.C., A.C.; data curation, A.G., R.R.; writing—review and editing, A.G, M.P., R.R.; supervision, M.P., A.G.; funding acquisition, A.G., M.P., R.R. All authors have read and agreed to the published version of the manuscript.

Funding: R.R. was supported by Associazione Italiana Ricerca sul Cancro (AIRC IG 2019–ID.23151), M.P. was supported by Associazione Italiana Ricerca sul Cancro (AIRC IG 2019–ID. 18493); A.G. was supported by Fondazione Cariplo (grant n° 2016-0570); S.R. was supported by Fondazione Umberto Veronesi fellowship.

Institutional Review Board Statement: The study was conducted according to the guidelines of the Declaration of Helsinki, and approved by the local animal ethics committee (OPBA, Organismo Preposto al Benessere degli Animali, Università degli Studi di Brescia, Italy), authorization n. 457/2015 issued by the Italian Ministry of Health.

Informed Consent Statement: Not applicable.

Data Availability Statement: Data is contained within the article or Supplementary Material.

Conflicts of Interest: The authors declare no conflict of interest.

References

1. Moreau, P.; Kumar, S.K.; Miguel, J.S.; Davies, F.; Zamagni, E.; Bahlis, N.; Ludwig, H.; Mikhael, J.; Terpos, E.; Schjesvold, F.; et al. Treatment of relapsed and refractory multiple myeloma: Recommendations from the International Myeloma Working Group. *Lancet Oncol.* **2021**, *22*, e105–e118. [CrossRef]
2. Ribatti, D.; Vacca, A. New insights in anti-angiogenesis in multiple myeloma. *Int. J. Mol. Sci.* **2018**, *19*, 2031. [CrossRef]
3. Ribatti, D.; Nico, B.; Vacca, A. Importance of the bone marrow microenvironment in inducing the angiogenic response in multiple myeloma. *Oncogene* **2006**, *25*, 4257–4266. [CrossRef] [PubMed]
4. Kumar, S.; Witzig, T.E.; Timm, M.; Haug, J.; Wellik, L.; Kimlinger, T.K.; Greipp, P.R.; Rajkumar, S.V. Bone marrow angiogenic ability and expression of angiogenic cytokines in myeloma: Evidence favoring loss of marrow angiogenesis inhibitory activity with disease progression. *Blood* **2004**, *104*, 1159–1165. [CrossRef]
5. Rajkumar, S.V.; Mesa, R.A.; Fonseca, R.; Schroeder, G.; Plevak, M.F.; Dispenzieri, A.; Lacy, M.Q.; Lust, J.A.; Witzig, T.E.; Gertz, M.A.; et al. Bone marrow angiogenesis in 400 patients with monoclonal gammopathy of undetermined significance, multiple myeloma, and primary amyloidosis. *Clin. Cancer Res. Off. J. Am. Assoc. Cancer Res.* **2002**, *8*, 2210–2216.
6. Jakob, C.; Sterz, J.; Zavrski, I.; Heider, U.; Kleeberg, L.; Fleissner, C.; Kaiser, M.; Sezer, O. Angiogenesis in multiple myeloma. *Eur. J. Cancer* **2006**, *42*, 1581–1590. [CrossRef]
7. Vacca, A.; Ribatti, D. Bone marrow angiogenesis in multiple myeloma. *Leukemia* **2006**, *20*, 193–199. [CrossRef]
8. Ronca, R.; Benkheil, M.; Mitola, S.; Struyf, S.; Liekens, S. Tumor angiogenesis revisited: Regulators and clinical implications. *Med. Res. Rev.* **2017**, *37*, 1231–1274. [CrossRef]
9. Presta, M.; Chiodelli, P.; Giacomini, A.; Rusnati, M.; Ronca, R. Fibroblast growth factors (FGFs) in cancer: FGF traps as a new therapeutic approach. *Pharmacol. Ther.* **2017**, *179*, 171–187. [CrossRef]
10. Sacco, A.; Federico, C.; Giacomini, A.; Caprio, C.; Maccarinelli, F.; Todoerti, K.; Favasuli, V.; Anastasia, A.; Motta, M.; Russo, D.; et al. Halting the FGF/FGFR axis leads to anti-tumor activity in Waldenstrom's macroglobulinemia by silencing MYD88. *Blood* **2020**. [CrossRef]
11. Ronca, R.; Ghedini, G.C.; Maccarinelli, F.; Sacco, A.; Locatelli, S.L.; Foglio, E.; Taranto, S.; Grillo, E.; Matarazzo, S.; Castelli, R.; et al. FGF trapping inhibits multiple myeloma growth through c-Myc degradation-induced mitochondrial oxidative stress. *Cancer Res.* **2020**, *80*, 2340–2354. [CrossRef] [PubMed]
12. Giacomini, A.; Ghedini, G.C.; Presta, M.; Ronca, R. Long pentraxin 3: A novel multifaceted player in cancer. *Biochim. Biophys. Acta Rev. Cancer* **2018**, *1869*, 53–63. [CrossRef]
13. Annese, T.; Ronca, R.; Tamma, R.; Giacomini, A.; Ruggieri, S.; Grillo, E.; Presta, M.; Ribatti, D. PTX3 modulates neovascularization and immune inflammatory infiltrate in a murine model of fibrosarcoma. *Int. J. Mol. Sci.* **2019**, *20*, 4599. [CrossRef] [PubMed]
14. Rezzola, S.; Ronca, R.; Loda, A.; Nawaz, M.I.; Tobia, C.; Paganini, G.; Maccarinelli, F.; Giacomini, A.; Semeraro, F.; Mor, M.; et al. The autocrine FGF/FGFR System in both skin and uveal melanoma: FGF trapping as a possible therapeutic approach. *Cancers* **2019**, *11*, 1305. [CrossRef]
15. Matarazzo, S.; Melocchi, L.; Rezzola, S.; Grillo, E.; Maccarinelli, F.; Giacomini, A.; Turati, M.; Taranto, S.; Zammataro, L.; Cerasuolo, M.; et al. Long pentraxin-3 follows and modulates bladder cancer progression. *Cancers* **2019**, *11*, 1277. [CrossRef] [PubMed]
16. Rodrigues, P.F.; Matarazzo, S.; Maccarinelli, F.; Foglio, E.; Giacomini, A.; Nunes, J.P.S.; Presta, M.; Dias, A.A.M.; Ronca, R. Long pentraxin 3-mediated fibroblast growth factor trapping impairs fibrosarcoma growth. *Front. Oncol.* **2018**, *8*, 472. [CrossRef]
17. Presta, M.; Foglio, E.; Schuind, A.C.; Ronca, R. Long pentraxin-3 modulates the angiogenic activity of fibroblast growth factor-2. *Front. Immunol.* **2018**, *9*, 2327. [CrossRef]

18. Ronca, R.; Giacomini, A.; Di Salle, E.; Coltrini, D.; Pagano, K.; Ragona, L.; Matarazzo, S.; Rezzola, S.; Maiolo, D.; Torrella, R.; et al. Long-pentraxin 3 derivative as a small-molecule fgf trap for cancer therapy. *Cancer Cell* **2015**, *28*, 225–239. [CrossRef]
19. Giacomini, A.; Taranto, S.; Rezzola, S.; Matarazzo, S.; Grillo, E.; Bugatti, M.; Scotuzzi, A.; Guerra, J.; Di Trani, M.; Presta, M.; et al. Inhibition of the FGF/FGFR system induces apoptosis in lung cancer cells via c-Myc downregulation and oxidative stress. *Int. J. Mol. Sci.* **2020**, *21*, 9376. [CrossRef]
20. Castelli, R.; Giacomini, A.; Anselmi, M.; Bozza, N.; Vacondio, F.; Rivara, S.; Matarazzo, S.; Presta, M.; Mor, M.; Ronca, R. Synthesis, structural elucidation, and biological evaluation of NSC12, an orally available fibroblast growth factor (FGF) ligand trap for the treatment of FGF-dependent lung tumors. *J. Med. Chem.* **2016**, *59*, 4651–4663. [CrossRef]
21. Basile, A.; Moschetta, M.; Ditonno, P.; Ria, R.; Marech, I.; De Luisi, A.; Berardi, S.; Frassanito, M.A.; Angelucci, E.; Derudas, D.; et al. Pentraxin 3 (PTX3) inhibits plasma cell/stromal cell cross-talk in the bone marrow of multiple myeloma patients. *J. Pathol.* **2013**, *229*, 87–98. [CrossRef]
22. Ribatti, D.; Gualandris, A.; Bastaki, M.; Vacca, A.; Iurlaro, M.; Roncali, L.; Presta, M. New model for the study of angiogenesis and antiangiogenesis in the chick embryo chorioallantoic membrane: The gelatin sponge/chorioallantoic membrane assay. *J. Vasc. Res.* **1997**, *34*, 455–463. [CrossRef]
23. Corsini, M.; Moroni, E.; Ravelli, C.; Grillo, E.; Presta, M.; Mitola, S. In situ DNA/protein interaction assay to visualize transcriptional factor activation. *Methods Protoc.* **2020**, *3*, 80. [CrossRef] [PubMed]
24. Ronca, R.; Benzoni, P.; Leali, D.; Urbinati, C.; Belleri, M.; Corsini, M.; Alessi, P.; Coltrini, D.; Calza, S.; Presta, M.; et al. Antiangiogenic activity of a neutralizing human single-chain antibody fragment against fibroblast growth factor receptor 1. *Mol. Cancer Ther.* **2010**, *9*, 3244–3253. [CrossRef]
25. Lavazza, C.; Carlo-Stella, C.; Giacomini, A.; Cleris, L.; Righi, M.; Sia, D.; Di Nicola, M.; Magni, M.; Longoni, P.; Milanesi, M.; et al. Human CD34+ cells engineered to express membrane-bound tumor necrosis factor-related apoptosis-inducing ligand target both tumor cells and tumor vasculature. *Blood* **2010**, *115*, 2231–2240. [CrossRef] [PubMed]
26. Belleri, M.; Coltrini, D.; Righi, M.; Ravelli, C.; Taranto, S.; Chiodelli, P.; Mitola, S.; Presta, M.; Giacomini, A. β-galactosylceramidase deficiency causes bone marrow vascular defects in an animal model of Krabbe disease. *Int. J. Mol. Sci.* **2019**, *21*, 251. [CrossRef] [PubMed]

Article

Anti-Angiogenic Properties of Ginsenoside Rg3 Epimers: In Vitro Assessment of Single and Combination Treatments

Maryam Nakhjavani [1,2], Eric Smith [1,2,*], Kenny Yeo [1,2], Helen M. Palethorpe [3], Yoko Tomita [1,2,4], Tim J. Price [2,4], Amanda R. Townsend [2,4] and Jennifer E. Hardingham [1,2]

[1] Molecular Oncology, Basil Hetzel Institute, The Queen Elizabeth Hospital, Woodville South, SA 5011, Australia; maryam.nakhjavani@adelaide.edu.au (M.N.); a1811332@student.adelaide.edu.au (K.Y.); yoko.tomita@sa.gov.au (Y.T.); jennifer.hardingham@adelaide.edu.au (J.E.H.)
[2] Adelaide Medical School, University of Adelaide, Adelaide, SA 5005, Australia; timothy.price@sa.gov.au (T.J.P.); amanda.townsend@sa.gov.au (A.R.T.)
[3] Centre for Cancer Biology, University of South Australia and SA Pathology, Adelaide, SA 5000, Australia; helen.palethorpe@unisa.edu.au
[4] Oncology Unit, The Queen Elizabeth Hospital, Woodville South, SA 5011, Australia
* Correspondence: eric.smith@adelaide.edu.au; Tel.: +61-8-8222-6142

Simple Summary: Angiogenesis is a critical step in tumour progression and metastasis. The application of current inhibitors of angiogenesis is accompanied by adverse effects. Therefore, there is a need for developing better treatments. *Panax ginseng* is a traditional herbal medicine that has been used by humans for thousands of years. 20(S) ginsenoside-Rg3 and 20(R) ginsenoside-Rg3 are two structurally similar molecules extracted from this plant, with distinct mechanisms of action. In this research, a combination of both of these molecules was optimised (C3) to inhibit angiogenesis, in lab settings. The results showed the role of C3 as a novel anti-angiogenic drug.

Abstract: Tumour angiogenesis plays a key role in tumour growth and progression. The application of current anti-angiogenic drugs is accompanied by adverse effects and drug resistance. Therefore, finding safer effective treatments is needed. Ginsenoside Rg3 (Rg3) has two epimers, 20(S)-Rg3 (SRg3) and 20(R)-Rg3 (RRg3), with stereoselective activities. Using response surface methodology, we optimised a combination of these two epimers for the loop formation of human umbilical vein endothelial cell (HUVEC). The optimised combination (C3) was tested on HUVEC and two murine endothelial cell lines. C3 significantly inhibited the loop formation, migration, and proliferation of these cells, inducing apoptosis in HUVEC and cell cycle arrest in all of the cell lines tested. Using molecular docking and vascular endothelial growth factor (VEGF) bioassay, we showed that Rg3 has an allosteric modulatory effect on vascular endothelial growth factor receptor 2 (VEGFR2). C3 also decreased the VEGF expression in hypoxic conditions, decreased the expression of aquaporin 1 and affected AKT signaling. The proteins that were mostly affected after C3 treatment were those related to mammalian target of rapamycin (mTOR). Eukaryotic translation initiation factor 4E (eIF4E)-binding protein 1 (4E-BP1) was one of the important targets of C3, which was affected in both hypoxic and normoxic conditions. In conclusion, these results show the potential of C3 as a novel anti-angiogenic drug.

Keywords: ginsenoside Rg3; response surface methodology; optimisation; epimer; angiogenesis

1. Introduction

Tumour angiogenesis is a critical step in tumour growth, survival, and metastasis. Several pro- and anti-angiogenic factors and signaling pathways contribute to regulate angiogenesis and facilitate tumour growth and metastasis [1–3]. The key driver of angiogenesis is the signaling of vascular endothelial growth factor receptor 2 (VEGFR2). VEGFR2 is activated upon interaction with its major ligand, VEGF. Hence, VEGF; VEGFR2; or the downstream signaling of VEGFR2, including PI3K/AKT, could be potential key targets in anti-angiogenesis drug development. Currently, the clinically approved anti-angiogenic agents are either antibodies against VEGF such as bevacizumab or small molecule tyrosine kinase inhibitors (TKIs). The administration of bevacizumab in advanced cancer patients could be accompanied by severe and sometimes fatal adverse effects, including hematological disorders, respiratory disorders, perforation and hemorrhage in the gastrointestinal system, and nervous system disorders [4]. TKIs also cause hematological and non-hematological events that may limit the application of treatment [5]. Furthermore, the administration of current anti-angiogenic treatments may also be limited because of drug resistance [6]. Therefore, developing effective less-toxic treatments is a fundamental effort for improving patient outcomes and it is the main aim of this research.

Epimers of ginsenoside Rg3 (Rg3), SRg3, and RRg3 are some of the most important pharmacologically active members of the ginsenosides family of chemicals extracted from *Panax ginseng* [7]. These molecules seem to be suitable anti-angiogenic candidates for drug development studies, because several studies have described their effects of inhibiting angiogenesis, and have shown their potential as anti-cancer agents (reviewed in [8,9]). Furthermore, in vitro and in vivo studies in animals and humans have shown tolerability and a low toxicity profile for these molecules (reviewed in [8,9]). These factors make Rg3 epimers intriguing candidates. In this regard, one important aspect of pharmacology of these epimers is their stereoselective anti-cancer action. We previously showed that these epimers have stereoselective activities for the inhibition of the migration and invasion of triple-negative breast cancer cell lines [10]. In addition, we showed that only SRg3 blocks the water transport function of aquaporin 1 (AQP1) [10], a protein that plays important roles in angiogenesis, tumour growth, and metastasis [11–13]. Furthermore, other studies have shown the stereoselectivity of these epimers on ion channels [14], the relaxation of the swine coronary artery [15], the anti-oxidant effect [16], promotion of immune system [17,18], and the inhibition of epithelial–mesenchymal transition [19]. Considering this stereoselective anti-cancer activity, these epimers should be considered as separate drugs that could be combined.

For the first time, in this research, the concentrations of these epimers in combination was optimised to yield the highest anti-angiogenic efficacy. The optimal combination was determined using response surface methodology (RSM), a statistical and experiment design modelling process, which aims at reducing the number of experiments and costs associated with the experiment design [20]. In recent years, RSM has gained popularity in drug design [21], drug interaction [22], and combination therapy in cancer treatment studies [23]. It describes a three-dimensional dose–response surface, measures drug interactions, and defines the optimised combination of two drugs [24]. In this study, the efficacy of the optimal combination of Rg3 epimers was confirmed in migration and proliferation assays in human and murine endothelial cells. The mode of cell death and several potential intracellular targets of this combination that play roles in angiogenesis were studied. These targets included the expression of VEGF, activation of VEGFR2, signaling of AKT downstream of the activation of VEGFR2, and expression of AQP1. Because of the essential role of hypoxia in driving angiogenesis in a rapidly growing tumour, the role of this combination was studied in both normoxic and hypoxic conditions. [25].

2. Materials and Methods
2.1. Reagents, Cell Lines, and Cell Culture

Human umbilical vein endothelial cell (HUVEC) and its media, endothelial cell growth medium-2 (EBM-2; Clonetics, Lonza, Belgium), were purchased from Lonza, Belgium. Murine endothelial cell lines, 2H-11 and 3B-11, and human triple-negative breast cancer cell line MDA-MB-231 were purchased from the American Type Culture Collection (Manassas, VA, USA) and maintained in Dulbecco's Modified Eagle Medium (DMEM; Life Technologies, Carlsbad, CA, USA), supplemented with 10% fetal bovine serum (Corning, Corning, NY, USA), 50 U/mL penicillin, and 50 µg/mL (Life Technologies). The cells were used within the first 10 passages. SRg3 (>98%) and RRg3 (>98%) (ChemFaces®, Wuhan, China) were dissolved in dimethyl sulfoxide (DMSO, HYBRI-MAX, Sigma-Aldrich, St. Louis, MO, USA). Aliquots of SRg3 and RRg3 at 6.5 and 12.7 mM, respectively, as the maximum concentrations of Rg3 epimers in aqueous media, were stored at −20 °C. The concentration of DMSO in the experiments did not exceed 0.8%, as described previously [10].

2.2. Response Surface Methodology (RSM)

To develop the RSM, the central composite design technique was employed with three levels, namely: low, mid, and high values corresponding to −1, 0, and +1, respectively, for the input parameters. The input parameters were the concentration of SRg3 and RRg3, which ranged from 0–100 µM for SRg3 and 0–50 µM for RRg3. Table 1 represents the values corresponding to low, mid, and high bounds of concentrations for the Rg3 epimers. The design matrix used in the RSM analysis is shown in Supplementary Table S1. To optimise the combination of concentrations, the RSM model reduced the total experiments to 13 iterations, with loop formation being the "main measurable target parameter".

Table 1. Low, mid, and high values used for response surface methodology (RSM) model.

Parameter	Index	Concentration (µM)		
		Lowest value (−1)	Centre Value (0)	Highest Value (+1)
SRg3	A	0	50	100
RRg3	B	0	25	50

Following optimising the combination, two other combinations were used to confirm the validity of the RSM model. These two combinations (C1 and C2) are as follows, which were tested along with the optimised combination (C3). Combination 1 (C1): SRg3 (12.5 µM) + RRg3 (6.25 µM). Combination 2 (C2): SRg3 (25 µM) + RRg3 (12.5 µM).

2.3. Proliferation Assay

A crystal violet assay was performed as previously described [26]. Briefly, cells were seeded at 800 cells per well of a 96-well plate and were cultured overnight. Single or combination concentrations of Rg3 epimers were added to the wells, and the absorbance was read at 595 nm at 3 time points, on days 0, 1, and 3, in order to assess the effect of the Rg3 epimers on the proliferation of the endothelial cell lines. The experiment included six replicates and the data are shown as mean ± standard deviation (SD).

2.4. Flow Cytometric Analysis of Cell Death

The cells were seeded at 5×10^4 cells per well on six-well plates overnight and were then exposed to Rg3 combinations for three days. Then, the samples were collected and stained using the Annexin-V-FLUOS staining kit (Roche Diagnostics, Mannheim, Germany), as previously described [10]. The samples were analysed in the BD FACSCanto II (BD Biosciences, San Jose, CA, USA) and FlowJo software, v 10.4 (FlowJo, LLC, Ashland, OR, USA). The experiment was performed in triplicate and the data are shown as mean ± SD.

2.5. Flow Cytometric Analysis of Cell Cycles

The cells were seeded at 5×10^4 cells per well on six-well plates, cultured overnight, and then exposed to C3 for 3 days. The cells were collected, fixed, stained, and analysed using BD FACSCanto II and FlowJo software, v10.4, as previously described [10]. The experiment was performed in triplicate and the data are shown as mean ± SD.

2.6. Migration Assay

A migration assay was performed based on the previously described method [27]. Briefly, HUVECs, 2H-11, and 3B-11 cells, either not pretreated or pretreated for 3 days with Rg3 epimers, were seeded in 96-well plates at 3.5×10^4, 1.2×10^4, and 4×10^4 cells per well, respectively, and were incubated overnight. A circular scratch was made in the cell monolayer. The area of the circular wound was measured based on a time of 0 and 10 h (murine endothelial cells), or 16 h (HUVEC), using ImageJ software (version 1.53a, National Health of Institute, Bethesda, MD, USA). The experiment included six replicates per treatment and the data are shown as mean ± SD.

2.7. Loop Formation Assay

A loop formation assay was optimised based on the cell proliferation index, viability, and cell number, and was performed as previously described [28]. Endothelial cells were seeded at 1.5×10^4 cells per well of a µ-plate (Ibidi, Martinsried, Germany) coated with Matrigel® (Corning) according to the manufacturer's protocol. The number of loops formed was counted at 16 h for HUVEC and 4 h for 2H-11 and 3B-11. The results are presented relative to the vehicle control. The experiment was performed in triplicate and the data are shown as mean ± SD.

2.8. Molecular Docking

For the molecular docking of Rg3 on the VEGF receptors, the SMILES structures of Rg3, sorafenib, and lenvatinib were obtained from PubChem. The crystal structure of VEGFR2 (2XIR and 3V2A) and VEGFR1 (5EX3) were from the protein data bank of NCBI (RCSB PDB). The UCSF Chimera program (version 1.15-mac64) and Autodock Vina algorithm (version 1-1-1-mac-catalina-64bit) were used to build the 3D structure of Rg3 and perform the molecular docking. The prediction of the Gibbs free energy of the protein-ligand binding was based on the flexible ligand docking simulations run within the docking grids on the interaction site of each protein, as previously described [10].

2.9. VEGFR2 Specific Interaction

To study the interaction between Rg3 epimers and VEGFR2, a VEGF bioassay kit (Promega, Madison, WI, USA) was used. It is a bioluminescent assay using KDR/NFAT-RE HEK293 cells. Upon activation of VEGFR2, intracellular signals triggered NFAT-RE-mediated luminescence. The experiment was performed according to the manufacturer's protocol. Briefly, the cells were seeded in white, flat-bottom 96-well assay plates (Delta Surface ™, Thermo Scientific, Roskilde, Denmark). Serial dilutions of SRg3 and RRg3 at final maximum concentrations of 100 and 50 µM were used alone or in combination with VEGF-A (recombinant VEGF, Promega, Madison, WI, USA) at a constant final concentration of 35 ng/mL (80% effective concentration). Bevacizumab (Avastin®, a maximum final concentration of 6 µg/mL) and VEGF-A (a maximum final concentration of 0.1 µg/mL) were used as the controls. The cells were incubated with the drugs for 6 h before a 10 min incubation with the Bio-Glo™ Reagent. Bioluminescence was measured using a FLUOstar Optima microplate reader (BMG LABTECH, Offenburg, Germany). The relative luminescence units (RLU) in each well were subtracted from the background. The experiment was performed in duplicate. GraphPad Prism (version 9.0.0 for Mac, GraphPad Software, San Diego, CA, USA, www.graphpad.com (accessed on 11 March 2021)) was used for plotting the dose–response curves (non-linear regression using log(inhibitor) vs. normalised response) and calculating the half inhibitory concentration (IC_{50}).

2.10. Quantitative PCR for the Expression of AQP1

The cells were seeded at 0.5×10^5 cells per well on six-well plates and were incubated overnight. Then, the cells were treated with Rg3 for 3 days at a normoxic (21% O_2) or hypoxic (0.1% O_2) condition. PureLink RNA mini kit (Life Technologies) was used to extract RNA and 20 ng RNA was used for reverse transcription using iScript cDNA Synthesis Kit (Bio-Rad Laboratories, Hercules, CA, USA). The duplex TaqMan Gene Expression Assays for aquaporin-1 (AQP1; Hs01028916_m1; Life Technologies) and the reference gene CCSER2 (HS00982799_mH, Life Technologies) was used in the study. Three biological replicates were used. Reactions were performed in triplicate and were analysed as previously described [10].

2.11. Enzyme-Linked Immunosorbent Assay (ELISA) for the Expression of VEGF-A

HUVEC and MDA-MB-231 cells were seeded on six-well plates at 1×10^5 cells per well on a 96-well plate. After overnight culture, the cells were exposed to C3 for three days. The expression of VEGF in these cells was compared in normoxic and hypoxic conditions. Following treatment, the supernatants were collected and centrifuged to pellet any debris. The cells were then lysed with a RIPA Lysis and Extraction Buffer (Pierce Biotechnology, Rockford, IL, USA) and the total protein was measured using Bio-Rad protein assay (Bio-Rad Laboratories). VEGF production was measured using the human VEGF-A ELISA Kit (RayBiotech, Norcross, GA, USA). The experiment was performed in duplicate, and the results are shown as mean $\pm$ SD.

2.12. Western Blotting for the Expression of Proteins Involved in Migration and Invasion

The total cell lysates were prepared and quantified as described above. Western blot was performed as previously described [28]. The anti-aquaporin-1 antibody [EPR20325] (ab219055, Abcam, Cambridge, UK, 1:1000) and goat anti-rabbit IgG H&L (ab6721, Abcam, 1:3000) were used as the primary and secondary antibodies, respectively. The experiments were repeated three times and the results are shown as mean $\pm$ SD.

2.13. AKT Pathway Phosphorylation Array

To assess the effect of Rg3 on the signaling of AKT, a Human/Mouse AKT Pathway Phosphorylation Array C1 (RayBiotech) was used. The HUVEC cells were pretreated with Rg3 or a vehicle (DMSO) for three days at normoxic and hypoxic conditions, and then the protein was collected using lysis buffer, protein inhibitor, and phosphatase inhibitor, as per the manufacturer's protocol. The protein concentration was determined using Bio-Rad Protein Assay Dye Reagent Concentrate (Bio-Rad Laboratories, Hercules, CA, USA). The density of each dot was measured using Image Lab™ Software (version 6.1). The results are shown as the mean $\pm$ SD of the two replicates.

2.14. Statistical Analysis

The results were analysed using parametric one-way or two-way analysis of variance using GraphPad Prism (version 9.0.0 for mac, GraphPad Software, San Diego, CA, USA, www.graphpad.com). The results are presented as mean $\pm$ SD for two to eight replicates, with $p < 0.05$.

3. Results

3.1. Optimisation of Concentration Combination of SRg3 and RRg3

The results of the response surface methodology modelling are depicted in Figure 1. Parameters A (SRg3), B (RRg3), and the combination of both (AB), all have significant effects (Figure 1a). Notably, AA is defined as a high concentration of SRg3, which is included as a reference. For further analysis, the Pareto chart analysis for the loop formation data (Figure 1b) reflects the effectiveness of each parameter and shows the critical parameters that needed to be investigated in this study. This chart shows that the concentration of SRg3 (A), RRg3 (B), and the combination of both drugs (AB) are key parameters playing a major role in the anti-angiogenic effects. The highest effect is sourced from SRg3, followed by the combination of both drugs and RRg3. Accordingly, the key parameter requiring optimisation is the combination of both drugs (AB), which shows a plausible efficacy and reduces the concentration needed of each if used singly. Both the Pareto analysis and standardised effect plots showed that the combination parameter is a key factor determining the efficacy of loop formation. By optimising the concentrations, the optimum region for a concentration of both drugs to give the minimum loop formation was identified, and is shown in a contour plot (Figure 1c) and surface plot (Figure 1d).

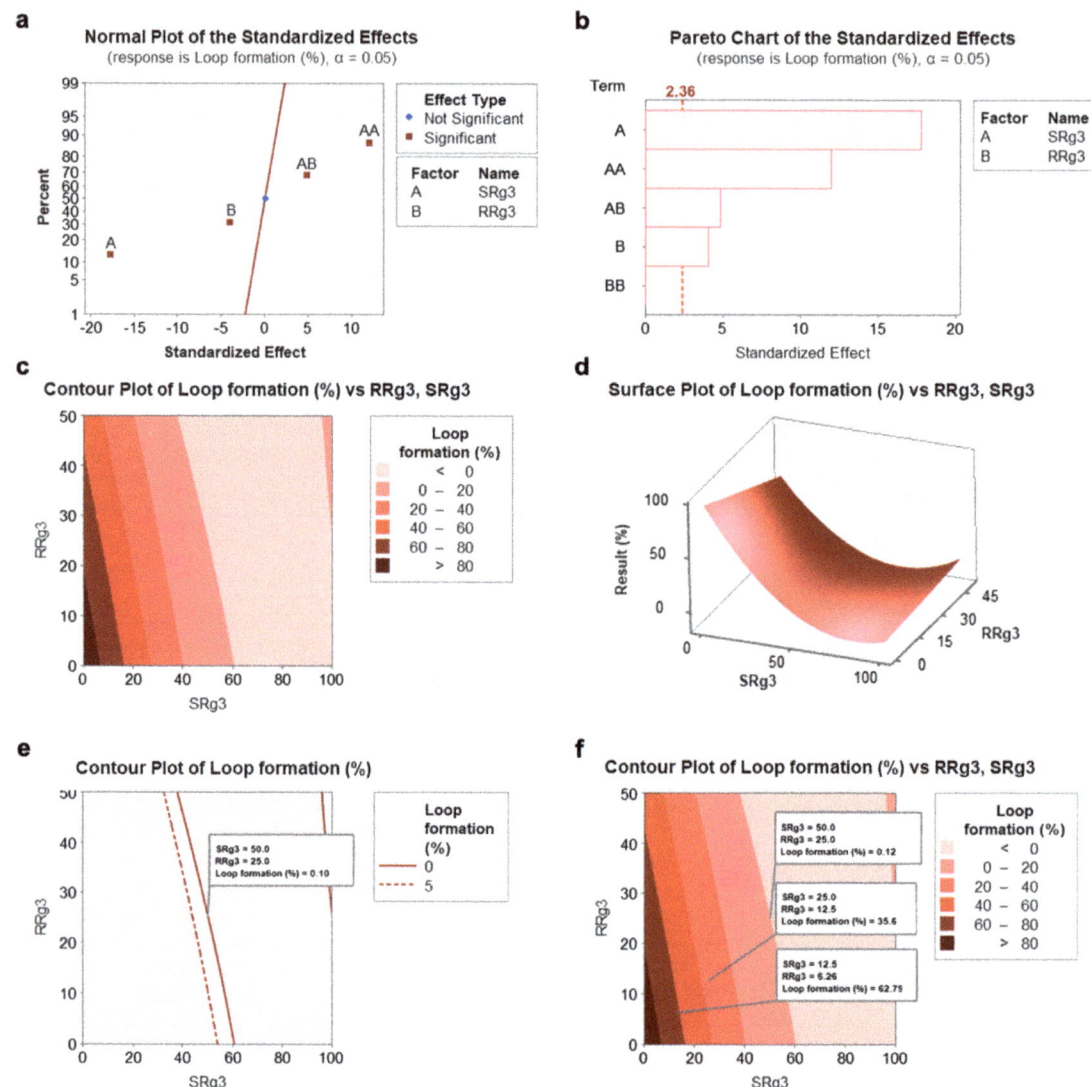

Figure 1. The calculated results of the response surface methodology developed using the central composite design technique for the optimisation of SRg3 and RRg3 drugs. (**a**) Standardised effect chart showing the critical parameters that need to be investigated, (**b**) Pareto chart analysis for loop formation data that reflects the effectiveness of each parameter, (**c**) contour plot, (**d**) surface plot for the percentage of loop formation following a combination of SRg3 or RRg3, (**e**) contour plot highlighting the area with the best efficacy for the combination, and (**f**) contour plot showing the predicted responses of two other combination treatments. A and B stand for SRg3 and RRg3, respectively, and AB represents the combination of SRg3 and RRg3. AA and BB show a mathematical expression of high concentrations of SRg3 and RRg3, respectively.

Notably, AA is defined as a high concentration of SRg3, which is included as a reference. For further analysis, the Pareto chart analysis for the loop formation data (Figure 1b) reflects the effectiveness of each parameter and shows the critical parameters that need to be investigated in this study. This chart shows that the concentration of SRg3 (A), RRg3 (B), and the combination of both drugs (AB) are key parameters playing a major role in the anti-angiogenic effects. The highest effect is sourced from SRg3, followed by the combination of both drugs and then RRg3. Accordingly, the key parameter requiring optimisation is the combination of both drugs (AB), which shows a plausible efficacy and reduces the concentration needed of each if used singly. Both the Pareto analysis and standardised effect plots showed that the combination parameter is a key factor determining the efficacy of loop formation. By optimising the concentrations, the optimum region for a concentration of both drugs to give the minimum loop formation was identified and is shown in a contour plot (Figure 1c) and surface plot (Figure 1d).

Accordingly, different areas, shown with different colours (Figure 1c), show the percentage of loop formation in response to the combination of concentrations of SRg3 and RRg3. As represented in Figure 1e, by narrowing down the identified region of the result to 0–5% loop formation, the response of the combination of 50 μM SRg3 + 25 μM RRg3 (C3) was minimised to 0.1%, in which loop formation was almost completely suppressed. Notably, a concentration of 50 μM SRg3 is a concentration that blocks AQP1 water channels [10], which, in combination with 25 μM RRg3, gives a minimum loop formation. To validate the results of this RSM model, two other combinations (C1 and C2) were considered and tested for loop formation. Figure 2a shows the results of the validation of the RSM model on HUVEC cells. As shown in Figure 1f, C1 and C2 are predicted to provide responses of 60–80% and 20–40% loop formation, respectively. In Figure 2a it is shown that C1 and C2 give mean responses of 74% and 22%, which is within the predicted regions in Figure 1f. Therefore, the identified concentration of C3 was used to conduct the rest of the experiments.

3.2. Effect of Rg3 on Loop Formation and Migration of Endothelial Cells

To show the effects of Rg3 epimers alone and in combination, loop formation and migration assays were performed at two-time points: on non-pretreated cells and three-day pretreated cells (Figure 2). In the non-pretreated state, HUVECs were the most sensitive of the three cell types to inhibitory effects of single and combination of Rg3 epimers, with C3 being the most effective combination to completely inhibit loop formation (Figure 2a–c) and cell migration (Figure 2d–f). In these cells, in this state, a dose–response relationship was observed for a single or combination of Rg3 epimers. The loop formation with RRg3 at 25 μM and 50 μM ($p = 0.0001$) was 76% and 50%, respectively. Loop formation with 50 μM SRg3 was inhibited by 74% ($p < 0.0001$) and was completely inhibited with 100 μM SRg3 ($p < 0.0001$).

With combinations of C1 and C2, loop formation was reduced to 74% ($p = 0.0348$) and 21% ($p < 0.0001$), while C3 completely inhibited loop formation ($p < 0.0001$). In this state, the murine 2H-11 and 3B-11 cell lines were less sensitive to the inhibitory effects of Rg3 and the treatment required more time to show an inhibitory action in these cell lines. 2H-11 was more sensitive to the effects of RRg3, and at 25 μM and 50 μM, loop formation was 63% and 45% ($p = 0.0023$), respectively (Figure 2b). However, only SRg3 inhibited loop formation in 3B-11. With 50 μM and 100 μM SRg3, 40% ($p = 0.0005$) and 21% ($p < 0.0001$) loop formation occurred, respectively (Figure 2c). Although the combinations did not significantly inhibit the loop formation of murine endothelial cell lines in the non-pretreated state, a dose–response pattern was observed with these treatments.

To study the time-dependency of the effects of Rg3, a three-day pretreatment was performed. Following this pretreatment of cells with Rg3, the inhibitory effects of treatment were exacerbated in all of the tested cells. In HUVEC, RRg3 at 25 and 50 μM inhibited loop formation by 35 and 70%, respectively ($p < 0.0001$), and RRg3 completely inhibited loop formation ($p < 0.0001$). With C1, only 35% loop formation occurred, and no loops formed with C2 and C3 ($p < 0.0001$). In the pretreated state, the effect of single and combination drugs increased in both murine cell lines. In 2H-11, 25 and 50 μM RRg3 decreased loop formation by an average of 59% ($p = 0.0004$) and 96% ($p < 0.0001$), respectively. SRg3 at 50 and 100 μM inhibited loop by 53% ($p < 0.0026$) and 83% ($p < 0.0001$), respectively. C1, C2 and C3, inhibited loop formation by 68%, 78% and 100% ($p < 0.0001$), respectively.

In pretreated 3B-11 cells, all single drugs inhibited loop formation by more than 90% ($p < 0.0001$), and C2 and C3 inhibited it by 73 and 100% ($p < 0.0001$), respectively. These results showed that Rg3 has a time- and dose-dependent effect on the inhibition of loop formation for endothelial cells.

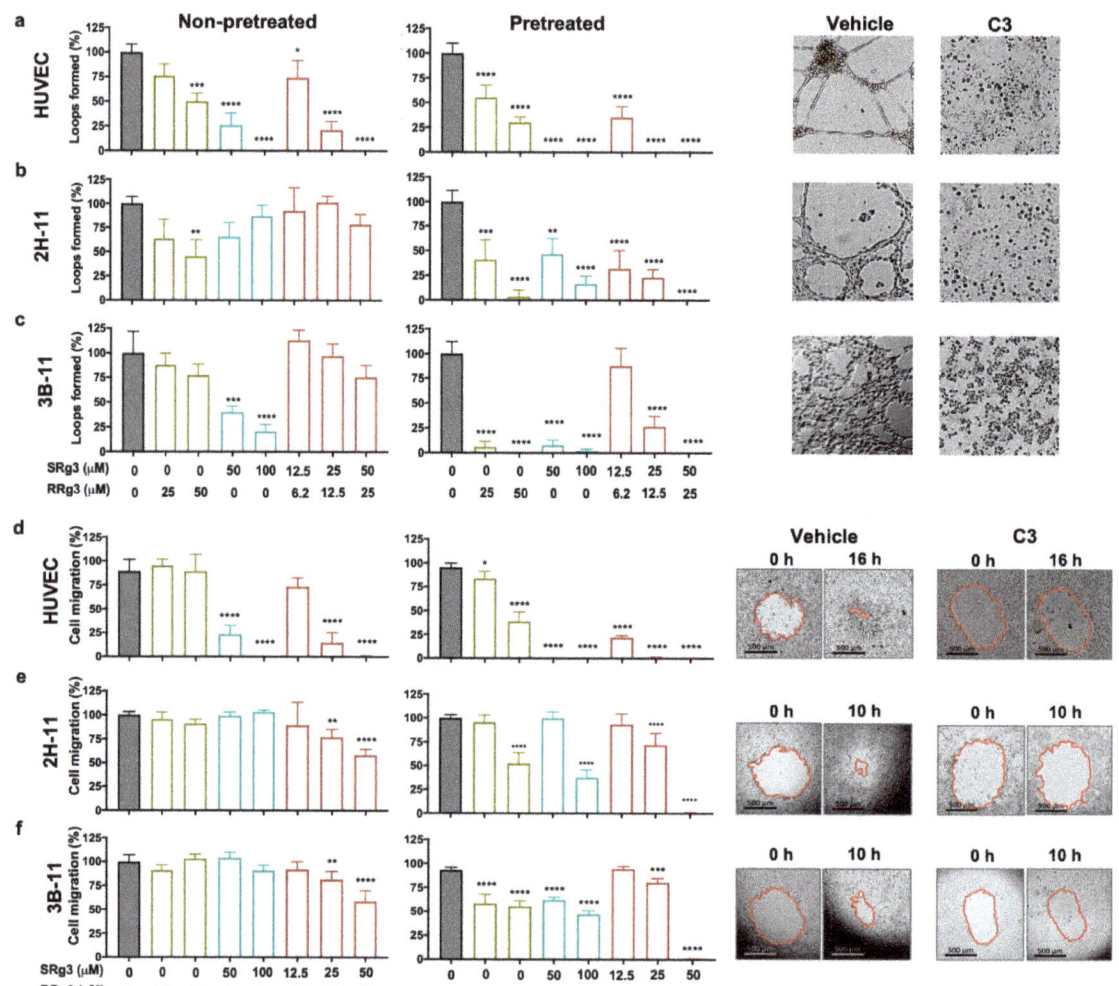

Figure 2. Effect of Rg3 epimers on loop formation (**a**–**c**) and migration (**d**–**f**) of human umbilical vein endothelial cell (HUVEC), 2H-11, and 3B-11 cells. Analysis of the loop formation and migration was performed at two timepoints; non-pretreated cells and 3-day pre-treated cells. Treatments are shown on the x axis, and (**a**–**c**) show the results of the loop formation in the HUVEC, 2H-11, and 3B-11 cell lines, respectively, at peak loop formation timepoints—16 h for HUVEC and 4 h for 2H-11 and 3B-11 cells. The experiments were done in triplicate and the results are presented as mean ± standard deviation (SD; $p < 0.05$). (**d**–**f**) show the results of cell migration in HUVEC, 2H-11 and 3B-11 cell lines, respectively. Results are presented as mean ± SD of 3 and 6 replicates for loop formation and migration assays, respectively ($p < 0.05$). The images represent the pre-treated cells. C3 represents a combination of 50 μM SRg3 + 25 μM RRg3. * $p < 0.05$, ** $p < 0.01$, *** $p < 0.001$ and **** $p < 0.0001$.

In the migration assay, the efficacy of single or a combination of Rg3 epimers was studied in non-pretreated or 3-day pretreated cells (Figure 2d–f). The trend of cells' response in this assay was similar to the results of loop formation assay. Similar to the inhibitory effects of Rg3 on loop formation, HUVEC was the most sensitive cell to the anti-migration effects of Rg3 (Figure 2d).

In non-pretreated HUVEC, only SRg3 inhibited cell migration by 66% and 80% for 50 and 100 µM, respectively ($p < 0.0001$; Figure 2d). C1, C2, and C3 inhibited loop formation dose-dependently by 16%, 75% ($p < 0.0001$), and 89% ($p < 0.0001$), respectively (Figure 2d). 2H-11 and 3B-11 were not sensitive to single epimers, but the C2 ($p < 0.001$) and C3 ($p < 0.0001$) significantly inhibited cell migration. A three-day exposure of the cells with the drugs increased the effects of single drugs. In HUVEC, the inhibitory effects of RRg3 increased and both concentrations of SRg3 completely inhibited migration ($p < 0.0001$). In 2H-11, the inhibitory effects of higher concentrations of SRg3 and RRg3 increased, while in 3B-11, the effects of all single epimers were increased. In both cell lines, C3 almost completely inhibited cell migration (Figure 2e,f). This experiment also showed time- and dose-dependent inhibition of migration by Rg3 epimers and confirmed the results of Rg3 in the loop formation assay.

3.3. Anti-Proliferative Effects of Rg3 in Endothelial Cells

As shown in Figure 3a, HUVEC was the most sensitive cell type to the anti-proliferative effects of Rg3. At equimolar concentrations, RRg3 was a less potent inhibitor of cell proliferation in HUVEC, while SRg3, C2, and C3 almost completely inhibited cell proliferation ($p < 0.0001$), and C1 had no significant inhibitory action. 2H-11 and 3B-11 were more sensitive to the combination of SRg3 and RRg3 compared with single epimers. In these two cell lines, C3 was the most effective inhibitor of cell proliferation (Figure 3a).

The induction of cell death was studied by staining the cells with annexin V and propidium iodide (PI; Figure 3b). In HUVEC cells, C2 and C3 induced about 29% ($p = 0.0003$) and 92% ($p < 0.0001$), respectively, cell death after three days of treatment. The cell death induced by C3 was associated with G0/G1 arrest in HUVEC ($p < 0.0001$; Figure 3c). Further studies showed that C2 and C3 induced the activation of caspase 3/7 in HUVEC by 536% and 980%, respectively (Figure 3d), with subsequent increases in the number of PI-positive cells (Figure 3e), consistent with late apoptosis. Interestingly, single epimers of Rg3 did not induce caspase activation in HUVEC. This shows that SRg3 and RRg3 in C2 and C3 combinations play a synergistic role in the induction of apoptosis in this cell.

In contrast, in murine 2H-11 and 3B-11 endothelial cells, C2 and C3 did not induce significant cell death (Figure 3b), while the cells were arrested in S phase (Figure 3c). Induction of cell cycle arrest in the S-phase was by 42% ($p = 0.0006$) and 63% ($p = 0.0001$) in 2H-11 and 3B-11, respectively. Therefore, it seems that the major mechanism of the inhibition of proliferation in these two cell lines is via the induction of cell cycle arrest.

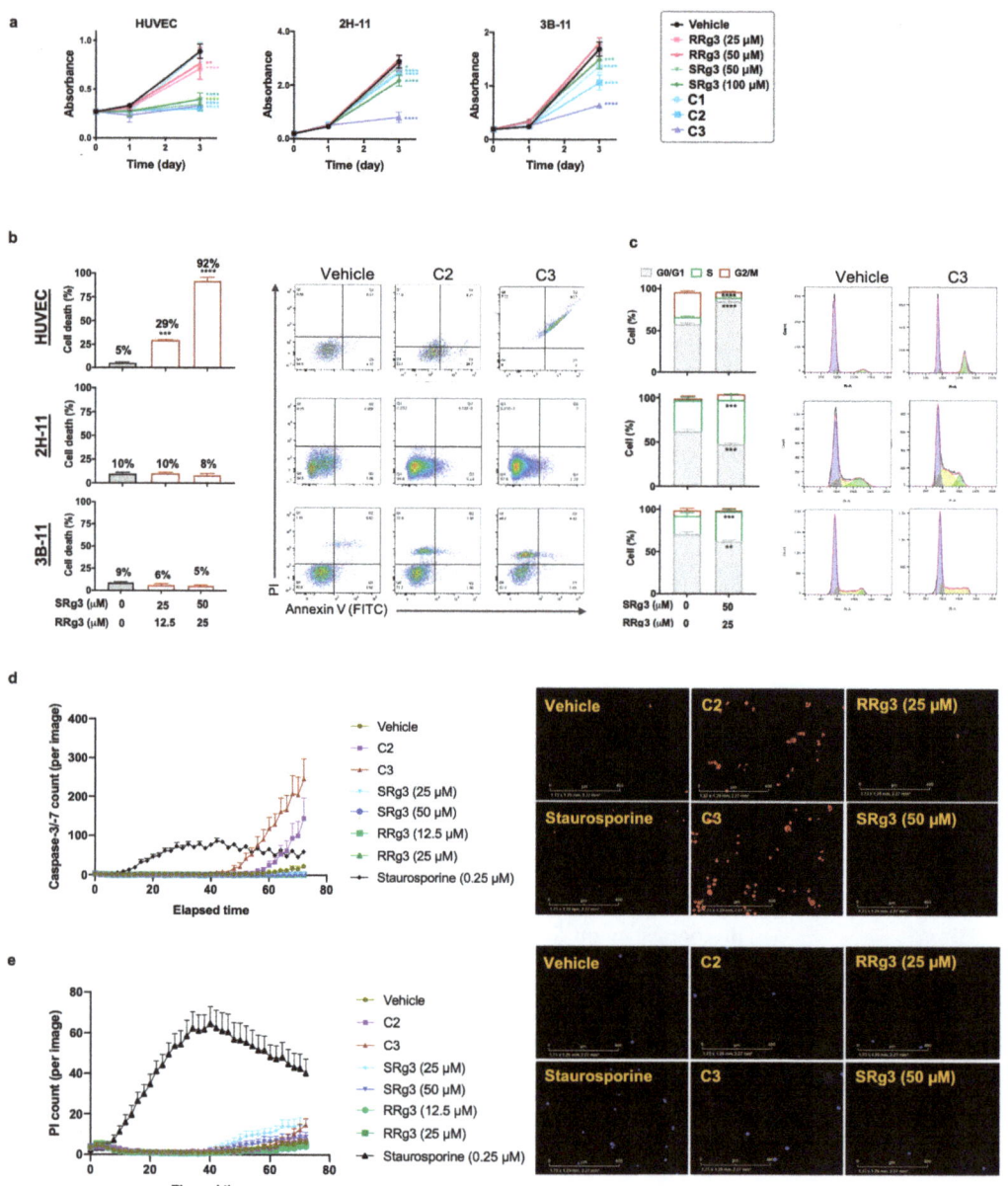

Figure 3. HUVEC, 2H-11, and 3B-11 cells were exposed to 0.8% dimethyl sulfoxide (DMSO) as a vehicle control or at concentrations of 25 and 50 µM RRg3; 50 and 100 µM SRg3; or three combinations of RRg3 + SRg3 at 6.2 + 12.5 (C1), 12.5 + 25 (C2), and 25 + 50 µM (C3). (**a**) The effect of single or combination Rg3 epimers on the proliferation of these cells in a three-day time frame. Each data point represents mean ± SD of six replicates. (**b**) The flow cytometric analysis of the induction of cell death and (**c**) cell cycle arrest in these cells by C2 and C3. Each data point represents mean ± SD of three replicates. (**d**) Activation of caspase 3/7, shown by red spots and (**e**) propidium iodide (PI) staining of cells shown by blue spots in HUVEC cells. Images are at 72 h and scale bars show 400 µm. Each data point represents mean ± SD of eight replicates. Statistical analyses were performed between the Rg3 and vehicle-treated cells ($p < 0.05$). * $p < 0.05$, ** $p < 0.01$, *** $p < 0.001$ and **** $p < 0.0001$.

3.4. The Effect of Rg3 on VEGF, VEGFR2, and Their Interaction

To further investigate the role of Rg3 epimers in angiogenesis, molecular docking was performed on two of the receptors VEGF—VEGFR1 and VEGFR2. As a comparable reference, molecular docking was also performed on two small-molecule TKIs—sorafenib and lenvatinib—which are known to interact with and inhibit VEGFR activity [29]. Human VEGFR2 includes an extracellular site with seven immunoglobulin (Ig)-like domains, and an intracellular tyrosine kinase domain, which are connected with a short transmembrane and a juxta-membrane domain (Figure 4a) [30]. The results of the molecular docking between Rg3 epimers and VEGFR1 and VEGFR2 predicted good binding scores between TKIs and the ATP-binding pocket of these receptors (Table 2). Molecular docking of SRg3 and RRg3 at this site of VEGFR2 predicted that both epimers have a strong binding with this site of the receptor, with scores of −9.0 and −8.9 kJ/mol, respectively. These scores are comparable with the binding scores of sorafenib (−9.9 kJ/mol) and lenvatinib (−9.1 kJ/mol; Table 2).

Table 2. Binding score (kJ/mol) of Rg3 epimers and growth factor receptors and the number of hydrogen binding (H-bond) predicted by Chimera program and Autodock vina algorithm.

Molecule	Binding Score (kJ/mol) (Number of H-Bonds)		
	VEGFR1	VEGFR2 [1]	VEGFR2 [2]
SRg3	4.8 (0)	−9.0 (8)	−7.2 (3)
RRg3	−7.4 (6)	−8.9 (5)	−7.0 (5)
Sorafenib	−4.9 (0)	−9.9 (0)	—
Lenvatinib	−8.9 (0)	−9.1 (0)	—

[1] Interaction with ATP-binding pocket; [2] interaction with vascular endothelial growth factor (VEGF)-binding site.

As shown in Figure 4a and summarised in Table 2, 8 and 5 H-bonds were predicted between VEGFR2 and the two epimers, SRg3 and RRg3, respectively. In both cases, Asn108, Asp180, Arg27, and Arg179 were suggested as potential H-bond residues. Although the ATP-binding cassette of VEGFR2 plays an important role in the activation of the receptor, the VEGF binding site in the extracellular side of the receptor is a key interaction site between VEGF and VEGFR2 to facilitate the intercellular signal transduction. Out of the 7 Ig-like domains, the first three domains, especially 2 and 3, mediate VEGF binding [31]. We performed a molecular docking on domains 2 and 3 of VEGFR2 and each of the Rg3 epimers. The results of this in silico study predicted binding scores of −7.2 and −7.0 kJ/mol for SRg3 and RRg3, respectively. These scores, together with the number of H-bond interactions with the receptor, were 3 and 5 H-bonds for SRg3 and RRg3, respectively, indicating strong binding. For both epimers, glycine, asparagine, and valine were the predicted amino acid residues to make H-bonds with each epimer at different positions of the domains, and provide affinity positions for H-bonds (Figure 4a).

To investigate the interaction between Rg3 and VEGFR2, in vitro, a VEGF bioassay was conducted. Figure 4b,c shows two dose–response curves of VEGF in a stimulatory state and a bevacizumab inhibitory state, respectively. VEGF, as the activator of the receptor, shows a stimulatory dose–response curve (Figure 4b) with a half effective concentration (EC_{50}) of 0.001 ng/mL (Table 3). The anti-VEGF monoclonal antibody, bevacizumab, antagonised the action of VEGF (Figure 4c) with the IC_{50} of 0.11 μg/mL (Table 3).

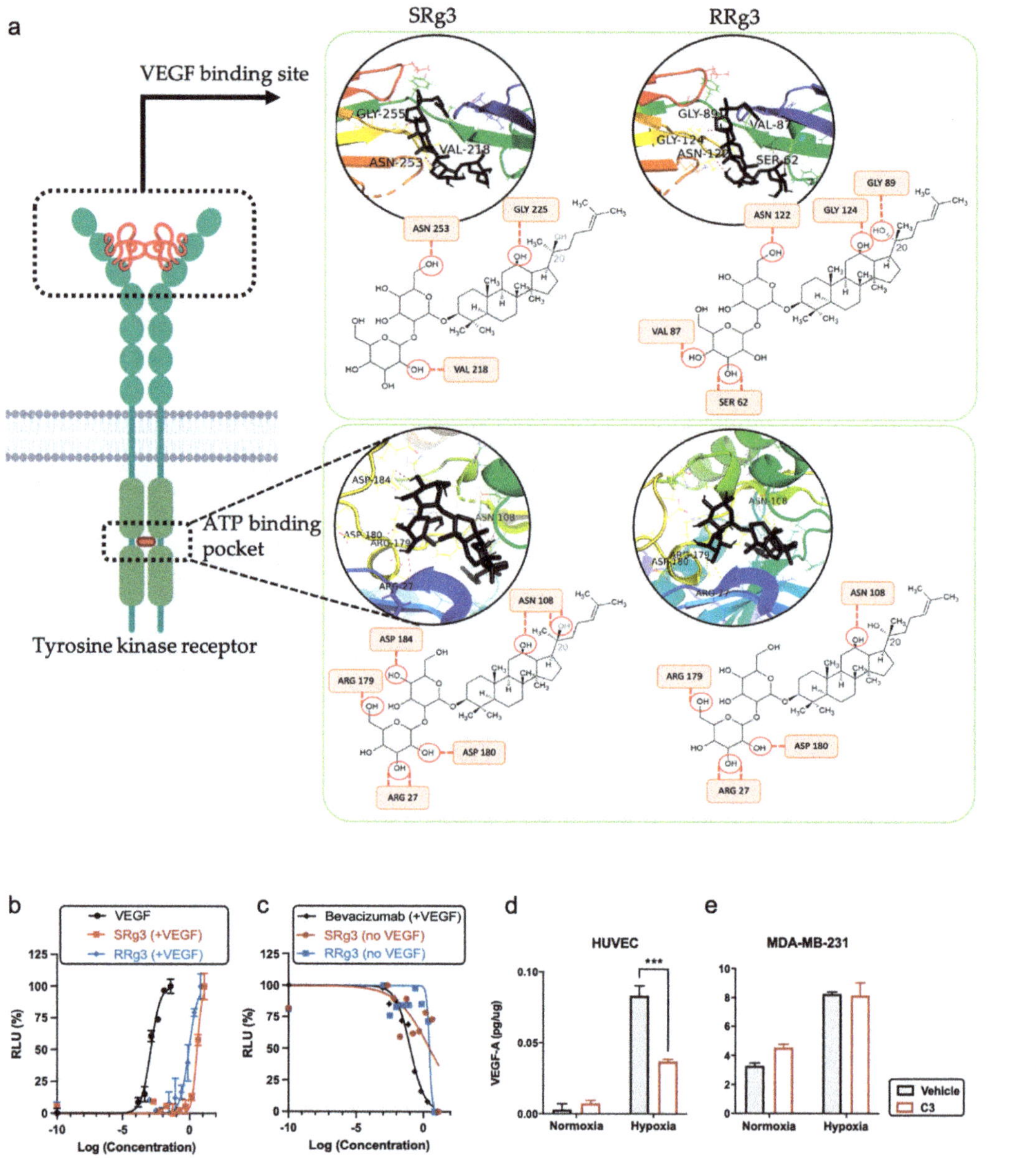

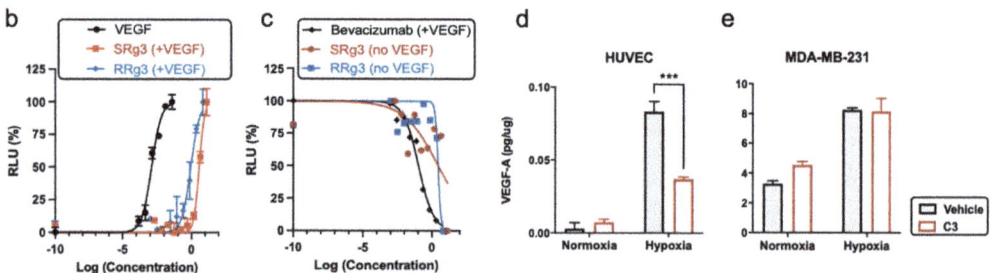

Figure 4. (**a**) A demonstration of the interaction between SRg3 and RRg3 (in black) with VEGFR2 at VEGF binding site or ATP-binding pocket. The interaction sites were predicted using molecular docking performed by AutoDock Vina algorithm. The predicted H-bonds between Rg3 and amino acid residues are shown with dashed lines. Dose–response curve of (**b**) VEGF, SRg3, and RRg3 in the presence of 35 ng/mL VEGF (stimulatory dose–response state) and (**c**) bevacizumab in the presence of 35 ng/mL VEGF, SRg3, and RRg3 alone (inhibitory dose–response state). Expression of VEGF in the presence of Rg3 in normoxic or hypoxic conditions in (**d**) HUVEC and (**e**) MDA-MB-231. The experiment was performed in duplicate, and the results are shown as mean ± SD, with $p < 0.05$. RLU—relative light units. *** $p < 0.001$.

To test the activity of Rg3 on VEGFR2, the bioassay was performed in two states: (i) in the presence of high levels (EC_{80}) of VEGF, which is a condition that encourages angiogenesis and highly activates VEGFR2, and (ii) in the absence of VEGF to test whether the molecules alone have any stimulatory or inhibitory effect on the receptor. In the presence of an EC_{80} value of VEGF (35 ng/mL), Rg3 epimers shifted the VEGF dose–response curve to the right. This means that Rg3 epimers reduced the efficacy of VEGF for the activation of VEGFR2. The EC_{50} values of SRg3 and RRg3 in this state were about 28 and 6.5 µM (Table 3). This means that RRg3 is almost four-fold more potent than SRg3 at reducing the efficacy of VEGF. SRg3, although less potent, more effectively shifted the VEGF dose–response curve to the right (Figure 4b).

Table 3. Calculated IC_{50} and EC_{50} values for VEGF (ng/mL), bevacizumab (µg/mL), SRg3 (µM), and RRg3 (µM) alone or in combination with 35 ng/mL VEGF, in interaction with VEGFR2. The experiment was performed in duplicate using the VEGF bioassay system (Promega) and was analysed using Prism software.

Compound	IC_{50}	EC_{50}	95% CI [1]	R Squared
VEGF	–	0.001	0.001–0.002	0.9781
Bevacizumab	0.11	–	0.08–0.15	0.9644
SRg3	21.23	–	3.25–8008	0.3391
RRg3	20.67	–	15.06–30.82	0.6963
SRg3 + VEGF	–	27.95	23.79–32.44	0.9670
RRg3 + VEGF	–	6.52	4.84–8.66	0.9411

[1] CI—confidence interval

To test whether the Rg3 epimers had any stimulatory effect on VEGFR2, the dose response curve was studied in the absence of VEGF. In this state, Rg3 epimers showed an almost steady response, except for the highest concentration (Figure 4c), and the regression analysis approach was not a plausible technique to fit a non-linear sigmoidal dose–response function, which resulted in poor R squared values (Table 3). However, the overall trend of their effect was inhibitory, as at the highest concentrations used (100 and 50 µM for SRg3 and RRg3, respectively), the response was minimised (Figure 4c). The decreased bioluminescence detected at the highest tested concentrations could be due to the cytotoxicity of the molecules, rather than the inhibitory effect of the drugs on the receptor.

Given the observed effects of Rg3 in this system and the results obtained from the molecular docking, it seems probable that Rg3 is an allosteric modulator of VEGFR2. In the absence of VEGF, as the primary ligand, SRg3 and RRg3 had a minimum activity on the receptor, while in the presence of VEGF, Rg3 epimers decreased the efficacy of VEGF-VEGFR2 interactions potentially by changing the conformation of the receptor. Furthermore, the efficacy of C3 on the VEGF expression in normoxic and hypoxic conditions in HUVEC and MDA-MB-231 cells was studied (Figure 4d,e). C3 did not have any significant effect on VEGF expression in MDA-MB-231 (Figure 5e), but in hypoxic HUVEC cells, when the expression of VEGF was significantly increased in vehicle-treated cells, Rg3 decreased this expression ($p = 0.0008$; Figure 4d).

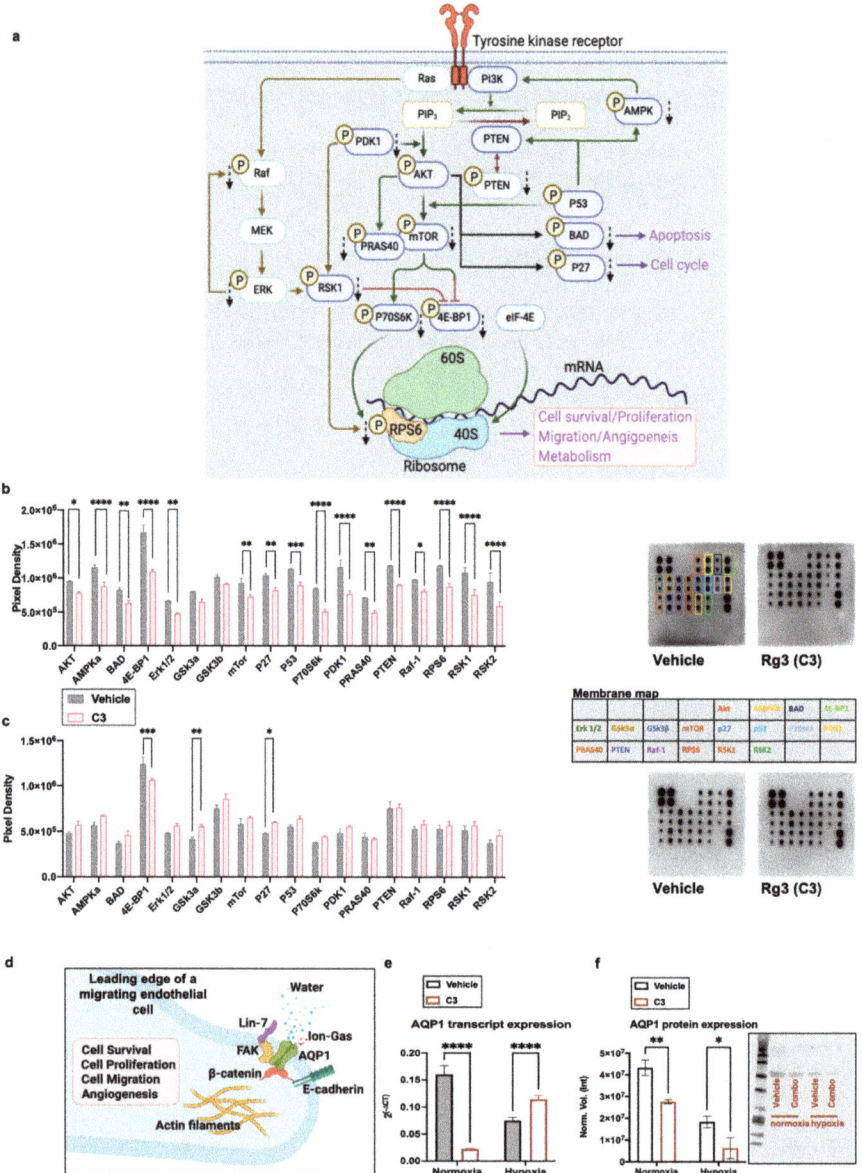

Figure 5. (**a**) A schematic diagram of the signaling of the downstream activation of a receptor tyrosine kinase, leading to cell survival, proliferation, migration, angiogenesis, and metabolism (created with BioRender.com (accessed on 11 March 2021)). The effects of C3 (50 µM SRg3 + 25 µM RRg3) on the phosphorylation of signaling proteins in PI3K/AKT signaling in HUVEC cells grown in (**b**) normoxia and (**c**) hypoxia (1% O_2) conditions. The cells were exposed with C3 for three days. The data show the mean ± SD of two replicates with $p < 0.05$. (**d**) A schematic diagram of the role of AQP1 localised in the leading edge of an endothelial cell facilitating cell migration (reviewed in [9]) (created with BioRender.com). The cells were exposed with C3 for thee days in normoxic or hypoxic (1% O_2) conditions. Expression levels of (**e**) AQP1 transcripts measured with QT-PCR or (**f**) proteins detected using the Western blotting technique in normoxic and hypoxic conditions are shown. The data represent the mean ± SD of three replicates with $p < 0.05$. * $p < 0.05$, ** $p < 0.01$, *** $p < 0.001$ and **** $p < 0.0001$.

3.5. Effect of Rg3 on the AKT Signalling Pathway and AQP1

Signaling of PI3K/AKT and its interaction with the Raf/MEK/ERK signal transduction pathway (Figure 5a) regulates several proteins controlling cell survival, proliferation, migration, and metabolism. To test whether C3 treatment has any effects on the signaling of AKT, a protein array was performed. C3 affected the phosphorylation of proteins downstream of the activation of AKT in both normoxia and hypoxia, although the effects in normoxia were more extensive (Figure 5). In normoxic conditions, Rg3 combined with C3 affected the phosphorylation of several proteins important in the signaling of AKT. The phosphorylation of AKT was decreased ($p = 0.017$) and regulators for the activation of AKT, including AMP-activated protein kinase (AMPK), phosphatase and tensin homolog (PTEN), and phosphoinositide-dependent kinase-1 (PDK1), were decreased ($p < 0.0001$) (Figure 5b). C3 also decreased the phosphorylation of the BCL2 associated agonist of cell death (BAD; $p = 0.0048$; Figure 5b), which plays important roles in AKT-mediated cell survival. C3 decreased the phosphorylation of cyclin-dependent kinase inhibitor 1B (p27^{kip1}; $p = 0.0015$; Figure 5b), hence keeping it in its active form, which could cause p27^{kip1}-mediated G1 arrest. With C3, the activation of p53 also decreased ($p = 0.0003$), which could also affect the activation of mTOR.

This experiment showed that C3 decreased the phosphorylation of mTOR ($p = 0.0045$), PRAS40 ($p = 0.0012$), P70S6K ($p < 0.0001$), 4E-BP1 ($p < 0.0001$), and RPS6 ($p < 0.0001$). These proteins play roles in the translation function of migrating cells. Furthermore, C3 decreased the phosphorylation of Raf ($p = 0.0187$), ERK 1/2 ($p = 0.0066$), RSK1, and RSK2 ($p < 0.0001$). In hypoxic conditions, the C3 affected proteins included 4E-BP1 ($p = 0.0003$), glycogen synthase kinase-3a (GSK3a; $p = 0.0086$), and p27^{kip1} ($p = 0.0323$).

AQP1 in combination with other proteins at the leading edge of a migrating cell facilitates cell migration (Figure 5d). We showed that in normoxic conditions, the levels of AQP1 transcript ($p < 0.0001$) and protein ($p = 0.0268$) were significantly decreased. In hypoxic conditions, although the transcript levels were increased, the protein levels were decreased ($p = 0.0195$; Figure 5e,f). In addition, in a mouse endothelial cell line, 3B-11, a significantly decreased expression of the AQP1 transcript was observed (Supplementary Figure S1). However, we did not detect any significant changes in the activation of focal adhesion kinase (FAK), as another player in the AQP1-facilitated migration in HUVECs (Supplementary Figure S2). Therefore, it seems that AQP1 is a more important protein in the C3-induced inhibition of migration in HUVECs. Furthermore, preliminary testing showed that in C2 treated HUVEC, the expression of the AQP1 transcript and protein and the activation of AKT were reduced (Supplementary Figure S3).

4. Discussion

A few studies have investigated the mechanisms of action of Rg3 as an anti-angiogenic agent. Keung et al. (2016) showed that RRg3 exerted its anti-angiogenic effects via an increased expression of hsa-miR-520h, which targeted ephrin type-B receptor 2 (EphB2) and EphB4 as a mediator of cancer migration and angiogenesis [32]. In addition, it was shown that an unspecified epimer of Rg3 (64 µM) decreased the protein and transcript expression of VEGF, basic fibroblast growth factor (b-FGF), matrix metalloproteinase-2 (MMP-2), and MMP-9 [33]. Because of the stereoselective activity of Rg3 epimers, for the first time, in this study, C3 was introduced as an optimised combination of SRg3 and RRg3 and a novel anti-angiogenic agent. This combination showed time- and dose-dependent anti-angiogenic properties in vitro. HUVECs were more sensitive to these effects of Rg3 than to the murine endothelial cell lines.

To further investigate the mechanisms involved in the anti-angiogenic properties of Rg3 and more specifically C3, (i) the effects of Rg3 on the VEGF–VEGFR2 interaction and (ii) the anti-angiogenic mechanisms Rg3 combination (C3) were studied. Molecular docking predicted good biding scores and VEGFR2, comparable to the binding of known TKIs. We showed that Rg3 has no stimulatory action on VEGFR2. The antiangiogenic effects observed by Rg3 are not comparable with a drug such as bevacizumab. Bevacizumab is a monoclonal antibody against VEGF, while Rg3 has several mechanisms, one of which is via the interaction with the activation of VEGFR2. For the first time, we showed that the interaction between Rg3 and VEGFR2 decreased the efficacy of VEGF on the system, working as an allosteric modulator. Allosteric modulators are of special interest in pharmacology. Since the introduction and successful treatment profile of benzodiazepines as allosteric ligands of γ-aminobutyric acid-A (GABA$_A$) receptors, versus the toxic direct-acting agonists of this receptor, much more attention has been paid to finding and registering allosteric drugs for various diseases (reviewed in [34]). Allosteric modulators offer several advantages over orthosteric ligands, such as subtype selectivity within receptor families and less adverse side effects [34,35]. Of special interest are tyrosine kinases, which play roles in several human diseases such as cancer. The ATP-binding pocket of kinases is a highly conserved part, and this results in a low selectivity and, consequently, off-target and side effects for the inhibitors designed for this target. Other types of inhibitors either bind at the ATP site extending into an adjacent allosteric pocket, specifically bind to the allosteric pockets near the ATP pocket, or bind to allosteric sites more remote from the ATP pocket [34]. The fact that Rg3 has been administered to humans without any reported serious side effects (reviewed in [8,9]) could be evidence for the safety of Rg3 allosterism. This research provided evidence of the anti-VEGFR2 action of Rg3 epimers as one of the anti-angiogenic mechanisms of these molecules. To confirm the interaction site of Rg3 with VEGFR2, further experiments are required in future research.

This research also provided evidence of the effectiveness of C3 in hypoxic conditions. This is especially important because of the importance of hypoxia in driving tumour invasiveness [36]. Hypoxia is a common feature of rapidly growing tumours, which affects tumour metabolism, metastasis, and resistance to chemotherapy [37], and is linked to a poor prognosis for several tumours (reviewed in [38]). Hypoxia leads to VEGF expression to encourage angiogenesis in the tumour. Increased VEGF expression promotes endothelial cell proliferation and migration, inhibits apoptosis in these cells, and facilitates the degradation of the extracellular matrix and endothelial cell migration and invasion [39]. Some solid tumours such as breast cancers overexpress VEGF and its receptors. This led to the development of anti-angiogenic drugs for these patients [40]. Our group is interested in developing novel treatments for breast cancer, in which VEGF expression is an independent prognostic factor and a possible target of treatment [41]. Both endothelial and breast cancer cells have an autocrine VEGF signaling pathway that supports angiogenesis and cancer progression [42]. Hence, we measured VEGF expression in both HUVEC and MDA-MB-231, in hypoxic and normoxic conditions, and showed that C3 significantly decreased VEGF expression only in hypoxic HUVEC cells and not in MDA-MB-231. Whether C3 has efficacy on other breast cancer cell lines should be further assessed. However, the responding cells, endothelial cells, play key roles in angiogenesis.

Downstream of the activation of VEGFR2, several signaling pathways are activated, including PI3K/AKT/mTOR signaling, the activation of which is one of the hallmark signaling pathways in cancer and angiogenesis [43,44]. Therefore, inhibitors of mTOR signaling have gained plenty of attention in cancer treatment. Currently, 70 trials of the inhibitors of mTOR signaling are recruiting in several tumour types, such as breast, lung, colorectal, and hematological tumours (https://clinicaltrials.gov/ (accessed on 11 March 2021)). In addition to several cellular functions, the activation of mTOR also plays roles in VEGF production and angiogenesis [45]. Inhibitors of this pathway, inhibiting either PI3K/mTOR or mTOR alone, show anti-angiogenic properties (reviewed in [46]). For example, rapamycin, an inhibitor of mTORC1, also inhibits VEGF production and angiogenesis [47]. As reviewed before, in leukemic and ovarian cancer models, Rg3 affected PI3K/AKT signaling [8]. Therefore, we examined the effects of C3 on this signaling pathway in normoxic and hypoxic conditions. In the normoxic condition, except GSK3, all other tested proteins were affected with C3. The proteins that showed more than 30% decreased phosphorylation were those that were related to mTORC1 function, including PRAS40 (a component and substrate of mTORC1); P70S6K, which phosphorylates and activates

RSP6, a component of 40S ribosomal subunit; and 4E-BP1, which, upon phosphorylation, releases eIF-4E, as one of the key components of ribosomal translation initiation for regulators of mTOR function including PDK1 and RSKs. PRAS40, when dephosphorylated, inhibits mTOR signaling, consequently decreasing ribosomal transcription via affecting the activation of 4E-BP1 and P70S6k, both of which play roles in tumour angiogenesis [48,49].

Decreased phosphorylation of 4E-BP1 was also observed both in hypoxic and normoxic HUVEC cells exposed to C3, thus implicating 4E-BP1 as having an important role in C3 mediated anti-angiogenic effects. In particular, as mTORC1, via mechanisms involving 4E-BP1, drives VEGF signaling in hypoxic conditions [50], it could be considered that the decreased expression of VEGF observed in these cells was mediated through mTORC1/4E-BP1. The leading edge of migrating cells is where many fundamental biological and biochemical processes occur to facilitate cell migration, including 4E-BP1, PRAS40, mRNAs, and translation initiation factors [51,52]. Therefore, one major mechanism of C3 could be via the inhibition of the translational function of mTOR.

Additionally, hypoxia-induced-endothelial cell proliferation requires functional mTOR complexes [53]. C3, via the decreased phosphorylation of 4E-BP1, could decrease the functionality of mTORC1 and hence play a contributing role in the decreased proliferation of these cells. C3, in hypoxic conditions, caused minor increased levels of p-p27^{Kip1}, which is a negative regulator of G1 cell cycle progression. It has been shown that GSK3 stabilises the levels of p27^{Kip1} and decreases cell proliferation [54] and hence, the observed minor increased levels of p27^{Kip1} could be a consequence of the deactivation of GSK3.

GSK3 is a constitutively active kinase, an activator of AKT, mTORC1, and mTORC2, which is in feedback and crosstalk with PI3K/AKT/mTOR. Phosphorylation on SER-21 (GSK3a) and SER-9 (GSK3b) is initiated by the growth factor activation of AKT/mTOR and inhibits GSK3 function [55], and C3 increased this phosphorylation to decrease the activation of this signaling. This could further decrease the functionality of mTORC1 and mTORC2. This is especially important, as mTORC2 is an essential cellular energy production element, which promotes cancer progression via lipid formation and fueling the PI3K/AKT/mTOR pathway [56]. It also plays roles in driving angiogenesis multiple myeloma, where mTORC2 inhibitors restrict angiogenesis in this tumour model [57]. mTORC2 is one of the molecular targets that is in advanced stages of translational application, and whether C3 has any inhibitory action on mTORC2 needs to be further investigated.

Some AQPs such as AQP1, AQP4, and AQP5 localise at the leading edge of migrating cells [58]. AQP1, for example, polarises at the leading edge, a phenomenon that is associated with an increased turnover of cell membrane protrusions and enhanced cell migration (reviewed in [12]). AQP1 was recognised as a pro-angiogenic factor [59], which, independent of VEGF, was required for the hypoxia-induced tube forming capacity of human retinal vascular endothelial cells [60]. AQP1-deficient cells were shown to have impaired migration and tube formation [61]. We have also shown that blockers of AQP1 impair angiogenesis [27,28]. In addition, using the molecular docking and oocyte swelling assay, we showed that Rg3 blocked AQP1 [10]. Therefore, it could be concluded that the blockage of AQP1 could contribute to the immediate inhibition of the loop formation observed. After a three-day pretreatment with C2 and C3, the protein expression of AQP1 was decreased. FAK, another important contributor to endothelial cell migration via VEGFR2-signalling or complexing with AQP1 (reviewed in [13]), did not seem to be involved in the mechanism of action of C3, which further highlights the role of AQP1 in this process.

5. Conclusions

In conclusion, we showed that Rg3 had a time- and dose-dependent inhibition of the migration and invasion of endothelial cells. The optimised combination of SRg3 and RRg3 inhibited the proliferation, migration, and invasion of endothelial cells. SRg3 and RRg3 potentiated each other's action in activating caspase 3/7 and inducing apoptosis, which was the major anti-angiogenic mechanism. This action was measured after 3 days of exposure with the treatment. Besides the induction of apoptosis, other inhibitory mechanisms were also involved that assisted with the anti-angiogenic action of Rg3. As our studies showed, these molecules were allosteric modulators of VEGFR2, and therefore potentially had far fewer off-target effects with less clinical side effects expected. A reduced expression of VEGF and AQP1, and decreased PI3K/AKT/mTOR signaling are suggested mechanisms of this drug. Further studies are needed to confirm the anti-angiogenic effects of C3 in vivo.

Supplementary Materials: The following are available online at https://www.mdpi.com/article/10.3390/cancers13092223/s1, Figure S1: Quantitative PCR for transcript expression of AQP1 in murine 3B-11 cell line following exposure with C3 for 3 days. Figure S2: Western blot analysis of activation of focal adhesion kinase (FAK) in HUVEC exposed to vehicle (V) or C3 (50 µM SRg3 + 25 µM RRg3), in normoxia or hypoxia conditions. Figure S3: HUVEC was exposed to vehicle (V) or C2 (25 µM SRg3 + 12.5 µM RRg3) for 3 days. Table S1: Design table developed for the RSM model.

Author Contributions: Conceptualization, M.N. and J.E.H.; methodology, M.N., E.S., H.M.P., K.Y. and Y.T.; software, M.N.; formal analysis, M.N., E.S. and K.Y.; investigation, M.N., E.S., J.E.H. and A.R.T.; resources, A.R.T., J.E.H. and T.J.P.; data curation, J.E.H. and A.R.T.; writing—original draft preparation, M.N.; writing—review and editing, E.S., H.M.P., Y.T., K.Y., A.R.T., T.J.P. and J.E.H.; visualization, M.N.; supervision, J.E.H. and A.R.T.; project administration, J.E.H.; funding acquisition, A.R.T., J.E.H. and T.J.P. All authors have read and agreed to the published version of the manuscript.

Funding: This research was funded by the Margaret Elcombe Hospital Research Foundation Research Grant.

Institutional Review Board Statement: Not applicable.

Informed Consent Statement: Not applicable.

Data Availability Statement: All of the data are available in the paper or as Supplementary Data.

Conflicts of Interest: The authors declare no conflict of interest.

References

1. Ollauri-Ibáñez, C.; Astigarraga, I. Use of antiangiogenic therapies in pediatric solid tumors. *Cancers* **2021**, *13*, 253. [CrossRef] [PubMed]
2. Maennling, A.E.; Tur, M.K.; Niebert, M.; Klockenbring, T.; Zeppernick, F.; Gattenlöhner, S.; Meinhold-Heerlein, I.; Hussain, A.F. Molecular targeting therapy against EGFR family in breast cancer: Progress and future potentials. *Cancers* **2019**, *11*, 1826. [CrossRef] [PubMed]
3. He, B.; Ganss, R. Modulation of the vascular-immune environment in metastatic cancer. *Cancers* **2021**, *13*, 810. [CrossRef] [PubMed]
4. Taugourdeau-Raymond, S.; Rouby, F.; Default, A.; Jean-Pastor, M.-J. Bevacizumab-induced serious side-effects: A review of the French pharmacovigilance database. *Eur. J. Clin. Pharmacol.* **2012**, *68*, 1103–1107. [CrossRef]
5. Hartmann, J.T.; Haap, M.; Kopp, H.-G.; Lipp, H.-P. Tyrosine kinase inhibitors-a review on pharmacology, metabolism and side effects. *Curr. Drug Metab.* **2009**, *10*, 470–481. [CrossRef]
6. Haibe, Y.; Kreidieh, M.; El Hajj, H.; Khalifeh, I.; Mukherji, D.; Temraz, S.; Shamseddine, A. Resistance mechanisms to anti-angiogenic therapies in cancer. *Front. Oncol.* **2020**, *10*. [CrossRef]
7. Yun, U.-J.; Lee, I.H.; Lee, J.-S.; Shim, J.; Kim, Y.-N. Ginsenoside Rp1, A ginsenoside derivative, augments anti-cancer effects of Actinomycin D via downregulation of an AKT-SIRT1 pathway. *Cancers* **2020**, *12*, 605. [CrossRef]
8. Nakhjavani, M.; Hardingham, J.E.; Palethorpe, H.M.; Tomita, Y.; Smith, E.; Price, T.J.; Townsend, A.R. Ginsenoside Rg3: Potential molecular targets and therapeutic indication in metastatic breast cancer. *Medicines* **2019**, *6*, 17. [CrossRef]
9. Nakhjavani, M.; Smith, E.; Townsend, A.R.; Price, T.J.; Hardingham, J.E. Anti-angiogenic properties of ginsenoside Rg3. *Molecules* **2020**, *25*, 4905. [CrossRef]
10. Nakhjavani, M.; Palethorpe, H.M.; Tomita, Y.; Smith, E.; Price, T.J.; Yool, A.J.; Pei, J.V.; Townsend, A.R.; Hardingham, J.E. Stereoselective anti-cancer activities of ginsenoside Rg3 on triple negative breast cancer cell models. *Pharmaceuticals* **2019**, *12*, 117. [CrossRef]
11. Nico, B.; Ribatti, D. Aquaporins in tumor growth and angiogenesis. *Cancer Lett.* **2010**, *294*, 135–138. [CrossRef]
12. Papadopoulos, M.; Saadoun, S.; Verkman, A. Aquaporins and cell migration. *Pflügers Arch. Eur. J. Physiol.* **2008**, *456*, 693–700. [CrossRef]
13. Tomita, Y.; Dorward, H.; Yool, A.J.; Smith, E.; Townsend, A.R.; Price, T.J.; Hardingham, J.E. Role of aquaporin 1 signalling in cancer development and progression. *Int. J. Mol. Sci.* **2017**, *18*, 299. [CrossRef]
14. Jeong, S.M.; Lee, J.-H.; Kim, J.-H.; Lee, B.-H.; Yoon, I.-S.; Lee, J.-H.; Kim, D.-H.; Rhim, H.; Kim, Y.; Nah, S.-Y. Stereospecificity of Ginsenoside Rg 3 Action on Ion Channels. *Mol. Cells* **2004**, *18*, 383–389.

15. Kim, J.-H.; Lee, J.-H.; Jeong, S.M.; Lee, B.-H.; Yoon, I.-S.; Lee, J.-H.; Choi, S.-H.; Kim, D.-H.; Park, T.-K.; Kim, B.-K. Stereospecific effects of ginsenoside Rg3 epimers on swine coronary artery contractions. *Biol. Pharm. Bull.* **2006**, *29*, 365–370. [CrossRef]
16. Wei, X.; Su, F.; Su, X.; Hu, T.; Hu, S. Stereospecific antioxidant effects of ginsenoside Rg3 on oxidative stress induced by cyclophosphamide in mice. *Fitoterapia* **2012**, *83*, 636–642. [CrossRef]
17. Wei, X.; Chen, J.; Su, F.; Su, X.; Hu, T.; Hu, S. Stereospecificity of ginsenoside Rg3 in promotion of the immune response to ovalbumin in mice. *Int. Immunol.* **2012**, *24*, 465–471. [CrossRef]
18. Wu, R.; Ru, Q.; Chen, L.; Ma, B.; Li, C. Stereospecificity of Ginsenoside Rg3 in the promotion of cellular immunity in hepatoma H22-bearing mice. *J. Food Sci.* **2014**, *79*, H1430–H1435. [CrossRef]
19. Kim, Y.-J.; Choi, W.-I.; Jeon, B.-N.; Choi, K.-C.; Kim, K.; Kim, T.-J.; Ham, J.; Jang, H.J.; Kang, K.S.; Ko, H. Stereospecific effects of ginsenoside 20-Rg3 inhibits TGF-β1-induced epithelial–mesenchymal transition and suppresses lung cancer migration, invasion and anoikis resistance. *Toxicology* **2014**, *322*, 23–33. [CrossRef]
20. Myers, R.H.; Montgomery, D.C.; Anderson-Cook, C.M. *Response Surface Methodology: Process and Product Optimization Using Designed Experiments*; John Wiley & Sons: Hoboken, NJ, USA, 2016.
21. Razura-Carmona, F.F.; Pérez-Larios, A.; González-Silva, N.; Herrera-Martínez, M.; Medina-Torres, L.; Sáyago-Ayerdi, S.G.; Sánchez-Burgos, J.A. Mangiferin-loaded polymeric nanoparticles: Optical characterization, effect of anti-topoisomerase I., and cytotoxicity. *Cancers* **2019**, *11*, 1965. [CrossRef]
22. Liou, J.-Y.; Tsou, M.-Y.; Ting, C.-K. Response surface models in the field of anesthesia: A crash course. *Acta Anaesthesiol. Taiwanica* **2015**, *53*, 139–145. [CrossRef] [PubMed]
23. Zafar, S.; Akhter, S.; Ahmad, I.; Hafeez, Z.; Rizvi, M.M.A.; Jain, G.K.; Ahmad, F.J. Improved chemotherapeutic efficacy against resistant human breast cancer cells with co-delivery of Docetaxel and Thymoquinone by Chitosan grafted lipid nanocapsules: Formulation optimization, in vitro and in vivo studies. *Colloids Surf. B Biointerfaces* **2020**, *186*, 110603. [CrossRef] [PubMed]
24. Lee, J.J.; Kong, M.; Ayers, G.D.; Lotan, R. Interaction index and different methods for determining drug interaction in combination therapy. *J. Biopharm. Stat.* **2007**, *17*, 461–480. [CrossRef] [PubMed]
25. Malfettone, A.; Silvestris, N.; Paradiso, A.; Mattioli, E.; Simone, G.; Mangia, A. Overexpression of nuclear NHERF1 in advanced colorectal cancer: Association with hypoxic microenvironment and tumor invasive phenotype. *Exp. Mol. Pathol.* **2012**, *92*, 296–303. [CrossRef] [PubMed]
26. Smith, E.; Palethorpe, H.; Tomita, Y.; Pei, J.; Townsend, A.; Price, T.; Young, J.; Yool, A.; Hardingham, J. The purified extract from the medicinal plant bacopa monnieri, bacopaside II, inhibits growth of colon cancer cells in vitro by inducing cell cycle arrest and apoptosis. *Cells* **2018**, *7*, 81. [CrossRef]
27. Tomita, Y.; Palethorpe, H.M.; Smith, E.; Nakhjavani, M.; Townsend, A.R.; Price, T.J.; Yool, A.J.; Hardingham, J.E. Bumetanide-derived aquaporin 1 inhibitors, AqB013 and AqB050 inhibit tube formation of endothelial cells through induction of apoptosis and impaired migration in vitro. *Int. J. Mol. Sci.* **2019**, *20*, 1818. [CrossRef]
28. Palethorpe, H.; Tomita, Y.; Smith, E.; Pei, J.; Townsend, A.; Price, T.; Young, J.; Yool, A.; Hardingham, J. The aquaporin 1 inhibitor bacopaside II reduces endothelial cell migration and tubulogenesis and induces apoptosis. *Int. J. Mol. Sci.* **2018**, *19*, 653. [CrossRef]
29. Jiao, Q.; Bi, L.; Ren, Y.; Song, S.; Wang, Q.; Wang, Y.-s. Advances in studies of tyrosine kinase inhibitors and their acquired resistance. *Mol. Cancer* **2018**, *17*, 1–12. [CrossRef] [PubMed]
30. Thieltges, K.M.; Avramovic, D.; Piscitelli, C.L.; Markovic-Mueller, S.; Binz, H.K.; Ballmer-Hofer, K. Characterization of a drug-targetable allosteric site regulating vascular endothelial growth factor signaling. *Angiogenesis* **2018**, *21*, 533–543. [CrossRef]
31. Brozzo, M.S.; Bjelić, S.; Kisko, K.; Schleier, T.; Leppänen, V.-M.; Alitalo, K.; Winkler, F.K.; Ballmer-Hofer, K. Thermodynamic and structural description of allosterically regulated VEGFR-2 dimerization. *Blood* **2012**, *119*, 1781–1788. [CrossRef]
32. Keung, M.-H.; Chan, L.-S.; Kwok, H.-H.; Wong, R.N.-S.; Yue, P.Y.-K. Role of microRNA-520h in 20 (R)-ginsenoside-Rg3-mediated angiosuppression. *J. Ginseng Res.* **2016**, *40*, 151–159. [CrossRef] [PubMed]
33. Li, J.-P.; Zhao, F.-L.; Yuan, Y.; Sun, T.-T.; Zhu, L.; Zhang, W.-Y.; Liu, M.-X. Studies on anti-angiogenesis of ginsenoside structure modification HRG in vitro. *Biochem. Biophys. Res. Commun.* **2017**, *492*, 391–396. [CrossRef] [PubMed]
34. Wenthur, C.J.; Gentry, P.R.; Mathews, T.P.; Lindsley, C.W. Drugs for allosteric sites on receptors. *Annu. Rev. Pharmacol. Toxicol.* **2014**, *54*, 165–184. [CrossRef] [PubMed]
35. Grover, A.K. Use of allosteric targets in the discovery of safer drugs. *Med Princ. Pract.* **2013**, *22*, 418–426. [CrossRef]
36. Muz, B.; de la Puente, P.; Azab, F.; Azab, A.K. The role of hypoxia in cancer progression, angiogenesis, metastasis, and resistance to therapy. *Hypoxia* **2015**, *3*, 83–92. [CrossRef]
37. Walsh, J.C.; Lebedev, A.; Aten, E.; Madsen, K.; Marciano, L.; Kolb, H.C. The clinical importance of assessing tumor hypoxia: Relationship of tumor hypoxia to prognosis and therapeutic opportunities. *Antioxid. Redox Signal.* **2014**, *21*, 1516–1554. [CrossRef]
38. Luo, D.; Liu, H.; Lin, D.; Lian, K.; Ren, H. The clinicopathologic and prognostic value of hypoxia-inducible factor-2α in cancer patients: A systematic review and meta-analysis. *Cancer Epidemiol. Prev. Biomark.* **2019**, *28*, 857–866. [CrossRef]
39. Ma, Q.; Reiter, R.J.; Chen, Y. Role of melatonin in controlling angiogenesis under physiological and pathological conditions. *Angiogenesis* **2020**, *23*, 91–104. [CrossRef]
40. Ramjiawan, R.R.; Griffioen, A.W.; Duda, D.G. Anti-angiogenesis for cancer revisited: Is there a role for combinations with immunotherapy? *Angiogenesis* **2017**, *20*, 185–204. [CrossRef]
41. Carpini, J.D.; Karam, A.K.; Montgomery, L. Vascular endothelial growth factor and its relationship to the prognosis and treatment of breast, ovarian, and cervical cancer. *Angiogenesis* **2010**, *13*, 43–58. [CrossRef]

42. Weigand, M.; Hantel, P.; Kreienberg, R.; Waltenberger, J. Autocrine vascular endothelial growth factor signalling in breast cancer. Evidence from cell lines and primary breast cancer cultures in vitro. *Angiogenesis* **2005**, *8*, 197–204. [CrossRef] [PubMed]
43. Masłowska, K.; Halik, P.K.; Tymecka, D.; Misicka, A.; Gniazdowska, E. The Role of VEGF receptors as molecular target in nuclear medicine for cancer diagnosis and combination therapy. *Cancers* **2021**, *13*, 1072. [CrossRef] [PubMed]
44. Tian, T.; Li, X.; Zhang, J. mTOR signaling in cancer and mTOR inhibitors in solid tumor targeting therapy. *Int. J. Mol. Sci.* **2019**, *20*, 755. [CrossRef] [PubMed]
45. Chen, M.C.; Hsu, W.L.; Chang, W.L.; Chou, T.C. Antiangiogenic activity of phthalides-enriched Angelica Sinensis extract by suppressing WSB-1/pVHL/HIF-1α/VEGF signaling in bladder cancer. *Sci. Rep.* **2017**, *7*. [CrossRef]
46. Karar, J.; Maity, A. PI3K/AKT/mTOR pathway in angiogenesis. *Front. Mol. Neurosci.* **2011**, *4*, 51. [CrossRef]
47. Guba, M.; von Breitenbuch, P.; Steinbauer, M.; Koehl, G.; Flegel, S.; Hornung, M.; Bruns, C.J.; Zuelke, C.; Farkas, S.; Anthuber, M. Rapamycin inhibits primary and metastatic tumor growth by antiangiogenesis: Involvement of vascular endothelial growth factor. *Nat. Med.* **2002**, *8*, 128–135. [CrossRef]
48. Mi, C.; Ma, J.; Wang, K.S.; Zuo, H.X.; Wang, Z.; Li, M.Y.; Piao, L.X.; Xu, G.H.; Li, X.; Quan, Z.S.; et al. Imperatorin suppresses proliferation and angiogenesis of human colon cancer cell by targeting HIF-1α via the mTOR/p70S6K/4E-BP1 and MAPK pathways. *J. Ethnopharmacol.* **2017**, *203*, 27–38. [CrossRef]
49. Saraswati, S.; Kumar, S.; Alhaider, A.A. α-santalol inhibits the angiogenesis and growth of human prostate tumor growth by targeting vascular endothelial growth factor receptor 2-mediated AKT/mTOR/P70S6K signaling pathway. *Mol. Cancer* **2013**, *12*, 1–18. [CrossRef]
50. Dodd, K.M.; Yang, J.; Shen, M.H.; Sampson, J.R.; Tee, A.R. mTORC1 drives HIF-1α and VEGF-A signalling via multiple mechanisms involving 4E-BP1, S6K1 and STAT3. *Oncogene* **2015**, *34*, 2239–2250. [CrossRef]
51. Herbert, S.P.; Costa, G. Sending messages in moving cells: mRNA localization and the regulation of cell migration. *Essays. Biochem.* **2019**, *63*, 595–606.
52. Willett, M.; Brocard, M.; Davide, A.; Morley, S.J. Translation initiation factors and active sites of protein synthesis co-localize at the leading edge of migrating fibroblasts. *Biochem. J.* **2011**, *438*, 217–227. [CrossRef]
53. Li, W.; Petrimpol, M.; Molle, K.D.; Hall, M.N.; Battegay, E.J.; Humar, R. Hypoxia-induced endothelial proliferation requires both mTORC1 and mTORC2. *Circ. Res.* **2007**, *100*, 79–87. [CrossRef]
54. Stein, J.; Milewski, W.M.; Hara, M.; Steiner, D.F.; Dey, A. GSK-3 inactivation or depletion promotes β-cell replication via down regulation of the CDK inhibitor, p27 (Kip1). *Islets* **2011**, *3*, 21–34. [CrossRef]
55. Hermida, M.A.; Kumar, J.D.; Leslie, N.R. GSK3 and its interactions with the PI3K/AKT/mTOR signalling network. *Adv. Biol. Regul.* **2017**, *65*, 5–15. [CrossRef]
56. Guri, Y.; Colombi, M.; Dazert, E.; Hindupur, S.K.; Roszik, J.; Moes, S.; Jenoe, P.; Heim, M.H.; Riezman, I.; Riezman, H. mTORC2 promotes tumorigenesis via lipid synthesis. *Cancer Cell* **2017**, *32*, 807–823. [CrossRef]
57. Lamanuzzi, A.; Saltarella, I.; Desantis, V.; Frassanito, M.A.; Leone, P.; Racanelli, V.; Nico, B.; Ribatti, D.; Ditonno, P.; Prete, M. Inhibition of mTOR complex 2 restrains tumor angiogenesis in multiple myeloma. *Oncotarget* **2018**, *9*, 20563. [CrossRef]
58. De Ieso, M.L.; Yool, A.J. Mechanisms of aquaporin-facilitated cancer invasion and metastasis. *Front. Chem.* **2018**, *6*, 135. [CrossRef]
59. Nicchia, G.P.; Stigliano, C.; Sparaneo, A.; Rossi, A.; Frigeri, A.; Svelto, M. Inhibition of aquaporin-1 dependent angiogenesis impairs tumour growth in a mouse model of melanoma. *J. Mol. Med.* **2013**, *91*, 613–623. [CrossRef]
60. Kaneko, K.; Yagui, K.; Tanaka, A.; Yoshihara, K.; Ishikawa, K.; Takahashi, K.; Bujo, H.; Sakurai, K.; Saito, Y. Aquaporin 1 is required for hypoxia-inducible angiogenesis in human retinal vascular endothelial cells. *Microvasc. Res.* **2008**, *75*, 297–301. [CrossRef] [PubMed]
61. Saadoun, S.; Papadopoulos, M.C.; Hara-Chikuma, M.; Verkman, A.S. Impairment of angiogenesis and cell migration by targeted aquaporin-1 gene disruption. *Nature* **2005**, *434*, 786–792. [CrossRef]

Article

Transient and Efficient Vascular Permeability Window for Adjuvant Drug Delivery Triggered by Microbeam Radiation

Sara Sabatasso [1,†], Cristian Fernandez-Palomo [1,†], Ruslan Hlushchuk [1], Jennifer Fazzari [1], Stefan Tschanz [1], Paolo Pellicioli [2], Michael Krisch [2], Jean A. Laissue [1] and Valentin Djonov [1,*]

[1] Institute of Anatomy, University of Bern, 3012 Bern, Switzerland; Sara.Sabatasso@hcuge.ch (S.S.); cristian.fernandez@ana.unibe.ch (C.F.-P.); ruslan.hlushchuk@ana.unibe.ch (R.H.); jennifer.fazzari@ana.unibe.ch (J.F.); stefan.tschanz@ana.unibe.ch (S.T.); jean-albert.laissue@pathology.unibe.ch (J.A.L.)
[2] Biomedical Beamline ID17, European Synchrotron Radiation Facility, 38043 Grenoble, France; paolo.pellicioli@esrf.fr (P.P.); krisch@esrf.fr (M.K.)
* Correspondence: valentin.djonov@ana.unibe.ch; Tel.: +41-31-631-84-32
† First Authors.

Citation: Sabatasso, S.; Fernandez-Palomo, C.; Hlushchuk, R.; Fazzari, J.; Tschanz, S.; Pellicioli, P.; Krisch, M.; Laissue, J.A.; Djonov, V. Transient and Efficient Vascular Permeability Window for Adjuvant Drug Delivery Triggered by Microbeam Radiation. *Cancers* **2021**, *13*, 2103. https://doi.org/10.3390/cancers13092103

Academic Editor: David Wong

Received: 31 March 2021
Accepted: 24 April 2021
Published: 27 April 2021

Publisher's Note: MDPI stays neutral with regard to jurisdictional claims in published maps and institutional affiliations.

Copyright: © 2021 by the authors. Licensee MDPI, Basel, Switzerland. This article is an open access article distributed under the terms and conditions of the Creative Commons Attribution (CC BY) license (https:// creativecommons.org/licenses/by/ 4.0/).

Simple Summary: One of the major challenges in the pharmacological treatment of solid tumours is ensuring that therapeutic concentrations of the agent reach and penetrate the tumour tissue. This is hampered by physiological barriers imposed by the aberrant and abnormal vessel structures of the tumours and high intratumoural pressure. We show that compound penetration into tumour tissue can be greatly enhanced by irradiating the tumour with an arrangement of discrete, synchrotron generated parallel X-rays in a range of 25–50 µm in width. This irradiation geometry induces a transient increase in vessel permeability in a time-dependent manner with a maximum between 45 min and 2 h after irradiation. The latter phenomenon was fully characterized in a vascular model of the developing chick embryo and termed "permeability window". The reported methodology could be considered as a potent and unique drug delivery system for combined tumour treatment. This will help to create new, more efficient treatment strategies against cancer and other vascular diseases.

Abstract: Background: Microbeam Radiation Therapy (MRT) induces a transient vascular permeability window, which offers a novel drug-delivery system for the preferential accumulation of therapeutic compounds in tumors. MRT is a preclinical cancer treatment modality that spatially fractionates synchrotron X-rays into micrometer-wide planar microbeams which can induce transient vascular permeability, especially in the immature tumor vessels, without compromising vascular perfusion. Here, we characterized this phenomenon using Chicken Chorioallantoic Membrane (CAM) and demonstrated its therapeutic potential in human glioblastoma xenografts in mice. Methods: the developing CAM was exposed to planar-microbeams of 75 Gy peak dose with Synchrotron X-rays. Similarly, mice harboring human glioblastoma xenografts were exposed to peak microbeam doses of 150 Gy, followed by treatment with Cisplatin. Tumor progression was documented by Magnetic Resonance Imaging (MRI) and caliper measurements. Results: CAM exposed to MRT exhibited vascular permeability, beginning 15 min post-irradiation, reaching its peak from 45 min to 2 h, and ending by 4 h. We have deemed this period the "permeability window". Morphological analysis showed partially fragmented endothelial walls as the cause of the increased transport of FITC-Dextran into the surrounding tissue and the extravasation of 100 nm microspheres (representing the upper range of nanoparticles). In the human glioblastoma xenografts, MRI measurements showed that the combined treatment dramatically reduced the tumor size by 2.75-fold and 5.25-fold, respectively, compared to MRT or Cisplatin alone. Conclusions: MRT provides a novel mechanism for drug delivery by increasing vascular transpermeability while preserving vessel integrity. This permeability window increases the therapeutic index of currently available chemotherapeutics and could be combined with other therapeutic agents such as Nanoparticles/Antibodies/etc.

Keywords: vascular permeability; drug delivery system; microbeam radiation therapy (MRT); Chicken Chorioallantoic Membrane (CAM); U-87 Glioblastoma

1. Introduction

Chemotherapy is one of the most suitable treatment options for cancer therapy. However, solid tumors' anatomical and physiological characteristics limit the exposure of all tumor cells to a sufficient concentration of such therapeutic agents [1,2]. For example, a compound must cross the vascular wall before it can affect the tumor tissue. In particular, the abnormal vasculature and the lack of a functional lymphatic network are tumor characteristics that lead to interstitial hypertension, which minimizes drug diffusion into the tumor core [3,4], ultimately diminishing the therapeutic potential of an anti-cancer compound.

Much work has focused on developing new strategies to overcome this blood-tumor barrier and improve the therapeutic potential of existing agents [5]. These strategies, which involve both pharmacological and physical approaches, include the following:

- Modulators of tumor blood flow reduce flow resistance through vasodilation and increase the blood pressure with vasoconstrictors, thereby also increasing the transvascular hydrostatic gradient [6]. Angiotensin II is a clinically proven agent in this group [7];
- Vascular normalization describes the correction of structural abnormalities by pruning immature branches, enhancing perivascular coverage, and reinstating the basal membrane [4]. This restores vascular functionality, in particular, the transportation of drugs to the tumor cells. Of all compounds used to achieve vascular normalization [8], VEGF inhibitors have been successful in clinical trials [9];
- Vascular permeabilization refers to the increase in capillary permeability due to the administration of inflammatory cytokines and vasomodulators, such as histamine, bradykinin, TNF-alpha, angiotensin II, botulinum neurotoxin and nitric oxide donors amongst others [5]. Some approaches use specific receptor-triggered endocytosis, i.e., employing the insulin-like growth factor 1 receptor to enable trafficking of compounds to the abluminal site [10];
- Overcoming the extracellular matrix (ECM) of tumors—extensive collagen networks are major obstacles for the penetration of therapeutic agents [11]. The use of collagen-degrading enzymes [12] or the downregulation of fibroblast activity [13] have shown great effectivity at improving the distribution of macromolecules;
- Hyperthermia is a simple, physical method that promotes drug delivery by increasing the local temperature of tissues to a range of 39–42 °C using tools such as microwaves, radiofrequency, and ultrasound. The induced capillary dilation increases perfusion and oxygenation, therefore enhancing the uptake and efficacy of chemotherapeutics [14];
- Ultrasound and microbubbles—the use of ultrasound in conjunction with intravenously administered microbubbles disrupts tight junction complexes and improves the delivery of chemotherapeutics in tumors [15,16]. Positive effects have been demonstrated in clinical studies of pancreatic cancer [17];
- Sonodynamic therapy is a novel, rapidly developing treatment based on preferential uptake of sonosensitizing compounds in tumor tissues and subsequent activation of the drug by high-intensity focused ultrasound. This strategy is minimally invasive and may be administered to deeply situated tumors [18,19].

Of all the drug-delivery systems mentioned above, more than a dozen have been approved by the Food and Drug Administration agency of the United States; however, most of them are mainly physical or only allow for topical application. This is not surprising since the permeability of the blood vessels is affected by the size and charge of the plasma components, making the delivery of macromolecules even more difficult compared to skin application [19]. As a result, there is a great need for a simple, precise, well-tolerated, and reliable drug delivery system to enhance the therapeutic potential of anti-cancer agents.

Synchrotron Microbeam Radiation Therapy (MRT) could be used as a novel drug delivery strategy that transiently enhances vessel permeability in tumors before drug administration. MRT has a unique vascular disruptive effect, where only the immature vessels are destroyed, while mature microvasculature is preserved [20,21]. MRT is based on the spatial fractionation of synchrotron-generated X-rays into arrays of micron-wide parallel, planar beamlets (25–100 µm), spaced 50–500 µm from center-to-center [22]. This generates a heterogenous dose deposition with tissue in the beam path receiving high (peak) doses of radiation (hGy) and the regions between microbeams (valley) receiving much lower doses (Gy). MRT has shown exceptional tumor control by reducing or even stopping tumor growth [23–29]. One potential mechanism of action involves MRT's preferential destruction of the immature dysregulated tumor vasculature, which decreases tumor blood volume, leading to necrosis [23,30]. Remarkably, normal tissues show extremely high tolerance to MRT, as has been observed in the brains of rodents [26,31–33], piglets [34], duck embryos [35], cerebella of suckling rat pups [36,37], weanling piglets [34,36], and in different types of normal tissues of mice after partial or total body MRT irradiation [38,39] (recently reviewed in [22]). This normal tissue sparing effect has been attributed to the preservation of mature microvasculature. Due to the spatial fractionation of MRT, vascular damage is confined to the beam path and, unlike the tumor, the minimally irradiated endothelial cells in the valley region can repair neighboring regions damaged by the microbeam [33]. The unique vascular disruptive effects of MRT have been demonstrated in the Chick Chorioallantoic Membrane (CAM) where the vascular properties resemble those of tumors [20]. The CAMs were exposed to peak doses of 200–300 Gy, which preferentially destroyed immature vessels (with the first subcellular changes occurring 15 min after exposure) while vascular integrity was maintained in the valley regions. This promoted the resolution of damaged regions and subsequent clearance of edema one hour post-irradiation with sustained capillary perfusion. Perfusion studies 6 h following MRT with FITC-dextran showed zones of intact, perfused capillaries in the valley (low-dose) region and vascular disruption and loss of perfusion limited to the microbeam path (high-dose) [20]. The reversible damage to the vasculature is attributable to the spatial fractionation of the incident beam, as homogenous dose delivery resulted in unresolved damage at doses hundreds of Gy below those delivered by MRT. The induction and subsequent resolution of edema suggest that MRT induces a transient increase in vascular permeability following MRT that could be exploited for therapeutic gain.

To further explore and characterize this observed period of transpermeability, we evaluated the effects of delivering MRT at a peak dose below the threshold in which vessel destruction was observed with the goal of preserving perfusion to the tumor and tumor-like vessels of (1) the vascularized CAM and (2) subcutaneous xenografts of human glioblastoma.

2. Materials and Methods

2.1. Animal Models

Two animal models were used for these experiments: the chick CAM and human U-87 Malignant Glioma xenotransplanted in BALB/c nude mice.

CAM: Fertilized chick eggs from a commercial hatchery were transferred to Petri dishes on the third day after fertilization following the shell-free culture method [40]. Embryos were maintained at 37 °C under a humidified atmosphere until day 12 of the embryonic development.

The CAM is the extraembryonic network of rapidly developing vasculature supporting respiration of the developing chicken embryo. Due to the ease of visualization and rapid development, the CAM model has been used extensively in the field of angiogenesis research with each stage of embryonic development corresponding with various stages of vascular maturation. This rapid vessel development also supports tumor grafts for the study of tumor dynamics, without the need for costly rodent models and eliminating ethical concerns [41]. The CAM has therefore become an indispensable model for the study of vessel development, and dynamics in particular, when testing anti-angiogenic therapies [42]. The versatility of the CAM as an experimental model has been extensively reviewed [43–45]. One of the major benefits of this model is that it can be maintained ex-ovo, permitting the real-time observation of vascular changes in response to various targeted treatments including radiation therapy (review by Mapanao et al., 2021 [46]). This made it an ideal system for visualizing MRT-induced changes in vascular permeability.

Human glioblastoma xenografts: U-87 Malignant Glioma cells (ECACC, Salisbury, UK) were cultured in D-MEM supplemented with 10% fetal calf serum and 1% antibiotics/antimycotics. Tumor cells (2×10^6 in 100 µL PBS per mouse) were implanted subcutaneously (sc) in the right flank of 61 male BALB/c nude mice, weighing about 20–22 g (Charles River Laboratories, Paris, France). There were four experimental groups: Control group (CO, $n = 13$), Cisplatin-treated group (CIS, $n = 17$), MRT group (MRT, $n = 17$), and double-treated group (MRT + Cisplatin, $n = 14$). A dose of 10 mg/kg BW of Cisplatin (Cisplatin-Teva®, Teva Pharma AG, Basel, Switzerland) was administered via the tail-vein 40 min after MRT.

2.2. Synchrotron Microbeam Irradiation

The irradiations of both animal models were performed at the ID17 Biomedical Beamline of the European Synchrotron Radiation Facility (ESRF) in Grenoble, France.

CAM: CAMs at day 12 of development were irradiated with a 1×1 cm array of 51 microbeams of 25 µm in width on average, spaced by 200 µm from their centers. To achieve this array configuration, we used a multi-slit "Archer" collimator with alternating Au and Al foils and fixed geometry [47,48]. A wiggler gap of 40 mm delivered an X-ray spectrum configuration of 93.4 keV mean energy, and 74.9 keV peak energy. Peak-entry doses were estimated at 75 Gy according to our Monte Carlo computation. A radiochromic film (GafChromic® radiochromic film type HD-810, ISP Corporation, Wayne, NJ, USA) was laid over the surface of the CAM prior to the irradiation, to visualize the microbeam paths and distinguish between the irradiated and non-irradiated parts.

Glioblastoma xenografts: Mice were anesthetized with an intraperitoneal injection of xylazine/ketamine (0.1/1% in saline solution buffer, 10 µL/g body weight), then placed on their left flank on a horizontal surface. The tumors were irradiated seventeen days after tumor cell inoculation when their volumes averaged 200 mm^3 (calculated by the formula: V(mm^3) = $4/3 \times \pi \times a \times b \times c$, where a, b and c are the length, the width, and the height of the tumor, measured by digital caliper). The tumors were irradiated unidirectionally with a skin-entrance dose of 150 Gy, using 50 µm wide microbeams and 200 µm on-center distance. This configuration was achieved by the ESRF-made multi-slit collimator made of tungsten carbide [49]. A wiggler gap of 24.8 mm delivered an X-ray spectrum of 104.2 keV mean energy, and 87.7 keV peak energy. The average dose rate was 12,000 Gy/s.

2.3. Vascular Permeability Assay with FITC-Dextran

Microscopic observations were made up to 48 h after MRT with video documentation. For the assay, we used FITC-dextran of MW 2×10^6 Daltons, (Fluorescein isothiocyanate from Sigma, Taufkirchen, Germany). FITC-dextran has a Stroke Ratio of 270 Angstrom [50], which converts to approximately 27 nm.

We also employed red fluorescent polystyrene microspheres (FluoSpheres™ Carboxylate-Modified Microspheres, Cat #: F8810, ThermoFisher, Waltham, MA, USA). The microspheres have a diameter of 100 nm with a coefficient variation in size of 5% [51]. Their surfaces are pre-coated with a high density of carboxylic acids, which endows the microspheres with a highly charged and relatively hydrophilic surface layer. The surface charges range between 0.1 and 2.0 mEq/g, which makes them stable to a relatively high concentration of electrolytes (max. 1 M univalent salt) and prevents agglomerates [52]. We selected this microsphere because the highly charged surface reduces their attraction to cells, which makes them ideal for studying vascular permeability.

CAM: CAMs were either left untreated or intravenously injected with FITC-dextran and red fluorescent microspheres. Moreover, we applied 1.0 µg of recombinant VEGF-A165 protein (Peprotech, London, UK) on the surface of 7 CAMs, and compared them against VEGF-untreated CAMs. A semi-quantitative analysis of the extravasated FITC-dextran was performed in vivo in the CAM. The blood flow and FITC-dextran extravasation were monitored every 15 min. The intensity of perivascular FITC-dextran was evaluated according to the score presented in the caption of Figure 1. The video sequences documented the presence and site of the microbeam stripes, the optically empty zones, the damaged medium- and large-sized vessels, the damage to the capillary network, and extravasation of the fluorescent probes.

Glioblastoma Xenograft: Forty-five minutes after irradiation, we performed a vascular permeability assay in a selected group of mice; MRT + Cisplatin ($n = 4$), MRT alone ($n = 7$), Cisplatin alone ($n = 4$), and control ($n = 4$). A solution of 3% FITC-dextran in sterile saline was injected (0.3 mL) into the tail vein of mice. In all cases, tumors were harvested approximately 30 min after injection of the fluorescent compound and fixed in 2% paraformaldehyde.

2.4. Semi-Thin Serial Sectioning and Transmission Electron Microscopy (TEM) of CAM

The sites of interest from CAM samples were harvested and fixed in 2.5% (v/v) glutaraldehyde solution buffered with 0.03 M potassium phosphate (pH 7.4, 370 mOsm), post-fixed in 1% OsO_4 (buffered with 0.1 M sodium cacodylate (pH 7.4, 340 mOsm)), dehydrated in ethanol, and embedded in epoxy resin. Thousands of 0.8 µm-thick serial sections, perpendicularly to the direction of the beam propagation, were prepared with glass knives and stained with toluidine blue. The serial sections were then viewed, and images captured at different magnifications using a light microscope (Leica, Leitz DM, Morrisville, NC, USA), equipped with a Leica DFC480 camera. For transmission electron microscopy, 80 to 90 nm-thick sections were prepared and mounted on copper grids coated with Formvar (polyvinyl formal; Fluka, Buchs, Switzerland). They were stained with lead citrate and uranyl acetate and viewed in a Philips EM-400 electron microscope [53].

2.5. Immunostaining and Analysis of Glioblastoma Xenograft

Sections of tumor blood vessels were dewaxed, rehydrated, and subjected to heat-induced epitope retrieval (Dako S1699). Endogenous proteins were blocked with 5% BSA in PBS for 30 min at room temperature, followed by the primary antibody (rabbit anti-mouse CD31, Abcam, Eugene, OR, USA), diluted 1:20 in 1% BSA/PBS and incubated for 48 h at 4 °C in a humidified chamber. After washing with PBS-Tween 20 for 30 min, sections were incubated with goat anti-rabbit Alexa Fluor 594 (Invitrogen, Carlsbad, CA, USA) secondary antibody at a dilution of 1:200 in 10% FCS/PBS for 30 min at room temperature. Quantification of the extravasated probe was performed, based on pictures taken with a confocal microscope (LSM Zeiss Meta, Caochen, Germany). For each experimental group, three to seven pictures per tumor were taken with three-vessel areas (VA) measured per picture. The overall fluorescence (OF) and the intravascular fluorescence (IF) of FITC-dextran were quantified using the ImageJ software. The extravascular fluorescence (EF) was defined as OF-IF. The vascular permeability index was calculated as EF/VA.

2.6. Magnetic Resonance Imaging (MRI) of Glioma Xenografts

A selected group of mice underwent MRI on days 0, 5, 13, 20, and 27 post-treatment to monitor tumor growth. Each group had 5 mice, including the control group. MRI was performed with a 4.7 Tesla Scan (Avance III console, Bruker, Ettlingen, Germany) at the "Institut des Neurosciences" in Grenoble, France. The animals were subjected to an anatomical T1/T2 scan, and to a permeability MRI after an intravenous injection of gadolinium-labelled albumin (Gd-Albumin, BioPAL) via the caudal vein.

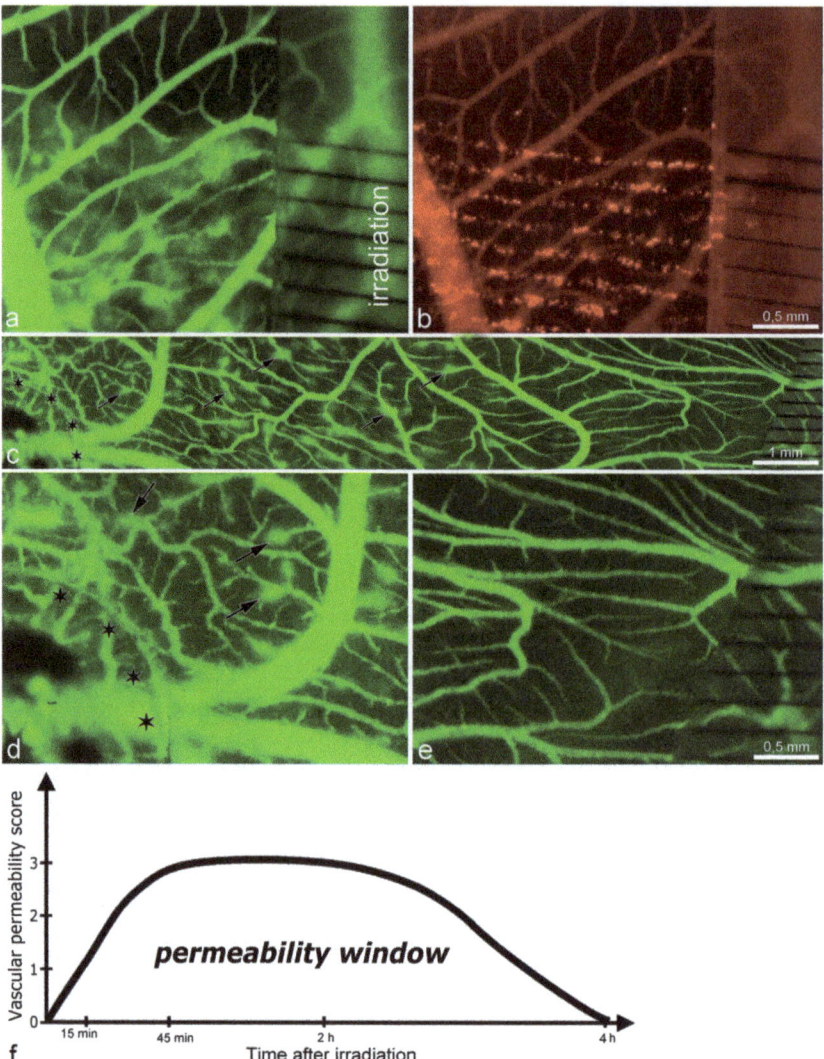

Figure 1. Images from intravital microscopy showing increased vascular permeability in CAM after exposure to MRT peak doses of 75 Gy; (**a,b**): normal CAM vasculature; (**c–e**): vasculature after VEGF treatment. Note: (**a**) forty-five minutes after exposure to MRT, the vascular permeability is increased, as demonstrated by the extravasation of FITC-dextran (green-fluorescent halos around the blood vessels). Conversely, in (**b**), the microspheres did not diffuse into the surrounding tissue but remained affixed as red-fluorescent dots along the microbeams path. Left side: (**c**) at the site of VEGF application (asterisks indicate the edge of the Thermanox® coverslip), the vascular permeability increased as early as 10 min after MRT, as shown by the halos of extravasated FITC-dextran (some marked by arrows). Right side: in the non-treated zone, no such signs of increased vascular permeability were observed simultaneously. Images (**d,e**): parts of (**c**) at higher magnification. Image (**f**): schematic representation of the vascular permeability window after 75 Gy of MRT. The score: (0) = no FITC extravasation; (1) = small non-confluent FITC "halo" surrounding the capillaries; (2) = FITC "halos" start merging but they are not completely confluent; (3) = the "halos" are completely confluent. In (**a–e**), black stripes on the radiochromic film indicate the path of the microbeams.

3. Results

3.1. Microbeams Induced a Transient Vascular "Permeability Window" in CAM without Impairing Tissue Perfusion

After irradiating the CAM with microbeam entrance peak-doses of 75 Gy, there was an increase in transpermeability without vascular destruction and preservation of vascular perfusion. The vascular permeability assay revealed that only FITC-dextran (~27 nm) extravasated into the surrounding tissue, while the larger microspheres (100 nm) remained stuck along the microbeam paths (Figure 1a,b). Successive semi-quantitative evaluation (every 15 min) showed that the extravasation of FITC-dextran was transient, detected from 15 min until it ended at 4 h after irradiation (Figure 1f, Video S1 in Supplementary Materials).

In addition, we administered VEGF to the top of the CAM prior to irradiation to induce neovascularization. The goal was to simulate the tumor microenvironment, which normally has high amounts of VEGF. Then, we compared the vascular effects caused by MRT in the VEGF-treated (Figure 1c, left side) and non-treated areas (Figure 1c, right side). We found that the vascular transpermeability occurred earlier (10 min after irradiation) in the VEGF-induced neovasculature (higher magnification in Figure 1d) than in the VEGF-untreated vasculature (higher magnification in Figure 1e). This suggests that immature vessels are more sensitive to the microbeams and show an earlier onset of vascular transpermeability.

3.2. Time Course of the Structural Changes in the CAM during the Vascular "Permeability Window"

Fifteen minutes after microbeam radiation of 75 Gy, the CAM thickness increased transiently to approximately three times its regular size. This was assessed by comparing it against the recovered CAM 4 h post-irradiation (Figure 2a,d). The acute increase in size is likely attributed to the development of edema underneath the capillary plexus (Figure 2b). The ultrastructural analysis revealed a discontinuous luminal surface of the microvessels, with rarefication of the endothelial cytoplasm resulting in fissures and gaps. The increased permeability was evidenced by the presence of FITC-dextran dots in the endothelial cell wall of the microvessels, as well as in the extravascular space (Figure 2c(c^1,c^2)). Conversely, 4 h after irradiation, the endothelial cells showed restored integrity, which was accompanied by only single holes and solitary FITC-dextran depositions in the endothelium (Figure 2f(f^1)). These observations suggest that microbeams of 75 Gy increased the vascular transpermeability without long-lasting damage to CAM vasculature.

3.3. Microbeams also Induced Vascular Permeability in Human U-87 Glioblastoma Xenografts

To determine whether microbeams promote vascular permeability in a human glioblastoma xenograft mouse model, we compared one group treated with 150 Gy (peak-entry dose) of microbeams with an unirradiated tumor control group (Figure 3). We observed clear extravasation of FITC-dextran in the irradiated tumors 45 min post-irradiation (Figure 3d). Conversely, the fluorescent compound remained intravascular in the control group (Figure 3c). The permeability index revealed a two-fold increase in transpermeability following MRT relative to the unirradiated control (Figure 3e). At the ultrastructural level, no extravasation of the fluorescent probe was observed in control tumors; FITC-dextran dots remained in the lumen (Figure 3f,h). However, in microbeam-treated tumors, FITC-dextran was observed in the extravascular space together with partially disintegrated endothelial cells (Figure 3g,i). These results confirm that MRT can also induce vascular permeability in this mammalian tumor model, and thus, vascular permeability is not restricted to the CAM (avian).

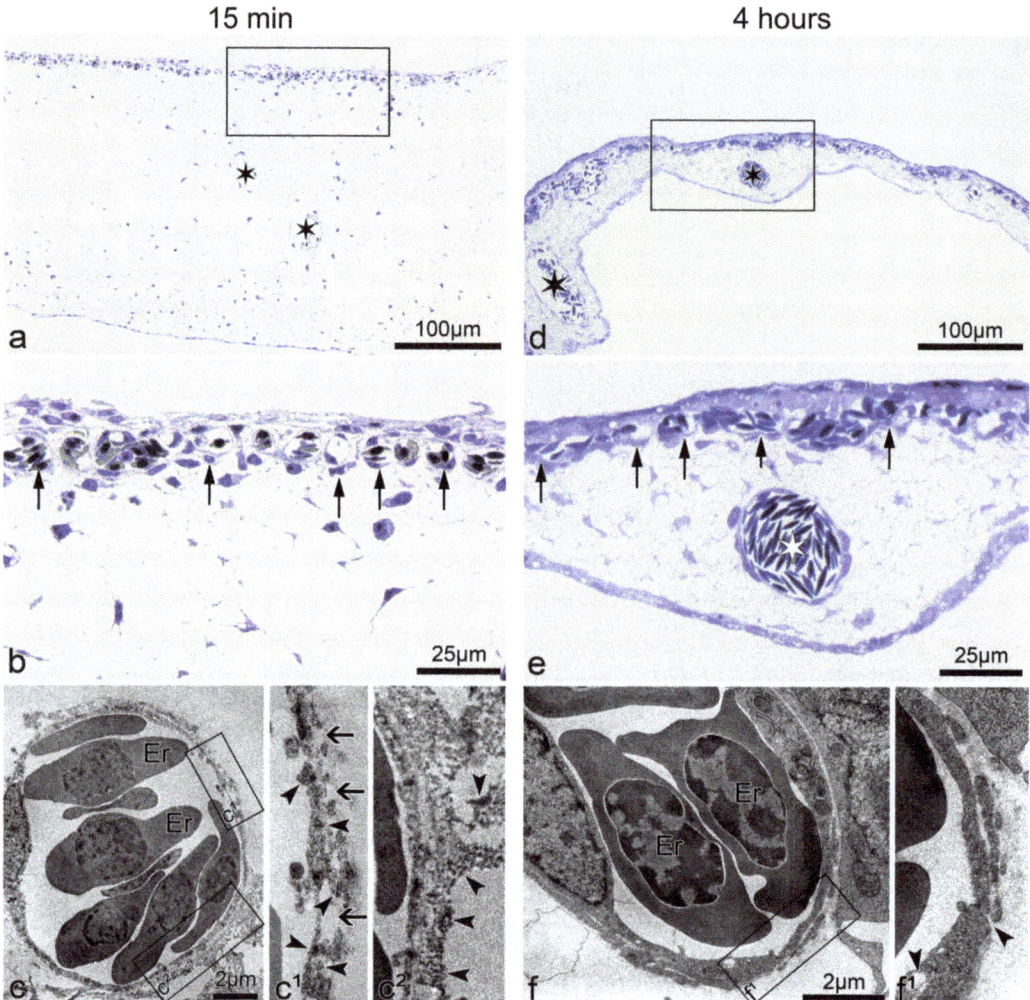

Figure 2. Morphological alteration of CAM vasculature 15 min and 4 h post-75 Gy of MRT. Images (**a,b**): semithin section of CAM fifteen minutes after MRT exposure: Irradiated CAM is enlarged (edematous). The capillary vessels (arrows) appear almost normal at light microscopy. (**c**(c^1,c^2)): Ultrastructure of CAM samples shown in (**a,b**) reveal a discontinuous endothelium with gaps and fissures (arrows). Those are most likely responsible for the increased permeability, as demonstrated by the presence of FITC-dextran dots (arrowheads) in the endothelial cell wall (c^1) as well as in the abluminal space (c^2). Images (**d,e**): four hours after 75 Gy of MRT, the CAM thickness decreased, thus almost reverting to the normal morphology. The capillary plexus (arrows) and supplying vessels (white asterisk) appear perfused and intact in semithin sections. Images (**f**(f^1)): four hours after microbeam exposure, the capillaries regained their normal ultrastructure, as evidenced by the nearly normal endothelial cells. Only occasional vacuoles and fissures were present (arrowheads). Images (**b**,c^1,c^2,**e**,f^1) are higher magnifications of the rectangles in (**a,c,d,f**), respectively. Er = erythrocyte.

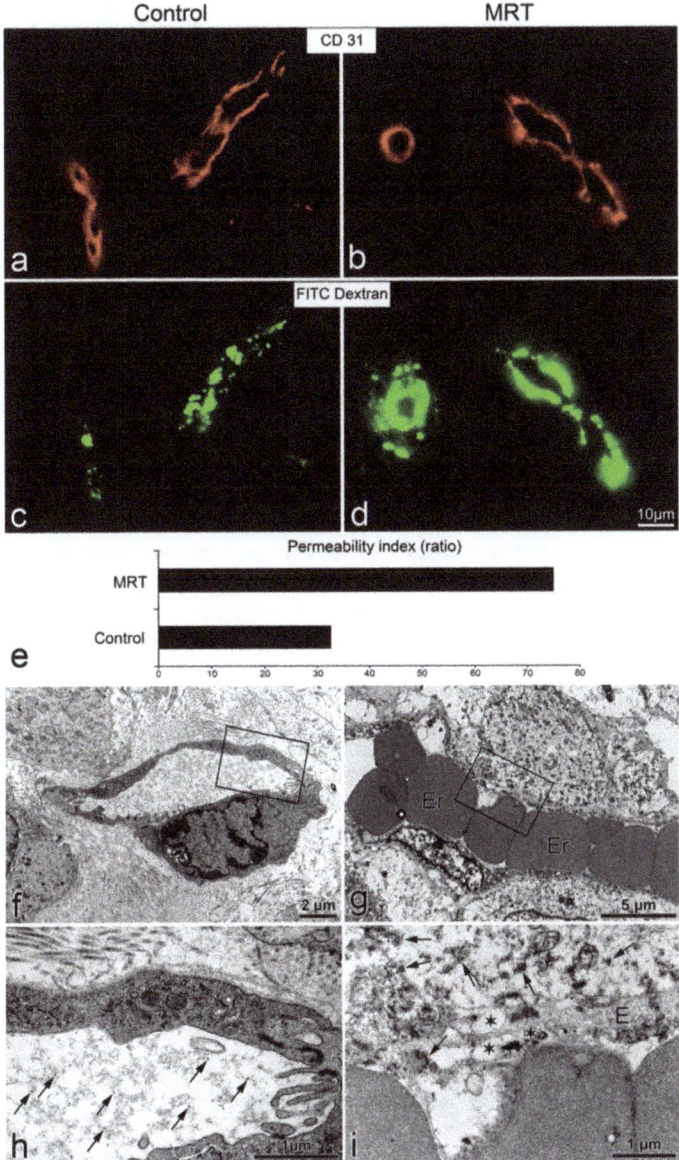

Figure 3. MRT-induced vascular permeability in mice harbouring the human U-87 glioblastoma xenograft. Fluorescence microscopy for CD31 and FITC-dextran in a control tumor (**a,c**); tumor post-MRT (**b,d**). There was no extravasation of the green FITC-dextran in the control tumor (**c**), while in the MRT-treated tumor (**d**), 45 min after 150 Gy, a bright halo of green fluorescence was visible. Image (**e**): graph showing the quantification of the vascular permeability in controls and MRT-treated tumors as the ratio of extravasated FITC-dextran fluorescent area/vessel area. The ultrastructure of tumor vessels was normal in controls (**f,h**), with no extravasation of FITC-dextran (intraluminal dextran as dark dots indicated by arrows). Conversely, in treated tumors (**g,i**), an extravasated fluorescent probe material was observed as dark dots (arrows) in the extravascular space; the disrupted endothelium contained multiple vacuoles of different sizes, indicated by asterisks. Er = erythrocyte. Images (**h,i**) are higher magnifications of the rectangles in (**f,g**), respectively.

3.4. Using the MRT-Induced Vascular Permeability to Enhance the Delivery of Cisplatin

To exploit the MRT-induced "permeability window", the adjuvant Cisplatin was administered in conjunction with MRT in mice bearing glioblastoma xenografts. Cisplatin is known to have efficacy against glioblastoma in vitro but a poor clinical response as a single agent and in combination with radiotherapy [54]. This is primarily due to poor penetration across the blood–brain barrier [55] and dose-limiting cytotoxicity [56]. Microbeam radiation therapy was delivered to the tumors 17 days after cell inoculation, and Cisplatin was administered 40 min after irradiation. Tumors in the control group began to grow exponentially 2 days after treatment (Figure 4). Differences between the treatment groups started on day 13, with the fastest-growing tumors belonging to the Cisplatin group, followed by those treated with MRT alone. In contrast, tumors treated with the combination of MRT + Cisplatin remained unchanged until approximately 22 days after treatment, when their growth rate began to abate slowly. In a second experimental trial, tumor growth measurements performed with Magnetic Resonance Imaging (MRI) on days 0, 5, 13, 20 and 27 after treatment yielded tumor volumes comparable to those measured with the digital caliper; tumor volumes decreased in the same order: Control > Cisplatin alone > MRT alone > MRT + Cisplatin (Figure 5a). Accordingly, images of tumor progression (Figure 5b) show the best treatment results on animals subjected to MRT + Cisplatin in comparison with the other experimental groups; with a 2.75-fold decrease in comparison with MRT alone, and a 5.25-fold decrease compared to Cisplatin alone. These results show that the administration of adjuvant Cisplatin can take advantage of the MRT-induced vascular permeability.

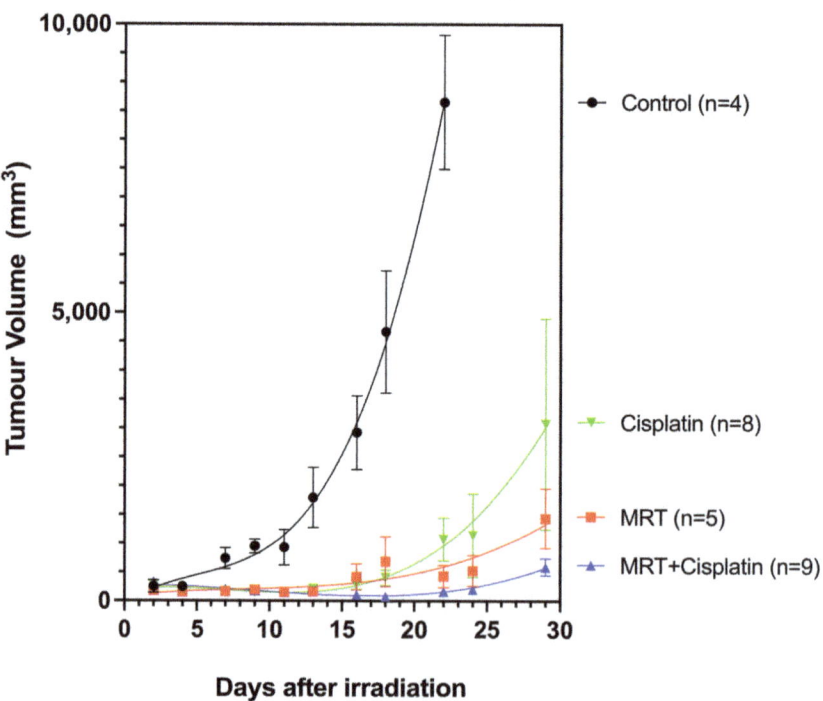

Figure 4. Growth of U-87 glioblastoma xenograft. Groups are unirradiated Controls ($n = 4$), Cisplatin ($n = 8$), MRT ($n = 5$), and Double Treatment (MRT + Cis) ($n = 9$). The tumors were measured with a digital caliper every second day.

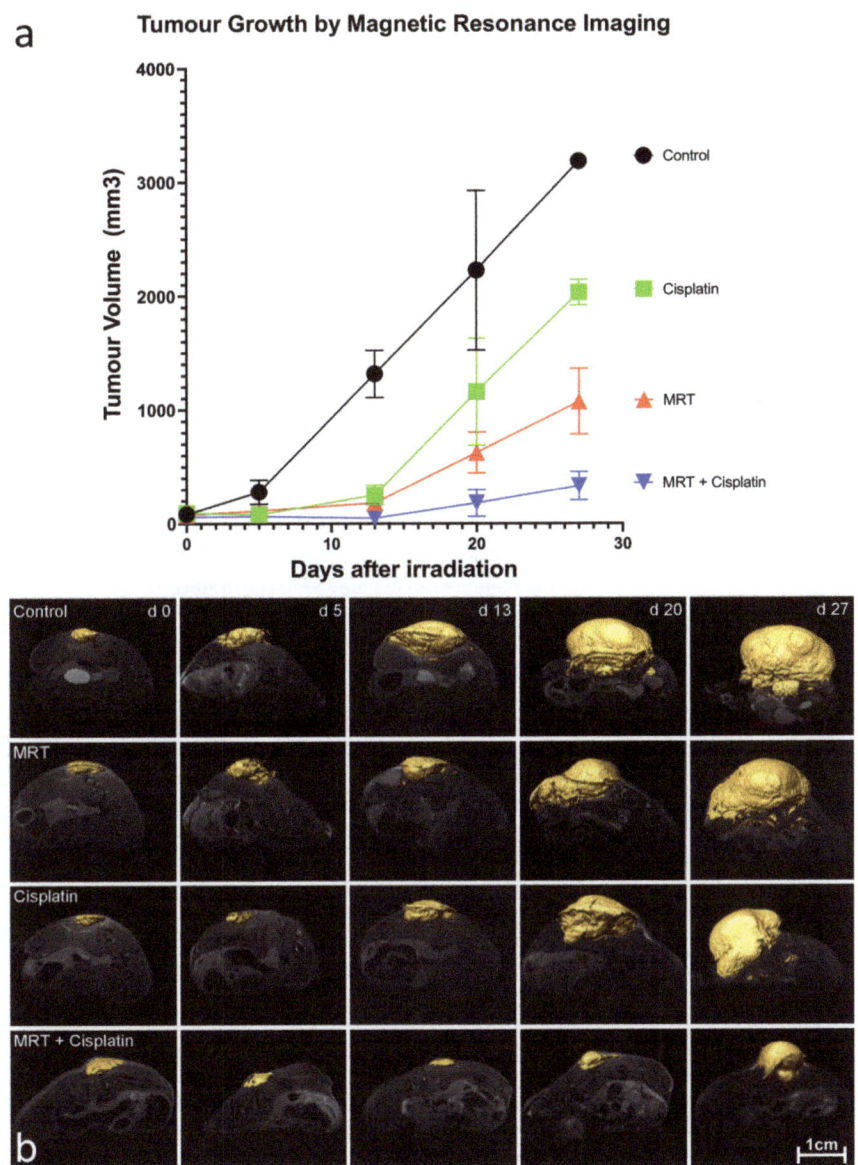

Figure 5. U-87 glioblastoma growth followed by MRI. Image (**a**) shows the tumor volume growth measured by MRI. Image (**b**): MRI images of the tumor progression for each animal group. Control ($n = 3$), Cisplatin ($n = 4$), 150 Gy MRT ($n = 7$), and Double Treatment (150 Gy MRT + Cis) ($n = 5$).

4. Discussion

One of the main challenges in cancer treatment is the delivery of sufficient quantities of chemotherapeutic agents and nanocarriers to tumors. For instance, many compounds do not extravasate into normal tissue but passively cross leaky tumor capillaries in a process referred to as Enhanced Permeability and Retention (EPR) [57]. This EPR has been shown to allow the passage of molecules from 40 to 70 kDa [58]. However, a retrospective study reported that, unfortunately, only 0.7% (median) of the administered agents were delivered to a solid tumor [59]. Furthermore, the chemotherapeutic compounds are often small molecules with a short half-life, and multiple applications or higher doses are needed to increase their therapeutic impact [60]. This can result in severe negative side effects and possible drug resistance [61]. Many of the approaches mentioned in the introduction have different limitations, and only a few are actually applied clinically. For example, modulators of blood flow and vascular normalization have a short half-life, making it difficult to estimate in advance the dose and timing of the drug administration [62]. Furthermore, systemic administration of inflammatory cytokines and vasomodulators for vascular permeabilization are associated with high, whole-body toxicity. For this reason, they are used in the clinic only for isolated limb perfusion, e.g., for the treatment of sarcomas and melanomas [63]. Therefore, to improve the therapeutic index of the wide range of potent anti-cancer agents, we need a precise, accurate, and well-tolerated drug delivery system that can increase their delivery into tumors while minimizing damage to normal tissue.

The promising new strategy for the enhancement of vessel permeability presented here may overcome the obstacles of the blood–tumor barrier (BTB). Low-dose MRT is a very simple and highly effective physical solution that does not rely on the use of any carriers. It induces a transient, vascular permeability window in CAM, beginning 15 min after MRT (75 Gy) and ending at 4 h. The extravasation of FITC-dextran (MW 2×10^6 Dalton, size of 27 nm) was visible as green "clouds" diffusing between high-dose microbeam regions, which indicates a penetration depth of a few hundreds of micrometers. At the same time, larger microspheres of 100 nm were constrained to the beam path (Figure 1a,b,f, and Supplementary Video S1), indicating that the transpermeability of particles following MRT is size-dependent.

Previously, it has been shown that diffusion of the small fluorescent probe, sulforhodamine B (0.58 kDa), across the blood–tumor barrier in 9 L gliosarcoma was only induced when peak doses were delivered at 1000 Gy and persisted up to 30 days post-irradiation [31]; meanwhile, diffusion of a larger 70 kDa FITC-dextran was not observed at any dose or time-point. However, the permeability observed by these authors was at much later time-points, with measurements beginning only 12 h following MRT (instead of at 15 min as we present in this manuscript), and the permeability was likely a consequence of endothelial destruction caused by the 1000 Gy peak dose [64]. Furthermore, an MRT scheme of 2 cross-fired arrays, each delivering a peak dose of 400 Gy (total dose of 800 Gy to the tumor), was effective in the destruction of tumor tissue but also induced edema in normal brain tissue, causing the majority of animal deaths in that study [30]. Although both studies acknowledge the use of MRT to increase permeability, the late time points of observation missed the effective "permeability window" that we report in this manuscript. Moreover, these studies used high peak doses that triggered the consequent vascular damage, while we show evidence that the delivery of lower MRT peak doses (<150 Gy) can enhance drug delivery to the tumor without inducing endothelial damage nor subsequent pathology. This is evidenced in the ultrastructure observations of endothelial cells from irradiated CAMs and glioma xenografts (Figures 2 and 3). Four hours after MRT, the vascular integrity was almost completely restored, and the permeability window closed (Figure 1).

Many tumors secrete high concentrations of VEGF, which is known to increase the hyperpermeability in already existing microvessels and at the same time induce neoangiogenesis in rapidly growing tumors [65,66]. Specifically, VEGF promotes vascular permeability by disrupting endothelial cell contacts. Exogenous VEGF administration to the CAM induced the rapid permeability of CAM vessels following MRT in comparison to MRT administered alone, decreasing the onset time of the permeability window from 15 min to 10 min (Figure 1c–e). This additive effect is a promising attribute of tumors producing a high amount of VEGF when they are treated by microbeams. MRT itself at higher peak doses has been shown to induce VEGF expression in normal and tumor tissue in the brain over time, contributing to brain edema [30]. However, lower doses of MRT (150 Gy), as used here, have been shown to induce a vascular normalization effect, with increased pericyte coverage and resolution of hypoxia within 2 weeks of treatment [28]. In previous studies, we reported that MRT at higher doses, in the range of 300–400 Gy, had a preferential vascular destructive effect in both the chick CAM vasculature [20] and the murine model of melanoma [29]. It seems that after the destruction of capillaries, the width of the microbeam is essential for the prediction of the grade of restoration of their integrity. In another study, we partially amputated the ventral half of the caudal fin of zebrafish to induce regeneration and the development of new, immature vasculature that mimics that seen in a tumor. After regeneration, we irradiated the regenerating ventral (immature) and undamaged, dorsal (mature) compartments with 25, 50, 100, 200, and 800 μm wide beamlets [21]. The restoration of vascular defects was observed when the beamlets were up to 100 μm wide, but not when they were 200 or 800 μm wide. It has been shown that one of the major mechanisms of the action of MRT is mediated by the induction of vascular toxicity on immature, or tumoral microcirculation (reviewed in [22]). The geometry of MRT, together with the applied doses, appears to be essential for the differential endothelial disintegration, which may be species-specific.

In the second part of this study, we exploited this "vascular permeability window" to enhance the delivery of co-adjuvant Cisplatin into U87 human glioblastoma xenografts in mice. First, it was confirmed that MRT could also induce a vascular permeability window in the glioblastoma xenograft model, as shown by the extravasated FITC-dextran 45 min after 150 Gy of MRT (Figure 3). Secondly, we showed the treatment efficacy of combining 150 Gy MRT + Cisplatin in two separate trials, where tumor measurements were performed with a digital caliper or more accurately with MRI, in each respective trial. The combined treatment of 150 Gy MRT + Cisplatin achieved the best tumor control among all treatment groups (Figures 4 and 5).

However, despite showing vascular permeability in glioblastoma xenografts with FITIC dextran (Figure 3), the present survival study had the limitation of not allowing for the measurement of the accumulated cisplatin in the tumor. Future mechanistic studies should include this variable to confirm the vascular permeability hypothesis.

5. Conclusions

In conclusion, we have confirmed that low doses of 75 and 150 Gy (relative to the MRT field) increase the vessel permeability in the chick CAM and a glioma xenograft model, respectively. Besides the effect of MRT on tumor growth, the results suggest that a preceding exposure to microbeams may render tumors more accessible to drug delivery. The MRT-induced vascular permeability observed in glioma xenografts could be exploited for the treatment of other intracranial tumors for which the transport of chemotherapeutic agents through the blood-brain barrier is difficult and limits treatment success [67]. Finally, the transient vascular permeability induced by MRT could also be applied to the delivery of drugs and/or agents other than chemotherapeutics, such as nanoparticles, antibodies, or vectors for the treatment of tumors or other pathologies.

Supplementary Materials: The following is available online at https://www.mdpi.com/article/10.3390/cancers13092103/s1, Video S1: Vascular Permeability in CAM.

Author Contributions: Conceptualization, J.A.L. and V.D.; methodology, S.S.; software, P.P.; validation, C.F.-P. and R.H.; formal analysis, C.F.-P.; investigation, S.S., J.A.L. and V.D.; data curation, S.S.; writing—original draft preparation, S.S., C.F.-P. and J.F.; writing—review and editing, C.F.-P., R.H., J.F., S.T., P.P., M.K., J.A.L. and V.D.; visualization, R.H. and S.T.; supervision, R.H. and V.D.; project administration, V.D.; funding acquisition, V.D. and C.F.-P. All authors have read and agreed to the published version of the manuscript.

Funding: This work received financial support from the Swiss National Science Foundation (31003A_176038) and Swiss Cancer Research Foundation (KFS-4281-08-2017) granted to Valentin Djonov; also by the Bernische Krebsliga (Grant number 190) awarded to Cristian Fernandez-Palomo.

Institutional Review Board Statement: Experiments involving Balb/c mice were performed under the Swiss Animal Experimentation numbers FR210/08 and FR21/09.

Informed Consent Statement: Not applicable.

Data Availability Statement: Data are available from the corresponding author upon request.

Acknowledgments: The authors have dedicated this work to their colleague Elke Bräuer-Krisch, who sadly passed away in 2018. Elke was a respected, essential, and enthusiastic promoter of Microbeam Radiation Therapy and a key partner in preparing the way towards its clinical application.

Conflicts of Interest: The authors declare no potential conflict of interest.

References

1. Boateng, F.; Ngwa, W. Delivery of Nanoparticle-Based Radiosensitizers for Radiotherapy Applications. *Int. J. Mol. Sci.* **2019**, *21*, 273. [CrossRef]
2. Holback, H.; Yeo, Y. Intratumoral drug delivery with nanoparticulate carriers. *Pharm. Res.* **2011**, *28*, 1819–1830. [CrossRef] [PubMed]
3. Chauhan, V.P.; Martin, J.D.; Liu, H.; Lacorre, D.A.; Jain, S.R.; Kozin, S.V.; Stylianopoulos, T.; Mousa, A.S.; Han, X.; Adstamongkonkul, P.; et al. Angiotensin inhibition enhances drug delivery and potentiates chemotherapy by decompressing tumour blood vessels. *Nat. Commun.* **2013**, *4*. [CrossRef] [PubMed]
4. Jain, R.K.; Tong, R.T.; Munn, L.L. Effect of Vascular Normalization by Antiangiogenic Therapy on Interstitial Hypertension, Peritumor Edema, and Lymphatic Metastasis: Insights from a Mathematical Model. *Cancer Res.* **2007**, *67*, 2729–2735. [CrossRef] [PubMed]
5. Marcucci, F.; Corti, A. How to improve exposure of tumor cells to drugs—Promoter drugs increase tumor uptake and penetration of effector drugs. *Adv. Drug Deliv. Rev.* **2012**, *64*, 53–68. [CrossRef]
6. Li, C.J.; Miyamoto, Y.; Kojima, Y.; Maeda, H. Augmentation of tumour delivery of macromolecular drugs with reduced bone marrow delivery by elevating blood pressure. *Br. J. Cancer* **1993**, *67*, 975–980. [CrossRef]
7. Nagamitsu, A.; Greish, K.; Maeda, H. Elevating blood pressure as a strategy to increase tumor-targeted delivery of macromolecular drug SMANCS: Cases of advanced solid tumors. *Jpn. J. Clin. Oncol.* **2009**, *39*, 756–766. [CrossRef]
8. Ojha, T.; Pathak, V.; Shi, Y.; Hennink, W.; Moonen, C.; Storm, G.; Kiessling, F.; Lammers, T. Pharmacological and Physical Vessel Modulation Strategies to Improve EPR-mediated Drug Targeting to Tumors. *Adv. Drug Deliv. Rev.* **2017**, *119*, 44–60. [CrossRef]
9. Willett, C.G.; Boucher, Y.; di Tomaso, E.; Duda, D.G.; Munn, L.L.; Tong, R.T.; Chung, D.C.; Sahani, D.V.; Kalva, S.P.; Kozin, S.V.; et al. Direct evidence that the VEGF-specific antibody bevacizumab has antivascular effects in human rectal cancer. *Nat. Med.* **2004**, *10*, 145–147. [CrossRef]
10. Lajoie, J.M.; Shusta, E.V. Targeting receptor-mediated transport for delivery of biologics across the blood-brain barrier. *Annu. Rev. Pharmacol. Toxicol.* **2015**, *55*, 613–631. [CrossRef]
11. Khawar, I.A.; Kim, J.H.; Kuh, H.-J. Improving drug delivery to solid tumors: Priming the tumor microenvironment. *J. Control Release* **2015**, *201*, 78–89. [CrossRef]
12. Magzoub, M.; Jin, S.; Verkman, A.S. Enhanced macromolecule diffusion deep in tumors after enzymatic digestion of extracellular matrix collagen and its associated proteoglycan decorin. *FASEB J.* **2008**, *22*, 276–284. [CrossRef]
13. Unemori, E.N.; Amento, E.P. Relaxin modulates synthesis and secretion of procollagenase and collagen by human dermal fibroblasts. *J. Biol. Chem.* **1990**, *265*, 10681–10685. [CrossRef]
14. Li, L.; ten Hagen, T.L.; Bolkestein, M.; Gasselhuber, A.; Yatvin, J.; van Rhoon, G.C.; Eggermont, A.M.M.; Haemmerich, D.; Koning, G.A. Improved intratumoral nanoparticle extravasation and penetration by mild hyperthermia. *J. Control Release* **2013**, *167*, 130–137. [CrossRef] [PubMed]
15. Dimcevski, G.; Kotopoulis, S.; Bjånes, T.; Hoem, D.; Schjøtt, J.; Gjertsen, B.T.; Biermann, M.; Molven, A.; Sorbye, H.; McCormack, E.; et al. A human clinical trial using ultrasound and microbubbles to enhance gemcitabine treatment of inoperable pancreatic cancer. *J. Control Release* **2016**, *243*, 172–181. [CrossRef] [PubMed]
16. Kooiman, K.; Roovers, S.; Langeveld, S.A.G.; Kleven, R.T.; Dewitte, H.; O'Reilly, M.A.; Escoffre, J.-M.; Bouakaz, A.; Verweij, M.D.; Hynynen, K.; et al. Ultrasound-Responsive Cavitation Nuclei for Therapy and Drug Delivery. *Ultrasound Med. Biol.* **2020**, *46*, 1296–1325. [CrossRef]
17. Kotopoulis, S.; Dimcevski, G.; Gilja, O.H.; Hoem, D.; Postema, M. Treatment of human pancreatic cancer using combined ultrasound, microbubbles, and gemcitabine: A clinical case study. *Med. Phys.* **2013**, *40*, 072902. [CrossRef] [PubMed]
18. Pandey, A.; Kulkarni, S.; Mutalik, S. Liquid metal based theranostic nanoplatforms: Application in cancer therapy, imaging and biosensing. *Nanomedicine* **2020**, *26*, 102175. [CrossRef]
19. Yang, R.; Wei, T.; Goldberg, H.; Wang, W.; Cullion, K.; Kohane, D.S. Getting Drugs Across Biological Barriers. *Adv. Mater.* **2017**, *29*, 1606596. [CrossRef] [PubMed]
20. Sabatasso, S.; Laissue, J.A.; Hlushchuk, R.; Graber, W.; Bravin, A.; Bräuer-Krisch, E.; Corde, S.; Blattmann, H.; Gruber, G.; Djonov, V. Microbeam radiation-induced tissue damage depends on the stage of vascular maturation. *Int. J. Radiat. Oncol. Biol. Phys.* **2011**, *80*, 1522–1532. [CrossRef]

21. Brönnimann, D.; Bouchet, A.; Schneider, C.; Potez, M.; Serduc, R.; Bräuer-Krisch, E.; Graber, W.; Von Gunten, S.; Laissue, J.A.; Djonov, V. Synchrotron microbeam irradiation induces neutrophil infiltration, thrombocyte attachment and selective vascular damage in vivo. *Sci. Rep.* **2016**, *6*, 33601. [CrossRef]
22. Fernandez-Palomo, C.; Fazzari, J.; Trappetti, V.; Smyth, L.; Janka, H.; Laissue, J.; Djonov, V. Animal Models in Microbeam Radiation Therapy: A Scoping Review. *Cancers* **2020**, *12*, 527. [CrossRef]
23. Dilmanian, F.A.; Button, T.M.; Le Duc, G.; Zhong, N.; Peña, L.A.; Smith, J.A.L.; Martinez, S.R.; Bacarian, T.; Tammam, J.; Ren, B. Response of rat intracranial 9L gliosarcoma to microbeam radiation therapy. *Neuro-Oncology* **2002**, *4*, 26–38. [CrossRef]
24. Regnard, P.; Le Duc, G.; Bräuer-Krisch, E.; Troprès, I.; Siegbahn, E.A.; Kusak, A.; Clair, C.; Bernard, H.; Dallery, D.; Laissue, J.A.; et al. Irradiation of intracerebral 9L gliosarcoma by a single array of microplanar x-ray beams from a synchrotron: Balance between curing and sparing. *Phys. Med. Biol.* **2008**, *53*, 861–878. [CrossRef] [PubMed]
25. Bouchet, A.; Boumendjel, A.; Khalil, E.; Serduc, R.; Brauer, E.; Siegbahn, E.A.; Laissue, J.A.; Boutonnat, J. Chalcone JAI-51 improves efficacy of synchrotron microbeam radiation therapy of brain tumors. *J. Synchrotron Radiat.* **2012**, *19*, 478–482. [CrossRef]
26. Laissue, J.A.; Geiser, G.; Spanne, P.O.; Dilmanian, F.A.; Gebbers, J.O.; Geiser, M.; Wu, X.Y.; Makar, M.S.; Micca, P.L.; Nawrocky, M.M.; et al. Neuropathology of ablation of rat gliosarcomas and contiguous brain tissues using a microplanar beam of synchrotron-wiggler-generated X rays. *Int. J. Cancer* **1998**, *78*, 654–660. [CrossRef]
27. Miura, M.; Blattmann, H.; Bräuer-Krisch, E.; Bravin, A.; Hanson, A.L.; Nawrocky, M.M.; Micca, P.L.; Slatkin, D.N.; Laissue, J.A. Radiosurgical palliation of aggressive murine SCCVII squamous cell carcinomas using synchrotron-generated X-ray microbeams. *Br. J. Radiol.* **2006**, *79*, 71–75. [CrossRef] [PubMed]
28. Griffin, R.J.; Koonce, N.A.; Dings, R.P.M.; Siegel, E.; Moros, E.G.; Bräuer-Krisch, E.; Corry, P.M. Microbeam Radiation Therapy Alters Vascular Architecture and Tumor Oxygenation and is Enhanced by a Galectin-1 Targeted Anti-Angiogenic Peptide. *Radiat. Res.* **2012**, *177*, 804–812. [CrossRef]
29. Potez, M.; Fernandez-Palomo, C.; Bouchet, A.; Trappetti, V.; Donzelli, M.; Krisch, M.; Laissue, J.; Volarevic, V.; Djonov, V. Synchrotron Microbeam Radiation Therapy as a New Approach for the Treatment of Radioresistant Melanoma: Potential Underlying Mechanisms. *Int. J. Radiat. Oncol. Biol. Phys.* **2019**, *105*, 1126–1136. [CrossRef]
30. Bouchet, A.; Lemasson, B.; Le Duc, G.; Maisin, C.; Bräuer-Krisch, E.; Siegbahn, E.A.; Renaud, L.; Khalil, E.; Rémy, C.; Poillot, C.; et al. Preferential effect of synchrotron microbeam radiation therapy on intracerebral 9l gliosarcoma vascular networks. *Int. J. Radiat. Oncol. Biol. Phys.* **2010**, *78*, 1503–1512. [CrossRef]
31. Serduc, R.; Vérant, P.; Vial, J.C.; Farion, R.; Rocas, L.; Rémy, C.; Fadlallah, T.; Brauer, E.; Bravin, A.; Laissue, J.; et al. In vivo two-photon microscopy study of short-term effects of microbeam irradiation on normal mouse brain microvasculature. *Int. J. Radiat. Oncol. Biol. Phys.* **2006**, *64*, 1519–1527. [CrossRef] [PubMed]
32. Serduc, R.R.; Christen, T.; Laissue, J.A.; Farion, R.R.; Bouchet, A.; van der Sanden, B.; Segebarth, C.; Brauer-Krisch, E.; Le Duc, G.G.; Bravin, A.; et al. Brain tumor vessel response to synchrotron microbeam radiation therapy: A short-term in vivo study. *Phys. Med. Biol.* **2008**, *53*, 3609–3622. [CrossRef] [PubMed]
33. Slatkin, D.N.; Spanne, P.; Dilmanian, F.A.; Gebbers, J.O.; Laissue, J.A. Subacute neuropathological effects of microplanar beams of x-rays from a synchrotron wiggler. *Proc. Natl. Acad. Sci. USA* **1995**, *92*, 8783–8787. [CrossRef] [PubMed]
34. Laissue, J.A.; Blattmann, H.; Di Michiel, M.; Slatkin, D.N.; Lyubimova, N.; Guzman, R.; Michiel, D.; Zimmermann, A.; Birrer, S.; Bey, T.; et al. The weaning piglet cerebellum: A surrogate for tolerance to MRT (microbeam radiation therapy) in paediatric neuro-oncology. *Proc. SPIE* **2001**, 65–73. [CrossRef]
35. Dilmanian, F.A.; Morris, G.M.; Le Duc, G.; Huang, X.; Ren, B.; Bacarian, T.; Allen, J.C.; Kalef-Ezra, J.; Orion, I.; Rosen, E.M.; et al. Response of avian embryonic brain to spatially segmented x-ray microbeams. *Cell. Mol. Biol.* **2001**, *47*, 485–493.
36. Laissue, J.A.; Blattmann, H.; Wagner, H.P.; Grotzer, M.A.; Slatkin, D.N. Prospects for microbeam radiation therapy of brain tumours in children to reduce neurological sequelae. *Dev. Med. Child Neurol.* **2007**, *49*, 577–581. [CrossRef]
37. Laissue, J.A.; Lyubimova, N.; Wagner, H.-P.; Archer, D.W.; Slatkin, D.N.; Di Michiel, M.; Nemoz, C.; Renier, M.; Brauer, E.; Spanne, P.O.; et al. Microbeam Radiation Therapy. 6 October 1999, Volume 3770. Available online: https://www.spiedigitallibrary.org/conference-proceedings-of-spie/3770/1/Microbeam-radiation-therapy/10.1117/12.368185.short?SSO=1 (accessed on 30 April 2020).
38. Potez, M.; Bouchet, A.; Wagner, J.; Donzelli, M.; Bräuer-Krisch, E.; Hopewell, J.W.; Laissue, J.; Djonov, V. Effects of Synchrotron X-Ray Micro-beam Irradiation on Normal Mouse Ear Pinnae. *Int. J. Radiat. Oncol. Biol. Phys.* **2018**, *101*, 680–689. [CrossRef]
39. Smyth, L.M.L.; Donoghue, J.F.; Ventura, J.A.; Livingstone, J.; Bailey, T.; Day, L.R.J.; Crosbie, J.C.; Rogers, P.A.W. Comparative toxicity of synchrotron and conventional radiation therapy based on total and partial body irradiation in a murine model. *Sci. Rep.* **2018**, *8*, 12044. [CrossRef]
40. Djonov, V.G.; Galli, A.B.; Burri, P.H. Intussusceptive arborization contributes to vascular tree formation in the chick chorio-allantoic membrane. *Anat. Embryol.* **2000**, *202*, 347–357. [CrossRef] [PubMed]
41. Ribatti, D.; Nico, B.; Perra, M.T.; Longo, V.; Maxia, C.; Annese, T.; Piras, F.; Murtas, D.; Sirigu, P. Erythropoietin is involved in angiogenesis in human primary melanoma. *Int. J. Exp. Pathol.* **2010**, *91*, 495–499. [CrossRef] [PubMed]
42. Ribatti, D.; Vacca, A.; Roncali, L.; Dammacco, F. The chick embryo chorioallantoic membrane as a model for in vivo research on anti-angiogenesis. *Curr. Pharm. Biotechnol.* **2000**, *1*, 73–82. [CrossRef]
43. Ribatti, D. The chick embryo chorioallantoic membrane (CAM): A multifaceted experimental model. *Mech. Dev.* **2016**, *141*, 70–77. [CrossRef]

44. DeBord, L.C.; Pathak, R.R.; Villaneuva, M.; Liu, H.-C.; Harrington, D.A.; Yu, W.; Lewis, M.T.; Sikora, A.G. The chick chorioallantoic membrane (CAM) as a versatile patient-derived xenograft (PDX) platform for precision medicine and preclinical research. *Am. J. Cancer Res.* **2018**, *8*, 1642–1660.
45. Chu, P.-Y.; Koh, A.P.-F.; Antony, J.; Huang, R.Y.-J. Applications of the Chick Chorioallantoic Membrane as an Alternative Model for Cancer Studies. *Cells Tissues Organs* **2021**, 1–16. [CrossRef] [PubMed]
46. Mapanao, A.K.; Che, P.P.; Sarogni, P.; Sminia, P.; Giovannetti, E.; Voliani, V. Tumor grafted—Chick chorioallantoic membrane as an alternative model for biological cancer research and conventional/nanomaterial-based theranostics evaluation. *Expert Opin. Drug Metab. Toxicol.* **2021**, 1–22. [CrossRef]
47. Bräuer-Krisch, E.; Bravin, A.; Lerch, M.; Rosenfeld, A.; Stepanek, J.; Di Michiel, M.; Laissue, J.A. MOSFET dosimetry for microbeam radiation therapy at the European Synchrotron Radiation Facility. *Med. Phys.* **2003**, *30*, 583–589. [CrossRef] [PubMed]
48. Archer, D.W. Collimator for Producing an Array of Microbeams. 1997. Available online: https://www.surechembl.org/document/US-5771270-A (accessed on 22 April 2021).
49. Bräuer-Krisch, E.; Requardt, H.; Brochard, T.; Berruyer, G.; Renier, M.; Laissue, J.A.; Bravin, A. New technology enables high precision multislit collimators for microbeam radiation therapy. *Rev. Sci. Instrum.* **2009**, *80*, 074301. [CrossRef] [PubMed]
50. FITC-Labelled Polysaccharides. Sigma-Aldrich. 2020. Available online: https://www.sigmaaldrich.com/technical-documents/articles/chemistry/fluorescently-labeled-dextrane.html (accessed on 13 April 2021).
51. FluoSpheres Fluorescent Microspheres—Product Information 2005 by Molecular Probes. Available online: https://tools.thermofisher.com/content/sfs/manuals/mp05000.pdf (accessed on 30 April 2020).
52. Working with FluoSpheres Fluorescent Microspheres—Product Information 2004 by Molecular Probes. Available online: http://tools.thermofisher.com/content/sfs/manuals/mp05001.pdf (accessed on 30 April 2020).
53. Djonov, V.; Schmid, M.; Tschanz, S.A.; Burri, P.H. Intussusceptive angiogenesis: Its role in embryonic vascular network formation. *Circ. Res.* **2000**, *86*, 286–292. [CrossRef] [PubMed]
54. Pérez, J.E.; Fritzell, S.; Kopecky, J.; Visse, E.; Darabi, A.; Siesjö, P. The effect of locally delivered cisplatin is dependent on an intact immune function in an experimental glioma model. *Sci. Rep.* **2019**, *9*, 5632. [CrossRef]
55. Jacobs, S.; McCully, C.L.; Murphy, R.F.; Bacher, J.; Balis, F.M.; Fox, E. Extracellular fluid concentrations of cisplatin, carboplatin, and oxaliplatin in brain, muscle, and blood measured using microdialysis in nonhuman primates. *Cancer Chemother. Pharmacol.* **2010**, *65*, 817–824. [CrossRef]
56. McWhinney, S.R.; Goldberg, R.M.; McLeod, H.L. Platinum Neurotoxicity Pharmacogenetics. *Mol. Cancer Ther.* **2009**, *8*, 10–16. [CrossRef]
57. Maeda, H.; Wu, J.; Sawa, T.; Matsumura, Y.; Hori, K. Tumor vascular permeability and the EPR effect in macromolecular therapeutics: A review. *J. Control Release* **2000**, *65*, 271–284. [CrossRef]
58. Azzi, S.; Hebda, J.K.; Gavard, J. Vascular permeability and drug delivery in cancers. *Front. Oncol.* **2013**, *3*, 211. [CrossRef]
59. Wilhelm, S.; Tavares, A.J.; Dai, Q.; Ohta, S.; Audet, J.; Dvorak, H.F.; Chan, W.C.W. Analysis of nanoparticle delivery to tumours. *Nat. Rev. Mater.* **2016**, *1*, 16014. [CrossRef]
60. Liu, J.; Li, M.; Luo, Z.; Dai, L.; Guo, X.; Cai, K. Design of nanocarriers based on complex biological barriers in vivo for tumor therapy. *Nano Today* **2017**, *15*, 56–90. [CrossRef]
61. Huang, P.; Wang, D.; Su, Y.; Huang, W.; Zhou, Y.; Cui, D.; Zhu, X.; Yan, D. Combination of Small Molecule Prodrug and Nanodrug Delivery: Amphiphilic Drug–Drug Conjugate for Cancer Therapy. *J. Am. Chem. Soc.* **2014**, *136*, 11748–11756. [CrossRef]
62. Jain, R.K. Normalizing tumor microenvironment to treat cancer: Bench to bedside to biomarkers. *J. Clin. Oncol.* **2013**, *31*, 2205–2218. [CrossRef] [PubMed]
63. Eggermont, A.M.; Schraffordt Koops, H.; Liénard, D.; Kroon, B.B.; van Geel, A.N.; Hoekstra, H.J.; Lejeune, F.J. Isolated limb perfusion with high-dose tumor necrosis factor-alpha in combination with interferon-gamma and melphalan for nonresectable extremity soft tissue sarcomas: A multicenter trial. *J. Clin. Oncol.* **1996**, *14*, 2653–2665. [CrossRef]
64. Nakata, H.; Yoshimine, T.; Murasawa, A.; Kumura, E.; Harada, K.; Ushio, Y.; Hayakawa, T. Early blood-brain barrier disruption after high-dose single-fraction irradiation in rats. *Acta Neurochir.* **1995**, *136*, 82–87. [CrossRef] [PubMed]
65. Kowanetz, M.; Ferrara, N. Vascular Endothelial Growth Factor Signaling Pathways: Therapeutic Perspective. *Clin. Cancer Res.* **2006**, *12*, 5018–5022. [CrossRef] [PubMed]
66. Lange, C.; Storkebaum, E.; de Almodóvar, C.R.; Dewerchin, M.; Carmeliet, P. Vascular endothelial growth factor: A neurovascular target in neurological diseases. *Nat. Rev. Neurol.* **2016**, *12*, 439–454. [CrossRef] [PubMed]
67. Faraji, A.H.; Wipf, P. Nanoparticles in cellular drug delivery. *Bioorganic Med. Chem.* **2009**, *17*, 2950–2962. [CrossRef] [PubMed]

Article

Diagnostic Value of VEGF-A, VEGFR-1 and VEGFR-2 in Feline Mammary Carcinoma

Catarina Nascimento [1], Andreia Gameiro [1], João Ferreira [2], Jorge Correia [1] and Fernando Ferreira [1,*]

[1] CIISA—Centro de Investigação Interdisciplinar em Sanidade Animal, Faculdade de Medicina Veterinária, Universidade de Lisboa, Avenida da Universidade Técnica, 1300-477 Lisboa, Portugal; catnasc@fmv.ulisboa.pt (C.N.); agameiro@fmv.ulisboa.pt (A.G.); jcorreia@fmv.ulisboa.pt (J.C.)

[2] Instituto de Medicina Molecular, Faculdade de Medicina, Universidade de Lisboa, 1649-028 Lisboa, Portugal; hjoao@medicina.ulisboa.pt

[*] Correspondence: fernandof@fmv.ulisboa.pt; Tel.: +351-21-365-2800 (ext. 431234)

Citation: Nascimento, C.; Gameiro, A.; Ferreira, J.; Correia, J.; Ferreira, F. Diagnostic Value of VEGF-A, VEGFR-1 and VEGFR-2 in Feline Mammary Carcinoma. *Cancers* **2021**, *13*, 117. https://doi.org/10.3390/cancers13010117

Received: 29 November 2020
Accepted: 18 December 2020
Published: 1 January 2021

Publisher's Note: MDPI stays neutral with regard to jurisdictional claims in published maps and institutional affiliations.

Copyright: © 2021 by the authors. Licensee MDPI, Basel, Switzerland. This article is an open access article distributed under the terms and conditions of the Creative Commons Attribution (CC BY) license (https://creativecommons.org/licenses/by/4.0/).

Simple Summary: Feline mammary carcinoma (FMC) is the third most common neoplasia in the cat, showing a highly malignant behavior, with both HER2-positive and triple negative (TN) subtypes presenting worse prognosis than luminal A and B subtypes. Furthermore, FMC has become a reliable cancer model for the study of human breast cancer, due to the similarities of clinicopathological, histopathological, and epidemiological features among the two species. Therefore, the identification of novel diagnostic biomarkers and therapeutic targets is needed to improve the clinical outcome of these patients. The aim of this study was to assess the potential of the VEGF-A/VEGFRs pathway, in order to validate future diagnostic and checkpoint-blocking therapies. Results showed that serum VEGF-A, VEGFR-1, and VEGFR-2 levels were significantly higher in cats with HER2-positive and TN normal-like tumors, presenting a positive association with its tumor-infiltrating lymphocytes expression, suggesting that these molecules may serve as promising non-invasive diagnostic biomarkers for these subtypes.

Abstract: Vascular endothelial growth factor (VEGF-A) plays an essential role in tumor-associated angiogenesis, exerting its biological activity by binding and activating membrane receptors, as vascular endothelial growth factor receptor 1 and 2 (VEGFR-1, VEGFR-2). In this study, serum VEGF-A, VEGFR-1, and VEGFR-2 levels were quantified in 50 cats with mammary carcinoma and 14 healthy controls. The expression of these molecules in tumor-infiltrating lymphocytes (TILs) and in cancer cells was evaluated and compared with its serum levels. Results obtained showed that serum VEGF-A levels were significantly higher in cats with HER2-positive and Triple Negative (TN) Normal-Like subtypes, when compared to control group ($p = 0.001$, $p = 0.020$). Additionally, serum VEGFR-1 levels were significantly elevated in cats presenting luminal A, HER2-positive and TN Normal-Like tumors ($p = 0.011$, $p = 0.048$, $p = 0.006$), as serum VEGFR-2 levels ($p = 0.010$, $p = 0.046$, $p = 0.005$). Moreover, a positive interaction was found between the expression of VEGF-A, VEGFR-1, and VEGFR-2 in TILs and their serum levels ($p = 0.002$, $p = 0.003$, $p = 0.003$). In summary, these findings point to the usefulness of VEGF-A and its serum receptors assessment in clinical evaluation of cats with HER2-positive and TN Normal-Like tumors, suggesting that targeted therapies against these molecules may be effective for the treatment of these animals, as described in human breast cancer.

Keywords: feline mammary carcinoma; VEGF-A; VEGFR-1; VEGFR-2; non-invasive biomarkers; angiogenesis

1. Introduction

Human breast cancer is the most diagnosed cancer and the leading cause of cancer-related death in women worldwide [1], being a heterogeneous disease driven by five distinct molecular profiles (Luminal A, Luminal B, HER2-positive, Triple-Negative Normal and Basal-Like) [2,3]. In parallel, the feline mammary carcinoma (FMC) is a very common neoplasia associated with local recurrence and distant metastasis, resulting in a high mortality rate [4], being HER2-positive and Triple Negative (TN) the most aggressive subtypes [5,6]. Furthermore, FMC has become a reliable cancer model for the study of human breast cancer, due to the similarities of clinicopathological, histopathological and epidemiological features among the two species [7–9]. Therefore, the development of new approaches allowing the early detection and appropriate therapeutic strategies and follow-up of cats with mammary carcinoma becomes crucial.

Angiogenesis, the formation of new blood vessels, is a hallmark of cancer and is fundamental to supply the high metabolic demands in nutrients and oxygen of cancer cells, leading to a rapid tumor growth and metastatic dissemination [10,11]. Accordingly, cancer cells and stromal cells are able to produce and release mediators of angiogenesis, such as the vascular endothelial growth factor A (VEGF-A) [12–14]. VEGF-A is a glycoprotein (45 kDa) that is highly conserved among mammalian species, being expressed by different cell populations, as tumor infiltrating lymphocytes (TILs), macrophages, platelets and cancer cells, promoting capillary network growth and vascular permeability, allowing cancer cells to migrate to distinct organs [15–17]. In humans, there are four distinct VEGF-A isoforms with 121, 165, 189, and 206 amino acids, as a result of alternative mRNA splicing, with $VEGF_{165}$ being the predominant isoform [13,18,19]. Several studies in human breast cancer have shown that VEGF-A overexpression is present in tumors with aggressive phenotype, such as HER2-positive and TN subtypes [20,21], being associated with poor prognosis and shorter disease-free survival (DFS) and overall survival (OS) [11,16]. Nevertheless, to show its biological activity, this angiogenic cytokine needs to bind to specific class-III-membrane tyrosine kinase receptors expressed on endothelial cells, as the vascular endothelial growth factor receptor 1 and 2 (VEGFR-1/Flt-1; VEGFR-2/KDR/Flk-1) [22], both having seven extracellular immunoglobulin homology domains, a transmembrane domain and an intracellular region with a tyrosine kinase domain, leading to distinct biological effects [23]. Accordingly, VEGFR-1 is more related with the pathological angiogenesis, while VEGFR-2 is involved in physiological and pathological angiogenesis [13]. In humans, the interaction between VEGF-A and VEGFR-2 is the most relevant for angiogenesis in solid tumors [12], as VEGFR-2 binds to all VEGF-A isoforms [24]. Activated VEGFR-2 promotes the activation of the PLC-γ, PKC-Raf-1-MEK-MAP kinase and PI3K-AKT pathways, as a signaling towards cell proliferation and endothelial cell survival [24,25].

Furthermore, it has been described that the secretion of soluble forms of VEGFR-1 (sVEGFR-1) and VEGFR-2 (sVEGFR-2) in the extracellular matrix displayed high affinity to VEGF-A. These isoforms are considered a natural defense strategy against malignant cells, exhibiting antiangiogenic, anti-edema and anti-inflammatory effects [13,22]. Accordingly, a low sVEGFR-1/VEGF-A ratio was associated with higher tumor malignancy and poor prognosis [13].

The discovery of antitumor immunotherapies targeting tumor-induced angiogenesis (e.g., VEGF-A, VEGFR-2) have been proposed as a universal therapeutic strategy to improve the clinical outcome of patients with several solid tumor types, as breast cancer [18,26]. Studies demonstrated that a humanized monoclonal antibody that bind to all soluble VEGF-A isoforms, bevacizumab, inhibit angiogenesis and tumor growth, promoting significant improvements in DFS of patients with breast cancer [10,27]. However, an increased overall survival (OS) could not be demonstrated, leading to a bevacizumab's approval withdrew by Food and Drug Administration (FDA) after two years following its initial approval, whereas the European Medicines Agency (EMA) maintained their approval [10,11]. Furthermore, several novel and potent VEGFR-1 and VEGFR-2 antagonists are being evaluated in clinical trials, showing promising results [24,28]. In cat, although Michishita et al. (2016) demonstrated that bevacizumab suppressed tumor growth in a xenograft model, suggesting its potential therapeutic effect for FMC [29], the role of VEGF-A in angiogenesis and its biological effects in feline mammary carcinoma is still poorly documented. Therefore, the aim of this study was to: (i) quantify and compare the serum VEGF-A, VEGFR-1 and VEGFR-2 levels between cats with distinct mammary carcinoma subtypes and healthy controls; (ii) test for associations between serum levels and clinicopathological features; (iii) evaluate the VEGF-A, VEGFR-1 and VEGFR-2 expression in TILs and cancer cells of feline spontaneous mammary carcinomas and (iv) screen for correlations between serum levels and expression levels of VEGF-A, VEGFR-1 and VEGFR-2 in TILs and cancer cells.

2. Results

2.1. Serum VEGF-A, VEGFR-1 and VEGFR-2 Levels Are Significantly Elevated in Cats with HER2-Positive and TN Normal-Like Mammary Carcinoma

Cats with mammary carcinoma were stratified according to their tumor subtype and serum VEGF-A, VEGFR-1 and VEGFR-2 levels were measured and compared with control group. Results showed that cats with HER2-positive and TN Normal-Like mammary carcinoma displayed higher serum VEGF-A levels than control group (1748.6 ± 3558.4 pg/mL vs. 0.0 pg/mL, $p = 0.001$; 1881.9 ± 2927.9 pg/mL vs. 0.0 pg/mL, $p = 0.020$; respectively, Figure 1A). Furthermore, cats presenting Luminal A, HER2-positive and TN Normal-Like mammary carcinoma subtypes revealed higher serum VEGFR-1 levels, comparing with healthy group (10197.4 ± 17679.4 pg/mL vs. 0.0 pg/mL, $p = 0.011$; 3068.9 ± 4935.5 pg/mL vs. 0.0 pg/mL, $p = 0.048$; 11527.6 ± 12845.4 vs. 0.0 pg/mL, $p = 0.006$; respectively, Figure 1B), as well as serum VEGFR-2 levels (2033.4 pg/mL ± 3550.7 vs. 0.0 pg/mL, $p = 0.010$; 502.3 ± 1091.8 pg/mL vs. 0.0 pg/mL, $p = 0.046$; 2023.6 ± 2416.0 pg/mL vs. 0.0 pg/mL, $p = 0.005$; respectively, Figure 1C).

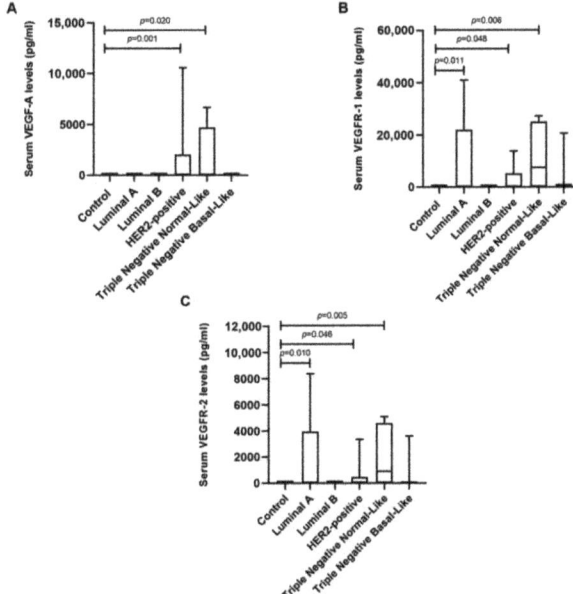

Figure 1. Serum vascular endothelial growth factor A (VEGF-A) levels are significantly increased in cats with HER2-positive and TN Normal-like tumors, while the vascular endothelial growth factor receptor 1 (VEGFR-1) and 2 (VEGFR-2) are significantly elevated in cats with luminal A, HER2-positive and TN Normal-like mammary carcinomas. (**A**) Box plot analysis of serum VEGF-A, (**B**) VEGFR-1 and (**C**) VEGFR-2 levels in the control group and in cats with mammary carcinoma grouped according to their molecular subtype.

In addition, a positive correlation was identified between serum VEGF-A and both VEGFR-1 (r = 0.567, p = 0.0001) and VEGFR-2 levels (r = 0.591, p = 0.0001), and also between serum VEGFR-1 and VEGFR-2 levels (r = 0.973, p = 0.0001).

2.2. Higher Serum VEGFR-1 and VEGFR-2 Levels are Correlated with the Administration of Contraceptives and Low-Grade Feline Mammary Carcinomas

A statistical analysis was performed between the serum VEGF-A, VEGFR-1 and VEGFR-2 levels in cats with mammary carcinoma and the studied clinicopathological features (Table 1). Although, no significant associations were found between serum VEGF-A levels and the recorded clinicopathologic parameters, serum VEGFR-1 and VEGFR-2 levels were positively associated with contraceptive administration (p = 0.026 and p = 0.042, respectively, Figure 2A,B) and tumors of lower malignancy grade (p = 0.037 and p = 0.046, respectively, Figure 2C,D).

Table 1. Statistical associations between serum VEGF-A, VEGFR-1 and VEGFR-2 levels and clinicopathological parameters examined in cats with mammary carcinoma (mean values ± standard deviation).

Clinicopathological Feature	Number of Animals (%)	VEGF-A (pg/mL)	p	VEGFR-1 (pg/mL)	p	VEGFR-2 (pg/mL)	p
Age							
<8 years old	4 (8.0%)	2643.5 ± 5287.0	0.483	2442.6 ± 4885.2	0.425	337.0 ± 674.1	0.58
8–12 years old	26 (52.0%)	159.5 ± 628.1		3771.3 ± 9414.6		754.1 ± 1838.2	
>12 years old	20 (40.0%)	738.4 ± 2042.5		5565.3 ± 11,514.6		963.3 ± 2288.2	
Spayed							
No	24 (48.0%)	1470.3 ± 2914.7	0.075	3996.3 ± 8062.8	0.39	644.0 ± 1443.7	0.537
Yes	25 (50.0%)	0		5757.9 ± 11,325.2		1117.1 ± 2271.9	
Unknown	1 (2.0%)						
Contraceptive administration							
No	21 (42.0%)	660.9 ± 2364.0	0.188	1077.3 ± 2740.1	0.026	140.4 ± 352.6	0.042
Yes	23 (46.0%)	882.9 ± 1900.5		8156.8 ± 12,291.0		1512.1 ± 2413.4	
Unknown	6 (12.0%)						
Multiple tumors							
Negative	19 (38.0%)	476.5 ± 1667.7	0.188	6572.4 ± 11,690.2	0.846	1217.0 ± 2286.1	0.701
Positive	31 (62.0%)	989.7 ± 2377.6		3602.3 ± 8396.6		621.3 ± 1617.4	
Lymph node status							
Negative	31 (62.0%)	1102.7 ± 2557.0	0.155	4817.9 ± 9291.5	0.345	840.3 ± 1742.9	0.432
Positive	16 (32.0%)	0		4842.9 ± 10,931.0		929.7 ± 2235.5	
Unknown	3 (6.0%)						
Stage							
I	11 (22.0%)	1753.6 ± 3736.3	0.502	4766.1 ± 8059.3	0.606	882.4 ± 1572.0	0.688
II	7 (14.0%)	138.5 ± 339.3		5632.5 ± 11,354.2		993.5 ± 2198.0	
III	27 (54.0%)	387.6 ± 1312.0		3609.1 ± 10,349.2		663.2 ± 2051.1	
IV	5 (10.0%)	0		6914.8 ± 10,335.0		1213.0 ± 1774.6	
Tumor size							
≤2 cm	26 (52.0%)	835.0 ± 2604.1	0.67	6024.2 ± 11,286.1	0.374	1140.8 ± 2239.7	0.5
>2 cm	24 (48.0%)	467.9 ± 1405.1		2754.6 ± 7646.5		452.9 ± 1386.5	
Tumor malignancy grade							
I	2 (4.0%)	5286.9 ± 7476.9	0.198	20,094.3 ± 14,600.2	0.037	3591.6 ± 3172.7	0.046
II	6 (12.0%)	0		1899.8 ± 4653.5		278.8 ± 683.0	
III	42 (84.0%)	480.0 ± 1526.4		3278.2 ± 9626.1		776.5 ± 1888.2	
Tumor necrosis							
Negative	11 (22.0%)	1358.2 ± 3725.1	0.587	8079.7 ± 14,010.3	0.227	1640.2 ± 2907.7	0.182
Positive	39 (78.0%)	415.3 ± 1549.8		2801.0 ± 8408.6		461.7 ± 1551.2	

Table 1. Cont.

Clinicopathological Feature	Number of Animals (%)	VEGF-A (pg/mL)	p	VEGFR-1 (pg/mL)	p	VEGFR-2 (pg/mL)	p
Tumor lymphatic invasion							
Negative	43 (86.0%)	544.3 ± 2112.3	0.956	4537.3 ± 10,320.0	0.098	820.5 ± 2011.0	0.117
Positive	7 (14.0%)	941.9 ± 2307.1		0		0	
Lymphocytic infiltration							
Negative	16 (32.0%)	881.2 ± 2932.7	0.466	5173.4 ± 9837.5	0.316	901.9 ± 1818.4	0.523
Positive	33 (66.0%)	485.1 ± 1669.0		3292.4 ± 9802.5		609.8 ± 1949.4	
Unknown	1 (2.0%)						
Tumor ulceration							
Negative	43 (86.0%)	682.0 ± 2286.4	0.073	3720.8 ± 9483.9	0.116	704.6 ± 1861.9	0.094
Positive	7 (14.0%)	161.5 ± 1020.3		4626.6 ± 11,316.9		656.7 ± 2151.0	
Metastasis							
No	22 (44%)	535.5 ± 198.8	0.89	5740.6 ± 11,041.4	0.165	1093.1 ± 2233.9	0.269
Yes	28 (56%)	747.6 ± 2810.5		3412.8 ± 8447.5		595.1 ± 1526.0	

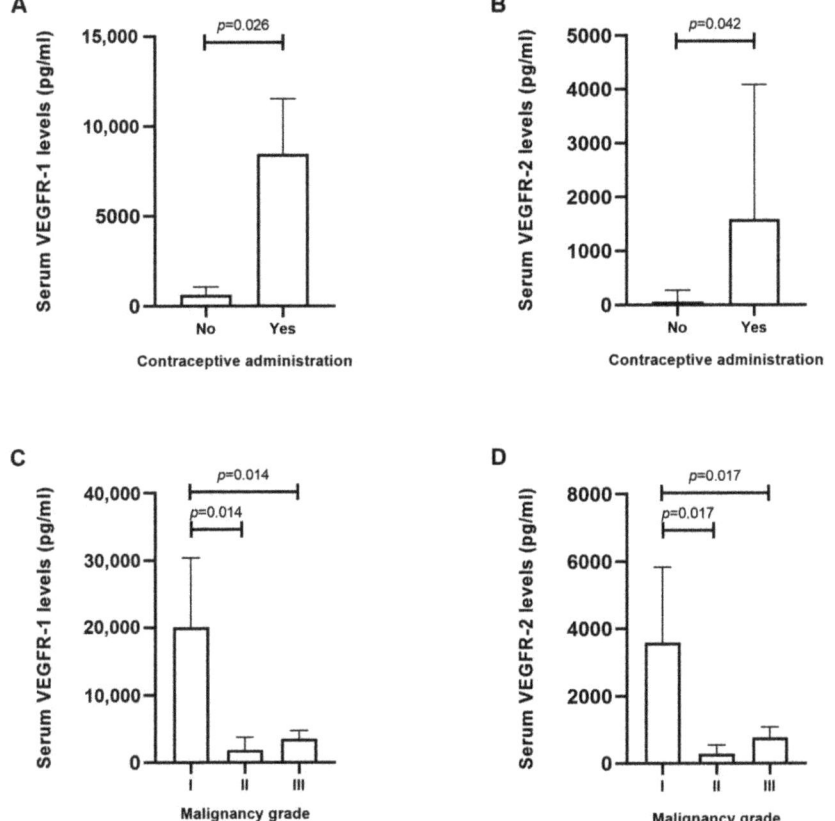

Figure 2. Serum levels of vascular endothelial growth factor receptor 1 (VEGFR-1) and receptor 2 (VEGFR-2) are positively correlated with the use of contraceptives and lower-malignancy tumors. (**A,B**) Box-plot analysis showing the mean ± SEM of serum VEGFR-1 and VEGFR-2 levels and its correlation with the use of contraceptive drugs and (**C,D**) tumor malignancy grade.

2.3. Serum VEGF-A, VEGFR-1 and VEGFR-2 Levels Are Positively Associated with Their Expression in Tumor Infiltrating Lymphocytes

Regarding the above results, the expression of VEGF-A, VEGFR-1 and VEGFR-2 was analyzed in cancer cells and in tumor infiltrating lymphocytes (TILs). Accordingly, the immunostaining analysis of cancer cells revealed that 95% (70% weak positive; 25% strong positive), 19% (17% weak positive; 2% strong positive) and 19% (19% weak positive; 0% strong positive) of tumors showed a positive score for VEGF-A, VEGFR-1 and VEGFR-2, respectively. In addition, 51% (33% weak positive; 18% strong positive), 22% (22% weak positive; 0% strong positive) and 24% (21% weak positive; 3% strong positive) of the tumors showed a positive IHC staining in TILs for VEGF-A, VEGFR-1 and VEGFR-2. Moreover, VEGF-A (Figure 3A,B) and VEGFR-1 expression (Figure 3C,D) was detected in cytoplasm of both cell types, while VEGFR-2 expression (Figure 3E,F) was found in cytoplasm and nucleus.

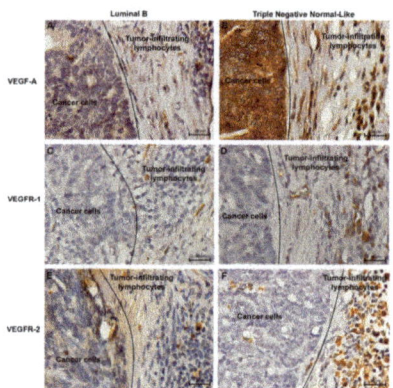

Figure 3. Representative images of immunohistochemical staining of vascular endothelial growth factor (VEGF-A), vascular endothelial growth factor receptor 1 (VEGFR-1) and receptor 2 (VEGFR-2) in tumor infiltrating lymphocytes and cancer cells of feline mammary carcinomas. Luminal B subtype graded as TILs negative for (**A**) VEGF-A, (**C**) VEGFR-1 and (**E**) VEGFR-2. Triple Negative Normal-Like subtype with a TILs-positive score for (**B**) VEGF-A, (**D**) VEGFR-1 and (**F**) VEGFR-2. Original magnification 400×.

Results also revealed that serum VEGF-A levels were significantly higher in cats showing a strong positive VEGF-A expression in TILs, in comparison to those with a weak positive ($p = 0.003$) or negative ($p = 0.003$) score (Figure 4A). Furthermore, a positive association was found between weak positive VEGFR-1 and VEGFR-2 expressions in TILs and their correspondent serum levels ($p = 0.002$, Figure 4B; $p = 0.002$, Figure 4C). No significant correlations were found between serum VEGF-A, VEGFR-1, or VEGFR-2 levels and the expression of these proteins in cancer cells ($p = 0.712$, $p = 0.235$, $p = 0.218$, respectively, data not shown). In addition, the expression of VEGFR-2 in TILs was associated with high serum VEGF-A (Figure 4D) and VEGFR-1 (Figure 4E) levels.

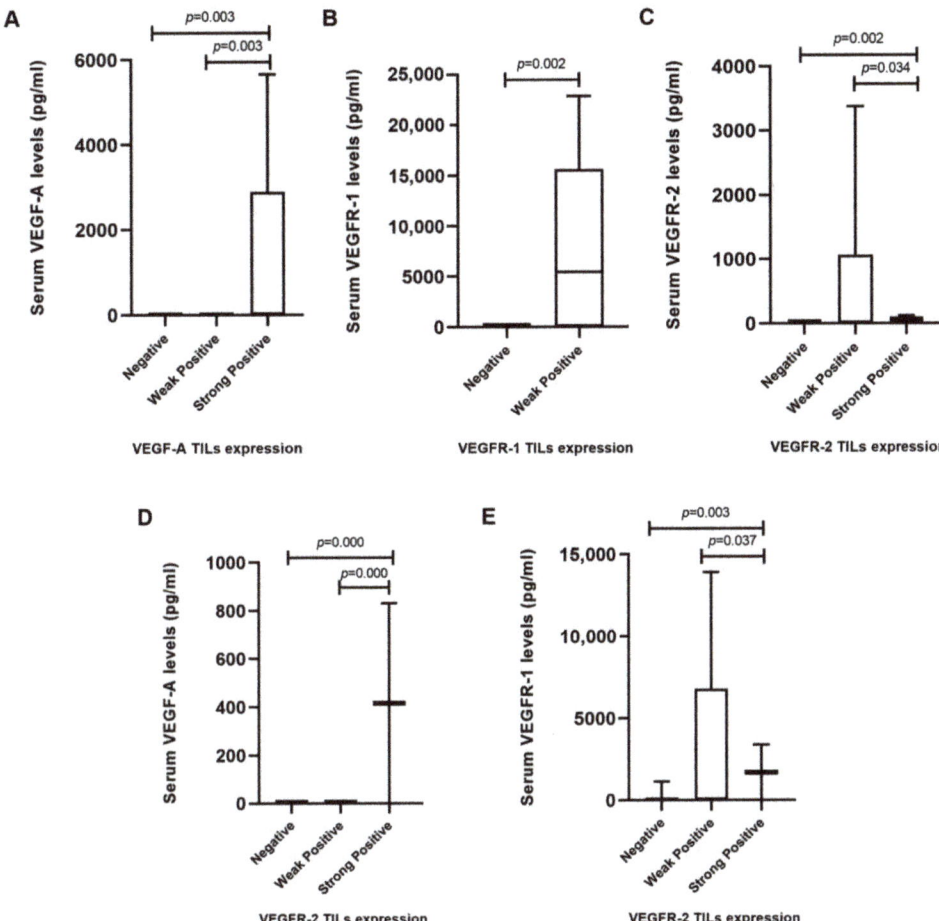

Figure 4. Serum levels and IHC scores of vascular endothelial growth factor (VEGF-A), vascular endothelial growth factor receptor 1 (VEGFR-1) and receptor 2 (VEGFR-2) in tumor infiltrating lymphocytes (TILs) of cats with mammary carcinoma. (**A**) Cats with tumors and a positive score for TILs showed higher serum VEGF-A, (**B**) VEGFR-1 and (**C**) VEGFR-2 levels in comparison with cats that showed a negative score for TILs. (**D**) Box plot diagrams showing that queens with mammary carcinomas scored as VEGFR-2-positive TILs had high serum VEGF-A and (**E**) VEGFR-1 levels.

3. Discussion

Feline mammary carcinoma shows a highly malignant behavior and a poor prognosis, particularly, the HER2-positive and triple negative subtypes, becoming challenging to treat due to a lack of specific targets [9,30]. Furthermore, angiogenesis is one of the key mechanisms involved in cancer progression, which is controlled by several growth factors secreted by tumor and stromal cells, with VEGF-A being the most potent angiogenic factor [31,32]. Therefore, in this study, the serum levels and tissue expression of VEGF-A and its receptors, VEGFR-1 and VEGFR-2, were evaluated in cats with mammary carcinoma, in order to improve diagnostic tools and therapeutic strategies.

The results showed that serum VEGF-A levels were significantly higher in cats with more aggressive mammary carcinoma subtypes, i.e., HER2-positive and TN normal-like, in accordance with previous studies in human breast cancer [19,33–35]. Furthermore, several studies have shown elevated serum VEGFR-1 and VEGFR-2 levels in breast cancer patients, when compared to healthy controls [22,36,37]. Accordingly, the results obtained in this study, revealed that cats showing luminal A, HER2-positive, and TN normal-like tumor subtypes presented higher serum VEGFR-1 and VEGFR-2 levels than control group. This phenomenon might be explained as a compensatory mechanism for high serum VEGF-A levels. Indeed, serum VEGFR-1 and VEGFR-2 receptors can bind to all VEGF-A isoforms, being considered as natural antagonists by decreasing VEGF-A biological activity and its availability for the membrane-bound receptors [13,22,38]. Moreover, a possible reason for the elevated serum VEGFR-1 and VEGFR-2 levels found in cats with luminal A subtype may be related with ulceration. Indeed, all luminal A tumors were ulcerated, suggesting the development of inflammation and consequently the presence of the soluble forms of VEGFR-1 and VEGFR-2, in order to exert anti-inflammatory activities [13]. Furthermore, the results obtained also demonstrated that increased serum VEGFR-1 and VEGFR-2 levels were associated with low-grade tumors, supporting a defense mechanism of these molecules in initial tumor phases against pathological angiogenesis [38]. However, these results were observed in only two animals, with more studies being necessary to better understand this mechanism. Moreover, elevated serum VEGFR-1 and VEGFR-2 levels were also correlated with contraceptive administration. Accordingly, studies in human breast cancer demonstrated that oestrogen and progesterone influence both VEGFR-1 and VEGFR-2 [39,40]. In addition, significant correlations were found between serum VEGF-A levels and serum VEGFR-1 and VEGFR-2 levels, in accordance with that described for human breast cancer [22,36].

The immunohistochemical analysis revealed cytoplasmic immunoreactivity for VEGF-A and VEGFR-1 and cytoplasmic and nuclear staining pattern for VEGFR-2 in cancer cells and TILs, which is consistent with the results of earlier reports [4,14,18,20]. Furthermore, results demonstrated that cats with strong positive VEGF-A expression in TILs showed higher serum VEGF-A levels than cats with a weak positive or negative VEGF-A expression, suggesting an effective endocrine mechanism for the release of serum VEGF-A from stromal cells to the bloodstream. Accordingly, as part of tumor microenvironment, TILs are able to release VEGF-A and inflammatory cytokines, showing immunosuppressive effects [41], including the formation of new blood vessels by acting on endothelial cells [42] and enhancing the inflammatory processes by increasing hypoxia inducible factor 1-alpha (HIF-1α) and VEGF-A synthesis [13]. Moreover, high serum VEGF-A levels were also associated with an intense VEGFR-2 reactivity, suggesting that serum VEGF-A also contributes to the activation of VEGFR-2 [18]. Further, it was identified an association between higher serum VEGFR-1 levels and its weak positive expression in TILs in FMC samples. This finding may be related with VEGFR-1 secretion in tumor microenvironment as a soluble isoform (sVEGFR-1) generated by alternative splicing. Accordingly, Orecchia et al. (2003) demonstrated that sVEGFR-1 present in tumor microenvironment may also play a protumoral action through the stimulation of endothelial cell adhesion and chemotaxis [13,43]. The same could be predicted for VEGFR-2. Finally, sVEGFR-1 can interact with VEGFR-2 abrogating its activity [13]. Whether this provides a compensation mechanism to counteract the concurrently elevated levels of VEGFR-2 that we observed in TILs infiltrating FMCs remains to be established.

4. Materials and Methods

4.1. Animal Population and Sample Collection

Fifty animals with spontaneous mammary carcinoma that underwent mastectomy and fourteen healthy queens presenting for elective ovariohysterectomy were recruited from Small Animal Hospital of the Faculty of Veterinary Medicine/ULisbon and private clinics around Lisbon. Tumor samples were collected in accordance with the EU Directive 2010/63/EU and all procedures involving the manipulation of animals were consented by the owners. All mammary lesions were embedded in paraffin after fixation in 10% buffered formalin (pH 7.2) during 24–48 h. Serum samples of the same animals were prepared by centrifugation of the fresh blood samples at 1500 g for 20 min at 4 °C and then aliquoted and stored at −80 °C.

For each animal enrolled in the study, the following clinicopathological characteristics were recorded: age, breed, reproductive status, contraceptive administration, treatment performed (none, surgery or surgery plus chemotherapy), number, location and size of tumor lesions, histopathological classification, malignancy grade, presence of tumor necrosis, lymphatic invasion, lymphocytic infiltration, cutaneous ulceration, regional lymph node involvement, stage of the disease (TNM system), DFS and OS. The mean age at diagnosis was 11.8 years (range 7–18 years), while the mean size of the primary lesions was 2.7 cm (range 0.3–7 cm). The DFS was 8.9 ± 1.1 months ($n = 46$; 95% CI: 6.8–11.1 months) and the OS was 13.8 ± 1.3 months ($n = 49$; 95% CI: 11.1–16.5 months).

Regarding the molecular-based subtyping of FMC [8,9], cats were stratified in five groups: Luminal A ($n = 9$), Luminal B ($n = 17$), HER2-positive ($n = 11$), TN Normal-Like ($n = 5$) and TN Basal-Like ($n = 8$).

The homology between human and feline VEGF-A_{121}, VEGF-A_{165} and VEGF-A_{165b}, is 90.8%, 94.2% and 94.1%, respectively (UniProt, accession numbers: *Homo sapiens* P15692-9, P15692-4, P15692-8; *Felis catus* Q95LQ4). Considering the VEGF receptors, the comparison between human and feline VEGFR-1 and VEGFR-2 revealed a homology of 87.8% and 93.2%, respectively (UniProt, accession numbers: *Homo sapiens* P17948, P35968; *Felis catus* M3WIL9, M3WBW2).

4.2. Quantification of Serum VEGF-A, VEGFR-1 and VEGFR-2 Levels

The assessment of serum VEGF-A, VEGFR-1 and VEGFR-2 levels was performed using the commercially VEGF (DY293B), VEGF R1/Flt-1 (DY321B) and VEGF R2/KDR (DY357) DuoSet ELISA kits (R&D Systems, Minneapolis, USA), and following the manufacturer's instructions. The absolute levels of VEGF-A, VEGFR-1 and VEGFR-2 were determined using standard curves (four parameter logistic) run on each ELISA plate. Briefly, the capture antibodies (100 µL/well) were incubated on 96-well plates overnight, at room temperature (RT). On the next day, plates were washed three times (3×400 µL/well) phosphate buffered saline (PBS)-Tween 0.05%) and coated with 300 µL/well of PBS/BSA blocking agent (1%, w/v), at RT for 60 min. After another washing step, serum samples previously diluted (1:20) were added (100 µL/well) and incubated during 2 h at RT. Antigen-antibody complexes were washed (3×400 µL/well PBS-Tween 0.05%) and 100 µL/well of detection antibodies were added and incubated for 2 h at RT. Then, after three washes with 400 µL/well of PBS-Tween 0.05%, 100 µL/well of streptavidin-HRP were added and incubated for 20 min at RT, avoiding placing the microplate in direct light. Afterwards a further washing step (3×400 µL/well PBS-Tween 0.05%), 100 µL/well of substrate solution (1:1 mixture of H_2O_2 and tetramethylbenzidine) were added and incubated for 20 min at RT in the dark, followed by a stop solution (50 µL/well of 2 N H_2SO_4). A microplate reader was used to measure the optical density at 450 nm and 570 nm (Fluostar Optima Microplate Reader, BMG, Ortenberg, Germany). Standards and negative controls were run on each ELISA plate.

4.3. Immunohistochemistry Staining and Evaluation

Immunohistochemistry was done on 3 µm thickness sections of FMC samples (Microtome Leica RM2135, Newcastle, UK). Deparaffinization, rehydration and antigen retrieval were performed using a PT-Link module (Dako, Agilent, Santa Clara, CA, USA), by boiling glass slides in Antigen Target Retrieval Solution pH 9 from Dako, during 20 min at 96 °C. Then, slides were cooled for 30 min at RT and rinsed twice for 5 min in distilled water. Thereafter, sections were blocked with Peroxidase Block Novocastra Solution (Novacastra, Leica Biosystems, Newcastle, UK) during 15 min at RT, followed by two washing steps with PBS pH 7.4, and Protein Block Novocastra Solution (Leica Biosystems) during 10 min. After two washes with PBS for 5 min, tissue slides were incubated with the primary antibodies (Table 2). After incubation, each tissue section was washed with PBS 2x for 5 min and subsequently treated with the Post-Primary Reagent (Leica Biosystems) for 30 min at RT and with the Novolink Polymer (Leica Biosystems) for 30 min. Afterwards, sections were stained with DAB Chromogen Solution (Leica Biosystems) for 5 min and nuclei were counterstained with Gills hematoxylin (Merck, NJ, USA). Slides were dehydrated in an ethanol gradient and mounted with Entellan mounting medium. Human and feline kidney tissues were used as negative and positive controls. Positive and negative control samples were included in each slide run.

Table 2. Primary antibodies and their conditions of use.

Monoclonal Antibody	Reference	Dilution	Incubation Time and Temperature
Anti-VEGF	Clone VG1 (Novus Biologicals)	1:50	60′ at RT
Anti-VEGFR1/Flt-1	Clone CL0345 (Novus Biologicals)	1:200	60′ at RT
Anti-VEGFR2/KDR/Flk-1	Clone EIC (Novus Biologicals)	1:10	120′ at RT plus 4 °C overnight

RT—Room Temperature.

The staining of VEGF-A, VEGFR-1 and VEGFR-2 in tumor-infiltrating lymphocytes and cancer cells was assessed manually by two independent pathologists. TILs were identified by their characteristic morphology and scored according to the International TILs Working Group 2014 [44]. Furthermore, cancer cells were evaluated in whole tumor sections with 200–400× magnification. The percentage of positive staining cells was scored using a 4-point scale: 0 (<10%), 1 (10–25%), 2 (26–50%) and 3 (>50%) and the staining intensity was graded as: 0 (no staining), 1+ (weak), 2+ (moderate), and 3+ (strong). The percentage of positive cells and intensity score were then added to obtain a final IHC score [20]. IHC scores of 0–3 were defined as negative, 4–5 as weak positive and 6 as strong positive (Table 3).

Table 3. Scoring criteria of immunostaining assay for VEGF-A, VEGFR-1 and VEGFR-2.

Percentage of Positive Staining Cells		Staining Intensity	
Score	Interpretation	Score	Interpretation
0	<10%	0	No staining
1	10–25%	1	Weak
2	26–50%	2	Moderate
3	>50%	3	Strong
Total score (0–6): Score of positive staining cells + intensity score			
0–3: Negative			
4–5: Weak Positive			
6: Strong Positive			

4.4. Statistical Analysis

Statistical analysis was performed using the SPSS software version 25.0 (IBM, Armonk, NY, USA), while the GraphPad Prism version 8.1.2 (GraphPad Software, CA, USA) was used to plot the graphs. The non-parametric Mann–Whitney U test and the Kruskal–Wallis test were carried out to analyze the differences among groups of continuous variables. Correlations between variables were performed using the Spearman's rank coefficient. Outliers with more than three standard deviations were removed from analysis. Results with a p-value < 0.05 were deemed to be statistically significant.

5. Conclusions

In conclusion, cats with HER-2 positive and TN Normal-Like mammary carcinoma subtypes showed more elevated serum VEGF-A, VEGFR-1, and VEGFR-2 levels than healthy animals, suggesting that these molecules may serve as promising non-invasive diagnostic biomarkers for these subtypes. Furthermore, circulating VEGF-A together with its receptors was positively associated with its expression in TILs, indicating that, besides hypoxia, inflammation is another mechanism that leads to cancer progression via VEGF-A/VEGFRs signaling. Altogether, the similarities found between FMC and human breast cancer further validate the utility of the cat as a valuable model for comparative oncology studies.

Author Contributions: Conceptualization, C.N. and F.F.; methodology, C.N. and F.F.; formal analysis, C.N.; investigation, C.N., A.G., J.C., J.F., and F.F.; supervision, F.F.; funding acquisition, C.N. and F.F.; project administration, F.F.; writing—original draft preparation, C.N. and F.F.; writing—review and editing, C.N., A.G., J.C., J.F., and F.F. All authors have read and agreed to the published version of the manuscript.

Funding: This research was funded by Fundação para a Ciência e a Tecnologia (Portugal), through the projects PTDC/CVT-EPI/3638/2014 and UIDB/00276/2020. C.N. is receipt of a PhD fellowship from University of Lisbon (ref.C00191r) and A.G. is receipt of a PhD fellowship from Fundação para a Ciência e a Tecnologia (ref. SFRH/BD/132260/2017).

Institutional Review Board Statement: Ethical review and approval were waived for this study, as there was no interference with animal well-being.

Informed Consent Statement: Not applicable.

Data Availability Statement: The data presented in this study are available on request from the corresponding author. The data are not publicly available as their containing information that could compromise the privacy of future research.

Acknowledgments: The authors would like to thank Maria Soares for the feline samples and clinical database and to Lucília Monteiro from Hospital Egas Moniz for the human kidney samples.

Conflicts of Interest: The authors declare no conflict of interest. The funders had no role in the design of the study; in the collection, analyses, or interpretation of data; in the writing of the manuscript, or in the decision to publish the results.

References

1. Bray, F.; Ferlay, J.; Soerjomataram, I.; Siegel, R.L.; Torre, L.A.; Jemal, A. Global cancer statistics 2018: GLOBOCAN estimates of incidence and mortality worldwide for 36 cancers in 185 countries. *CA Cancer J. Clin.* **2018**, *68*, 394–424. [CrossRef] [PubMed]
2. Eliyatkin, N.; Yalcin, E.; Zengel, B.; Aktaş, S.; Vardar, E. Molecular classification of breast carcinoma: From traditional, old-fashioned way to a new age, and a new way. *J. Breast Health* **2015**, *11*, 59–66. [CrossRef] [PubMed]
3. Dai, X.; Li, T.; Bai, Z.; Yang, Y.; Liu, X.; Zhan, J.; Shi, B. Breast cancer intrinsic subtype classification, clinical use and future trends. *Am. J. Cancer Res.* **2015**, *5*, 2929–2943. [PubMed]
4. Chen, B.; Lin, S.J.H.; Li, W.T.; Chang, H.W.; Pang, V.F.; Chu, P.Y.; Lee, C.C.; Nakayama, H.; Wu, C.H.; Jeng, C.R. Expression of HIF-1α and VEGF in feline mammary gland carcinomas: Association with pathological characteristics and clinical outcomes. *BMC Vet. Res.* **2020**, *16*, 1–10. [CrossRef] [PubMed]
5. Soares, M.; Ribeiro, R.; Carvalho, S.; Peleteiro, M.; Correia, J.; Ferreira, F. Ki-67 as a prognostic factor in feline mammary carcinoma: What is the optimal cutoff value? *Vet. Pathol.* **2016**, *53*, 37–43. [CrossRef] [PubMed]
6. Nascimento, C.; Urbano, A.C.; Gameiro, A.; Correia, J.; Ferreira, F. Serum PD-1/PD-L1 levels, tumor expression and PD-L1 somatic mutations in HER2-positive and triple negative normal-like feline mammary carcinoma subtypes. *Cancers* **2020**, *12*, 1386. [CrossRef]
7. Urbano, A.C.; Nascimento, C.; Soares, M.; Correia, J.; Ferreira, F. Clinical relevance of the serum CTLA-4 in cats with mammary carcinoma. *Sci. Rep.* **2020**, *10*, 1–11. [CrossRef]
8. Soares, M.; Madeira, S.; Correia, J.; Peleteiro, M.; Cardoso, F.; Ferreira, F. Molecular based subtyping of feline mammary carcinomas and clinicopathological characterization. *Breast* **2016**, *27*, 44–51. [CrossRef]
9. Soares, M.; Correia, J.; Peleteiro, M.C.; Ferreira, F. St Gallen molecular subtypes in feline mammary carcinoma and paired metastases—Disease progression and clinical implications from a 3-year follow-up study. *Tumor Biol.* **2016**, *37*, 4053–4064. [CrossRef]
10. Garcia, J.; Hurwitz, H.I.; Sandler, A.B.; Miles, D.; Coleman, R.L.; Deurloo, R.; Chinot, O.L. Bevacizumab (Avastin®) in cancer treatment: A review of 15 years of clinical experience and future outlook. *Cancer Treat. Rev.* **2020**, *86*, 102017. [CrossRef]
11. Gullo, G.; Eustace, A.J.; Canonici, A.; Collins, D.M.; Kennedy, M.J.; Grogan, L.; Breathhnach, O.; McCaffrey, J.; Keane, M.; Martin, M.J.; et al. Pilot study of bevacizumab in combination with docetaxel and cyclophosphamide as adjuvant treatment for patients with early stage HER-2 negative breast cancer, including analysis of candidate circulating markers of cardiac toxicity: ICORG 08–10 trial. *Ther. Adv. Med. Oncol.* **2019**, *11*, 1–9.
12. Arai, R.J.; Petry, V.; Hoff, P.M.; Mano, M.S. Serum levels of VEGF and MCSF in HER2+ / HER2- breast cancer patients with metronomic neoadjuvant chemotherapy. *Biomark. Res.* **2018**, *6*, 1–6. [CrossRef] [PubMed]

13. Ceci, C.; Atzori, M.G.; Lacal, P.M.; Graziani, G. Role of VEGFs/VEGFR-1 signaling and its inhibition in modulating tumor invasion: Experimental evidence in different metastatic cancer models. *Int. J. Mol. Sci.* **2020**, *21*, 1388. [CrossRef] [PubMed]
14. Cîmpean, A.M.; Raica, M.; Suciu, C.; Tătucu, D.; Sârb, S.; Mureşan, A.M. Vascular endothelial growth factor A (VEGF A) as individual prognostic factor in invasive breast carcinoma. *Rom. J. Morphol. Embryol.* **2008**, *49*, 303–308. [PubMed]
15. Sahana, K.R.; Akila, P.; Prashant, V.; Chandra, B.S.; Suma, M.N. Quantitation of vascular endothelial growth factor and interleukin-6 in different stages of breast cancer. *Rep. Biochem. Mol. Biol.* **2017**, *6*, 32–38.
16. Salven, P.; Perhoniemi, V.; Tykkä, H.; Mäenpää, H.; Joensuu, H. Serum VEGF levels in women with a benign breast tumor or breast cancer. *Breast Cancer Res. Treat.* **1999**, *53*, 161–166. [CrossRef]
17. Shibuya, M. Vascular endothelial growth factor (VEGF) and its receptor (VEGFR) signaling in angiogenesis: A crucial target for anti- and pro-angiogenic therapies. *Genes Cancer* **2011**, *2*, 1097–1105. [CrossRef]
18. Koukourakis, M.I.; Limberis, V.; Tentes, I.; Kontomanolis, E.; Kortsaris, A.; Sivridis, E.; Giatromanolaki, A. Serum VEGF levels and tissue activation of VEGFR2/KDR receptors in patients with breast and gynecologic cancer. *Cytokine* **2011**, *53*, 370–375. [CrossRef]
19. Foekens, J.A.; Peters, H.A.; Grebenchtchikov, N.; Look, M.P.; Meijer-van Gelder, M.E.; Geurts-Moespot, A.; Van der Kwast, T.H.; Sweep, C.G.J.; Klijn, J.G.M. High tumor levels of vascular endothelial growth factor predict poor response to systemic therapy in advanced breast cancer. *Cancer Res.* **2001**, *61*, 5407–5414.
20. Ragab, H.M.; Shaaban, H.M.; El Maksoud, N.A.; Radwan, S.M.; Elaziz, W.A.; Hafez, N.H. Expression of vascular endothelial growth factor protein in both serum samples and excised tumor tissues of breast carcinoma patients. *Int. J. Cancer Res.* **2016**, *12*, 152–161. [CrossRef]
21. Dent, S.F. The role of VEGF in triple-negative breast cancer: Where do we go from here? *Ann. Oncol.* **2009**, *20*, 1615–1617. [CrossRef] [PubMed]
22. Thielemann, A.; Baszczuk, A.; Kopczyński, Z.; Kopczyński, P.; Grodecka-Gazdecka, S. Clinical usefulness of assessing VEGF and soluble receptors sVEGFR-1 and sVEGFR-2 in women with breast cancer. *Ann. Agric. Environ. Med.* **2013**, *20*, 293–297. [PubMed]
23. Simons, M.; Gordon, E.; Claesson-Welsh, L. Mechanisms and regulation of endothelial VEGF receptor signalling. *Nat. Rev. Mol. Cell Biol.* **2016**, *17*, 611–625. [CrossRef] [PubMed]
24. Dang, Y.Z.; Zhang, Y.; Li, J.P.; Hu, J.; Li, W.W.; Li, P.; Wei, L.C.; Shi, M. High VEGFR1/2 expression levels are predictors of poor survival in patients with cervical cancer. *Medicine* **2017**, *96*, 1–6. [CrossRef] [PubMed]
25. Takahashi, T.; Ueno, H.; Shibuya, M. VEGF activates protein kinase C-dependent, but Ras-independent Raf-MEK-MAP kinase pathway for DNA synthesis in primary endothelial cells. *Oncogene* **1999**, *18*, 2221–2230. [CrossRef] [PubMed]
26. Wang, R.; Chen, S.; Huang, L.; Zhou, Y.; Shao, Z. Monitoring serum VEGF in neoadjuvant chemotherapy for patients with triple-negative breast cancer: A new strategy for early prediction of treatment response and patient survival. *Oncologist* **2019**, *24*, 753–761. [CrossRef]
27. Kut, C.; Mac Gabhann, F.; Popel, A.S. Where is VEGF in the body? A meta-analysis of VEGF distribution in cancer. *Br. J. Cancer* **2007**, *97*, 978–985. [CrossRef]
28. Golfmann, K.; Meder, L.; Koker, M.; Volz, C.; Borchmann, S.; Tharun, L.; Dietlein, F.; Malchers, F.; Florin, A.; Büttner, R.; et al. Synergistic anti-angiogenic treatment effects by dual FGFR1 and VEGFR1 inhibition in FGFR1-amplified breast cancer. *Oncogene* **2018**, *37*, 5682–5693. [CrossRef]
29. Michishita, M.; Ohtsuka, A.; Nakahira, R.; Tajima, T.; Nakagawa, T.; Sasaki, N.; Arai, T.; Takahashi, K. Anti-tumor effect of bevacizumab on a xenograft model of feline mammary carcinoma. *J. Vet. Med. Sci.* **2016**, *78*, 685–689. [CrossRef]
30. Cannon, C. Cats, cancer and comparative oncology. *Vet. Sci.* **2015**, *2*, 111–126. [CrossRef]
31. Dumond, A.; Pagès, G. Neuropilins, as relevant oncology target: Their role in the tumoral microenvironment. *Front. Cell Dev. Biol.* **2020**, *8*, 1–10. [CrossRef] [PubMed]
32. Fujii, T.; Hirakata, T.; Kurozumi, S.; Tokuda, S.; Nakazawa, Y.; Obayashi, S.; Yajima, R.; Oyama, T.; Shirabe, K. VEGF-A is associated with the degree of TILs and PD-L1 expression in primary breast cancer. *In Vivo* **2020**, *34*, 2641–2646. [CrossRef] [PubMed]
33. Linderholm, B.K.; Hellborg, H.; Johansson, U.; Elmberger, G.; Skoog, L.; Lehtiö, J.; Lewensohn, R. Significantly higher levels of vascular endothelial growth factor (VEGF) and shorter survival times for patients with primary operable triple-negative breast cancer. *Ann. Oncol.* **2009**, *20*, 1639–1646. [CrossRef] [PubMed]
34. Konecny, G.E.; Meng, Y.G.; Untch, M.; Wang, H.J.; Bauerfeind, I.; Epstein, M.; Stieber, P.; Vernes, J.M.; Gutierrez, J.; Hong, K.; et al. Association between HER-2/neu and vascular endothelial growth factor expression predicts clinical outcome in primary breast cancer patients. *Clin. Cancer Res.* **2004**, *10*, 1706–1716. [CrossRef] [PubMed]
35. Ali, E.M.; Sheta, M.; El Mohsen, M.A. Elevated serum and tissue VEGF associated with poor outcome in breast cancer patients. *Alexandria J. Med.* **2011**, *47*, 217–224. [CrossRef]
36. Toi, M.; Bando, H.; Ogawa, T.; Muta, M.; Hornig, C.; Weich, H.A. Significance of vascular endothelial growth factor (VEGF)/soluble VEGF receptor-1 relationship in breast cancer. *Int. J. Cancer* **2002**, *98*, 14–18. [CrossRef]
37. Zajkowska, M.; Lubowicka, E.; Malinowski, P.; Szmitkowski, M.; Ławicki, S. Plasma levels of VEGF-A, VEGF B, and VEGFR-1 and applicability of these parameters as tumor markers in diagnosis of breast cancer. *Acta Biochim. Pol.* **2018**, *65*, 621–628. [CrossRef]
38. Zarychta, E.; Rhone, P.; Bielawski, K.; Rosc, D.; Szot, K.; Zdunska, M.; Ruszkowska-Ciastek, B. Elevated plasma levels of tissue factor as a valuable diagnostic biomarker with relevant efficacy for prediction of breast cancer morbidity. *J. Physiol. Pharmacol.* **2018**, *69*, 921–931.

39. Garvin, S.; Nilsson, U.W.; Dabrosin, C. Effects of oestradiol and tamoxifen on VEGF, soluble VEGFR-1, and VEGFR-2 in breast cancer and endothelial cells. *Br. J. Cancer* **2005**, *93*, 1005–1010. [CrossRef]
40. Botelho, M.; Soares, R.; Alves, H. Progesterone in breast cancer angiogenesis. *SM J. Reprod. Health Infertil.* **2015**, *1*, 1–3.
41. Aguilar-Cazares, D.; Chavez-Dominguez, R.; Carlos-Reyes, A.; Lopez-Camarillo, C.; Hernadez de la Cruz, O.N.; Lopez-Gonzalez, J.S. Contribution of angiogenesis to inflammation and cancer. *Front. Oncol.* **2019**, *9*, 1–10. [CrossRef] [PubMed]
42. Angelo, L.S.; Kurzrock, R. Vascular endothelial growth factor and its relationship to inflammatory mediators. *Clin. Cancer Res.* **2007**, *13*, 2825–2830. [CrossRef] [PubMed]
43. Orecchia, A.; Lacal, P.M.; Schietroma, C.; Morea, V.; Zambruno, G.; Failla, C.M. Vascular endothelial growth factor receptor-1 is deposited in the extracellular matrix by endothelial cells and is a ligand for the α5β1 integrin. *J. Cell Sci.* **2003**, *116*, 3479–3489. [CrossRef] [PubMed]
44. Salgado, R.; Denkert, C.; Demaria, S.; Sirtaine, N.; Klauschen, F.; Pruneri, G.; Wienert, S.; Van den Eynden, G.; Baehner, F.L.; Penault-Llorca, F.; et al. The evaluation of tumor-infiltrating lymphocytes (TILS) in breast cancer: Recommendations by an International TILS working group 2014. *Ann. Oncol.* **2015**, *26*, 259–271. [CrossRef]

Article

Resveratrol Promotes Tumor Microvessel Growth via Endoglin and Extracellular Signal-Regulated Kinase Signaling Pathway and Enhances the Anticancer Efficacy of Gemcitabine against Lung Cancer

San-Hai Qin [†], Andy T. Y. Lau [†], Zhan-Ling Liang, Heng Wee Tan, Yan-Chen Ji, Qiu-Hua Zhong, Xiao-Yun Zhao and Yan-Ming Xu *

Laboratory of Cancer Biology and Epigenetics, Department of Cell Biology and Genetics, Shantou University Medical College, Shantou, Guangdong 515041, China; sanhaiqin@stu.edu.cn (S.-H.Q.); andytylau@stu.edu.cn (A.T.Y.L.); 16zlliang@stu.edu.cn (Z.-L.L.); hwtan@stu.edu.cn (H.W.T.); joyceycji@stu.edu.cn (Y.-C.J.); 17qhzhong@stu.edu.cn (Q.-H.Z.); 18xyzhao@stu.edu.cn (X.-Y.Z.)
* Correspondence: amyymxu@stu.edu.cn; Tel.: +86-754-8890-0437
† These authors contributed equally to this work.

Received: 14 March 2020; Accepted: 13 April 2020; Published: 15 April 2020

Abstract: The synergistic anticancer effect of gemcitabine (GEM) and resveratrol (RSVL) has been noted in certain cancer types. However, whether the same phenomenon would occur in lung cancer is unclear. Here, we uncovered the molecular mechanism by which RSVL enhances the anticancer effect of GEM against lung cancer cells both in vitro and in vivo. We established human lung adenocarcinoma HCC827 xenografts in nude mice and treated them with GEM and RSVL to detect their synergistic effect in vivo. Tumor tissue sections from nude mice were subjected to hematoxylin and eosin staining for blood vessel morphological observation, and immunohistochemistry was conducted to detect CD31-positive staining blood vessels. We also established the HCC827-human umbilical vein endothelial cell (HUVEC) co-culture model to observe the tubule network formation. Human angiogenesis antibody array was used to screen the angiogenesis-related proteins in RSVL-treated HCC827. RSVL suppressed the expression of endoglin (ENG) and increased tumor microvessel growth and blood perfusion into tumor. Co-treatment of RSVL and GEM led to more tumor growth suppression than treatment of GEM alone. Mechanistically, using the HCC827-HUVEC co-culture model, we showed that RSVL-suppressed ENG expression was accompanied with augmented levels of phosphorylated extracellular signal-regulated kinase (ERK) 1/2 and increased tubule network formation, which may explain why RSVL promoted tumor microvessel growth in vivo. RSVL promoted tumor microvessel growth via ENG and ERK and enhanced the anticancer efficacy of GEM. Our results suggest that intake of RSVL may be beneficial during lung cancer chemotherapy.

Keywords: resveratrol; gemcitabine; endoglin; ERKs; microvessel growth; lung cancer

1. Introduction

Lung cancer is one of the worldwide malignancies with the highest incidence and mortality, which has been ranked number one in terms of morbidity and accounts for 11.6% of the total incidence of cancers [1]. Gemcitabine (GEM), a first-line chemotherapeutic agent for advanced non-small cell lung cancer (NSCLC), is commonly used but not very effective as a single agent, and therefore it is important to find ways to enhance its therapeutic efficacy.

Angiogenesis is crucial during the development of human lung cancer, in which cancer cells secrete angiogenic factors and induce neovascularization to establish tumor vascular network, providing nutrients required for cell expansion, facilitating their growth, proliferation, and metastasis.

Traditionally, anti-angiogenic therapy should be a very logical strategy for lung cancer therapy [2]. However, it should also be noted that because the tumor vasculatures are abnormal and consist of chaotic labyrinth of malformed and destabilized vessels that are structurally and functionally impaired, the tumor microenvironment is therefore not only hypoxic and acidic but also is surrounded by high interstitial pressure, which acts as a pathologic barrier for drug delivery into tumors and leads to a notably reduced therapeutic effect [3]. Traditional anti-angiogenic approaches often cause extreme hypoxia in tumors and eventually lead to increased drug resistance, local invasion, and more distant metastasis [4]. In addition, angiogenesis compensatory pathways and alternative modes of tumor vascularization, such as vascular co-option or mimicry, may also play roles in resistance to anti-angiogenesis therapy [5]. Suffice to say, the clinical efficacies of current angiogenesis inhibitors are limited and some of them are totally invalid or unacceptably toxic [6], and therefore new, safer, and more effective agents are urgently needed.

Resveratrol (trans-3,5,4'-trihydroxystilbene, RSVL) is a well-known plant-derived natural polyphenolic compound that widely presents in grapes, berries, peanuts, and is abundant in red wine, exerting extensive bioactivities including antioxidative, anticancer, antiaging, anti-inflammatory, and other effects [7–11]. However, opposing results were obtained among studies that showed that RSVL possesses both anti- and pro-angiogenesis effects, depending on model systems and circumstances [12]. Combination of RSVL with chemotherapeutic drugs was found to enhance the efficacy of these drugs, such as the fact that (1) RSVL sensitizes pancreatic cancer MIA-PaCa-2 cells to chemotherapeutic agents such as docetaxel, mitoxantrone, 5-fluorouracil, cisplatin, and oxaliplatin [13]; (2) RSVL decreases Rad51 expression and sensitizes cisplatin-resistant MCF-7 breast cancer cells [14]; (3) RSVL markedly potentiates the effect of sorafenib in hepatocellular carcinoma Hep3b cells [15]; (4) RSVL enhances the apoptotic and oxidant effects of paclitaxel in DBTRG glioblastoma cells [16]; and (5) RSVL sensitizes colorectal cancer HCT116 and HT-29 cells to doxorubicin [17]. However, whether RSVL can potentiate the effect of GEM in lung cancer is unclear.

In this study, we investigated whether RSVL could enhance the anticancer effect of GEM against human lung cancer cells both in vitro and in vivo. We established a HCC827-human umbilical vein endothelial cell (HUVEC) co-culture model and examined the effect of RSVL on tubule network formation in vitro. We also performed hematoxylin and eosin (HE) staining for blood vessel morphology observation and immunohistochemistry (IHC) to detect CD31-positive staining blood vessels in tumor tissue sections of nude mice with HCC827 xenografts, which were conducted to examine the effect of RSVL in tumor microvessel growth in vivo. Mechanistically, the present work showed that the downregulated protein expression level of endoglin (ENG) and the activation of extracellular signal-regulated kinase (ERK) signaling pathway play important roles in RSVL-promoted tumor microvessel growth, leading to increased blood perfusion and drug delivery into tumor and thereby resulting in enhanced anticancer effect of GEM. The implications of our findings suggest the potential clinical applications of RSVL to enhance the anticancer efficacy of anticancer drugs against lung cancer.

2. Results

2.1. RSVL in Combination with GEM Showed No Synergistic Effects on HCC827 Cancer Cells Cultured In Vitro

To determine whether RSVL, a stilbene, might have a role in the treatment of lung cancer in combination with GEM, we firstly investigated the effect of RSVL on the proliferation of HCC827 lung cancer cell line to determine a suitable concentration. We undertook BEAS-2B bronchial epithelial cells and human umbilical vein endothelial cells (HUVEC) as normal controls. Upon treatment of these cells with RSVL, it can be seen that RSVL at the concentration of 5–10 µM showed negligible cytotoxicity to HCC827 lung cancer cells, BEAS-2B, and HUVEC at 24 h (Figure 1A). Because the concentration at 10 µM was relatively low and non-cytotoxic, and more importantly, RSVL at 10 µM was water-soluble and may have reached the indicated concentration in vivo, thus RSVL was used in most of the subsequent experiments at the concentration of 10 µM. Next, we investigated whether

RSVL can potentiate the effect of GEM against HCC827 lung cancer cells cultured in vitro, however, there was no significant difference in cell viability between GEM treated alone and GEM combined with RSVL on HCC827 cells (Figure 1B).

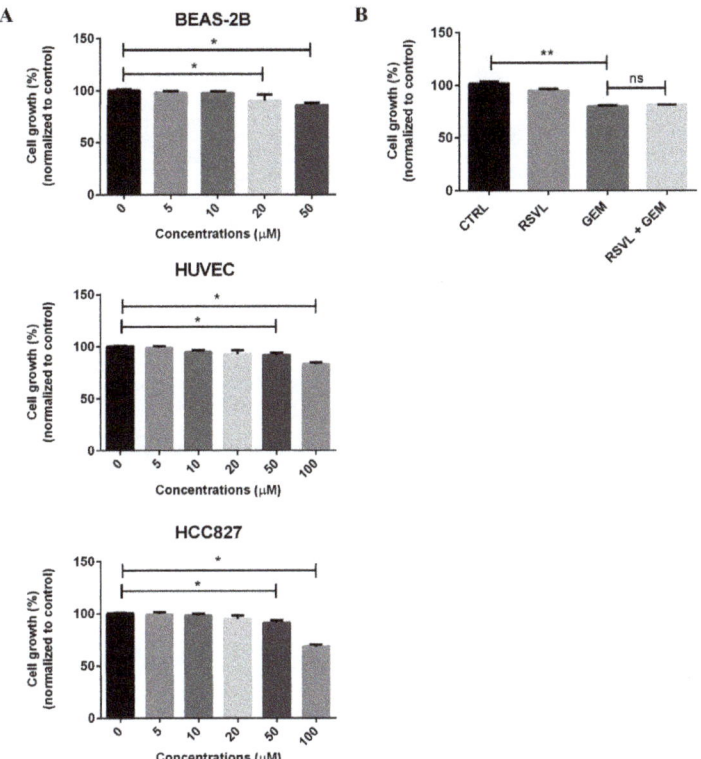

Figure 1. Resveratrol (trans-3,5,4′-trihydroxystilbene, RSVL) in combination with gemcitabine (GEM) showed no synergistic cytotoxic effects on HCC827 cancer cells cultured in vitro alone. (**A**) BEAS-2B, human umbilical vein endothelial cell (HUVEC), and HCC827 cells were treated with various concentrations of RSVL. (**B**) HCC827 cells were sham-exposed or treated with 10 μM RSVL and/or 1 μM GEM. After 24 h, the cell viability was measured by naphthol blue black (NBB) staining assay. The percentage of viability was plotted as 100% for control (no treatment of RSVL or GEM). Results are expressed as mean ± SD of triplicate samples, and reproducibility was confirmed in three separate experiments. * ($p \leq 0.05$), ** ($p \leq 0.01$), ns (not significant).

2.2. RSVL Enhanced the Anticancer Efficacy of GEM in HCC827 Lung Cancer Bearing Nude Mice

From the information above, we can see that there was no observable synergistic effect of RSVL in GEM-treated HCC827 cancer cell culture in vitro. However, we wondered whether this was due to the simplicity of the experimental design (only a monolayer of cancer cell culture in a 24-well plate) because in reality the tumor microenvironment is so complex and many cell–cell interactions are actually involved. For this reason, we examined the therapeutic potential of RSVL and GEM either alone or in combination on the growth of transplanted HCC827 human lung cancer cells in nude mice. The experimental protocol is depicted in Figure 2A. Briefly, HCC827 cells were subcutaneously inoculated into the right flanks of nude mice. After 7 days, we randomized the animals into four groups and started the treatment following the experimental protocol. Tumors were measured twice a week, and after administration of 25 days, mice were sacrificed and tumors were excised surgically and weighed, and then were fixed in 4% formaldehyde solution for further study. Compared with GEM treated alone, the combination of the two agents was more effective in reducing the tumor burden. The tumors in the group of combination grew slower, appearing with lower volume and weight, as well as a lower tumor growth rate (Figure 2B–E). These results showed that RSVL enhanced the anticancer efficacy of GEM against HCC827 lung cancer in vivo in xenograft-bearing nude mice.

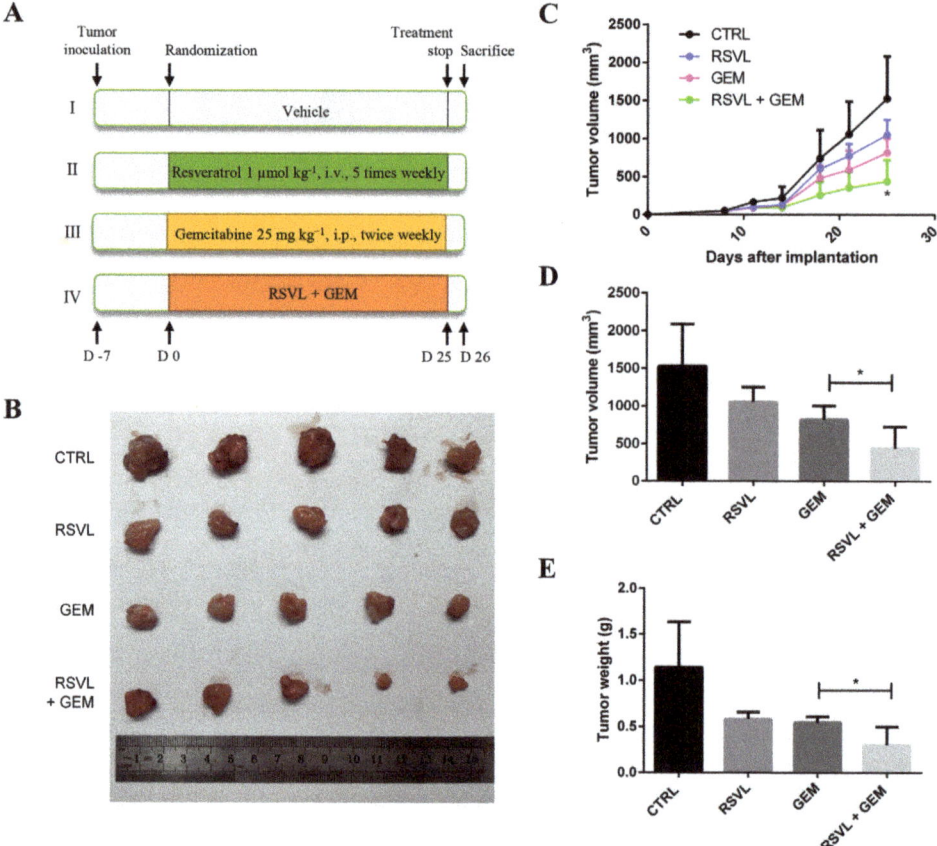

Figure 2. RSVL enhanced the anticancer efficacy of GEM in HCC827 lung cancer xenograft-implanted nude mice. (**A**) Schematic representation of the experimental protocol as described in the Materials and Methods section. A total of four mice groups were used. Group I was administrated with vehicle (100 μL, i.v. injection, five times weekly) and phosphate-buffered saline (100 μL, i.p. injection, twice weekly), group II was administrated with RSVL (1 μmol kg^{-1}, i.v. injection, five times weekly), group III was administrated with GEM (25 mg kg^{-1}, i.p. injection, twice weekly), and group IV was administrated with RSVL (1 μmol kg^{-1}, five times weekly by i.v. injection) and GEM (25 mg kg^{-1}, twice weekly by i.p. injection). (**B**) Image showing the excised tumor nodules from the above mice. (**C**) tumor volume measurement upon implantation of HCC827 cells in nude mice. (**D**) Comparison of tumor volumes at the last measurement. (**E**) Comparison of tumor weights at the last measurement. Values are mean ± SD and * ($p \leq 0.05$) as compared with GEM-treated group alone.

2.3. RSVL Increased Microvessel Growth and Promoted Blood Perfusion into Tumor in Lung Cancer Xenograft Mice

From the above results, it is quite intriguing that RSVL enhanced the anticancer efficacy of GEM against HCC827 lung cancer in vivo but not in vitro. To answer this question, we made tumor tissues from nude mice into sections and performed HE staining for the morphology observation. The results showed that there were more tumor microvessels and bloodstream in RSVL or combined treatment groups as compared with control or groups treated with GEM alone (Figure 3A,B). The results of immunohistochemistry (IHC) assay also indicated increased CD31-positive staining blood vessels in RSVL or combined treatment groups (Figure 3C,D), suggesting that RSVL increases tumor microvessel growth and promotes blood perfusion into tumor in lung cancer-transplanted nude mice.

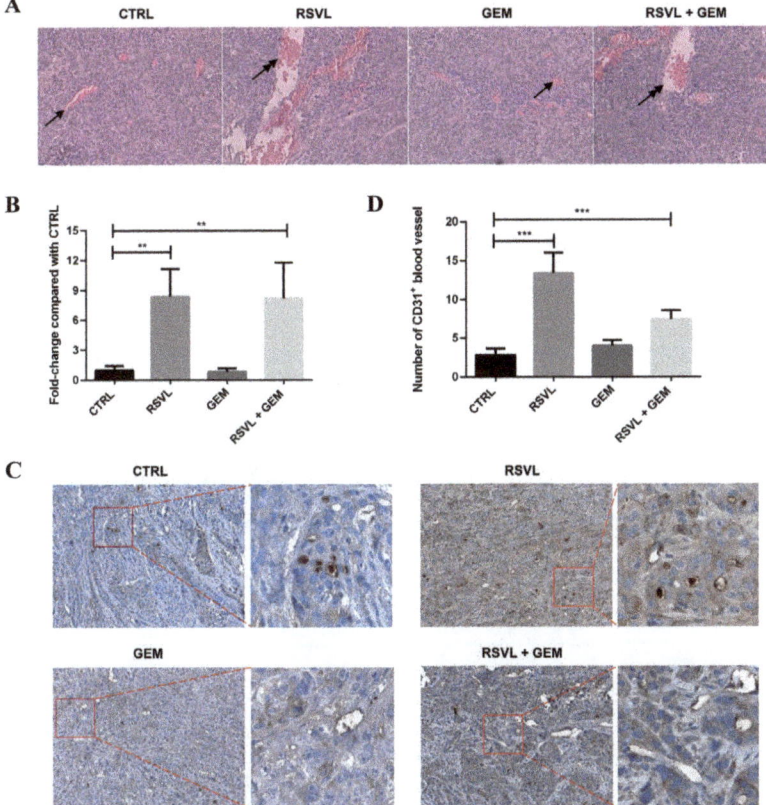

Figure 3. RSVL increased microvessel growth and promoted blood perfusion into tumor of lung cancer xenograft bearing nude mice. (**A**) Representative images of hematoxylin and eosin (HE) staining. Blood vessels formed in xenografts were indicated by arrows. Upon RSVL and RSVL + GEM treatment, more blood vessels, lacunae (indicated with double head arrows), as well as red-dyed blood cells can be seen. (**B**) The degree of blood perfusion in each group was quantitated and expressed as relative ratio, setting 1 for control. (**C**) Representative images of immunohistochemistry (IHC) staining for CD31 protein expression. Enlarged view is also shown on the right of each image. Upon RSVL and RSVL + GEM treatment, more CD31-positive staining blood vessels can be observed. (**D**) The number of CD31-positive staining blood vessels in each group was quantitated and presented. ** ($p \leq 0.01$), *** ($p \leq 0.001$).

2.4. RSVL Promoted Tubule Network Formation in HCC827-HUVEC Co-Culture Model

To confirm our findings in vivo, we used enhanced green fluorescent protein (EGFP) stably-expressing HUVEC that was cultured with HCC827 cells to establish the HCC827-HUVEC co-culture model, and observed tubule network formations in vitro under the fluorescent microscope after RSVL treatment. At the end, photos were captured and analyzed. Experimental protocol is depicted in Figure 4A. After treatment with 10 µM RSVL for 24 h, the HCC827-HUVEC co-culture model showed better tubular formation, appearing to have a higher percentage of elongation, tubules, and junctions formed, whereas HUVEC alone/BEAS-2B-HUVEC co-culture model showed no significant difference in tubule network formation (Figure 4B,C). The above evidence suggests that RSVL enhanced the anticancer efficacy of GEM against lung cancer in vivo, which may be explained by the promoted microvessel growth and blood perfusion in tumor, increasing the concentration of GEM in the surrounding interstitial space, thereby enhancing its anticancer efficacy against lung cancer.

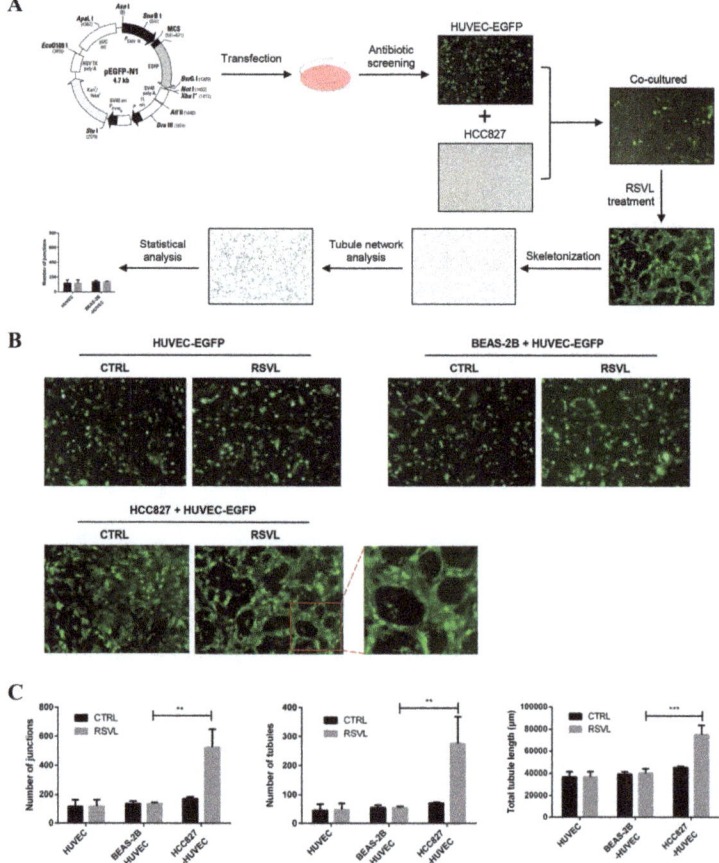

Figure 4. RSVL promoted tubule network formation in the HCC827-HUVEC co-culture model. (**A**) Schematic representation of experimental protocol as described in the Materials and Methods section. (**B**) Representative images taken under fluorescent microscope (10× magnification). HCC827-HUVEC appeared to have a more obvious tubule network formation than BEAS-2B-HUVEC or HUVEC alone. (**C**) The corresponding number of junctions, number of tubules, and total tubule length of the images of (**B**) were quantified and compared. ** ($p \leq 0.01$), *** ($p \leq 0.001$).

2.5. RSVL Suppressed both the mRNA and Protein Levels of ENG in HCC827 Lung Cancer Cells, and also Decreased the Protein Level of ENG in Tumor Tissues from HCC827 Xenograft Mice

Without any clues on which angiogenic factors might possibly be involved in the above phenomenon, we resolved to screen the differential expression of 55 angiogenesis-related protein targets in HCC827 cancer cells after RSVL treatment by using the human angiogenesis array. According to the screening results, one of the downregulated targets, endoglin (ENG), caught our attention (Figure 5). We performed qPCR and Western blotting again, finding that RSVL suppressed both mRNA and protein levels of ENG in HCC827 cells cultured in vitro (Figure 6A,B), which is consistent with the angiogenesis array screening result. Next, we conducted IHC to detect the protein levels of ENG in tumor tissue sections from HCC827 xenograft-bearing nude mice. Notably, ENG-positive staining became weaker in RSVL alone and RSVL + GEM groups as compared with control or groups treated with GEM alone (Figure 6C,D), which suggested that RSVL also decreased the protein level of ENG in vivo.

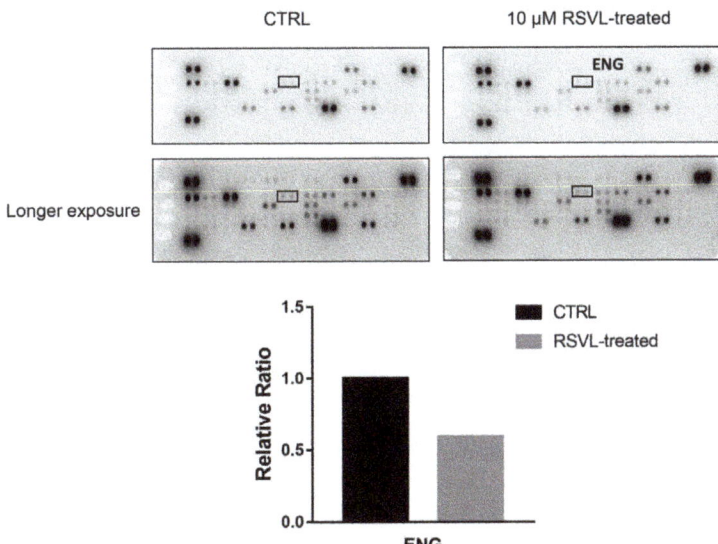

Figure 5. Human angiogenesis antibody array analysis identified endoglin (ENG) as one of the downregulated protein targets in 10 µM RSVL-treated HCC827. After 24 h of RSVL treatment, HCC827 cells were lysed and protein extract (300 µg) were used for angiogenesis array analysis. Array spots were visualized in accordance with the manufacturer's instructions. The intensity of the spot was measured as described in the Materials and Methods section. The graph shows the relative ratios of ENG protein expression in cells, setting 1 for control (CTRL).

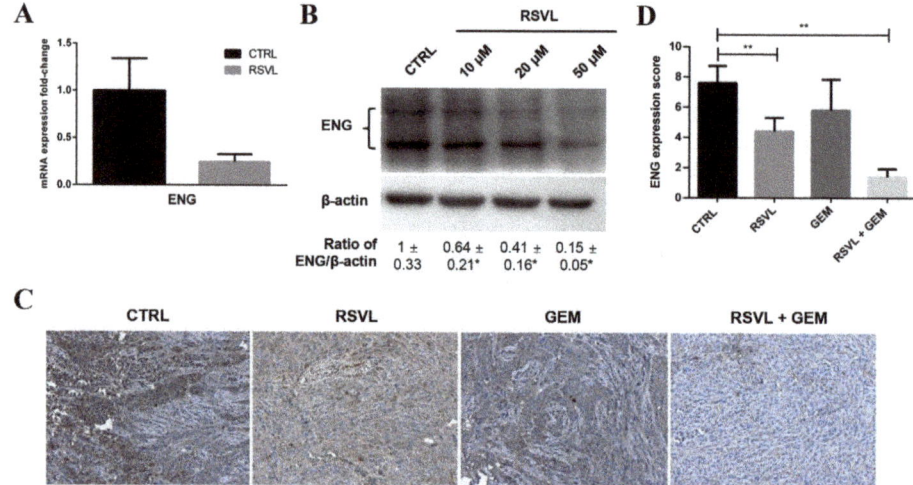

Figure 6. RSVL suppressed ENG expression both in vitro and in vivo. (**A**) Fold change of ENG mRNA expression in HCC827 cells cultured in vitro after treatment with 10 µM RSVL for 24 h. (**B**) RSVL suppressed the protein expression of ENG in a dose-dependent manner after treatment of HCC827 cells with RSVL from 10 to 50 µM for 24 h. (**C**) Representative images of IHC for ENG in tumor tissues from HCC827 xenograft nude mice. Positive ENG staining became weaker in RSVL group and in the combined treatment group as compared with the control or GEM alone group. (**D**) Comparsion for IHC score of ENG among each group in nude mice. IHC score of ENG decreased in RSVL group and in the combined treatment group as compared with control. * ($p \leq 0.05$), ** ($p \leq 0.01$).

2.6. ENG Was Crucial in RSVL-Promoted Microvessel Growth

As indicated in previous sections, ENG expressions were suppressed by RSVL both in vitro and in vivo, which suggested ENG may play an important role in RSVL-promoted microvessel growth. Thus, we focused on the study of ENG.

First, we transfected ENG-small interfering RNA (siRNA) into HCC827 cells to knockdown the endogenous ENG, then mixed them with HUVEC to establish the HCC827-HUVEC co-culture model and observed the tubule network formation under the fluorescent microscope. The knockdown efficiency of ENG- or control (CTRL)-siRNA was validated (Figure S1). After knockdown of ENG, HUVEC showed better tubular formation, appearing to have a higher percentage of elongation, tubules, and junctions formed as compared with CTRL-siRNA group (Figure 7A,B), which indicated the fact that knockdown of ENG can promote tubule network formation in HCC827-HUVEC co-culture model. We also constructed the plasmid of pcDNA3.1(+)-ENG-mCherry and transfected it into HCC827 cells to overexpress ENG, then cultured them with HUVEC to establish the HCC827-HUVEC co-culture model and observed the tubule network formations under fluorescence microscope. However, after overexpression of ENG, HUVEC showed no differences in tubule network formations compared with the control vector group both in the presence or absence of RSVL for 24 h (Figure 7C–E), suggesting that overexpression of ENG can inhibit tubule network formation induced by RSVL in HCC827-HUVEC co-culture model. Thus, it can be seen that ENG plays a crucial role in RSVL-promoted tumor microvessel growth, of which the expression levels of ENG are negatively correlated with tubule network formations in the HCC827-HUVEC co-culture model.

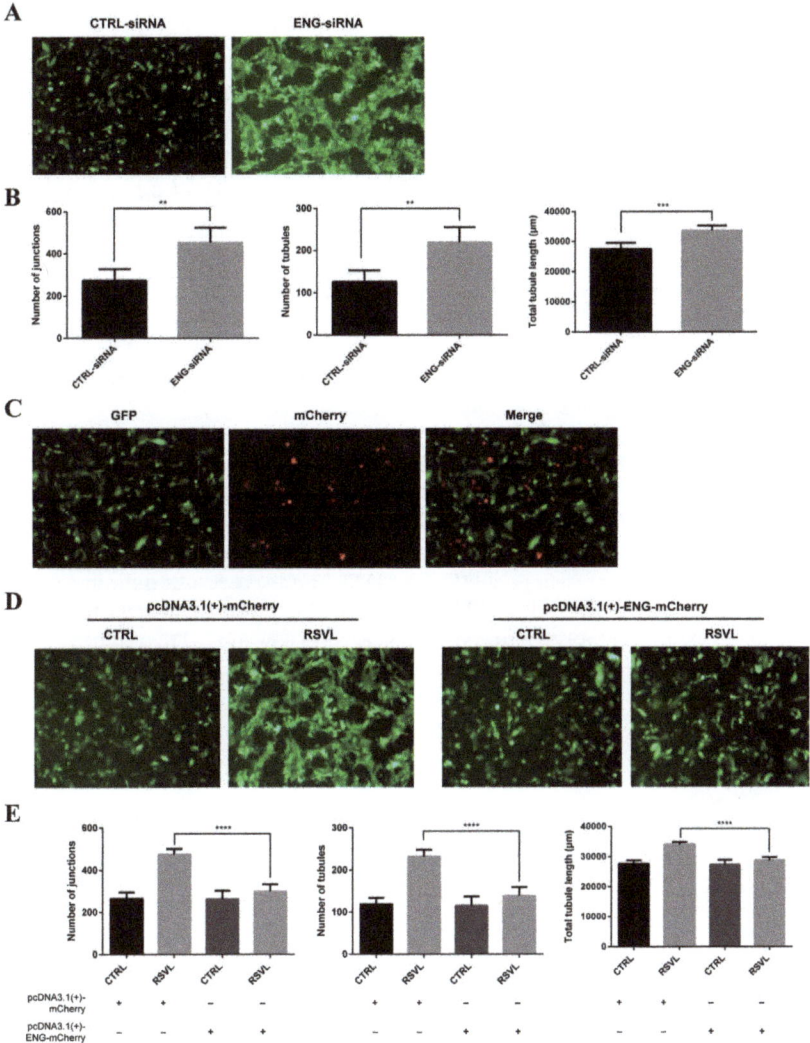

Figure 7. ENG is crucial in RSVL-promoted tubule network formation in HCC827-HUVEC co-culture model. (**A**) Representative images taken under the fluorescent microscope (10×) after knockdown of ENG. Upon transfection of HCC827 cells with ENG-siRNA for 24 h, HCC827 and HUVEC-EGFP cells were seeded at a ratio of 1:1 and co-cultured for 24 h, wherein tubule network formations were captured. HCC827-HUVEC showed better tubular formation, appearing to have a higher percentage of elongation, tubules, and junctions formed after ENG knockdown. (**B**) Quantification of tubule network formation in (**A**). Number of junctions (left), number of tubules (middle), and total tubule length (right). (**C**) Representative images of pcDNA3.1(+)-ENG-mCherry group taken under fluorescence microscope (10×). (**D**) Representative images after overexpression of ENG (10×). Upon transfection of HCC827 cells with pcDNA3.1(+)-mCherry or pcDNA3.1(+)-ENG-mCherry for 24 h, HCC827 and HUVEC-EGFP cells were seeded at a ratio of 1:1 and co-cultured for 24 h, wherein tubule network formations were observed and images of each group were captured. (**E**) Quantification of tubule network formation in (**D**). Number of junctions (left), number of tubules (middle), total tubule length (right). ** ($p \leq 0.01$), *** ($p \leq 0.001$), **** ($p \leq 0.0001$).

2.7. ENG and ERK Signaling Pathway Played Important Roles in RSVL-Promoted Tumor Microvessel Growth

Some reports demonstrated that the ERK signaling pathway is involved in endothelial cell growth and migration [18,19]. To investigate the roles of ERK signaling pathway in RSVL-promoted tumor microvessel growth, we performed Western blotting to detect the levels of phosphorylated ERK1/2 in HCC827-HUVEC co-culture model.

First, we treated HCC827 cells with RSVL and found that levels of p-ERK1/2 decreased in a dose-dependent manner from 10 to 50 µM (Figure 8A). Then, we manipulated the ENG protein levels by knockdown or overexpression of ENG in HCC827 cells co-cultured with HUVEC, and found that the levels of p-ERK1/2 increased after knockdown of ENG (Figure 8B, left panel) whereas they decreased after overexpression of ENG (Figure 8C, left panel) in the HCC827-HUVEC co-culture models. Of note, the increase of the levels of p-ERK1/2 was likely contributed to by the HUVEC in the co-culture model as we can see that the level of p-ERK1/2 was increased in the co-culture situation (Figure 8B, left panel), but decreased in HCC827 alone (Figure 8B, right panel) (which further supports our data in Figures 5 and 8A showing that downregulation of ENG by RSVL caused decrease of p-ERK1/2 level in HCC827 cells). Therefore, the ENG expression level in HCC827 cells corresponded with the degree of ERK1/2 activation in HUVEC in the HCC827-HUVEC co-culture model.

Furthermore, we also blocked the phosphorylation of ERKs by treatment with MEK inhibitor PD0325901. We found that HUVEC showed obvious tubular formations stimulated by RSVL treatment, whereas HUVEC showed impaired tubule network formations after PD0325901 treatment, no matter whether RSVL was present or not (Figure 8D,E). The above results indicated that the ENG expression level negatively corresponded with the degree of ERK1/2 activation in the HCC827-HUVEC co-culture model. Treatment of RSVL reduced the ENG expression and led to the attenuation of ERK phosphorylation inhibition, resulting in increased tumor microvessel growth, which can be suppressed by PD0325901. Collectively, the ERK signaling pathway may play an important role in RSVL-promoted tumor microvessel growth.

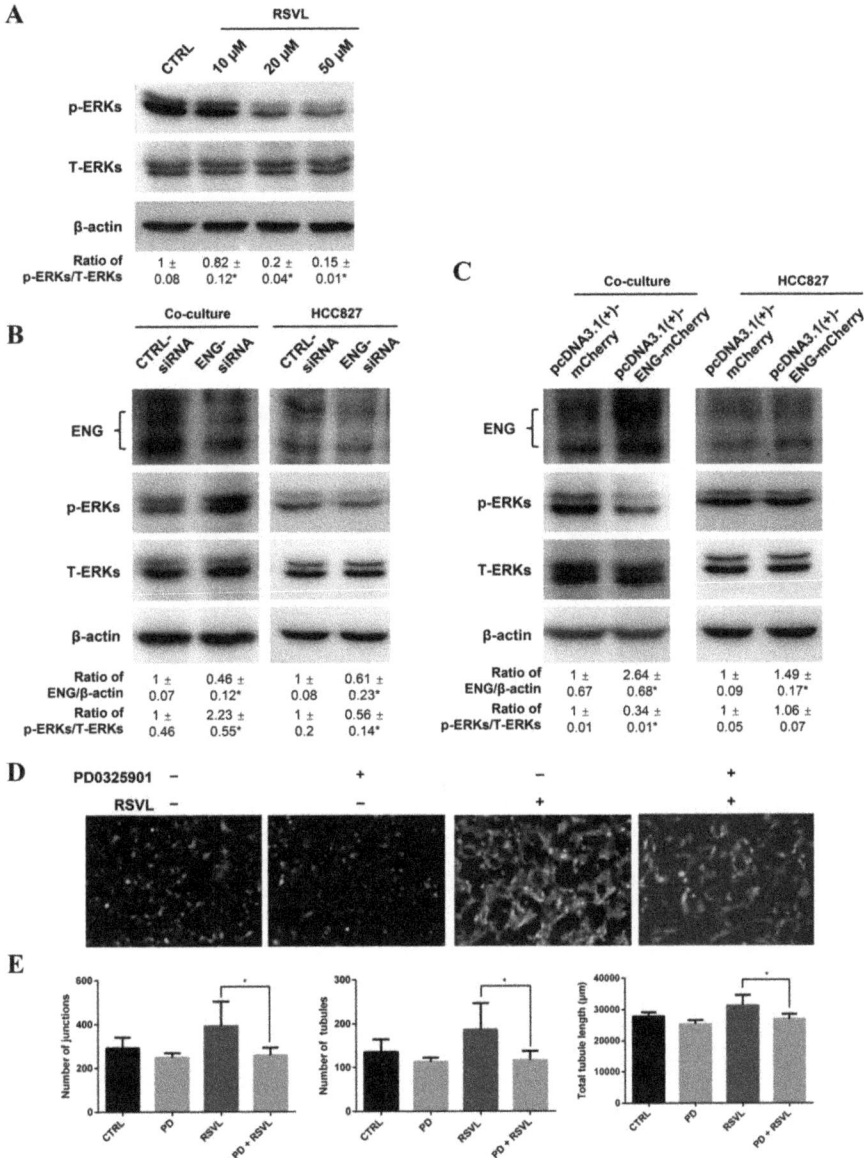

Figure 8. ENG and extracellular signal-regulated kinase (ERK) signaling pathway played important roles in RSVL-promoted tubule network formation in the HCC827-HUVEC co-culture model. (**A**) RSVL (10 to 50 µM for 24 h) inhibited the phosphorylation of ERKs in a dose-dependent manner in HCC827 cells. (**B**) Knockdown of ENG resulted in enhanced ERK1/2 activation in the HCC827-HUVEC co-culture model.

(**C**) Overexpression of ENG resulted in suppressed ERK1/2 activation in the HCC827-HUVEC co-culture model. After transfected with ENG-siRNA (50 nM) or pcDNA3.1(+)-ENG-mCherry (3 µg) for 24 h in HCC827 cells, HCC827 and HUVEC-EGFP cells were seeded at a ratio of 1:1 to establish the co-culture model, and after 24 h of incubation, cell pellets were harvested and subjected to Western blot analyses. (**D**) Representative images taken under fluorescence microscope (10×). HCC827 and HUVEC-EGFP cells were seeded at a ratio of 1:1, and after 24 h of co-culture, cells were pretreated with PD0325901 for 1 h, then were treated with 10 µM RSVL for 24 h and tubule network formations were captured. HUVEC-EGFP showed better tubular formation after RSVL treatment, which can be abrogated by PD0325901 pretreatment. (**E**) Quantification of tubule network formation in (**D**). Number of junctions (left), number of tubules (middle), and total tubule length (right). * ($p \leq 0.05$).

2.8. Data from Online Databases Suggest Increased Expression of ENG May Be Negatively Correlated to the Survival of Lung Cancer Patients

Lastly, to correlate our findings more clinically, we analyzed the ENG mRNA levels in patients from an online database (The Cancer Genome Atlas (TCGA) Provisional 2015) and found that a high percentage (3.72%–10.24%) of lung cancer patients had altered ENG gene expression, and in most cases, ENG was up-regulated instead of down-regulated. There were more lung cancer patients with adenocarcinoma that showed increased ENG expression than patients with squamous cell carcinoma (Figure 9A). Increased expression of ENG was negatively-correlated with the survival of lung cancer patients. However, when we performed the survival analysis based on different cancer histology, we found that the ENG expression specifically had a strong impact on the survival of patients with adenocarcinoma but not those with squamous cell carcinoma (Figure 9B). These results indicated that a high level of ENG may be an elevated risk factor in the development of lung cancer, and thus decreased ENG induced by RSVL will be favorable for cancer prevention, especially for lung adenocarcinoma prevention.

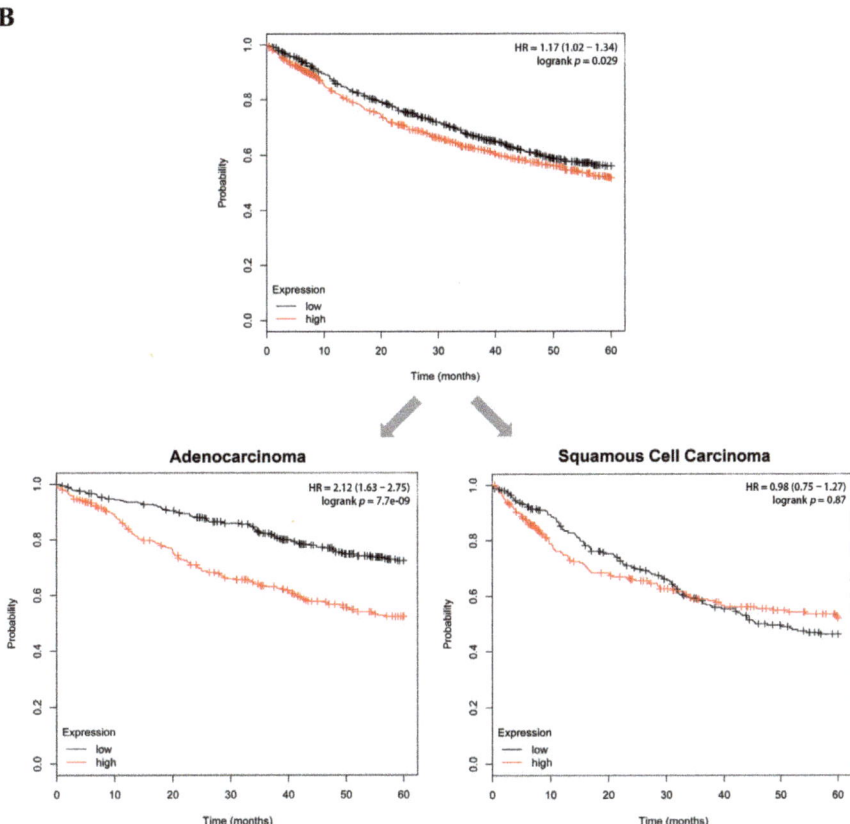

Figure 9. Altered expression of ENG was found in lung cancer patients. (**A**) Clinical datasets from The Cancer Genome Atlas (TCGA) were utilized to analyze the mRNA level of ENG in lung cancer patients with lung adenocarcinoma or lung squamous cell carcinoma. For each of the datasets, the mRNA expression z-score threshold was set at ± 1.5. (**B**) Kaplan–Meier plotter database was utilized to assess the correlation between ENG expression and survival of lung cancer patients.

3. Discussion

In recent years, many natural products have been recognized as anticancer agents, and previous studies have indicated the synergistic anticancer effect of GEM and RSVL in certain cancer types, such as pancreatic cancer and ovarian carcinoma [20,21]. However, whether the same phenomenon would occur in lung cancer is unclear. Here, we delineated for the very first time the molecular mechanism by which RSVL enhances chemosensitivity and the critical role of ENG in GEM-treated human lung adenocarcinoma cell line HCC827.

ENG (also known as CD105) is a cell membrane glycoprotein mainly expressed in endothelial cells and that is overexpressed in tumor-associated vascular endothelium, which functions as an accessory component of the transforming growth factor-β (TGF-β) receptor complex and is involved in vascular development and remodeling. In solid malignancies, ENG is almost exclusively expressed in endothelial cells of both peri- and intra-tumoral blood vessels and on tumor stromal components. Several studies have defined the role of ENG as a powerful marker to quantify intratumoral microvessel density (IMVD) in solid and hematopoietic tumors, including breast, prostate, cervical, colorectal, and non-small cell lung cancer (NSCLC), and in multiple myeloma and hairy cell leukemia. Quantification of tumor microvessel density, as determined by immunohistochemical staining for ENG, is a significant indicator of poor prognosis in patients with selected solid neoplasias including NSCLC, cervical cancer, prostate cancer, and breast carcinoma [22]. Data from online databases suggest that increased expression of ENG is negatively correlated with the survival of lung cancer patients. In this case, the fact that RSVL decreased ENG level *in vitro* and *in vivo* in our model systems here may suggest the beneficial therapeutic role of RSVL in conjunction with chemotherapy (such as GEM) for lung cancers.

Lee and Blobe [23] reported that β-arrestin2 binding to ENG causes the internalization of ENG and simultaneous accumulation of ENG and β-arrestin2 in endocytic vesicles, which antagonized TGF-β-mediated ERK signaling, altered the subcellular distribution of activated ERK, and inhibited endothelial cell migration in a manner dependent on the ability of ENG to interact with β-arrestin2. Moreover, ENG impedes endothelial cell growth through sustained inhibition of ERK-induced c-Myc and cyclinD1 expression in a TGF-β-independent manner, by which ENG augments growth-inhibition by targeting ERK and key downstream mitogenic substrates [19]. In our current study, as summarized in Figure 10, RSVL inhibited the expression of ENG in HCC827 lung cancer cells, corresponding with decreased ENG in the surrounding interstitial space and microvessels in tumor tissues by direct physical contact or in a paracrine manner. Subsequently, the ERK signaling towards endothelial growth and migration was activated, which contributed to enhanced tubule network formation and microvessel growth. Downregulation of ENG plays a crucial role in RSVL-promoted tumor microvessel growth, which leads to increased blood perfusion and drug delivery into tumor, thereby resulting in an enhanced anticancer effect of GEM.

The participation of cancer cells, as well as growth factors released by tumor cells and endothelium–extracellular matrix (ECM) interactions are highlighted in tumor angiogenesis, as well as the physical contacts and the paracrine actions that are the keys to endothelial cell (EC) differentiation [24,25]. Many cancer cell lines, such as hepatocellular carcinoma HepG2 and human neuroblastoma SK-N-SH, possess the ability to induce EC morphological changes, whereas for normal cells, such as human embryonic kidney HEK-293, liver cell L0-2, human fiber cell IMR90, and human smooth muscle cell VSMC, also possess the ability to induce EC morphological changes. Compared to Matrigel model, a common model for studying tubule network formation in vitro, the co-culture model simulates the real angiogenic microenvironment in the human body, which allows direct interactions between cancer cells and ECs, thus facilitating the study of factors and signaling pathways governing blood vessel formation in cancer [26,27]. In this study, we used the HCC827-HUVEC co-culture model to evaluate the effect of RSVL on tumor angiogenesis and its possible mechanism.

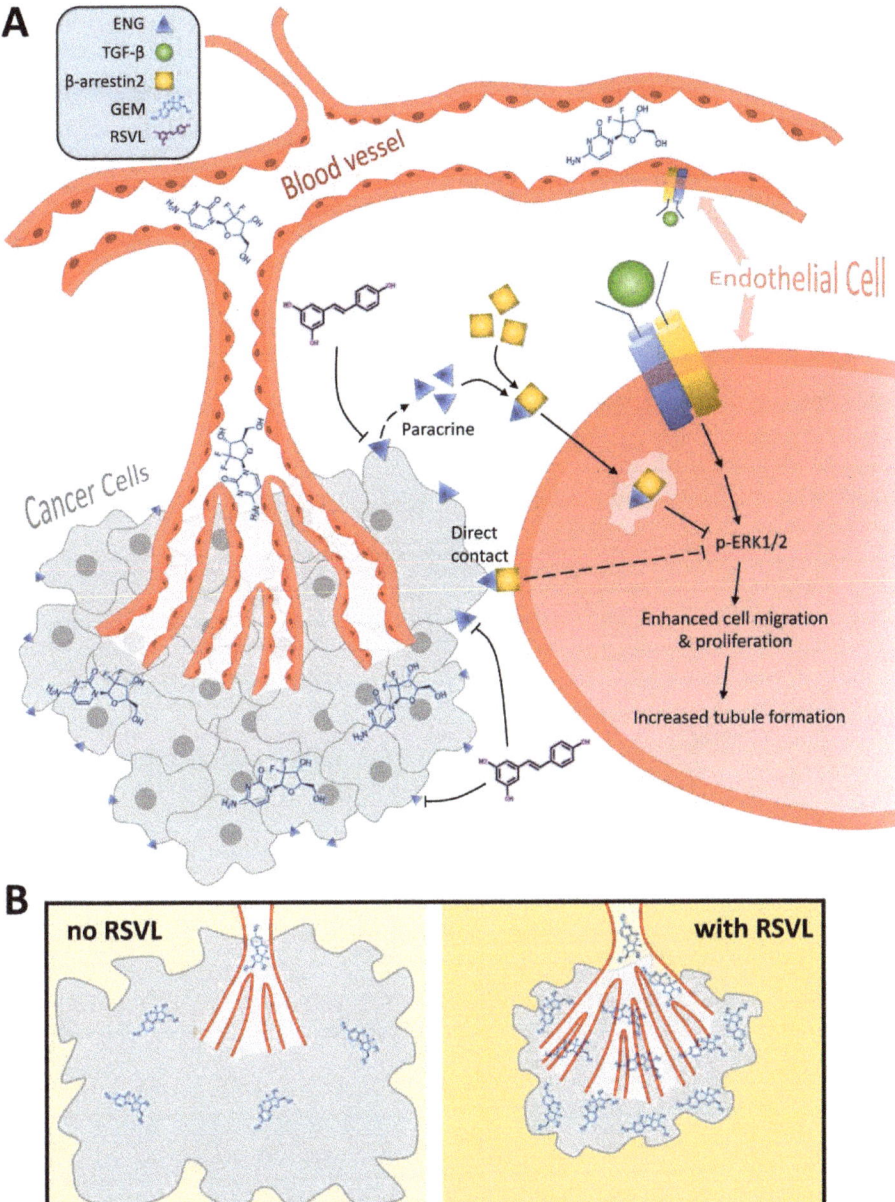

Figure 10. The putative molecular mechanism by which RSVL enhances tubule network formation and increases microvessel growth (**A**), leading to increased blood perfusion and drug delivery into tumor and thereby resulting in enhanced anticancer effect of GEM (**B**). The implications of our findings suggest the potential clinical applications of RSVL to enhance the anticancer efficacy of anticancer drugs against lung cancer.

The concentrations of RSVL in most anticancer studies are often beyond 20 µM and above, some of them even reach 100 µM [28–32]. However, due to the low water solubility and bioavailability of RSVL, plasma levels as high as above 20 µM may not be physiologically attainable in humans [33]. Moreover, the high concentration of RSVL may be cytotoxic to normal cells. In contrast, the concentration we used in our experiment (10 µM) is relatively low and non-cytotoxic to BEAS-2B, HUVEC, and HCC827 lung cancer cell lines, and more importantly, RSVL at 10 µM is water-soluble and may reach the indicated concentration in serum and even achieve higher levels of drug accumulation in tumor tissues, which in turn will inform the need for dietary advice on the intake of RSVL for the patients undergoing chemotherapy. Nevertheless, our findings indicate that RSVL may have the potential to augment the therapeutic efficacy of anticancer drugs and suggest that consumption of RSVL may be beneficial during cancer therapy. Further study of RSVL in combination of other well-known anticancer drugs (besides GEM) is warranted, for attesting whether similar phenomenon would occur.

4. Materials and Methods

4.1. Reagents

RSVL (98%) was purchased from Sigma-Aldrich (St. Louis, MO, USA). GEM and PD0325901 were purchased from Selleck Chemicals (Shanghai, China). All other general chemicals were purchased from Sigma-Aldrich and GE Healthcare (Uppsala, Sweden). Antibodies used for immunoblotting and immunohistochemical staining were purchased from Santa Cruz Biotechnology (Santa Cruz, CA, USA) and Cell Signaling Technology (Danvers, MA, USA).

4.2. Cell Lines and Culture Conditions

All cell lines used were purchased from the American Type Culture Collection (ATCC) (Rockville, MD). The human lung adenocarcinoma cell line HCC827 (CRL-2868) harbors an acquired mutation in the EGFR tyrosine kinase domain (exon 19, del E746-A750), and this mutation has been verified through gene sequencing. Cells were routinely grown in RPMI-1640 complete medium containing 10% FBS, 100 U/mL penicillin, and 100 µg/mL streptomycin at 37 °C in an atmosphere of 5% CO_2/95% air, as recommended by ATCC. The normal human bronchial epithelial cell (BEAS-2B) and human umbilical vein endothelial cell (HUVEC) were cultured in standard culture conditions, as recommended by the ATCC.

4.3. Plasmids, Small Interfering RNAs (siRNAs), and Transfection

The pEGFP-N1 plasmid was from Clontech (Mountain View, CA). The pcDNA3.1(+)-ENG-mCherry encoding plasmid and siRNAs were synthesized by IGE Biotechnology Ltd. (Guangzhou, China). Cells were transfected with the above expression plasmids, with either siRNA duplexes against ENG- (5'-AGAAAGAGCUUGUUGCGCA-3') or CTRL-siRNA (a scrambled sequence that will not lead to the specific degradation of any known cellular mRNA) [34], using LipofectAMINE 2000 (Invitrogen, Carlsbad, CA).

4.4. Cell Proliferation Assay

The effects of RSVL on proliferation of cells were determined by naphthol blue black (NBB) staining assay. Briefly, HCC827, BEAS-2B, or HUVEC cells (5×10^4 per well) were seeded in triplicate in a 24-well plate. After culturing for 24 h, the cells were treated with various concentrations of RSVL and were further incubated for 24 h; cells were then fixed with 10% formalin for 8 min and stained with 0.05% NBB solution for 30 min, the wells were washed by distilled water for three times, 50 mM NaOH was added to each well, and the absorbance of the cell suspension was measured at 595 nm using a 96-well multiscanner (Thermo Scientific Multiskan FC, Thermo Scientific, USA).

4.5. HCC827-HUVEC Co-Culture Model Establishment and Image Analysis

After transfection with pEGFP-N1 and antibiotic screening by G418, HUVEC with stable EGFP expression were constructed and cultured with HCC827 cells to establish the HCC827-HUVEC co-culture model following the method described previously [27]. Briefly, HCC827 and HUVEC-EGFP cells were harvested by trypsin digestion and mixed in ratios of 1:1 before seeding them in 12-well plates (3×10^5 per well). BEAS-2B cells co-cultured with HUVEC-EGFP cells or HUVEC-EGFP cells cultured alone were served as controls. There were three replicates in each group, and after treatment with RSVL, images were captured under a ZEISS Observer A1 inverted fluorescence microscope (ZEISS, Germany) and analyzed with Angiogenesis Analyzer software with ImageJ plugin. Quantification of tubule network formation was obtained by averaging the number of junctions, number of tubules, and total branching lengths.

4.6. In Vivo Xenograft Mouse Model Establishment and Treatment

Athymic nude mice were purchased from Beijing Vital River Laboratory Animal Technology Co., Ltd. Animals were maintained under "specific pathogen-free" conditions and had free access to food and water. All animal studies were conducted according to guidelines approved by the Shantou University Medical College Institutional Animal Care and Use Committee. The ethic code is SUMC2019-321. HCC827 cells (2×10^6 in 200 µL RPMI) were injected s.c. into the right flank of mouse, and tumor growth was monitored. Approximately 1 week after tumor cell inoculation, when the HCC827 xenografts were growing to suitable size, mice were randomly assigned to four groups ($n = 5$) and treated with vehicle, RSVL (1 µmol kg^{-1}, five times weekly by i.v. injection), GEM (25 mg kg^{-1}, twice weekly by i.p. injection), or both. Mice were weighed and tumors were measured by caliper twice weekly. Tumor volume was calculated according to the following formula: tumor volume (mm^3) = 0.5 × length × width^2. After administration of 25 days, when tumors reached about 1000 mm^3 total volume, mice were sacrificed, tumors were excised surgically and weighed, and then were fixed in 4% formaldehyde solution for further study.

4.7. RNA Isolation and Conditions for Quantitative Real-Time RT-PCR

Total RNA was extracted with Trizol reagent (Life Technology, NY, USA) and was reverse-transcribed into cDNA using the GoScript RT reagent mix (Promega Corporation, Madison, USA) according to the manufacturer's instructions. For RT-qPCR, 10 ng of cDNA template was amplified by the appropriate primer set in a reaction contained RT2 SYBR Green ROX qPCR Mastermix (Qiagen-SABiosciences, USA). The real-time PCR assay was performed in an ABI QS5 Real-Time PCR Detection System (Applied Biosystems, CA, USA). Primers used in quantitative real-time RT-PCR were designed by using the RTPrimerDB database (www.rtprimerdb.org) [35] or Primer-BLAST (Primer3 and NCBI, Bethesda, USA), and were synthesized by Beijing Genomics Institute (Shenzhen, Guangdong, China; the primer sequences are available upon request). β-actin was used as a reference gene and the relative gene expressions were calculated using the comparative C_T method ($2^{-\Delta\Delta C_T}$ method), as described previously [36].

4.8. The Human Angiogenesis Array Screening

The Human Angiogenesis Array Kit (R&D Systems, Minneapolis, MN) was used to detect the relative levels of expression of 55 angiogenesis-related proteins according to the manufacturer's instructions (The Human Angiogenesis Array coordinates are shown in Table S1). Briefly, after treatment with RSVL or Dulbecco's phosphate-buffered saline (as control), HCC827 cells were lysated and the supernatants were collected. The membranes were blocked with Array Buffer in advance, then equal amounts of protein supernatants were incubated with reconstituted Detection Antibody Cocktail and hybridized with diluted Streptavidin-HRP, then Chemi Reagent Mix was added and exposed using a gel imaging and analysis system (Tenan, China). Integrated optical density (IOD) in each spot of the array was analyzed using Gel-Pro analyzer 4, and was compared with corresponding signals on different arrays to determine the relative change in angiogenesis-related proteins.

4.9. Western Blotting Analysis

Extracted protein was resolved by 10% SDS-PAGE and transferred to polyvinylidene difluoride membrane. The membrane was then probed with various primary antibodies followed by incubations with appropriate secondary antibodies and subjected to Enhanced Chemiluminescence Western Blotting Detection Reagents (GE Healthcare, PA, USA), as described previously [37]. Antibodies used for Western blot and IHC were purchased from Santa Cruz Biotechnology (Santa Cruz, CA), Cell Signaling Technology (Danvers, MA), and Sigma-Aldrich, with the following dilutions: ENG (sc-20072; Santa Cruz Biotechnology), 1:1000 for Western blot, and 1:100 for IHC; CD31 (sc-13537; Santa Cruz Biotechnology), 1:100 for IHC; p-ERK1/2 Thr^{202}/Tyr^{204} (4370; Cell Signaling Technology), 1:1000; ERK1/2 (9102; Cell Signaling Technology), 1:1000; and β-actin (A2228; Sigma-Aldrich), 1:10,000. Whole blot showing all the bands with all molecular weight markers on the Western can be found in Figure S2.

4.10. HE Staining and Immunohistochemistry

Tumor tissues from HCC827 xenograft-implanted mice were fixed by formalin and embedded by paraffin, then were made into 4 μm thick sections and stained with hematoxylin and eosin (HE) for histo-morphological evaluation. CD31-positive staining blood vessels as well as ENG protein levels were detected by immunohistochemistry (IHC). CD31 (1:100) and ENG (1:100) were used as primary antibodies. Sections were scanned as digitalized images using Perkin Elmer Mantra Quantitative Pathology Workstation (PerkinElmer, MA, USA). Tumor blood perfusion ratio, which is defined as the ratio of red-dyed blood cell pixel to the total image pixel, was used to quantify blood perfusion in tumor tissue. The number of CD31-positive staining blood vessels was assessed by counting the vascular structures in five high power fields (HPFs, 200× magnification) and then averaging the counts of the five fields. The IHC staining score of each section was determined through assessment, as described previously [38].

4.11. Clinical Database and Statistical Analysis

The connections between ENG mRNA expression and lung cancer were investigated using cBioPortal database (www.cbioportal.org) [39]. Datasets of lung cancer patients were obtained from The Cancer Genome Atlas (TCGA) studies. Expressions of ENG with z-score ≥ +1.5 or ≤ −1.5 were considered significantly altered. The z-score is calculated as ((expression in tumor sample − mean expression in normal sample) ÷ standard deviation of expression in normal sample). Survival analysis of lung cancer patients was performed using the Kaplan–Meier plotter database (http://kmplot.com) [40]. SPSS version 23.0 was used for all of the statistical analysis in the study. All quantitative data are expressed as means ± SD, as indicated. Student's *t*-test or one-way ANOVA was used for statistical analysis. A probability of $p \leq 0.05$ was used as the criterion for statistical significance.

5. Conclusions

In summary, the results from the current study showed that RSVL enhanced the anticancer efficacy of GEM against HCC827 lung cancer in vivo. We discovered a new molecular mechanism in which ENG and ERK signaling pathway played important roles in RSVL-promoted tumor microvessel growth and blood perfusion into tumor, which resulted in enhanced anticancer effect of GEM. Thus, intake of RSVL may be beneficial during lung cancer chemotherapy.

Supplementary Materials: The following are available online at http://www.mdpi.com/2072-6694/12/4/974/s1, Table S1. Human Angiogenesis Array coordinates, Figure S1. Knockdown efficiency of ENG-small interfering RNA. (A) Expression of ENG after transfection with 50 nM of ENG-siRNA for 0–36 h. (B) Expression of ENG after transfection with 0–100 nM of ENG-siRNA for 24 h. (C) Expression of ENG after transfection with 0–100 nM of CTRL-siRNA for 24 h. * ($p \leq 0.05$), Figure S2. Whole blot showing all the bands with all molecular weight markers on the Western.

Author Contributions: Conceptualization, Y.-M.X. and A.T.Y.L.; data curation, S.-H.Q., Z.-L.L., H.W.T., and Y.-C.J.; formal analysis, S.-H.Q., Z.-L.L., H.W.T., Y.-M.X., and A.T.Y.L.; funding acquisition, Y.-M.X. and A.T.Y.L.;

investigation, S.-H.Q., Z.-L.L., H.W.T., Y.-M.X., and A.T.Y.L.; methodology, S.-H.Q., H.W.T., Y.-M.X., and A.T.Y.L.; project administration, Y.-M.X. and A.T.Y.L.; resources, Y.-M.X. and A.T.Y.L.; validation, S.-H.Q., Z.-L.L., Y.-C.J., Q.-H.Z., and X.-Y.Z.; visualization, S.-H.Q., Z.-L.L., and H.W.T.; writing—original draft, S.-H.Q., Y.-M.X., and A.T.Y.L.; writing—review and editing, Y.-M.X. and A.T.Y.L.. All authors read and approved the final version of the manuscript.

Funding: This work was supported by the grants from the National Natural Science Foundation of China (no. 31771582 and 31271445), the Guangdong Natural Science Foundation of China (no. 2017A030313131), the "Thousand, Hundred, and Ten" project of the Department of Education of Guangdong Province of China, the Basic and Applied Research Major Projects of Guangdong Province of China (2017KZDXM035 and 2018KZDXM036), the "Yang Fan" Project of Guangdong Province of China (Andy T. Y. Lau-2016; Yan-Ming Xu-2015), and the "Young Innovative Talents" Project of Guangdong Province of China (2019KQNCX034).

Acknowledgments: We would like to thank members of the Lau And Xu laboratory for critical reading of this manuscript.

Conflicts of Interest: The authors declare no conflict of interest.

References

1. Bray, F.; Ferlay, J.; Soerjomataram, I.; Siegel, R.L.; Torre, L.A.; Jemal, A. Global cancer statistics 2018: GLOBOCAN estimates of incidence and mortality worldwide for 36 cancers in 185 countries. *CA Cancer J. Clin.* **2018**, *68*, 394–424. [CrossRef] [PubMed]
2. Malapelle, U.; Rossi, A. Emerging angiogenesis inhibitors for non-small cell lung cancer. *Expert Opin. Emerg. Drugs* **2019**, *24*, 71–81. [CrossRef] [PubMed]
3. Park, J.S.; Kim, I.K.; Han, S.; Park, I.; Kim, C.; Bae, J.; Oh, S.J.; Lee, S.; Kim, J.H.; Woo, D.C.; et al. Normalization of tumor vessels by Tie2 activation and Ang2 Inhibition enhances drug delivery and produces a favorable tumor microenvironment. *Cancer Cell* **2017**, *31*, 157–158. [CrossRef] [PubMed]
4. Paez-Ribes, M.; Allen, E.; Hudock, J.; Takeda, T.; Okuyama, H.; Vinals, F.; Inoue, M.; Bergers, G.; Hanahan, D.; Casanovas, O. Antiangiogenic therapy elicits malignant progression of tumors to increased local invasion and distant metastasis. *Cancer Cell* **2009**, *15*, 220–231. [CrossRef]
5. Pinto, M.P.; Sotomayor, P.; Carrasco-Avino, G.; Corvalan, A.H.; Owen, G.I. Escaping antiangiogenic therapy: Strategies employed by cancer cells. *Int. J. Mol. Sci.* **2016**, *17*, 1489. [CrossRef]
6. Petrovic, N. Targeting angiogenesis in cancer treatments: Where do we stand? *J. Pharm. Pharm. Sci.* **2016**, *19*, 226–238. [CrossRef]
7. Awais, W.; Gao, K.; Caixia, J.; Feilong, Z.; Guihua, T.; Ghulam, M.; Jianxin, C. Significance of resveratrol in clinical management of chronic diseases. *Molecules* **2017**, *22*, 1329.
8. Dvorakova, M.; Landa, P. Anti-inflammatory activity of natural stilbenoids: A review. *Pharm. Res.* **2017**, *124*, 126–145. [CrossRef]
9. Ferramosca, A.; Giacomo, M.D.; Zara, V. Antioxidant dietary approach in treatment of fatty liver: New insights and updates. *World J. Gastroenterol.* **2017**, *23*, 4146–4157. [CrossRef]
10. Kursvietiene, L.; Staneviciene, I.; Mongirdiene, A.; Bernatoniene, J. Multiplicity of effects and health benefits of resveratrol. *Medicina (Kaunas Lith.)* **2016**, *52*, 148–155.
11. Sawda, C.; Moussa, C.; Turner, R.S. Resveratrol for Alzheimer's disease. *Ann. N. Y. Acad. Sci.* **2017**, *1403*, 142. [CrossRef] [PubMed]
12. Chen, Y.; Tseng, S.H. Review. Pro- and anti-angiogenesis effects of resveratrol. *In Vivo (Athensgreece)* **2007**, *21*, 365–370.
13. McCubrey, J.A.; Abrams, S.L.; Lertpiriyapong, K.; Cocco, L.; Steelman, L.S. Effects of berberine, curcumin, resveratrol alone and in combination with chemotherapeutic drugs and signal transduction inhibitors on cancer cells-Power of nutraceuticals. *Adv. Biol. Regul.* **2018**, *67*, 190–211. [CrossRef] [PubMed]
14. Leon-Galicia, I.; Diaz-Chavez, J.; Albino-Sanchez, M.; Garcia-Villa, E.; Bermudez-Cruz, R.; Garcia-Mena, J.; Herrera, L.; García-Carrancá, A.; Gariglio, P. Resveratrol decreases Rad51 expression and sensitizes cisplatin-resistant MCF-7 breast cancer cells. *Oncol. Rep.* **2018**, *39*, 3025–3033. [CrossRef] [PubMed]
15. Bahman, A.A.; Abaza, M.S.I.; Khoushiash, S.I.; Al-Attiyah, R.J. Sequence-dependent effect of sorafenib in combination with natural phenolic compounds on hepatic cancer cells and the possible mechanism of action. *Int. J. Mol. Med.* **2018**, *42*, 1695–1715. [CrossRef] [PubMed]

16. Öztürk, Y.; Günaydın, C.; Yalçın, F.; Nazıroğlu, M.; Braidy, N. Resveratrol enhances apoptotic and oxidant effects of paclitaxel through TRPM2 channel activation in DBTRG glioblastoma cells. *Oxidative Med. Cell. Longev.* **2019**, *2019*, 1–13.
17. Khaleel, S.A.; Al-Abd, A.M.; Ali, A.A.; Abdel-Naim, A.B. Didox and resveratrol sensitize colorectal cancer cells to doxorubicin via activating apoptosis and ameliorating P-glycoprotein activity. *Sci. Rep.* **2016**, *6*, 36855. [CrossRef]
18. Wang, J.; He, L.; Huwatibieke, B.; Liu, L.; Lan, H.; Zhao, J.; Li, Y. Ghrelin stimulates endothelial cells angiogenesis through extracellular regulated protein kinases (ERK) signaling pathway. *Int. J. Mol. Sci.* **2018**, *19*, 2530. [CrossRef]
19. Pan, C.C.; Bloodworth, J.C.; Mythreye, K.; Lee, N.Y. Endoglin inhibits ERK-induced c-Myc and cyclin D1 expression to impede endothelial cell proliferation. *Biochem. Biophys. Res. Commun.* **2012**, *424*, 620–623. [CrossRef]
20. Dutta, S.; Mahalanobish, S.; Saha, S.; Ghosh, S.; Sil, P.C. Natural products: An upcoming therapeutic approach to cancer. *Food Chem. Toxicol.* **2019**, *128*, 240–255. [CrossRef]
21. Lichota, A.; Gwozdzinski, K. Anticancer activity of natural compounds from plant and marine environment. *Int. J. Mol. Sci.* **2018**, *19*, 3533. [CrossRef]
22. Fonsatti, E.; Altomonte, M.; Nicotra, M.R.; Natali, P.G.; Maio, M. Endoglin (CD105): A powerful therapeutic target on tumor-associated angiogenetic blood vessels. *Oncogene* **2003**, *22*, 6557–6563. [CrossRef]
23. Lee, N.Y.; Blobe, G.C. The interaction of endoglin with beta-arrestin2 regulates transforming growth factor-beta-mediated ERK activation and migration in endothelial cells. *J. Biol. Chem.* **2007**, *282*, 21507–21517. [CrossRef]
24. Kleinman, H.K.; Martin, G.R. Matrigel: Basement membrane matrix with biological activity. *Semin. Cancer Biol.* **2005**, *15*, 378–386. [CrossRef]
25. Khodarev, N.N.; Yu, J.; Labay, E.; Darga, T.; Brown, C.K.; Mauceri, H.J.; Yassari, R.; Gupta, N.; Weichselbaum, R.R. Tumour-endothelium interactions in co-culture: Coordinated changes of gene expression profiles and phenotypic properties of endothelial cells. *J. Cell Sci.* **2003**, *116*, 1013–1022. [CrossRef]
26. Benton, G.; Arnaoutova, I.; George, J.; Kleinman, H.K.; Koblinski, J. Matrigel: From discovery and ECM mimicry to assays and models for cancer research. *Adv. Drug Deliv. Rev.* **2014**, *79–80*, 3–18. [CrossRef]
27. Chiew, G.G.Y.; Fu, A.; Perng Low, K.; Qian Luo, K. Physical supports from liver cancer cells are essential for differentiation and remodeling of endothelial cells in a HepG2-HUVEC co-culture model. *Sci. Rep.* **2015**, *5*, 10801. [CrossRef]
28. Kwon, S.H.; Choi, H.R.; Kang, Y.A.; Park, K.C. Depigmenting effect of resveratrol is dependent on FOXO3a activation without SIRT1 activation. *Int. J. Mol. Sci.* **2017**, *18*, 1213. [CrossRef]
29. Oi, N.; Yuan, J.; Malakhova, M.; Luo, K.; Li, Y.; Ryu, J.; Zhang, L.; Bode, A.M.; Xu, Z.; Li, Y.; et al. Resveratrol induces apoptosis by directly targeting Ras-GTPase-activating protein SH3 domain-binding protein 1. *Oncogene* **2015**, *34*, 2660–2671. [CrossRef]
30. Mao, Q.Q.; Bai, Y.; Lin, Y.W.; Zheng, X.Y.; Qin, J.; Yang, K.; Xie, L.P. Resveratrol confers resistance against taxol via induction of cell cycle arrest in human cancer cell lines. *Mol. Nutr. Food Res.* **2010**, *54*, 1574–1584. [CrossRef]
31. Zhang, C.; Lin, G.; Wan, W.; Li, X.; Zeng, B.; Yang, B.; Huang, C. Resveratrol, a polyphenol phytoalexin, protects cardiomyocytes against anoxia/reoxygenation injury via the TLR4/NF-kappaB signaling pathway. *Int. J. Mol. Med.* **2012**, *29*, 557–563. [CrossRef]
32. Oi, N.; Jeong, C.H.; Nadas, J.; Cho, Y.Y.; Pugliese, A.; Bode, A.M.; Dong, Z. Resveratrol, a red wine polyphenol, suppresses pancreatic cancer by inhibiting leukotriene A(4)hydrolase. *Cancer Res.* **2010**, *70*, 9755–9764. [CrossRef]
33. Gambini, J.; Ingles, M.; Olaso, G.; Lopez-Grueso, R.; Bonet-Costa, V.; Gimeno-Mallench, L.; Mas-Bargues, C.; Abdelaziz, K.M.; Gomez-Cabrera, M.C.; Vina, J.; et al. Properties of resveratrol: In vitro and in vivo studies about metabolism, bioavailability, and biological effects in animal models and humans. *Oxidative Med. Cell. Longev.* **2015**, *2015*, 837042. [CrossRef]
34. Pardali, E.; Van der Schaft, D.W.J.; Wiercinska, E.; Gorter, A.; Hogendoorn, P.C.W.; Griffioen, A.W.; Ten Dijke, P. Critical role of endoglin in tumor cell plasticity of Ewing sarcoma and melanoma. *Oncogene* **2011**, *30*, 334–345. [CrossRef]

35. Lefever, S.; Vandesompele, J.; Speleman, F.; Pattyn, F. RTPrimerDB: The portal for real-time PCR primers and probes. *Nucleic Acids Res.* **2009**, *37*, D942–D945. [CrossRef]
36. Schmittgen, T.D.; Livak, K.J. Analyzing real-time PCR data by the comparative CT method. *Nat. Protoc.* **2008**, *3*, 1101–1108. [CrossRef]
37. Liang, Z.L.; Wu, D.D.; Yao, Y.; Yu, F.Y.; Yang, L.; Tan, H.W.; Hylkema, M.N.; Rots, M.G.; Xu, Y.M.; Lau, A.T.Y. Epiproteome profiling of cadmium-transformed human bronchial epithelial cells by quantitative histone post-translational modification–enzyme-linked immunosorbent assay. *J. Appl. Toxicol.* **2017**, *38*, 888–895. [CrossRef]
38. Xiao, Y.-S.; Zeng, D.; Liang, Y.-K.; Wu, Y.; Li, M.-F.; Qi, Y.-Z.; Wei, X.-L.; Huang, W.-H.; Chen, M.; Zhang, G.-J. Major vault protein is a direct target of Notch1 signaling and contributes to chemoresistance in triple-negative breast cancer cells. *Cancer Lett.* **2019**, *440–441*, 156–167. [CrossRef]
39. Gao, J.; Aksoy, B.A.; Dogrusoz, U.; Dresdner, G.; Gross, B.; Sumer, S.O.; Sun, Y.; Jacobsen, A.; Sinha, R.; Larsson, E.; et al. Integrative analysis of complex cancer genomics and clinical profiles using the cBioPortal. *Sci. Signal.* **2013**, *6*, pl1. [CrossRef]
40. Lánczky, A.; Nagy, Á.; Bottai, G.; Munkácsy, G.; Szabó, A.; Santarpia, L.; Győrffy, B. miRpower: A web-tool to validate survival-associated miRNAs utilizing expression data from 2178 breast cancer patients. *Breast Cancer Res. Treat.* **2016**, *160*, 439–446. [CrossRef]

© 2020 by the authors. Licensee MDPI, Basel, Switzerland. This article is an open access article distributed under the terms and conditions of the Creative Commons Attribution (CC BY) license (http://creativecommons.org/licenses/by/4.0/).

Article

Retinol-Binding Protein 4 Accelerates Metastatic Spread and Increases Impairment of Blood Flow in Mouse Mammary Gland Tumors

Diana Papiernik [1], Anna Urbaniak [1], Dagmara Kłopotowska [1], Anna Nasulewicz-Goldeman [1], Marcin Ekiert [2,3], Marcin Nowak [4], Joanna Jarosz [1], Monika Cuprych [1], Aleksandra Strzykalska [1], Maciej Ugorski [4], Rafał Matkowski [2,3] and Joanna Wietrzyk [1,*]

1. Department of Experimental Oncology, Hirszfeld Institute of Immunology and Experimental Therapy, Polish Academy of Sciences, 53-114 Wroclaw, Poland; diana.papiernik@gmail.com (D.P.); anna.urbaniak@hirszfeld.pl (A.U.); dagmara.klopotowska@hirszfeld.pl (D.K.); anna.nasulewicz-goldeman@hirszfeld.pl (A.N.-G.); joanna.jarosz@hirszfeld.pl (J.J.); monika.cuprych@hirszfeld.pl (M.C.); s.aleksandra0@wp.pl (A.S.)
2. Division of Surgical Oncology and Clinical Oncology, Department of Oncology, Wroclaw Medical University, 50-367 Wroclaw, Poland; marcin.ekiert@umed.wroc.pl (M.E.); rafal.matkowski@umed.wroc.pl (R.M.)
3. Wroclaw Comprehensive Cancer Center, 53-413 Wroclaw, Poland
4. Faculty of Veterinary Medicine, Wroclaw University of Environmental and Life Sciences, 50-375 Wroclaw, Poland; marcin.nowak@upwr.edu.pl (M.N.); maciej.ugorski@upwr.edu.pl (M.U.)
* Correspondence: joanna.wietrzyk@hirszfeld.pl

Received: 17 February 2020; Accepted: 4 March 2020; Published: 7 March 2020

Abstract: Retinol-binding protein 4 (RBP4) is proposed as an adipokine that links obesity and cancer. We analyzed the role of RBP4 in metastasis of breast cancer in patients and in mice bearing metastatic 4T1 and nonmetastatic 67NR mammary gland cancer. We compared the metastatic and angiogenic potential of these cells transduced with *Rbp4* (4T1/RBP4 and 67NR/RBP4 cell lines). Higher plasma levels of RBP4 were observed in breast cancer patients with metastatic tumors than in healthy donors and patients with nonmetastatic cancer. Increased levels of RBP4 were observed in plasma, tumor tissue, liver, and abdominal fat. Moreover, the blood vessel network was highly impaired in mice bearing 4T1 as compared to 67NR tumors. RBP4 transductants showed further impairment of blood flow and increased metastatic potential. Exogenous RBP4 increased lung settlement by 67NR and 4T1 cells. In vitro studies showed increased invasive and clonogenic potential of cancer cells treated with or overexpressing RBP4. This effect is not dependent on STAT3 phosphorylation. RBP4 enhances the metastatic potential of breast cancer tumors through a direct effect on cancer cells and through increased endothelial dysfunction and impairment of blood vessels within the tumor.

Keywords: RBP4; metastasis; breast cancer; angiogenesis; endothelial dysfunction; STAT3; VEGF; endothelin-1

1. Introduction

Recent studies utilizing the model of mouse mammary gland cancer 4T1, reflecting a basal-like phenotype (in human: negative for nuclear estrogen receptor (ER) α, progesterone receptor (PR) and human epidermal growth factor receptor 2 (HER2), i.e., triple-negative, and positive for epidermal growth factor receptor (EGFR)) [1–4], have shown a predominant role of endothelium damage during the metastatic process of these cells [5–7]. For instance, it is evidenced that endothelial dysfunction in the lungs, which was assessed as decreased activity and phosphorylation of endothelial nitric oxide synthase (eNOS) resulting in a low nitric oxide (NO) production state, was an early event in breast cancer pulmonary metastasis. These processes precede the onset of a phenotypic switch in the lung endothelium toward a mesenchymal phenotype (EndMT), which is parallel to the appearance of the first pulmonary metastatic colonies [7]. Therefore, therapeutic strategies that aim to normalize endothelial dysfunction can decrease the metastatic potential of this type of breast cancer [8–10].

Apart from its involvement in cancer development, endothelial dysfunction plays an important role in the development of cardiovascular diseases and atherosclerosis. Moreover, in type 2 diabetes mellitus, endothelial dysfunction and insulin resistance often coexist at the earliest stage of atherosclerosis with elevation of serum retinol-binding protein 4 (RBP4), a specific retinol transporter in the blood [11]. It is documented that RBP4 induces inflammation of endothelial cells in vitro. This action is due to the stimulation of proinflammatory molecules involved in leukocyte recruitment and their adherence to endothelium, and it is independent of retinol and the RBP4 membrane receptor STRA6 [12]. Endothelial inflammation induced by RBP4 is largely mediated by toll-like receptor 4 (TLR4), and in part, through the c-Jun N-terminal kinase (JNK) and p38 mitogen-activated protein kinase (MAPK) signaling pathways [13]. Moreover, in isolated aorta rings, RBP4 treatment significantly increased NO production, stimulating the PI3K/Akt/eNOS pathway [14].

RBP4, classified as adipokine [15], is proposed as the protein linking obesity and cancer [16]. Studies have shown various correlations between RBP4 plasma/tumor tissue levels and the development of certain types of cancer. For instance, Fei et al. reported lower RBP4 serum levels in patients with colon cancer than in healthy individuals [17]; on the other hand, Karunanithi et al. and Abola et al. have shown that elevated RBP4 is associated with colon cancer progression and liver metastasis [18,19]. There are also studies showing the importance of this protein in ovarian, renal, hepatocellular, oral squamous cell, and pancreatic cancer patients [20–24]. Such analyses conducted in patients with breast cancer suggested a link between elevated RBP4 and the risk of breast cancer [25]. The proposed mechanisms of RBP4 effects on cancer cells are dependent on the activation of the signaling receptor and transporter of retinol STRA6 by bound RBP4 and further transduction of the JAK2-STAT3 signaling cascade [18]. Other authors have shown that knockdown of RBP4 significantly reduces ovarian cancer cell migration and proliferation driven through the RhoA/Rock1 and extracellular signal-regulated kinase (ERK) pathways [26].

The aim of our studies was to explain the role of RBP4 protein in the growth and metastatic spread of two murine breast cancer isogenic cell lines (derived from a single tumor of BALB/c mouse); metastatic 4T1 and nonmetastatic 67NR, representing basal-like and luminal-like phenotype, respectively [2,4]. Moreover, because, in many cases, cancer is associated with the aging process (largely reviewed in [27]) and aging affects the metastatic phenotype of cancer cells as well as tumor angiogenesis [28–30], we decided to include in our studies young and aged animals. To the best of our knowledge, no previous studies have investigated the effect of RBP4 protein on the metastatic spread of cancer.

2. Results

2.1. Impaired Angiogenesis in 4T1 Metastatic Tumors Compared to That in Nonmetastatic 67NR Tumors

The 4T1 and 67NR cells from in vitro culture were injected orthotopically (ort.) into the mammary fat pad of 6–8-week-old mice. The mice were observed and tumor growth was measured. Kinetics of growth of 4T1 and 67NR tumors were similar, but only 4T1 cells formed lung metastatic foci (Figure 1A).

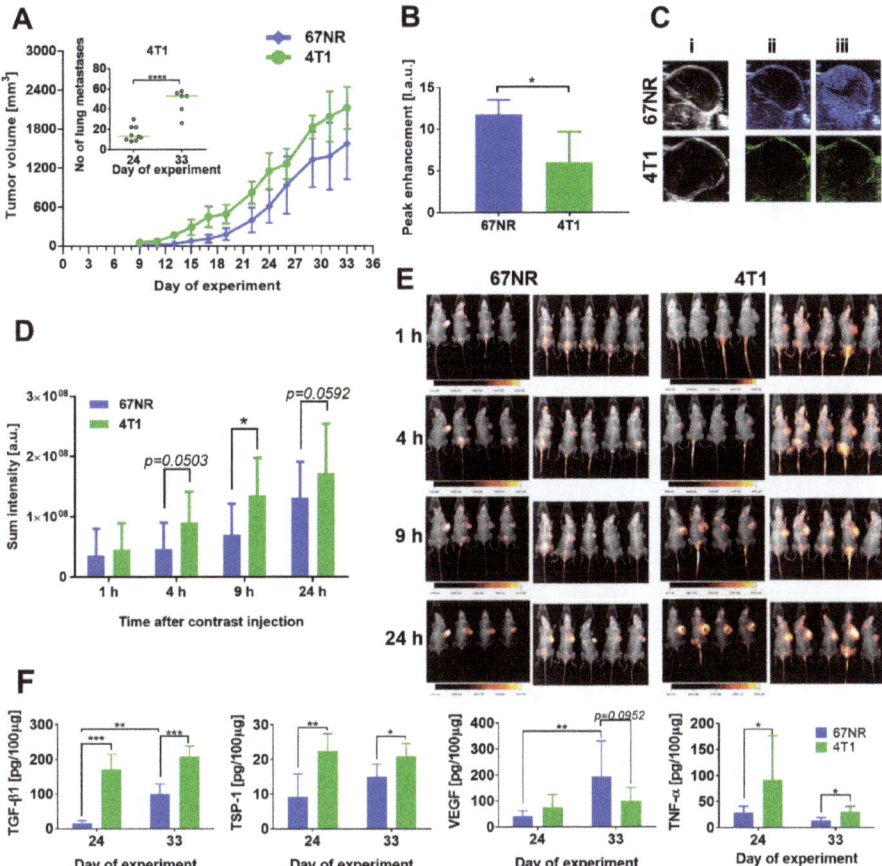

Figure 1. Basic characteristics of 67NR tumors compared to 4T1 mouse mammary gland tumors in 6- to 8-week-old mice. (**A**) Tumor growth kinetics of 67NR and 4T1 cancer and in inserted graph: number of metastases on 2 steps of 4T1 tumor progression. (**B**) Analysis of tumor blood perfusion: peak enhancement parameter. (**C**) Representative pictures of (**i**) Ultrasonography (USG) image, (**ii**) tumors before contrast agent injection, (**iii**) tumors maximally filled with the contrast agent. (**D**) Blood vessel permeability: fluorescence of IRDye® 800CW fluorescent dye PEG Contrast Agent. (**E**) Pictures of X-ray and fluorescence images of mice. (**F**) Tumor tissue level of transforming growth factor β1 (TGF-β1), thrombospondin 1 (TSP-1), vascular endothelial growth factor (VEGF), and tumor necrosis factor α (TNF-α). Data presented as mean ± SD or data for individual mice (insert in figure **A**). Number of mice per group: (**A**) 6–9; (**B**) 3–4; (**D**) 9; (**F**) 5. Statistical analysis: Tukey's multiple comparison test * $p < 0.05$, ** $p < 0.01$, *** $p < 0.001$, **** $p < 0.0001$.

Literature data [4] and our histopathological analyses confirmed that no cancer cells were detected in the lungs of 67NR tumor-bearing mice (Figure S1A). The 4T1 tumors implanted ort. demonstrated decreased blood flow (Figure 1B,C) and increased blood vessel permeability as compared to 67NR tumors (Figure 1D,E). Tumor tissue levels of transforming growth factor β1 (TGF-β1), thrombospondin 1 (TSP-1), and tumor necrosis factor α (TNF-α) were higher in 4T1 tumors than in 67NR tumors (Figure 1F). On the other hand, the level of vascular endothelial growth factor (VEGF) had a tendency to be lower in 4T1 tumors on day 33 than in 67NR tumors (Figure 1F). Plasma levels of soluble P-selectin, E-selectin, vascular cell adhesion molecule 1 (V-CAM-1), and insulin-like growth factor 1 (IGF-1) were similar or lower in mice bearing 4T1 tumors than in mice bearing 67NR tumors; however, plasma levels of soluble intercellular adhesion molecule 1 (sI-CAM-1), VEGF, and endothelin-1 (ET-1) were significantly higher in mice bearing 4T1 tumors than in mice bearing 67NR tumors (Figure 2).

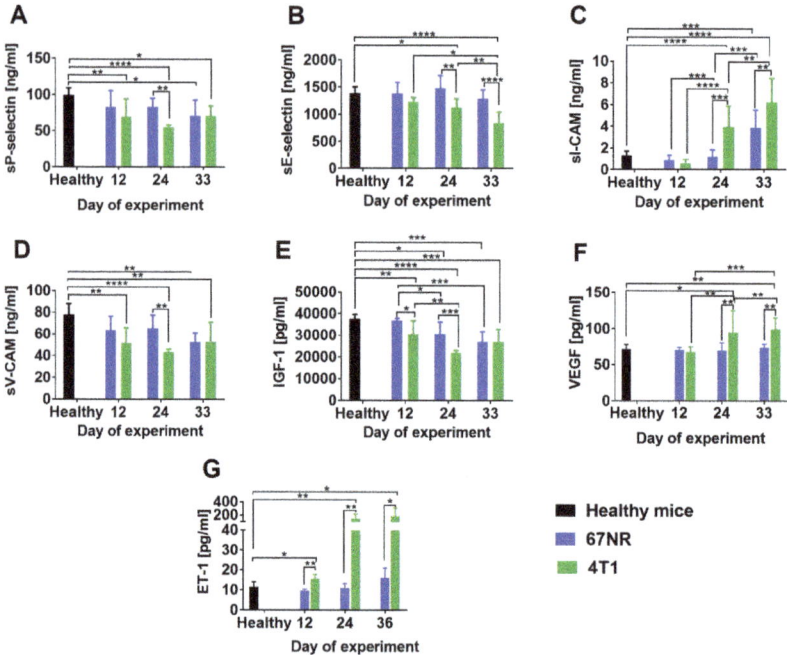

Figure 2. Plasma level of soluble proteins in 6- to 8-week-old mice bearing 67NR tumors compared to 4T1 mouse mammary gland tumors. (**A**) sP- and (**B**) sE-selectin, (**C**) intercellular adhesion molecule (sI-CAM) and (**D**) vascular cell adhesion molecule 1 (sV-CAM), (**E**) insulin-like growth factor 1 (IGF-1), (**F**) vascular endothelial growth factor (VEGF), and (**G**) endothelin-1 (ET-1). Data presented as mean ± SD. Number of mice per group: 5. Statistical analysis: Tukey's multiple comparison test * $p < 0.05$, ** $p < 0.01$, *** $p < 0.001$, **** $p < 0.0001$.

Figure S1B shows changes in basic blood morphological parameters during the progression of 67NR and 4T1 tumors. The 4T1 breast cancer-bearing mice exhibit higher levels of leukocytes (including lymphocytes, monocytes and granulocytes) than 67NR-bearing mice (Figure S1B).

In summary, metastatic 4T1 tumors show that impaired blood flow and blood vessels in these tumors are more permeable. In addition, higher TGF-β1, TSP-1, VEGF, and TNF-α levels are observed in 4T1 than in 67NR tumor tissue. Plasma level of sP-selectin, sE-selectin, sV-CAM and IGF-1 is decreased, whereas the level of sI-CAM, VEGF and ET-1 is elevated in mice bearing 4T1 as compared to 67NR tumors.

2.2. Increased RBP4 Protein Level in Young and Aged Mice and in Patients with Breast Cancer with Metastatic and Nonmetastatic Tumors

Young (6–8-week-old) and aged (1-year-old) mice were ort. injected with both cell lines (4T1 and 67NR). Plasma level of RBP4 protein significantly increased in young and aged 4T1 tumor-bearing mice starting from approximately 2 weeks after cell transplantation and was significantly higher than that in 67NR tumor-bearing mice. Although the plasma level of RBP4 was also increased in 67NR tumor-bearing mice, the increase was not significant in young mice and was significant in aged mice only on the last day of observation (Figure 3A,B, respectively).

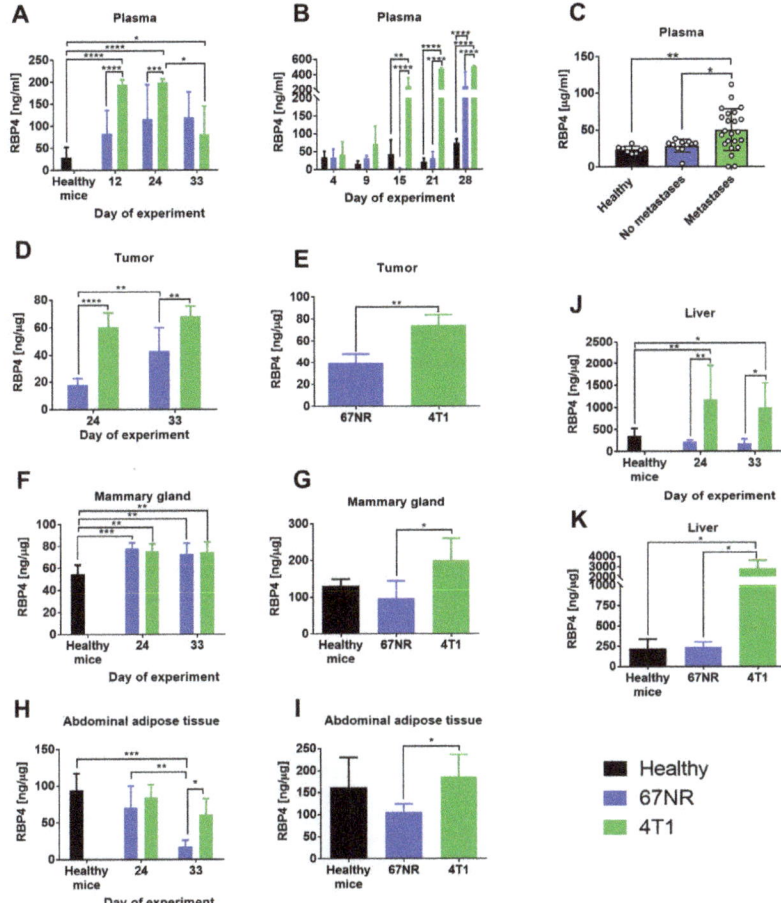

Figure 3. The level of RBP4 protein in plasma and various tissues from mice bearing nonmetastatic 67NR and metastatic 4T1 mammary gland cancer cells and in plasma from patients with breast cancer. Plasma from (**A**) young and (**B**) aged mice. (**C**) Plasma from patients with breast cancer. Tumors from (**D**) young and (**E**) aged mice. Mammary glands from (**F**) young and (**G**) aged mice. Abdominal adipose tissue from (**H**) young and (**I**) aged mice. Liver from (**J**) young and (**K**) aged mice. Data presented as mean ± SD or data for individual patients (Figure C). Number of mice per group: (**A**) 4–5; (**B**) 4 (2 healthy mice); (**D**) 5; (**E**) 4; (**F**) 5; (**G**) 4; (**H**) 4; (**I**) 3–4; (**J**) 5; (**K**) 3–4. Statistical analysis: Tukey's multiple comparison test * $p < 0.05$, ** $p < 0.01$, *** $p < 0.001$, **** $p < 0.0001$.

The plasma level of RBP4 in patients with breast cancer was significantly higher in specimens from metastatic cancers (lymph nodes or disseminated metastases) than those in healthy volunteers and patients without diagnosed metastases (Figure 3C). The level of RBP4 protein was elevated in the tumor tissue of young and aged 4T1 tumor-bearing mice as compared to that in 67NR tumor-bearing mice (Figure 3D,E) and in the liver tissue of 4T1 tumor-bearing mice as compared to that of control healthy mice (Figure 3J,K). In mammary gland adipose tissue from young mice, the level of RBP4 increased significantly in mice bearing both tumors as compared to that in healthy mice (Figure 3F), whereas in aged mice, the level of RBP4 increased only in mice bearing 4T1 cells (Figure 3G). In abdominal adipose tissue, we observed a higher level of RBP4 in mice bearing 4T1 cells than in mice bearing 67NR cells, but the level of this protein did not differ significantly as compared to that in healthy mice

(Figure 3H,I). The exception is its significant decrease in young mice bearing 67NR tumors at the end of observation (Figure 3H). In general, the levels of RBP4 were higher in old mice than in young mice.

The levels of transthyretin (TTR; complex formation of RBP4 with TTR prevents extensive loss of RBP4 by renal filtration [31,32]) in the plasma, tumor tissue, and mammary gland (day 24) were lower in 4T1 tumor-bearing mice than in 67NR tumor-bearing mice, and the TTR levels in the liver and abdominal tissue did not differ significantly between these groups of mice (Figure S2). Figure S3 presents the results of the plasma and tumor tissue levels of RBP4 (tumor tissue), TTR, ET-1, and IGF-1 in patients and healthy volunteers. The values did not differ significantly among healthy volunteers and the analyzed groups of patients.

In summary, the plasma level of RBP4 in young and old mice bearing metastatic 4T1 cancer is higher compared to mice with non-metastatic 67NR. Similar observations are made in patients' plasma. In addition, we observe a higher level of RBP4 in tumor, liver, and adipose tissue (young and old mice), and mammary glands (old mice) of 4T1-bearing mice when compared to 67NR.

2.3. Intravenous Injection of RBP4 Increases Settlement of Breast Cancer Cells in the Lungs

Figure S4A shows the kinetics of RBP4 protein in mice plasma after intravenous injection of 500 ng/mouse of RBP4. The highest concentration of RBP4 was observed between 15 min and 1 h, which then gradually decreased to the basal value after 8 h. Thus, up to 1 h, the plasma level of RBP4 reached a maximum and then started to decrease. Activation of endothelium measured as an increase of P-selectin expression on mouse endothelial cells in vitro (after incubation with 200 ng/mL of RBP4) reaches a maximum after 3 h of incubation and remains at a high level for up to 5 h (Figure S4B). On the basis of these results, we planned the following experiment (Figure 4).

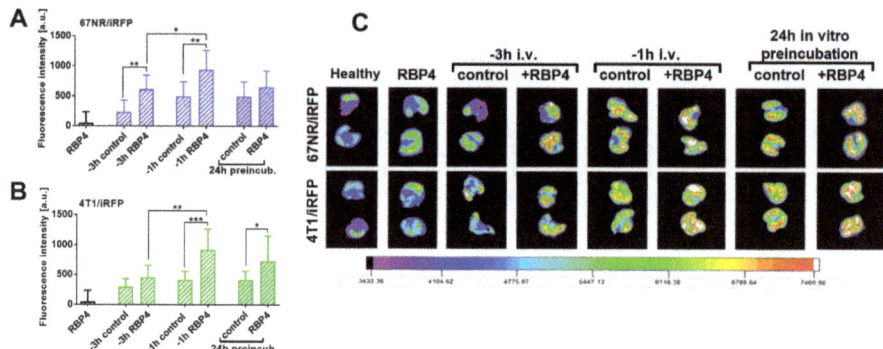

Figure 4. Lung fluorescence images after injection of 67NR/iRFP and 4T1/iRFP cells preceded by RBP4 protein administration. (**A**) 67NR/iRFP or (**B**) 4T1/iRFP cells were inoculated intravenously (i.v.). Three (-3 h) or one (-1 h) hour before cell inoculation, mice were administered i.v. with 500 ng/mouse of RBP4 or with saline. Alternatively, the cells were incubated with 200 ng/mL of RBP4 for 24 h and then inoculated i.v. (24 h preincubation). The lungs from healthy mice inoculated i.v. with saline were used as a reference for fluorescence measurements. Healthy mice injected with 500 ng/mouse of RBP4 were used as an additional control (RBP4). Lung fluorescence measurements were performed 48 h after cell inoculation. (**C**) Fluorescence of representative lungs from all groups is presented. No. of mice per group: healthy and RBP4 = 5, -3 h control = 6, remaining groups = 8–10. Data presented as mean ± SD. Statistical analysis: Unpaired t-test * $p < 0.05$, ** $p < 0.01$, *** $p < 0.001$.

The 67NR/iRFP or 4T1/iRFP cells injected i.v. 1 h after the administration of 500 ng/mouse RPB4 protein were stopped significantly in the lungs (Figure 4). This influence of RBP4 protein was also observed at 3 h after its injection, but only in the case of 67NR/iRFP cells (Figure 4A). In vitro 24 h preincubation of cancer cells with RBP4 significantly increased lung settlement of 4T1 cells (Figure 4B).

2.4. Increase in Metastatic Potential and Tumor Blood Vessel Impairment in Mice Bearing RBP4-transduced Cells

Both 67NR/RBP4 and 4T1/RBP4 cells with overexpression of RBP4 protein when transplanted ort. showed similar kinetics of tumor growth as compared to wild-type tumors (Figure S5A,B). Only tumors growing from 67NR/0 cells exhibited slow growth kinetics; therefore, the analyses of plasma and tissues and angiogenesis assessment were performed 10 days later, when tumor volumes of 67NR/0 tumors were comparable to those of wild-type and 67NR/RBP4 tumors (Figure S5A). Lung weight of mice bearing 67NR/RBP4 cells did not change as compared to that of controls; however, histopathological analysis revealed the presence of tumor cells in the lung tissue of 3/9 mice bearing 67NR/RBP4 cells (Figure 5A,B).

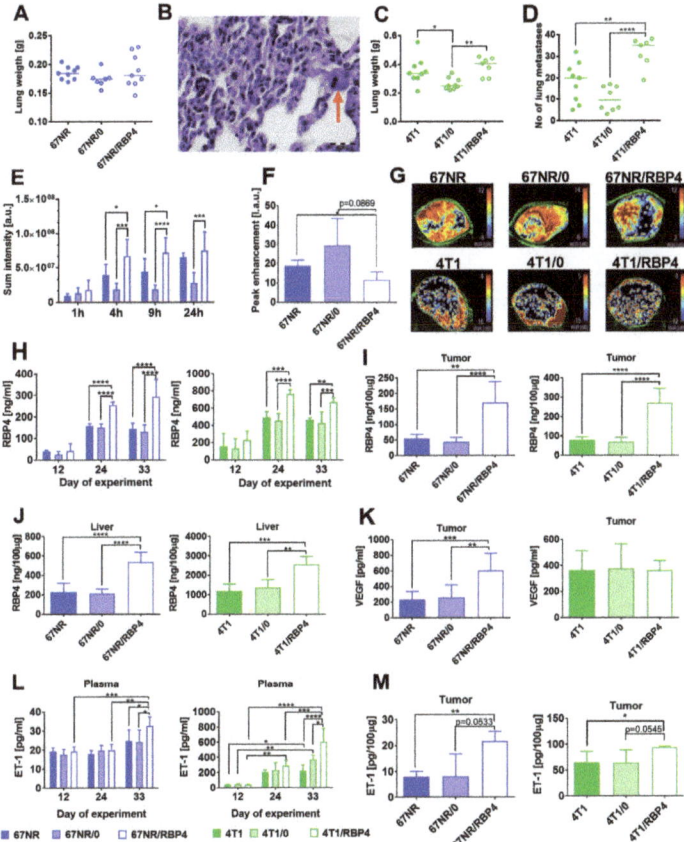

Figure 5. The effect of RBP4 overexpression on metastatic potential and angiogenesis of 67NR and 4T1 tumors. (**A**) Lung weight of 67NR tumor-bearing mice (N = 7–9) and (**B**) microphotograph of lung metastasis in mice bearing 67NR/RBP4 cells. Red arrow indicates epithelial cell with mitotic spindle. (**C**) Lung weight and (**D**) number of lung metastatic foci in mice bearing 4T1/RBP4 tumors (N = 7–9). (**E**) Blood vessel permeability in 67NR/RBP4 tumors (N = 4). (**F**) Peak enhancement in tumor tissue of mice bearing 67NR/RBP4 tumors (N = 3–4). (**G**) Representative pictures of wash in rate parameter. Concentration of RBP4 protein in (**H**) plasma (N = 3–5), (**I**) tumor tissue (N = 5), and (**J**) liver (N = 5). (**K**) VEGF in tumor tissue (N = 6–9). Concentration of endothelin-1 (ET-1) in (**L**) plasma (N = 3–4) and (**M**) tumor tissue (N = 3–4). Data presented as mean ± SD or data for individual measurements (Figures (**A**), (**C**), and (**D**)). Statistical analysis: Tukey's multiple comparison test. * $p < 0.05$, ** $p < 0.01$, *** $p < 0.001$, **** $p < 0.0001$.

The same analysis of lungs from mice with 67NR and 67NR/0 cells did not reveal the presence of cancer cells in the lung. Lung weight and the number of metastatic foci significantly increased in mice bearing 4T1/RBP4 cells as compared to those in mice bearing 4T1/0 cells (Figure 5C,D). Moreover, we observed the impairment of tumor tissue angiogenesis (increased blood vessel permeability and decreased blood flow) when mice were transplanted with 67NR/RBP4 tumors (Figure 5E–G) or to a lesser extent even with transplantation of 4T1/RBP4 cells (Figure S6). The RBP4 protein level was significantly increased in plasma (Figure 5H), tumor tissue (Figure 5I), and liver (Figure 5J) of mice bearing 67NR/RBP4 or 4T1/RBP4 cells as compared to that in mice inoculated with cells not overexpressing RBP4. The RBP4 protein level in mammary gland and abdominal adipose tissue of mice bearing cells overexpressing RBP4 protein did not differ significantly as compared to that in control animals (Figure S5A,B). The level of VEGF in tumor tissue increased significantly, but only in mice bearing 67NR/RBP4 cells (Figure 5K). ET-1 level in plasma (Figure 5L) and tumor tissue (Figure 5M) was significantly higher in mice bearing 67NR/RBP4 or 4T1/RBP4 cells than in appropriate controls.

In summary, overexpression of RBP4 causes impaired blood flow in tumors and an increase in vascular permeability. Elevated RBP4 levels are observed in tumors, liver and plasma of mice inoculated with *Rbp4*-transduced cells. Increased ET-1 level is noticed in plasma and tumors and VEGF in tumors of mice transplanted with RBP4-overexpressing cells.

2.5. RBP4 Increases the Invasive Potential of 67NR and 4T1 Mouse Mammary Gland Tumor Cells In Vitro

The 4T1 cell lysates from in vitro culture showed significantly higher level of RBP4 protein than 67NR cell lysates (Figure 6A).

Incubation of both cell lines with 200 ng/mL of RBP4 did not significantly influence the proliferation rate of cells (Figure 6B). The migration of 67NR cells through collagen and fibronectin and of 4T1 cells through collagen was enhanced by RBP4 (Figure 6C). Moreover, the adhesion of 4T1 cells was inhibited significantly after incubation with RBP4 (Figure 6D). However, the expression levels of E- and N-cadherin, CD44, CD29, CD61, CD162, CD51, CD24, and CD41 did not change significantly after incubation with RBP4 (Figure S7).

Because 67NR cells seem to be more sensitive to the effect of RBP4 in vitro, we also analyzed proliferation, colony formation, and migration using the 67NR/RBP4 cell line (Figure 6E–H). Proliferation of wild-type 67NR, 67NR/0, and 67NR/RBP4 cell lines did not differ significantly (Figure 6E). However the long-term colony formation assay showed significant improvement in the number of colonies formed for the 67NR/RBP4 cell line as compared to that for both 67NR and 67NR/0 cell lines (Figure 6F,G). Migratory properties of 67NR/RBP4 cells through collagen and fibronectin significantly increased as compared to that of 67NR/0 cells (Figure 6H). Overexpression of RBP4 did not significantly influence the sensitivity of 67NR/RBP4 cells to the selected anticancer agents (Figure 6I). Except for cisplatin, we observed the tendency of increased antiproliferative activity of cisplatin against 67NR/RBP4 cells as compared to that against 67NR/0 cells ($p = 0.0653$). We also assessed the effect of RBP4 overexpression on STAT3 phosphorylation in 67NR/RBP4 cells in vitro (as well as in 67NR/RBP4 and 4T1/RBP4 tumor cell lysates), and we did not observed significant differences between wild-type or empty vector-transduced cell lines and *Rbp4*-transduced cells (Figure S8A–C). The expression of VEGF did not differ between 67NR/RBP4 cell lines and control cell lines (Figure S8D).

Exogenous RBP4 added to cell culture, as well as transfection of cells with *Rbp4*, do not affect cell proliferation but increase their migration. Overexpression of RBP4 increases the clonogenic potential of cells and may sensitize cells to cisplatin.

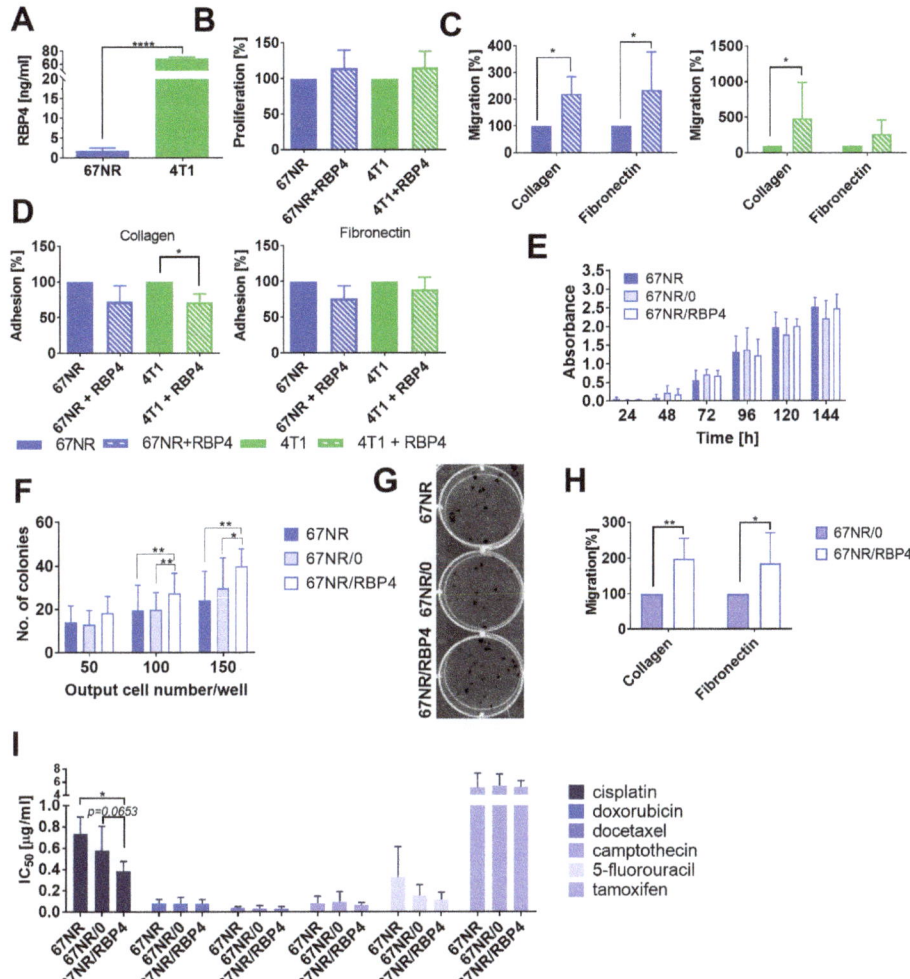

Figure 6. In vitro characteristics of the effect of RBP4 on 4T1 and 67NR mouse mammary gland cancer. (**A**) Comparison of RBP4 level in 4T1 and 67NR cell lysates using ELISA. (**B**–**D**) The effect of 24 h incubation with 200 ng/mL of RBP4 on (**B**) cell proliferation ($N = 4$), (**C**) migration ($N = 6$), and (**D**) adhesion ($N = 3$). (**E**) The proliferation of wild-type and transduced 67NR cells measured between 24 and 144 h ($N = 2$–6). (**F**) Number of colonies formed 14 days after seeding three different numbers of wild-type and transduced 67NR cells ($N = 3$ for 150 cells; $N = 8$ for 50 and 100 cells/well). (**G**) Representative image of colonies formed after seeding of 100 cells/well. (**H**) Migration of 67NR/0 and 67NR/RBP4 cell lines through collagen and fibronectin ($N = 4$–6). (**I**) The sensitivity of 67NR/RBP4 cells to commonly used anticancer drugs ($N = 3$–7). Statistical analysis: Tukey's multiple comparison test or unpaired t test; (**I**) Dunnett's multiple comparison test. * $p < 0.05$, ** $p < 0.01$, **** $p < 0.0001$.

3. Discussion

Although several studies have presented the analysis of the effect of RBP4 on several types of cancers, only one study has shown such research on patients with breast cancer [25]. In their case-control study, Jiao et al. showed that serum RBP4 levels were positively associated with breast cancer risk among patients with lower BMI (<25 kg/m^2) and that patients with ER- or PR-negative tumors possessed significantly higher serum levels of RBP4 [25]. A similar tendency (in the case of ER) was observed in our studies (Figure S3H). However, in contrast to the studies of Jiao et al., our results showed significant differences in the plasma levels of RBP4 between patients with metastatic and nonmetastatic tumors [25]. Moreover, our animal studies on two sister cell lines, nonmetastatic 67NR and metastatic 4T1, confirmed significantly higher RBP4 plasma levels in mice bearing metastatic tumors than in mice bearing nonmetastatic tumors. Moreover, the overexpression of RBP4 in cancer cells further increased the metastatic potential of the 4T1/RBP4 cell line, and for the nonmetastatic 67NR cell line [4], we could detect cancer cells in the lungs of 67NR/RBP4 tumor-bearing mice.

RBP4 protein is reported to induce endothelial inflammation through the stimulation of expression of proinflammatory molecules involved in leukocyte recruitment and adherence to the endothelium, including V-CAM-1, I-CAM-1, and E-selectin [12,13]. Endothelial inflammation and/or prolonged activation during obesity and cancer lead to endothelial dysfunction and are among the factors facilitating tumor progression and metastasis [15,33]. ET-1 is a useful and sensitive marker of endothelial dysfunction [34,35]. We observed an elevated level of ET-1 in 4T1 metastatic tumors as compared to that in 67NR tumor; moreover, the overexpression of RBP4 led to further elevation of its plasma level in mice bearing 67NR/RBP4 or 4T1/RBP4 tumors, indicating increasing endothelial dysfunction with increasing RBP4 expression in tumor cells. By using the model of bovine vascular aortic endothelial cells (BAECs), Takebayashi et al. showed that RBP4 inhibited insulin-stimulated secretion of ET-1 and induced NO production [14]. However, these interesting effects observed in their paper should be described as acute, whereas our studies showed a systematic increase in RBP4 levels in the plasma of mice, indicating chronic exposure of endothelium to its effects. Moreover, increased ET-1 levels were observed in our studies at later steps of tumor progression (24–33 days), i.e., after a significant increase of RBP4 plasma level (day 12).

The other soluble factor related to endothelial activation, namely sI-CAM-1, was elevated in plasma of mice bearing metastatic tumor as compared to that in mice bearing nonmetastatic mammary gland tumor. On the other hand, the plasma levels of sV-CAM-1, sE-selectin, and sP-selectin were decreased in metastatic tumors as compared to those in nonmetastatic tumors. The expression and shedding of all these molecules are enhanced in angiogenesis-associated diseases by angiogenic mediators released by tumor and inflammatory cells. In turn, these soluble molecules can stimulate neovascularization [36–39]. Two studies conducted in vitro showed that exogenous RBP4 can induce the expression and shedding of V-CAM, I-CAM, and E-selectin, and this activity was described to be realized through the activation of NADPH oxidase and nuclear factor κB (NF-κB) [12] or by TLR4 and in part by the JNK and p38 MAPK signaling pathways [13]. We cannot exclude the influence of various other factors/molecules whose expression was reported to differ between 4T1 and 67NR tumors on the expression of these proteins [4]. However, the final effect on angiogenesis is unambiguous, namely dysfunctional blood vessel network in highly metastatic tumors (4T1 and 4T1/RBP4) and increased blood vessel permeability with decreased blood flow in 67NR/RBP4 tumors overexpressing RBP4. The increased level of VEGF in the plasma of 4T1 tumor-bearing mice and in the tumor tissue of 67NR/RBP4 tumors also contributes to this final effect. VEGF, the main proangiogenic molecule, is responsible for excessive angiogenic response within the tumor tissue, and antiangiogenic therapies directed against this molecule or its receptors result in the normalization of the blood vessel network [40]. TGF-β, which was increased in the tumor tissue of 4T1 tumor-bearing mice, is another molecule known to affect tumor angiogenesis [41,42], and activated endothelial cells are also characterized by increased TGF-β production [43]. Therefore, the endothelium activated by RBP4 may also lead to the increased expression of this molecule. TGF-β is also the main activator of the epithelial–mesenchymal transition (EMT) process during cancer progression, in which epithelial cells break down their junctional structures, begin to express mesenchymal cell proteins, remodel their extracellular matrix, and migrate [44].

TNF-α (whose expression was higher in 4T1 tumors than in 67NR tumors) is known to be a promoter of invasion and metastasis through the activation of NF-κB signaling [45], and RBP4 can activate NF-κB [12]. Therefore, both RBP4 and NF-κB may lead to a synergistic increase in the tumor progression and metastasis process observed in our studies. The potential role of NF-κB signaling in the mechanism of the effects of RBP4 observed in our studies may be supported by the observation that NF-κB regulates the expression of VEGF (and thus tumor angiogenesis) [46]. The overexpression of RBP4 increased the level of VEGF in the tumor tissue of 67NR/RBP4, but not in the cell culture of these cells (Figure S8D). We can therefore assume that this effect is not dependent on the direct effect of RBP4 on cancer cells.

It is also known that RBP4 production is downregulated in human adipocytes by TNF-α [47]. Therefore, in our studies using wild-type tumors, the increased expression of TNF-α may be responsible for the observed lower levels of RBP4 in the plasma and abdominal adipose tissue of young mice in the last days of observation. Such effects were not observed in older mice bearing transduced cells as well as in aged mice bearing wild-type 4T1 tumors. Moreover, the plasma levels of RBP4 were higher in aged mice bearing cancer than in young mice. RBP4 expression increases during obesity, and a previous study suggested that the development of obesity leads to the increased expression of RBP4 by adipocytes [48]. Although RBP4 plasma levels did not differ significantly between healthy young 6-week-old (weighing about 20 g) and 52-week-old (about 25 g) female BALB/c mice, 4T1 tumor growth induced higher levels of RBP4 in the plasma of aged mice (about 200 ng/mL in young vs. 600 ng/mL in aged mice). Interestingly, the tumor tissue level of RBP4 did not differ between young and aged mice, but was again higher in mammary glands, abdominal adipose tissue, and liver of aged mice, similar to that observed in the plasma. Interestingly, tumors overexpressing RBP4 led to increased (as compared to that in wild-type or transduced with control vector cell lines) level of RBP4 only in the liver (besides tumor tissue and plasma) and not in the mammary gland or abdominal adipose tissue (Figure S5). Thompson et al. reported that hepatocytes are the main source of circulating RBP4 in mice and RBP4 produced by adipocytes may have a more important autocrine or paracrine function [49]. Recent studies reported that IL-6 is an important modulator of RBP4 production in the liver. Mohd et al. proposed a new mechanism involving peroxisome proliferator-activated receptor α (PPARα) and different CCAAT/enhancer binding protein (C/EBP) isoforms necessary for the regulation of *RBP4* gene expression in response to external stimuli, like IL-6, during physiological changes [50]. However, further research is required to understand the mechanisms by which growing tumors enhance RBP4 levels in other tissues and the difference observed between young and aged mice.

Apart from the influence of RBP4 in vitro on endothelial cells [12–14], RBP4 has also been reported to influence cancer cells [18,26]. Wang et al. showed that RBP4 can drive ovarian cancer cell migration and proliferation through the RhoA/Rock1 and ERK pathways [26]. Exogenous RBP4 and RBP4 overexpression resulted in increased migration of 4T1 or 67NR cells through collagen or fibronectin, but we did not observe any effect of RBP4 on E- and N-cadherins and other adhesion molecules analyzed (Figure S7). We also observed that 67NR/RBP4 cells possessed increased ability to form colonies from a single cell. On the other hand, an increase in MMP2 and MMP9 expression was observed in ovarian cancer cell lines in parallel with increased migratory potential of these cells [26], and downregulation of STRA6 or RBP4 in colon cancer cells decreased the fraction of cancer stem cells and tumor initiation frequency through mechanisms dependent on the activation of the STRA6 receptor by bound RBP4 and further transduction of the JAK2-STAT3 signaling cascade [18]. These mechanisms could also be important in our studies on breast cancer cells. Therefore, we analyzed the phosphorylation status of STAT3 in tumor tissue and cell culture and found that it did not change with the overexpression of RBP4 (Figure S8).

The abovementioned effects of RBP4 on metastasis and angiogenesis may therefore rely on the direct effects of RBP4 on cancer cells and endothelial cells. To show which of these effects prevail, we conducted studies by performing intravenous injection of cancer cells. Our initial research showed that endothelial cell activation in vitro (measured as expression of P-selectin) was the highest after 3 h, and at the same time, the plasma level of RBP4 injected i.v. persisted at the highest level between 15 and 60 min and then rapidly diminished (Figure S4). Therefore, we assumed that injecting RBP4 1 h prior to the injection of cancer cells may allow to observe the combined effects of RBP4 on cancer cells and endothelial cells. On the other hand, when cancer cells were injected 3 h after RBP4 injection, only the effect of RBP4 on the endothelial cells could be observed. The incubation of cancer cells for 24 h before intravenous injection should at least represent the effect of RBP4 on cancer cells. All these experimental schedules resulted in the increase of lung settlement by cancer cells, indicating that the effect of RBP4 on both cancer and endothelial cells is important. Moreover, the highest number of cancer cells in the lungs were observed in the experiment where both effects occurred: injection of cancer cells 1 h after RBP4 injection.

Increased in vitro sensitivity of 67NR/RBP4 cells to cisplatin (reduced IC_{50} value) is possibly the only beneficial property of RBP4 observed in our studies. It should be emphasized that increased levels of RBP4 in patients with diabetes are considered as a marker of renal tubular dysfunction [51]. In addition, high levels of RBP4 were also observed in patients with kidney graft dysfunction [52]. Other authors have also shown that cisplatin increases RBP4 expression in mice by inducing kidney damage [53]. Similar relationships were observed for the platinum-based drug LA-12 in both rats and patients [54]. However, our initial in vitro studies indicate that RBP4 alone does not adversely affect the sensitivity of cancer cells to anticancer drugs, including those based on platinum or other drugs that lead to kidney damage and may cause an increase in RBP4. Further research is needed to confirm these observations.

4. Materials and Methods

4.1. Cell Lines and Cell Culture

Mouse mammary gland cancer cell line 67NR was obtained from Barbara Ann Karmanos Cancer Institute (Detroit, MI, USA) and 4T1 cell line from ATTC (Rockville, MD, USA). Variants of these cell lines transduced with near-infrared fluorescent protein iRFP670 (KC991142.1)–67NR/iRFP and 4T1/iRFP–and RBP4 protein (NM_011255.3)–67NR/RBP4 and 4T1/RBP4–as well as cells with empty vector–67NR/0 and 4T1/0–were produced using the pRRL-cppt-CMV-ires-puro-PRE-sin lentiviral vector kindly provided as part of the lentivirus system by Dr. Didier Trono (Ecole Polytechnique Fédérale de Lausanne, Lausanne, Switzerland). The efficacy of transduction was presented in Figure S4C (iRFP670) and Figure S5C (RBP4).

For lentivirus production and packaging, Lenti-X™ 293FT cells (Clontech, Mountain View, CA, USA) were cotransfected at 60% confluence with 20 µg pRRL-cppt-CMV-RBP4-ires-puro-PRE-sin, 10 µg pMDL-g/p-RRE, 5 µg pRSV-REV, and 5 µg pMk-VSVG (D. Trono, École Polytechnique Fédérale de Lausanne, Lausanne, Switzerland) using polyethylenimine (Polysciences Inc., Warrington, PA, USA) at a concentration of 1 mg/mL dissolved in phosphate-buffered saline (PBS). The virus-containing supernatant was concentrated 100× on an Amicon Ultra-15K:100.000 (Millipore, Billerica, MA, USA). The 4T1 and 67NR cells (2.5×10^4) were transduced with the concentrated virus stock by centrifugation (2460× g) at 24 °C for 2.5 h.

67NR cells were cultured in high-glucose Dulbecco's Modified Eagle's Medium (DMEM; Thermo Fisher Scientific, Waltham, MA, USA) supplemented with 10% fetal bovine serum (FBS) + Fe, 2 mM L-glutamine, and 1% MEM. 4T1 cells were cultured in the 1:1 mixture of RPMI1640 + Opti-MEM medium with 5% FBS (Thermo Fisher Scientific), 2 mM L-glutamine, 4.5 g/L glucose, 1 mM sodium pyruvate (all from Sigma-Aldrich, St. Louis, MO, USA). Both culture media were supplemented with 100 µg/mL streptomycin and 100 U/mL penicillin (both from Polfa Tarchomin S.A., Warszawa, Poland). Culture media for transfected cells were supplemented with puromycin (8 µg/mL for 67NR cells and 1 µg/mL for 4T1 cells; Thermo Fisher Scientific). The Lenti-X™ 293FT cell line was maintained in high-glucose DMEM (Gibco, Scotland, UK) supplemented with 1% MEM (Sigma-Aldrich), 100 U/mL penicillin, 100 mg/mL streptomycin (both from Polfa Tarchomin S.A, Warszawa, Poland), 1 mM sodium pyruvate, 5% FBS (HyClone, Logan, UT, USA), and 6 mm L-glutamine.

4.2. In Vivo Experiments

The experiments were carried out on 6-week-old (about 20 g), 16-week-old (about 22 g), and 52-week-old (about 25 g) female BALB/c mice, under protocol Nos. 46/2013, 44/2016, 75/2017, and 09/2018 approved by the Local Ethical Committee for Experiments on Animals in Wroclaw, Poland. All experiments were conducted in accordance with the Directive of the European Parliament and Council No. 2010/63/EU on the protection of animals used for scientific purposes. Mice were obtained from the Animal Facility of the Experimental Medicine Center of the Medical University of Bialystok, Poland. The mice were maintained under the conditions of a 12-h day/night cycle with unrestricted access to food and drinking water.

4.2.1. Cell Transplantation

The cells (67NR, 67NR/0, 67NR/RBP4: 2×10^6 cells/mouse; 4T1, 4T1/0, 4T1/RBP4: 0.2×10^6 cells/mouse) from in vitro culture were injected orthotopically (ort.) into the mammary fat pad. After orthotopic cell injection, the mice were observed, and their body weight and tumor growth were measured. Tumor volume [mm^3] was calculated according to the Formula (1):

$$TV = \frac{1}{2} \times a^2 \times b \quad (1)$$

where *TV*—tumor volume; *a*—shorter diameter; *b*—longer diameter.

At 1–3 time points (Table 1), the mice were euthanized and blood, tumor, lungs, liver, abdominal visceral adipose tissue [55], and the tissue of the healthy mammary gland from the site opposite to the tumor location site were harvested for further analyses.

Table 1. Detailed characteristics of animal experiments.

Cell Line	Route of Transplantation	Age of Mice (weeks)	Time Points of Euthanasia—Days after Cell Transplantation
67NR, 4T1	ort.	6	12, 24, 33, "26"
67NR, 4T1	ort.	52	28 (4, 9, 15, 21) *
67NR, 67NR/0, 67NR/RBP4, 4T1, 4T1/0, 4T1/RBP4	ort.	16	12, 24, 33 **, "26"
67NR/iRFP and 4T1/iRFP	i.v.	16	48 h after cell transplantation

* in brackets: days of blood collection from zygomatic vein; ** 67NR/0 cell line: last time point of euthanasia 43 day; "26"—day of angiogenesis assessment; except for 67NR/0: the day when tumor reached the volume of 1000 mm^3, comparable to 67NR and 67NR/RBP4 on day 26; ort.—orthotopically; i.v.—intravenously.

Both 67NR/iRFP and 4T1/iRFP cells were injected intravenously (i.v.) into the lateral tail vein in the number of 0.6×10^6 cells/mouse and mice were euthanized 48 h after cell transplantation. One or three hours before intravenous cell transplantation, the mice were injected i.v. with 500 ng/mouse of RBP4 (RBP4 Recombinant Mouse Protein, His Tag; Thermo Fisher Scientific) or with 67NR/iRFP or 4T1/iRFP cells preincubated for 24 h before transplantation with 200 ng/mL of RBP4. The details of experiments are summarized in Table 1. As a control, healthy BALB/c mice of the corresponding age were used in selected analyses.

4.2.2. Tumor Angiogenesis Assessment

To compare tumor angiogenesis between 67NR and 4T1 cell lines and in mice bearing cells overexpressing RBP4 on day 26 (or in the case of slowly growing 67NR/0 cells, when tumors reached volume of 1000 mm^3), two methods were used.

Tumor blood perfusion analysis was performed using intravenous injection of MicroMarker™ Contrast Agent by the Vevo2100 ultrasound imaging system (VisualSonics, Ontario, Canada) as described previously [10]. The analysis of the received data was carried out using the VevoLab and VevoCQ software (VisualSonics). Tumor perfusion was assessed on the basis of quantitative contrast analysis in the central part of the tumor at the pixel level by calculating the perfusion parameters related to the amplitude and time according to the fit of the curve algorithm.

To evaluate vascular permeability, mice were administered i.v. (1 nmol/mouse) the IRDye® 800CW fluorescent dye PEG Contrast Agent (LI-COR, Lincoln, NE, USA). IRDye® 800CW selectively accumulates within the tumor tissue through increased vascular permeability and impaired lymphatic drainage in the tumor. At 1, 4, 9 and 24 h after administration, fluorescence measurements were performed. For this purpose, the animals were anesthetized by infusing a continuous 3% isoflurane mixture in synthetic air. The animals were then placed in a chamber for visualization for small rodents of the In Vivo MS FX Pro system (Carestream Health Inc., Rochester, NY, USA), equipped with individual masks for providing an anesthetic. During fluorescence imaging, the following camera settings were used: $t = 30$ s, f-stop = 2.8, FOV = 200 mm, excitation wavelength: 760 nm, emission wavelength: 830 nm. In addition, X-ray pictures of the examined animals were taken to allow localization of tumor. The imaging was performed using the following camera settings: $t = 2$ min, f-stop = 5.57, FOV = 200. The obtained fluorescence images were analyzed using the Carestream MI SE software (Carestream Health Inc.) based on analyzed regions.

4.2.3. Lung Fluorescence Measurement

Measurement of fluorescence of lungs dissected during autopsy was performed using the In Vivo MS FX Pro system with coregistration of fluorescence and X-ray. The obtained fluorescence images were analyzed using the Carestream MI SE software based on analyzed regions as described previously [56].

4.2.4. Blood Morphological Analyses

Whole blood was collected in a tube containing low-molecular-weight heparin (LMWH) at 5000 IU/mL and then analyzed using the hematology analyzer Mythic 18 (C2 Diagnostics, Montpellier, France).

4.3. Plasma and Tumor Tissue from Patients with Breast Cancer

Approval was obtained from the Bioethical Commission at the Medical University in Wroclaw for studies on plasma and tumor tissue from patients with breast cancer and healthy donors (Approval No. 71/2017). Informed consent was obtained from persons participating in the study. All procedures were conducted in accordance with the institutional and international ethical standards. Blood samples and tumor samples were collected from July to September 2017 in Wroclaw Comprehensive Cancer Center, Poland, from 34 patients at various stages of breast cancer. In addition, blood from eight healthy donors was used as a control. Blood was collected from peripheral veins into heparin-containing tubes. Patients were divided into two groups according to the stage of breast cancer: nonmetastatic and patients with metastases (in the lymph nodes and patients with distant metastases) (Table 2).

Table 2. Characteristics of patients with breast cancer.

Patients	No.	Age: Median (min-max)	Diabetes	Tumor Diameter: Mean ± SD (mm)	Ki67: Median (min-max) [%]	ER+	PR+	HER2+
No metastases	10	62 (47–85)	0/10	25 ± 23	20 (1–50)	9/10	7/10	8/10
Metastases	24	57 (31–83)	5/24	32 ± 21	15 (3–60)	14/24	13/24	13/24

4.4. In Vitro Experiments

4.4.1. Cell Preparation to Evaluate Proliferation, Migration, Adhesion, and Integrin Expression after Incubation with RBP4

After 24 h culture of 67NR and 4T1 cells, the culture medium was changed to medium with 5% FBS. After further 24 h, the medium was removed and replaced with a fresh medium containing 5% FBS and RBP4 protein at a concentration of 200 ng/mL. After 24 h of incubation, the cell proliferation was assessed, or the cells were harvested using nonenzymatic Cell Dissociation Solution (Sigma-Aldrich). It was then neutralized by the addition of medium, and the density of the cells was counted. Cells prepared in this way were used for further tests:

- Proliferation

The sulforhodamine B (SRB) assay was performed as described previously [57] and the percentage of cell proliferation was calculated as follows Formula (2):

$$\% \ of \ proliferation = \left[100 \times \left(1 - \frac{Ab - Am}{Ak - Am}\right)\right] \qquad (2)$$

where:

Ab—absorbance value measured for cells treated with RBP4
Ak—absorbance value measured for untreated cells
Am—absorbance value measured for the culture medium

- Migration

Inserts (Transwell Permeable Supports 6.5 mm Insert, Corning Incorporated, New York, NY, USA) were coated with type IV collagen or fibronectin (both from Sigma-Aldrich) at a concentration of 10 µg/mL diluted in 2% acetic acid or water, respectively, and incubated overnight at 4 °C. Subsequently, the inserts were rinsed twice with PBS and blocked with 1% BSA (Bio-Rad, Laboratories, Hercules, CA, USA) for 1 h at 37 °C. After incubation, the inserts were rinsed again with PBS and 25,000 cells suspended in 250 µL DMEM were added. The inserts were placed in the wells with culture medium and left in an incubator at 37 °C for 8 h for 4T1 and 67NR cells to migrate on collagen and for 6 h for 67NR cells to migrate on fibronectin. After incubation, the inserts were rinsed twice with PBS to remove cells that were not migrated, and the cells were stained with 0.2% crystal violet in 20% MetOH and counted using an Olympus CX microscope (Olympus Europe Holding GmbH, Hamburg, Germany).

- Adhesion

Wells of 96-well plates were coated with 10 µg/mL of collagen or fibrinogen diluted in 2% acetic acid or water, respectively, and incubated overnight at 4 °C. The next day, the plates were washed with TSM buffer (2 × 300 µL), and 100 µL of 1% BSA/TSM solution was added to block nonspecific binding sites. The plates were incubated for 30 min at 37 °C. After incubation, the plates were washed again with TSM buffer (2 × 300 µL). The cells were then plated at a concentration of 5×10^5/mL in 50 µL of 0.5% BSA/TSM. The plates were incubated for 60 min at 37 °C. After incubation, the plates were washed three times with 300 µL of TSM buffer to remove nonadherent cells. Adherent cells were stained with a solution of 0.2% crystal violet in 20% MetOH in a volume of 50 µL per well. The plates were incubated for 30 min at 4 °C and then washed, and the cell suspension was diluted with 100 µL of 80% MetOH. The optical density of the samples was read using a Biotek Hybrid H4 reader (BioTek Instruments, Winooski, VT, USA) at 570 nm.

- Flow cytometry analysis

The cell pellet was suspended in PBS solution with the addition of 2% FBS. Cells were counted and $2.5–5 \times 10^5$ cells were stained with antibodies for 30 min at 4 °C in the dark, centrifuged, and suspended in PBS. The analysis was performed in a BD Fortessa cytometer using the Diva software (Becton Dickinson, East Rutherford, NJ, USA).

List of antibodies used for flow cytometry analysis: BV421 Rat Anti-Mouse CD162, BV421 Rat Anti-Mouse CD41, FITC Rat Anti-Mouse CD29, FITC Rat Anti-Mouse CD44, FITC Rat Anti-Mouse CD61, PE Rat Anti-Mouse CD24, PE Rat Anti-Mouse CD51, and PE P-selectin 62P (all from BD Biosciences, San Jose, CA, USA).

4.4.2. Proliferation and Migration Evaluation and Clonogenic Assay Using 67NR/RBP4 Cell Line

- Proliferation

The MTT assay was performed as described previously [57] with minor modifications. Briefly, cells were seeded at a density of 1500 cells per well in 96-well cell culture plates and maintained at 37 °C in 5% CO_2. After incubation (24, 48, 72, 96, 120, or 144 h), 20 µL of a 5 mg/mL solution of 3-(4,5-dimethylthiazol-2-yl)-2,5-diphenyltetrazolium bromide (Sigma-Aldrich, St. Louis, MO, USA) in PBS was added to each well. The cells were then incubated at 37°C for 4 h. Then, the medium was removed, and the cells were lysed by adding 200 µL/well of DMSO (Avantor Performance Materials, Gliwice, Poland) The resulting formazan crystals were dissolved in DMSO, and absorbance at 570 nm was measured using a Biotek Hybrid H4 reader (BioTek Instruments, Winooski, VT, USA).

- Proliferation of cells treated with anticancer agents

Cisplatin, doxorubicin, docetaxel, and 5-fluorouracil were purchased from Accord Healthcare Poland (Warsaw, Poland). Camptothecin and tamoxifen were purchased from Sigma-Aldrich (St. Louis, MO, USA). The effects of anticancer drugs on the cell growth of 67NR, 67NR/0, and 67NR/RBP4 cell lines were measured using the MTT assay as described above. Briefly, cells were seeded at a density of 1500 cells per well in 96-well cell culture plates 1 day prior to the assay and maintained at 37 °C in 5% CO_2. The cells were then treated with cisplatin, doxorubicin, 5-FU, or camptothecin at four concentrations in the range of 0.001 to 1 µg/mL, docetaxel at four concentrations in the range of 0.0001 to 0.1 µg/mL, and tamoxifen at four concentrations in the range of 0.01 to 10 µg/mL for 72 h. The solvent for camptothecin and tamoxifen (DMSO) used at the highest concentration (0.1%) in the assay did not cause any cytotoxicity. All compounds were diluted prior to use in culture medium to the required concentrations. The IC_{50} value was defined as the concentration required for half-maximal (50%) inhibition of cell growth as compared to the growth of untreated cells. The IC_{50} values were calculated based on Cheburator 0.4 software [58]. In each experiment, samples containing specific concentrations of the preparation were used in triplicate. The experiments were repeated 3–7 times.

- Migration

The inserts (prepared as described above) were placed in the wells with culture medium and left in an incubator at 37 °C for 6 h. After incubation, the cells were stained with RAL Diff-Quik kit (RAL Diagnostics, Martillac, France), rinsed twice with PBS to remove cells that had not migrated, and counted using an Olympus CX microscope (Olympus Europe Holding GmbH, Hamburg, Germany).

- Clonogenic assay

The viable cells were counted and seeded at a density of 50, 100, or 150 cells on the wells of a 6-well plate. After 7 days, the colonies were fixed and stained with 1% crystal violet/methanol (Sigma-Aldrich), documented with a ChemiDoc Imaging System (Bio-Rad Laboratories), and counted manually.

4.5. Tissue and Cell Lysate Preparation for ELISA and Western Blot

Frozen tissue was homogenized with an appropriate amount of RIPA buffer with a cocktail of phosphatase and protease inhibitors (all from Sigma-Aldrich, St. Louis, MO, USA) using Fast Prep®-24 MP Bio homogenizer (MP Biomedicals LLC, Santa Ana, CA, USA). The samples were then incubated on ice for 20 min and centrifuged (4 °C, 15 min, 12,000× g). The obtained supernatant was transferred to 1.5 mL tubes and again centrifuged.

Cells plated on culture dishes were rinsed twice with PBS, and 90 μL of RIPA buffer containing a cocktail of protease and phosphatase inhibitors was added to the cells. The cells were then harvested with scrapers, transferred to tubes, and incubated on ice for 20 min. After incubation, the tubes were centrifuged for 15 min at 4 °C at 10,000× g.

The obtained supernatants were transferred to new tubes and stored at −80°C for further analysis. Protein content were analyzed using the Bio-Rad Protein Assay kit (Bio-Rad Laboratories).

4.5.1. ELISA Tests

ELISA tests were performed according to the manufacturer's protocols. The result of the analysis was read using a Biotek Synergy H4 Hybrid reader (BioTek Instruments, Winooski, VT, USA) by measuring the absorbance at 450 nm. Standard curves were prepared, which were used to determine the concentration of test samples.

List of ELISA kits used (anti-mouse): ET-1 (Endothelin 1), I-CAM-1/CD54 (Intercellular Adhesion Molecule 1), IGF-1 (Insulin-like Growth Factor 1), RBP4 (Retinol Binding Protein 4, Plasma), SeLE (E-selectin), SeLP (P-Selectin), sV-CAM-1/CD106 (soluble Vascular Cell Adhesion Molecule 1), TNF-α (Tumor Necrosis Factor-Alpha), TSP-1 (Thrombospondin-1), TTR (Transthyretin), TGF-β1 (TGF-beta1 (Transforming Growth Factor-beta1) (all from Elabscience Biotechnology Co, Wuhan, China); VEGF-A (Vascular Endothelial Cell Growth Factor A) (from Thermo Fisher Scientific, Waltham, MA, USA or R&D Systems, MN, USA); InstantOne ELISA STAT3 (Total/Phospho), Invitrogen (Waltham, MA, USA).

4.5.2. Western Blot

List of anti-mouse protein antibodies used: anti-E-cadherin, anti-N-cadherin (both from Proteintech, Chicago, IL, USA), anti-RBP4 (Abcam, Cambridge, UK), and anti-β-actin-horse radish peroxidase (HRP) (Santa Cruz Biotechnology, Inc., Heidelberg, Germany). Equal amounts of protein (50 μg of cell culture lysates) were mixed with 4× Laemmli Sample Buffer (Bio-Rad Laboratories, Hercules, CA, USA). Then, the samples were separated in a 4–20% sodium dodecyl sulfate (SDS)-polyacrylamide gel and transferred to a polyvinylidene difluoride (PVDF) membrane (0.45 μm; Merck Millipore, city, state abbrev, USA). The membranes were blocked for 1 h at room temperature in 5% non-fat dry milk in 0.1% PBS/Tween-20 (PBST). Next, the membranes were washed (3 × 10 min) with 0.1% PBST and then incubated overnight at 4 °C with a primary antibody. After incubation, the membranes were washed (3 × 10 min) with 0.1% PBST and incubated for 1 h with the secondary mouse anti-rabbit immunoglobulin G (IgG)-HRP antibody (Santa Cruz Biotechnology Inc., Santa Cruz, CA, USA). The membranes after washing with 0.1% PBST were detected by the ECL method. Chemiluminescence was visualized using Image Station 4000MM PRO (Carestream Health Inc., Rochester, NY, USA). Densitometry analysis of the blots was performed using Carestream MI Software 5.0.6.20 (Carestream Health Inc., Rochester, NY, USA).

4.6. Statistical Analysis

Statistical analysis of the results was performed using GraphPad Prism 7. The data normality analysis was performed using the Shapiro-Wilk data normality test assuming the significance of the test for $p < 0.05$. Statistical analysis for normal distribution data was performed using the ANOVA test. When the ANOVA test showed significant differences between the groups under consideration, further analyses were performed using Tukey's test or Sidak's test for multiple comparisons. In the event that the data distribution differed from normal, the analysis was conducted using the Kruskal-Wallis test for multiple comparisons. In some cases, the Mann–Whitney test or the t test was applied depending on the data distribution. Differences between the groups were considered statistically significant at $p < 0.05$.

5. Conclusions

The RBP4 protein may be an important driver of metastasis and angiogenesis of breast tumors. It affects endothelial cells by increasing the symptoms of endothelial dysfunction/activation and dysfunctional tumor angiogenesis. Furthermore, the direct effect of RBP4 on cancer cells through increased migratory and colony-forming properties contributes to the final prometastatic effect. The effect of RBP4 on tumor tissue and cancer cells in this model is not dependent on STAT3 phosphorylation.

Supplementary Materials: The following are available online at http://www.mdpi.com/2072-6694/12/3/623/s1, Figure S1. Selected blood morphological parameters of mice bearing 67NR and 4T1 tumors. Figure S2. The level of TTR protein in plasma and various tissues from mice bearing nonmetastatic 67NR and metastatic 4T1 mammary gland cancer cells (young mice). Figure S3. Patients plasma and tumor tissue levels. Figure S4. Kinetics of plasma level of RBP4 and in vitro kinetics of endothelial cells activation after incubation with RBP4. Figure S5. The effect of RBP4 overexpression on kinetics of 67NR/RBP4 and 4T1/RBP4 tumor growth as well as the level of RBP4 in mammary gland and abdominal adipose tissue. Figure S6. The effect of RBP4 overexpression on angiogenesis of 67NR and 4T1 tumors. Figure S7. Expression of surface molecules after 24 h in vitro incubation of 67NR and 4T1 cells with RBP4 protein at a concentration of 200 ng/mL. Figure S8. STAT3 phosphorylation and VEGF expression in cell lines with overexpression of RBP4. STAT3 phosphorylation status in tumor lysates.

Author Contributions: All authors contributed to the work presented in this paper. All authors read and approved the final manuscript. D.P., A.U., D.K., and A.N.-G. were responsible for the design and execution of the experiments, data analysis, and revision of the manuscript; M.C., J.J., and A.S. performed experiments and data analysis; R.M. and M.E. collected patient samples; M.N. performed histological examinations; M.U. contributed to design and interpretation of the data; J.W. was responsible for the conception of the study, interpretation of results; and writing of the manuscript.

Funding: This study was supported by the National Centre for Research and Development under the Polish Strategical Framework Program STRATEGMED (grant coordinated by JCET-UJ No. STRATEGMED1/233226/11/NCBR/2015). The funding bodies did not participate in the design of the study; collection, analysis, and interpretation of data; or writing of the manuscript.

Conflicts of Interest: The authors declare no conflict of interest. The funders had no role in the design of the study; in the collection, analyses, or interpretation of data; in the writing of the manuscript, or in the decision to publish the results.

References

1. DuPré, S.A.; Redelman, D.; Hunter, K.W. The mouse mammary carcinoma 4T1: Characterization of the cellular landscape of primary tumours and metastatic tumour foci. *Int. J. Exp. Pathol.* **2007**, *88*, 351–360. [CrossRef] [PubMed]
2. Aslakson, C.J.; Miller, F.R. Selective events in the metastatic process defined by analysis of the sequential dissemination of subpopulations of a mouse mammary tumor. *Cancer Res.* **1992**, *52*, 1399–1405. [CrossRef] [PubMed]
3. Heppner, G.H.; Miller, F.R.; Shekhar, P.M. Nontransgenic models of breast cancer. *Breast Cancer Res.* **2000**, *2*, 331–334. [CrossRef] [PubMed]
4. Johnstone, C.N.; Smith, Y.E.; Cao, Y.; Burrows, A.D.; Cross, R.S.N.; Ling, X.; Redvers, R.P.; Doherty, J.P.; Eckhardt, B.L.; Natoli, A.L.; et al. Functional and molecular characterisation of EO771.LMB tumours, a new C57BL/6-mouse-derived model of spontaneously metastatic mammary cancer. *Dis. Model. Mech.* **2015**, *8*, 237–251. [CrossRef]
5. Buczek, E.; Denslow, A.; Mateuszuk, L.; Proniewski, B.; Wojcik, T.; Sitek, B.; Fedorowicz, A.; Jasztal, A.; Kus, E.; Chmura-Skirlinska, A.; et al. Alterations in NO- and PGI2- dependent function in aorta in the orthotopic murine model of metastatic 4T1 breast cancer: Relationship with pulmonary endothelial dysfunction and systemic inflammation. *BMC Cancer* **2018**, *18*, 582. [CrossRef]

6. Pacia, M.Z.; Mateuszuk, L.; Buczek, E.; Chlopicki, S.; Blazejczyk, A.; Wietrzyk, J.; Baranska, M.; Kaczor, A. Rapid biochemical profiling of endothelial dysfunction in diabetes, hypertension and cancer metastasis by hierarchical cluster analysis of Raman spectra. *J. Raman Spectrosc.* **2016**, *47*, 1310–1317. [CrossRef]
7. Smeda, M.; Kieronska, A.; Adamski, M.G.; Proniewski, B.; Sternak, M.; Mohaissen, T.; Przyborowski, K.; Derszniak, K.; Kaczor, D.; Stojak, M.; et al. Nitric oxide deficiency and endothelial–Mesenchymal transition of pulmonary endothelium in the progression of 4T1 metastatic breast cancer in mice. *Breast Cancer Res.* **2018**, *20*, 86. [CrossRef]
8. Porshneva, K.; Papiernik, D.; Psurski, M.; Nowak, M.; Matkowski, R.; Ekiert, M.; Milczarek, M.; Banach, J.; Jarosz, J.; Wietrzyk, J. Combination Therapy with DETA/NO and Clopidogrel Inhibits Metastasis in Murine Mammary Gland Cancer Models via Improved Vasoprotection. *Mol. Pharm.* **2018**, *15*, 5277–5290. [CrossRef]
9. Blazejczyk, A.; Switalska, M.; Chlopicki, S.; Marcinek, A.; Gebicki, J.; Nowak, M.; Nasulewicz-Goldeman, A.; Wietrzyk, J. 1-methylnicotinamide and its structural analog 1,4-dimethylpyridine for the prevention of cancer metastasis. *J. Exp. Clin. Cancer Res.* **2016**, *35*, 110. [CrossRef]
10. Porshneva, K.; Papiernik, D.; Psurski, M.; Łupicka-Słowik, A.; Matkowski, R.; Ekiert, M.; Nowak, M.; Jarosz, J.; Banach, J.; Milczarek, M.; et al. Temporal inhibition of mouse mammary gland cancer metastasis by CORM-A1 and DETA/NO combination therapy. *Theranostics* **2019**, *9*, 3919–3939. [CrossRef]
11. Park, S.E.; Kim, D.H.; Lee, J.H.; Park, J.S.; Kang, E.S.; Ahn, C.W.; Lee, H.C.; Cha, B.S. Retinol-binding protein-4 is associated with endothelial dysfunction in adults with newly diagnosed type 2 diabetes mellitus. *Atherosclerosis* **2009**, *204*, 23–25. [CrossRef] [PubMed]
12. Farjo, K.M.; Farjo, R.A.; Halsey, S.; Moiseyev, G.; Ma, J.X. Retinol-Binding Protein 4 Induces Inflammation in Human Endothelial Cells by an NADPH Oxidase- and Nuclear Factor Kappa B-Dependent and Retinol-Independent Mechanism. *Mol. Cell. Biol.* **2012**, *32*, 5103–5115. [CrossRef] [PubMed]
13. Du, M.; Martin, A.; Hays, F.; Johnson, J.; Farjo, R.A.; Farjo, K.M. Serum retinol-binding protein-induced endothelial inflammation is mediated through the activation of toll-like receptor 4. *Mol. Vis.* **2017**, *23*, 185–197. [PubMed]
14. Takebayashi, K.; Sohma, R.; Aso, Y.; Inukai, T. Effects of retinol binding protein-4 on vascular endothelial cells. *Biochem. Biophys. Res. Commun.* **2011**, *408*, 58–64. [CrossRef] [PubMed]
15. Jung, U.; Choi, M.-S.; Jung, U.J.; Choi, M.-S. Obesity and Its Metabolic Complications: The Role of Adipokines and the Relationship between Obesity, Inflammation, Insulin Resistance, Dyslipidemia and Nonalcoholic Fatty Liver Disease. *Int. J. Mol. Sci.* **2014**, *15*, 6184–6223. [CrossRef]
16. Noy, N.; Li, L.; Abola, M.V.; Berger, N.A. Is retinol binding protein 4 a link between adiposity and cancer? *Horm. Mol. Biol. Clin. Investig.* **2015**, *23*, 39–46. [CrossRef]
17. Fei, W.; Chen, L.; Chen, J.; Shi, Q.; Zhang, L.; Liu, S.; Li, L.; Zheng, L.; Hu, X. RBP4 and THBS2 are serum biomarkers for diagnosis of colorectal cancer. *Oncotarget* **2017**, *8*, 92254–92264. [CrossRef]
18. Karunanithi, S.; Levi, L.; DeVecchio, J.; Karagkounis, G.; Reizes, O.; Lathia, J.D.; Kalady, M.F.; Noy, N. RBP4-STRA6 Pathway Drives Cancer Stem Cell Maintenance and Mediates High-Fat Diet-Induced Colon Carcinogenesis. *Stem Cell Rep.* **2017**, *9*, 438–450. [CrossRef]
19. Abola, M.V.; Thompson, C.L.; Chen, Z.; Chak, A.; Berger, N.A.; Kirwan, J.P.; Li, L. Serum levels of retinol-binding protein 4 and risk of colon adenoma. *Endocr. Relat. Cancer* **2015**, *22*, L1–L4. [CrossRef]
20. Cheng, Y.; Liu, C.; Zhang, N.; Wang, S.; Zhang, Z. Proteomics Analysis for Finding Serum Markers of Ovarian Cancer. *Biomed. Res. Int.* **2014**, *2014*, 1–9. [CrossRef]
21. Sobotka, R.; Čapoun, O.; Kalousová, M.; Hanuš, T.; Zima, T.; Koštířová, M.; Soukup, V. Prognostic Importance of Vitamins A, E and Retinol-binding Protein 4 in Renal Cell Carcinoma Patients. *Anticancer Res.* **2017**, *37*, 3801–3806. [CrossRef] [PubMed]
22. El-Mesallamy, H.O.; Hamdy, N.M.; Zaghloul, A.S.; Sallam, A.M. Serum retinol binding protein-4 and neutrophil gelatinase-associated lipocalin are interrelated in pancreatic cancer patients. *Scand. J. Clin. Lab. Invest.* **2012**, *72*, 602–607. [CrossRef] [PubMed]
23. Wang, D.-D.; Zhao, Y.-M.M.; Wang, L.; Ren, G.; Wang, F.; Xia, Z.-G.G.; Wang, X.-L.L.; Zhang, T.; Pan, Q.; Dai, Z.; et al. Preoperative serum retinol-binding protein 4 is associated with the prognosis of patients with hepatocellular carcinoma after curative resection. *J. Cancer Res. Clin. Oncol.* **2011**, *137*, 651–658. [CrossRef]
24. Chen, Y.; Azman, S.N.; Kerishnan, J.P.; Zain, R.B.; Chen, Y.N.; Wong, Y.L.; Gopinath, S.C.B. Identification of host-immune response protein candidates in the sera of human oral squamous cell carcinoma patients. *PLoS ONE* **2014**, *9*, e109012. [CrossRef]

25. Jiao, C.; Cui, L.; Ma, A.; Li, N.; Si, H. Elevated serum levels of retinol-binding protein 4 are associated with breast cancer risk: A Case-Control study. *PLoS ONE* **2016**, *11*, e0167498. [CrossRef]
26. Wang, Y.; Wang, Y.; Zhang, Z. Adipokine RBP4 drives ovarian cancer cell migration. *J. Ovarian Res.* **2018**, *11*, 29. [CrossRef]
27. Piano, A.; Titorenko, V.I. The Intricate Interplay between Mechanisms Underlying Aging and Cancer. *Aging Dis.* **2015**, *6*, 56–75. [CrossRef]
28. Meehan, B.; Dombrovsky, A.; Lau, K.; Lai, T.; Magnus, N.; Montermini, L.; Rak, J. Impact of host ageing on the metastatic phenotype. *Mech. Ageing Dev.* **2013**, *134*, 118–129. [CrossRef]
29. Meehan, B.; Garnier, D.; Dombrovsky, A.; Lau, K.; D'Asti, E.; Magnus, N.; Rak, J. Ageing-related responses to antiangiogenic effects of sunitinib in atherosclerosis-prone mice. *Mech. Ageing Dev.* **2014**, *140*, 13–22. [CrossRef]
30. Klement, H.; St Croix, B.; Milsom, C.; May, L.; Guo, Q.; Yu, J.L.; Klement, P.; Rak, J. Atherosclerosis and vascular aging as modifiers of tumor progression, angiogenesis, and responsiveness to therapy. *Am. J. Pathol.* **2007**, *171*, 1342–1351. [CrossRef]
31. Zanotti, G.; Berni, R. Plasma Retinol-Binding Protein: Structure and Interactions with Retinol, Retinoids, and Transthyretin. *Vitam. Horm.* **2004**, *69*, 271–295. [CrossRef] [PubMed]
32. Naylor, H.M.; Newcomer, M.E. The structure of human retinol-binding protein (RBP) with its carrier protein transthyretin reveals an interaction with the carboxy terminus of RBP. *Biochemistry* **1999**, *38*, 2647–2653. [CrossRef]
33. Blazejczyk, A.; Papiernik, D.; Porshneva, K.; Sadowska, J.; Wietrzyk, J. Endothelium and cancer metastasis: Perspectives for antimetastatic therapy. *Pharmacol. Reports* **2015**, *67*, 711–718. [CrossRef] [PubMed]
34. Endemann, D.H.; Schiffrin, E.L. Endothelial Dysfunction. *J. Am. Soc. Nephrol.* **2004**, *15*, 1983–1992. [CrossRef] [PubMed]
35. Iglarz, M.; Clozel, M. Mechanisms of ET-1-induced endothelial dysfunction. *J. Cardiovasc. Pharmacol.* **2007**, *50*, 621–628. [CrossRef] [PubMed]
36. Giavazzi, R.; Chirivi, R.G.; Garofalo, A.; Rambaldi, A.; Hemingway, I.; Pigott, R.; Gearing, A.J. Soluble intercellular adhesion molecule 1 is released by human melanoma cells and is associated with tumor growth in nude mice. *Cancer Res.* **1992**, *52*, 2628–2630.
37. Gho, Y.S.; Kleinman, H.K.; Sosne, G. Angiogenic activity of human soluble intercellular adhesion molecule-1. *Cancer Res.* **1999**, *59*, 5128–5132.
38. Morbidelli, L.; Brogelli, L.; Granger, H.J.; Ziche, M. Endothelial cell migration is induced by soluble P-selectin. *Life Sci.* **1997**, *62*. [CrossRef]
39. Koch, A.E.; Halloran, M.M.; Haskell, C.J.; Shah, M.R.; Polverini, P.J. Angiogenesis mediated by soluble forms of E-selectin and vascular cell adhesion molecule-1. *Nature* **1995**, *376*, 517–519. [CrossRef]
40. Maj, E.; Papiernik, D.; Wietrzyk, J. Antiangiogenic cancer treatment: The great discovery and greater complexity (Review). *Int. J. Oncol.* **2016**, *49*, 1773–1784. [CrossRef]
41. Ferrari, G.; Cook, B.D.; Terushkin, V.; Pintucci, G.; Mignatti, P. Transforming growth factor-beta 1 (TGF-beta1) induces angiogenesis through vascular endothelial growth factor (VEGF)-mediated apoptosis. *J. Cell. Physiol.* **2009**, *219*, 449–458. [CrossRef] [PubMed]
42. Viñals, F.; Pouysségur, J. Transforming growth factor beta1 (TGF-beta1) promotes endothelial cell survival during in vitro angiogenesis via an autocrine mechanism implicating TGF-alpha signaling. *Mol. Cell. Biol.* **2001**, *21*, 7218–7230. [CrossRef]
43. Pintavorn, P.; Ballermann, B.J. TGF-β and the endothelium during immune injury. *Kidney Int.* **1997**, *51*, 1401–1412. [CrossRef] [PubMed]
44. Moustakas, A.; Heldin, C.-H. Signaling networks guiding epithelial-mesenchymal transitions during embryogenesis and cancer progression. *Cancer Sci.* **2007**, *98*, 1512–1520. [CrossRef] [PubMed]
45. Tang, D.; Tao, D.; Fang, Y.; Deng, C.; Xu, Q.; Zhou, J. TNF-Alpha Promotes Invasion and Metastasis via NF-Kappa B Pathway in Oral Squamous Cell Carcinoma. *Med. Sci. Monit. Basic Res.* **2017**, *23*, 141–149. [CrossRef] [PubMed]
46. Xie, T.-X.X.; Xia, Z.; Zhang, N.; Gong, W.; Huang, S. Constitutive NF-κB activity regulates the expression of VEGF and IL-8 and tumor angiogenesis of human glioblastoma. *Oncol. Rep.* **2010**, *23*, 725–732.
47. Sell, H.; Eckel, J. Regulation of retinol binding protein 4 production in primary human adipocytes by adiponectin, troglitazone and TNF-α [2]. *Diabetologia* **2007**, *50*, 2221–2223. [CrossRef]

48. Yang, Q.; Graham, T.E.; Mody, N.; Preitner, F.; Peroni, O.D.; Zabolotny, J.M.; Kotani, K.; Quadro, L.; Kahn, B.B. Serum retinol binding protein 4 contributes to insulin resistance in obesity and type 2 diabetes. *Nature* **2005**, *436*, 356–362. [CrossRef]
49. Thompson, S.J.; Sargsyan, A.; Lee, S.A.; Yuen, J.J.; Cai, J.; Smalling, R.; Ghyselinck, N.; Mark, M.; Blaner, W.S.; Graham, T.E. Hepatocytes are the principal source of circulating RBP4 in mice. *Diabetes* **2017**, *66*, 58–63. [CrossRef]
50. Mohd, M.A.; Ahmad Norudin, N.A.; Muhammad, T.S.T. Transcriptional regulation of retinol binding protein 4 by Interleukin-6 via peroxisome proliferator-activated receptor α and CCAAT/Enhancer binding proteins. *Mol. Cell. Endocrinol.* **2020**, *505*, 110702. [CrossRef]
51. Shimizu, H.; Negishi, M.; Shimomura, Y.; Mori, M. Changes in urinary retinol binding protein excretion and other indices of renal tubular damage in patients with non-insulin dependent diabetes. *Diabetes Res. Clin. Pract.* **1992**, *18*, 207–210. [CrossRef]
52. Hosaka, B.; Park, S.I.; Felipe, C.R.; Garcia, R.G.; Machado, P.G.P.; Pereira, A.B.; Tedesco-Silva, H.; Medina-Pestana, J.O. Predictive value of urinary retinol binding protein for graft dysfunction after kidney transplantation. *Transplant. Proc.* **2003**, *35*, 1341–1343. [CrossRef]
53. Hung, Y.C.; Huang, G.S.; Lin, L.W.; Hong, M.Y.; Se, P.S. Thea sinensis melanin prevents cisplatin-induced nephrotoxicity in mice. *Food Chem. Toxicol.* **2007**, *45*, 1123–1130. [CrossRef] [PubMed]
54. Bouchal, P.; Jarkovsky, J.; Hrazdilova, K.; Dvorakova, M.; Struharova, I.; Hernychova, L.; Damborsky, J.; Sova, P.; Vojtesek, B. The new platinum-based anticancer agent LA-12 induces retinol binding protein 4 in vivo. *Proteome Sci.* **2011**, *9*, 68. [CrossRef]
55. Kong, S.; Ruan, J.; Zhang, K.; Hu, B.; Cheng, Y.; Zhang, Y.; Yang, S.; Li, K. Kill two birds with one stone: Making multi-transgenic pre-diabetes mouse models through insulin resistance and pancreatic apoptosis pathogenesis. *PeerJ* **2018**, *2018*, e4542. [CrossRef]
56. Denslow, A.; Świtalska, M.; Jarosz, J.; Papiernik, D.; Porshneva, K.; Nowak, M.; Wietrzyk, J. Clopidogrel in a combined therapy with anticancer drugs—Effect on tumor growth, metastasis, and treatment toxicity: Studies in animal models. *PLoS ONE* **2017**, *12*, e0188740. [CrossRef]
57. Wietrzyk, J.; Chodyński, M.; Fitak, H.; Wojdat, E.; Kutner, A.; Opolski, A. Antitumor properties of diastereomeric and geometric analogs of vitamin D3. *Anticancer. Drugs* **2007**, *18*, 447–457. [CrossRef] [PubMed]
58. Nevozhay, D. Cheburator software for automatically calculating drug inhibitory concentrations from in vitro screening assays. *PLoS ONE* **2014**, *9*, e106186. [CrossRef] [PubMed]

© 2020 by the authors. Licensee MDPI, Basel, Switzerland. This article is an open access article distributed under the terms and conditions of the Creative Commons Attribution (CC BY) license (http://creativecommons.org/licenses/by/4.0/).

Review

Role of bFGF in Acquired Resistance upon Anti-VEGF Therapy in Cancer

Fatema Tuz Zahra, Md. Sanaullah Sajib and Constantinos M. Mikelis *

Department of Pharmaceutical Sciences, School of Pharmacy, Texas Tech University Health Sciences Center, Amarillo, TX 79106, USA; fatema.zahra@ttuhsc.edu (F.T.Z.); s.sajib@ttuhsc.edu (M.S.S.)
* Correspondence: constantinos.mikelis@ttuhsc.edu; Tel.: +1-806-414-9242; Fax: +1-806-356-4770

Simple Summary: Anti-angiogenic therapies targeting the vascular endothelial growth factor (VEGF) signaling are established in the arsenal of cancer treatments. Despite the expectations, their benefits are temporary in cancer patients, partly due to the compensatory function of other angiogenic growth factors. This review focuses on the role of basic fibroblast growth factor (bFGF), one of the highly implicated players in the emergence of resistance to anti-angiogenic approaches. Here, we summarize data from various tumor types where bFGF is upregulated after anti-angiogenic treatment, the molecular mechanisms involved, and we highlight the current status and future perspectives of multi-target anti-angiogenic drugs for cancer.

Abstract: Anti-angiogenic approaches targeting the vascular endothelial growth factor (VEGF) signaling pathway have been a significant research focus during the past decades and are well established in clinical practice. Despite the expectations, their benefit is ephemeral in several diseases, including specific cancers. One of the most prominent side effects of the current, VEGF-based, anti-angiogenic treatments remains the development of resistance, mostly due to the upregulation and compensatory mechanisms of other growth factors, with the basic fibroblast growth factor (bFGF) being at the top of the list. Over the past decade, several anti-angiogenic approaches targeting simultaneously different growth factors and their signaling pathways have been developed and some have reached the clinical practice. In the present review, we summarize the knowledge regarding resistance mechanisms upon anti-angiogenic treatment, mainly focusing on bFGF. We discuss its role in acquired resistance upon prolonged anti-angiogenic treatment in different tumor settings, outline the reported resistance mechanisms leading to bFGF upregulation, and summarize the efforts and outcome of combined anti-angiogenic approaches to date.

Keywords: bFGF; VEGF; angiogenesis; anti-angiogenic therapy; resistance; cancer

1. Introduction

Angiogenesis is the formation of new blood vessels from preexisting ones [1]. It is the outcome of a coordinated series of events, which takes place mostly during development and in certain occasions during adulthood. Angiogenic activity is controlled by a dynamic balance between growth factors and angiogenesis inhibitors. This balance is disrupted in a series of diseases, where dysregulated angiogenesis is primarily responsible or augments the progression of the disease [2]. Among the diseases where angiogenesis is abnormally increased, thus requiring pharmaceutical intervention, is cancer. Therapeutic endeavors against tumor angiogenesis are a field of intense scientific efforts since Judah Folkman's visionary observation and pioneering work in the 1970s [3]. The boost in the angiogenesis research field emerged a few years later with the isolation and identification of the two best-known growth factors, vascular endothelial growth factor (VEGF) [4–7] and basic fibroblast growth factor (bFGF or FGF2) [8], followed by the isolation of a series of heparin-binding growth factors shortly after [3]. To date, VEGF's isoforms and receptors have been the target for the majority of Food and Drug Administration-(FDA)-approved therapies for tumor angiogenesis blockade [9]. Current anti-angiogenic therapies targeting VEGF signaling pathways are classified as anti-VEGF monoclonal antibodies, VEGF-binding proteins, and VEGF receptor (VEGFR) tyrosine kinase inhibitors (TKIs) [10].

Targeting the tumor microenvironment has been considered an attractive approach for tumor therapy, because contrary to the very heterogeneous cancer cells, stromal cells are considered relatively homogeneous [11]. Preclinical studies with anti-VEGF approaches demonstrated promising results in tumor angiogenesis and permeability inhibition [12,13]. Shortly after, clinical trials with the anti-VEGF monoclonal antibody Bevacizumab as monotherapy or combination therapy were initiated, highlighting the benefit of anti-angiogenesis therapy as cancer treatment for many malignancies [14]. However, in most cases this benefit was assessed in terms of disease-free survival and not overall survival. Thus, with the exception of some indications, such as metastatic colorectal cancer, the final outcome of clinical trials has not met the expectations [15–17]. Bevacizumab was FDA-approved in February 2004 as a first line treatment for patients with metastatic carcinoma of the colon and rectum (CRC) in combination with 5-fluorouracil-based chemotherapy, and in 2006, it was approved as a second line treatment for patients with advanced or metastatic CRC after irinotecan with 5-fluorouracil-based chemotherapy [18]. To date, more than ten anti-angiogenic drugs, antibodies or tyrosine kinase inhibitors have been FDA-approved for the treatment of a variety of cancers including glioblastoma, lung, colorectal, renal and breast cancers [19]. However, despite the increasing number of anti-angiogenesis inhibitors and the several years of clinical experience since the approval of bevacizumab, the response to anti-VEGF therapies is still moderate and not outstanding. The reason is the ephemeral effects of anti-angiogenic drugs with limited prolongation of overall survival, which is only seen in some cancers [9].

There are several potential variants for the poor outcome of anti-angiogenic therapies in clinical practice, such as the stage of the primary tumor, the level of vessel maturation, differential VEGF expression, differentiated anti-angiogenic drug efficacy in the presence of chemotherapy and the differential genetic identity of tumor endothelial cells, to name a few [9,16]. Apart from the VEGF family, several other growth factors either mediate distinct functions of the angiogenic process or act synergistically [2]. One of the major reasons for the limited outcome of anti-angiogenic therapies is "evasive resistance", which refers to the alternative pathways that are activated upon the blockade of a specific angiogenesis pathway [20]. The outcome of evasive resistance, where the specific anti-angiogenic target remains inhibited, is adaptive response, which differs from the traditional drug resistance or intrinsic non-responsiveness, the other resistance mechanism, where the inhibition of the anti-angiogenic target is not achieved due to mutational alteration of the target or alterations in drug uptake and efflux [21].

Resistance to the VEGF/VEGFR signaling inhibitors has been attributed to the activation of alternative pro-angiogenic signaling pathways in the tumor or tumor microenvironment. A variety of other cell types, such as bone marrow-derived cells, fibroblasts and monocytes express a plethora of alternative angiogenic factors such as basic fibroblast growth factor (bFGF), angiopoietins, platelet-derived growth factor (PDGF) and epidermal growth factor (EGF), which can substitute for VEGF. Among these alternative growth factors, bFGF has been widely considered a major player in anti-angiogenic tumor resistance mechanisms, with other growth factors to follow [11,16,21]. In this review, we will discuss the role, preclinical, clinical evidence and molecular pathways triggered by bFGF-driven resistance to anti-VEGF therapy.

2. Basic Fibroblast Growth Factor (bFGF): A Pro-Angiogenic Growth Factor

The FGF family in mammals consists of 18 secreted glycoproteins [22], which signal through the FGF receptors (FGFRs). The FGFRs comprise four transmembrane receptor tyrosine kinases FGFR1, FGFR2, FGFR3 and FGFR4 which get auto-phosphorylated upon the binding of FGF members on different types of cells [23,24]. The extracellular domain of the FGFRs contains three immunoglobulin-like (Ig-like) domains, which present structural variability and thus ligand binding specificity due to alternative splicing [25,26]. The role of the FGF/FGFR family during development and adulthood is pivotal. During development it regulates mesoderm patterning and organogenesis [27,28] and in adults it regulates angiogenesis-related functions, such as wound healing [22]. Gain- or loss-of-function mutations of the FGFR family are driving forces of several pathological conditions, highlighting them as targets for pharmaceutical intervention [22]. In cancer, the FGF/FGFR family regulates cellular proliferation, differentiation, apoptosis, angiogenesis and inflammation through different mechanisms, including aberrant expression, mutation and gene amplification [29–32]. The classical FGF signaling can be transduced by RAS/MAPK, PI3K/Akt, Src tyrosine kinase and STAT pathways, which consist targets of current anti-cancer approaches [29,32,33].

Among the FGF family members, bFGF constitutes the prototypic and best characterized pro-angiogenic factor. The expression of bFGF is increased at sites of chronic inflammation [34–36], after tissue injury [37], and in different types of human cancers [38]. Among the members of the FGF1 subfamily, FGF1 can bind all FGFRs whereas bFGF has preference to the c isoforms of FGFR1, FGFR2 and FGFR3 [39,40]. Among the FGFRs, FGFR1, FGFR3 and less frequently FGFR2 are found in endothelial cells (ECs) with minimal or no expression of FGFR4 [26,41]. Upon binding with its receptors on ECs, bFGF can directly promote angiogenesis in vitro and in vivo [22,42,43]. In vivo, bFGF is able to induce neovascularization in a variety of animal models, such as the chick embryo chorioallantoic membrane (CAM) assay, the rodent cornea assay, the subcutaneous matrigel plug assay in mice, and the zebrafish yolk membrane assay [38,44]. bFGF can act on endothelial cells via a paracrine mode of action released by tumor stromal and inflammatory cells and/or by mobilization from the extracellular matrix (ECM). On the other hand, bFGF can also be produced endogenously by ECs and induce angiogenesis via autocrine, intracrine or paracrine manners [38,45]. However, bFGF deficiency, double FGF1 and bFGF deficiency, as well as bFGF overexpression did not lead to lethality due to vascular defects, which can be explained by the presence of compensatory mechanisms in the vascular system [22,38].

Several studies have confirmed the integration of angiogenesis and inflammation in a number of physiological and pathological conditions, including cancer [46–49]. bFGF-mediated angiogenesis can be promoted by inflammation [50]. Inflammatory cells can express bFGF and inflammatory mediators can activate the endothelium to synthesize and release bFGF, which in turn stimulates angiogenesis through an autocrine manner. The inflammatory response can also increase bFGF production and release by causing cell damage, fluid and plasma protein exudation, and hypoxia [51,52]. On the other hand, bFGF can amplify the inflammatory and angiogenic response by interacting with endothelial cells. Gene expression profiling has revealed a pro-inflammatory signature of bFGF-stimulated murine microvascular endothelial cells characterized by the up-regulation of pro-inflammatory cytokines/chemokines and their receptors, endothelial cell adhesion molecules, and members of the eicosanoid pathway [51]. Macrophages are a source of bFGF and express FGFRs. Monocytes/macrophages play a functional, non-redundant role in bFGF-mediated angiogenesis revealed from early recruitment of mononuclear phagocytes preceding blood vessel formation in bFGF-driven angiogenesis in the matrigel plug assay, while in tumors, increased bFGF regulates macrophage polarization [51,53]. Apart from the pro-inflammatory signature, bFGF also contributes to the increased expression of a variety of pro-angiogenic growth factors in the endothelium, including itself, VEGF and angiopoietin-2 (Ang2) [51,54–56]. Overall, bFGF contributes to the modulation of the neovascularization process triggered by growth factors via activating an autocrine loop of amplification of the angiogenic response and by paracrine activity exerted by endothelium-derived cytokines/chemokines on inflammatory cells [57].

3. bFGF in Cancer: A Prominent Resistance Mechanism upon Anti-Angiogenic Therapy

Targeting tumor-induced angiogenesis has mostly focused on the VEGF signaling pathway, and was implemented more than 15 years ago with the introduction of bevacizumab, a humanized, recombinant monoclonal antibody against VEGF-A [58]. By binding to circulating, soluble VEGF-A, bevacizumab inhibits its interaction with VEGFR2 and the activation of the downstream signaling pathways. Thus, it provides anti-tumor effectiveness by inhibiting angiogenesis and microvascular density, inducing the regression of newly formed vessels. An important and more recent goal of antiangiogenic therapies is vascular normalization. Normalizing the tumor vasculature renders the tumor susceptible for anti-cancer therapy or immunotherapy [59,60]. Despite the encouraging preclinical data for anti-VEGF therapy and the clinical success in other angiogenesis-related pathologies, such as age-related macular degeneration [61], the clinical outcome in cancer treatments did not meet the expectations. Bevacizumab has been approved since 2004 and is currently marketed in 134 countries worldwide for a number of solid tumors [60], thus there is an increasing number of studies denoting the upregulation of bFGF as an important resistance mechanism, contributing to the ephemeral nature of anti-angiogenic results, important examples of which we highlight below and are summarized in Table 1.

Table 1. Summary of clinical, preclinical and in vitro tumor studies demonstrating that anti-angiogenic inhibition induced basic fibroblast growth factor (bFGF) expression. The cancer type, anti-angiogenic treatment, effect in bFGF expression and observed outcomes of each study are presented. CD31: cluster of differentiation 31; SMA: smooth muscle actin; FGFR: FGF receptors; MMPs: matrix metalloproteinases; SPARC: secreted protein acidic and rich in cysteine; TIMPs: tissue inhibitors of metalloproteinases; PDGF: platelet-derived growth factor; VEGF: vascular endothelial growth factor; RIP-Tag2: rat insulin promoter-1 driven viral SV40 large T-antigen; HUVEC: human umbilical vein endothelial cells; PDGFR: platelet-derived growth factor receptor.

Cancer Type	Model Used	Treatment	Effect on bFGF	Observed Outcomes	References
Glioblastoma	Clinical	Bevacizumab	↑ bFGF in pericytes, endothelial and tumor cells	↓ Vessel density/no difference ↑ CD31(-)/SMA(+) pericytes ↑ MMPs ↑ VEGFR1 ↓ Akt	[62,63]
	Preclinical (U87)	Bevacizumab	↑ bFGF after 7 weeks	↑ Vascularity, cell proliferation ↑ HIF-2a, CA IX	[64]
	In vitro	Bevacizumab	↑ bFGF in U87 and NCS23 tumor cells	↑ Cell invasion ↑ MMP-2, MMP-9, MMP-12 ↑ Collagen IV, CXCL9 ↑ SPARC, TIMPS ↓ Laminin, integrin β_2, MMP-1	[64]
Head and neck squamous cell carcinoma	Preclinical (Tu138)	Bevacizumab	↑ bFGF, FGFR1-3	- Sustained angiogenesis ↑ PLCg2, FZD4, CX3CL1 ↑ ERK ↓ Endothelial apoptosis	[30]
Gastric cancer	Clinical/Preclinical (MKN45)/In vitro	Pazopanib	↓ FGFRP1 (in vitro)	↑ TWIST ↑ CYP2C19, TFF3, PLA2G2A ↓ EGLN2, MIR590, ↓ LCN2, TET1 ↑ Mesenchymal phenotype	[65]
	Preclinical (GXF97, MKN-45, MKN-28, 4-1ST, SC-08-JCK, SC-09-JCK, SCH, SC-10-JCK, NCI-N87)	Bevacizumab	↑ bFGF in bevacizumab-resistant tumor cells	↑ Vessel density ↑ Tumor volume	[66]

Table 1. Cont.

Cancer Type	Model Used	Treatment	Effect on bFGF	Observed Outcomes	References
Colorectal carcinoma	Clinical	Bevacizumab, fluorouracil, leucovorin, irinotecan (FLORFIRI+B)	↑ Plasma bFGF levels	↑ Resistance	[67]
	In vitro	VEGF RNAiBevacizumab	↑ bFGF in endothelial cells from colon tumors	↑ ANG1	[68]
Pancreatic cancer	Preclinical (RIP-Tag2 model)	VEGFR2-blocking antibodies	↑ bFGF in endothelial and tumor cells	↓ Vessel density ↑ Tumor hypoxia, HIF-1α ↑ FGF1, ANG1 ↑ EphA1, EphA2	[69]
Liver cancer	Preclinical (H22)/ In vitro (HUVEC, HEPG2)	Sorafenib	Potential bFGF increase (higher lenvatinib efficacy)	↑ PD1, CTLA-4, Tim-3 ↑ PD-L1 expression	[70]
Renal cell carcinoma	Clinical	Sunitinib	↑ Plasma bFGF levels	↑ HGF, IL-6, IL-8 ↑ PDGF1, ANG1	[71,72]
	In vitro (HUVEC)	Sunitinib	↑ bFGF efficacy, FGFR activation	↑ Angiogenesis	[73]
Breast cancer	Preclinical (E0771, MCaIV)	Anti-VEGF antibody	↑ bFGF in adipocyte-rich tumor periphery ↑ bFGF in cancer-associated fibroblasts	↑ IL-6, IL-12, CXCL1, TNFα ↓ Tumor vasculature ↑ Hypoxia	[74]
	Preclinical (T-47D)	Tet-regulated VEGF expression	↑ bFGF	↑ Tumor growth	[75]
Cervical carcinoma	Preclinical	Imatinib	↓ bFGF in cancer-associated fibroblasts	↓ PDGFR ↓ Angiogenesis ↓ Epithelial proliferation	[76]
Prostate cancer	Clinical	VEGF inhibitors	↑ FGF-FGFR in tumors	↑ Angiogenic pathways	[10,77]

3.1. Glioblastoma

Bevacizumab in combination with temozolomide has been approved for newly diagnosed and recurrent malignant glioma in the United States and other countries and provides the clinically meaningful prolongation of progression-free survival (PFS) and non-detrimental increase in overall survival (OS) [60,78]. In a case study, this treatment led to dramatic but transient tumor reduction, and tumor analysis upon recurrence demonstrated VEGF signaling blockade but upregulation of matrix metalloproteinases (MMPs) and sustained p44/42 phosphorylation, denoting the activation of compensatory mechanisms [62]. Immunohistochemical staining in four autopsied malignant gliomas showed increased proliferation in CD31(-)/SMA(+) pericytes around tumor vessels after bevacizumab treatment and no significant changes in the number of tumor vessels in initial and autopsied tumor vessels before and after bevacizumab administration. VEGF-A was present in all tumors at the initial surgery, but its expression was reduced after bevacizumab administration. Interestingly, bFGF and PDGF expression was increased in the endothelial cells, pericytes and tumor cells upon bevacizumab treatment, indicating that the inhibition of VEGF alone is not sufficient to maintain the inhibition of neovascularization due to resistance by bFGF and pericyte coverage by PDGF. The molecular mechanism of bFGF upregulation upon bevacizumab treatment, although not delineated, was speculated to be a result of negative feedback due to the continuous inhibition of the VEGF-driven angiogenic pathway [63].

In vitro, although bevacizumab was capable of sequestering the majority of the autocrine secretion of the highly VEGF-expressing U87 glioblastoma and NCS23 glioma stem cells, it induced invasion in a concentration dependent manner [64]. Moreover, it led to bFGF mRNA and protein upregulation in vitro and in vivo, which indicates the potential of glioblastoma cells to escape from antiangiogenic treatment. Consistent with this phenotype, further upregulation of invasion-related proteins, such as matrix metalloproteinases (MMP-2, MMP-9, MMP-12), secreted protein acidic and rich in cysteine (SPARC) and tissue inhibitors of metalloproteinases (TIMPs), allowed the cancer cells to invade into surrounding brain areas in the in vivo glioblastoma xenograft model. The upregulation of bFGF in the glioblastoma xenograft model was further responsible for the rapid increase in vascularity and cellular proliferation, denoting resistance development after the long-term antiangiogenic treatment. Mechanistically, bFGF upregulation was hypoxia-driven, since the hypoxia markers hypoxia-inducible factor 2a (HIF-2a) and carbonic anhydrase IX (CA IX) were also increased. In the U87 xenograft model, after short term (4 weeks) VEGF blockade, bFGF levels were not increased and microvessel density was significantly reduced, but as VEGF blockade continued (7 weeks) bFGF levels increased, similar to the in vitro study, along with microvessel density and tumor cell proliferation, indicating the reactivation of angiogenesis [64].

3.2. Head and Neck Squamous Cell Carcinoma (HNSCC)

bFGF upregulation appears to be an important resistance mechanism upon bevacizumab treatment in head and neck squamous cell carcinoma (HNSCC). Through an HNSCC xenograft model of acquired resistance to bevacizumab, it was demonstrated that bevacizumab-resistant tumors maintained angiogenesis and prevented endothelial apoptosis, despite the sequestration of VEGF. Whole genome microarray analysis revealed the upregulation of angiogenesis-related genes including bFGF, FGFR1-3, PLCg2, FZD4, CX3CL1 and CCL5 in the bevacizumab-resistant tumor cells. The fact that bevacizumab led to the overexpression of several members of the FGF/FGFR family, including bFGF and FGFR1-3, as well as the activation of downstream signaling effectors including PLCg1, PLCg2, AKT and ERK, strengthens the involvement of the FGF axis in bevacizumab-associated resistance in the HNSCC xenograft model. Co-targeting of the VEGF and FGF pathways led to the restoration of sensitivity to anti-VEGF therapy in bevacizumab-resistant tumors, demonstrating that the upregulation of FGF/FGFR autocrine signaling plays a crucial role in circumventing VEGF inhibition in bevacizumab-resistant tumor cells [30].

3.3. Gastric Cancer

In human gastric cancer xenograft models, bFGF expression was proposed as a biomarker for antitumor activity of bevacizumab. Refractory to bevacizumab treatment models presented high bFGF levels and the VEGF/bFGF ratio provided a more accurate correlation of sensitivity to bevacizumab, than VEGF expression itself [66]. Irrespective from its role in the vascular system, the deregulation of the FGFR pathway, through point mutations, gene fusions or ligand overexpression, has been recently considered an oncogenic driver for gastrointestinal stromal tumors [79]. It was recently reported that the higher response of MKN45 than SNU5 gastric cancer cells to Pazopanib, a tyrosine kinase inhibitor that targets VEGFR1-3, PDGFRα,β, c-KIT, FGFR1-4 and CSF1R, was due to the higher FGFR2 and FGFR3 expression. The sensitivity of MKN45 cells was higher in the in vivo compared to the in vitro settings, which was attributed to the lower expression of FGF-binding protein 1 (FGFBP1) in the in vitro setting. FGFBP1 mediates the release of bFGF from the extracellular matrix, thus highlighting the FGF signaling as an important mediator for pazopanib treatment. Although the MKN45 xenografts were initially responsive to pazopanib, they later transitioned to a mesenchymal-like phenotype, becoming more invasive and developing resistance, which led to tumor regrowth after drug withdrawal [65].

3.4. Colorectal Carcinoma

The stimulating role of bFGF on colorectal carcinoma cell invasion is long established [80]. Cytokine analysis in metastatic colorectal cancer patients undergoing a phase

II [67] clinical trial of bevacizumab and FLORFIRI+B treatment regimen revealed an increment of bFGF levels in the plasma of a subset of patient population during the emergence of resistance. The FLORFIRI+B regimen contained bevacizumab, irinotecan, bolus fluorouracil and leucovorin, followed by infusion of fluorouracil. Although the mean bFGF levels decreased after one cycle of FLORFIRI+B, they increased before and at the time of disease progression [67], indicating the participation of bFGF in resistance mechanisms. VEGF downregulation in endothelial cells isolated from tumors of colon cancer patients led to significant bFGF upregulation, further highlighting the impact of the tumor vascular endothelium in bFGF-dependent compensatory mechanisms [68].

3.5. Pancreatic Cancer

In a murine model of islet cell carcinogenesis, qRT-PCR analysis from total tumor mRNA revealed the upregulation of several FGF members, including bFGF, upon VEGFR2-blocking treatment, which was further confirmed by ELISA. Although bFGF was upregulated both in tumor cells and tumor endothelial cells, expression of FGFR1 and FGFR2 was not affected in this model. The trigger for bFGF upregulation was the increased levels of tumor hypoxia after VEGFR2 inhibition, which was also confirmed in the RIP-Tag2 tumor-derived βTC3 cell line under hypoxia in vitro. In the same model, and contrary to the in vivo data, the FGF1 levels remained unaffected [69].

3.6. Liver Cancer

Hepatocellular carcinoma is the most common type of liver cancer and occurs frequently in patients with liver cirrhosis or chronic liver diseases. Anti-VEGF treatment increases survival and is the standard-of-care for hepatocellular carcinoma (HCC), with sorafenib (VEGFR2, PDGFR, Raf1 inhibitor), lenvatinib (VEGFR, FGFR, c-Kit and RET inhibitor) and regorafenib (VEGFR2, Tie2 inhibitor) being common treatments [81,82]. The plasma levels of VEGF and bFGF in hepatocellular carcinoma patients are increased with the progression of the disease, upregulating PD-1 expression and inducing immune suppression [70].

Tumor vessel normalization, a major goal of anti-angiogenic treatments, was achieved in liver cancer with the combined inhibition of VEGFR and FGFR pathways. An elegant study demonstrated that combined VEGFR and FGFR inhibition potentiated the efficacy of anti-PD-1 treatment, inducing vessel normalization and antitumor efficacy [70].

3.7. Renal Cell Carcinoma

Renal cell carcinoma (RCC) is a highly vascularized tumor, thus tumor angiogenesis plays a critical role in the development of metastatic RCC. Several anti-angiogenic drugs have been approved for RCC treatment in the United States, including bevacizumab, sunitinib, pazopanib and sorafenib [19,83,84]. While angiogenesis targeting via VEGF blockade is the standard of care in metastatic RCC, around 20% of the patients do not respond to the treatment. For the rest, although they gain initial benefits from anti-angiogenic therapy, they eventually develop resistance between 6 and 15 months of treatment, which is attributed to revascularization, driven by the tumor microenvironment [85]. Sunitinib treatment of RCC patients led to an increase in serum bFGF levels, irrespective of the treatment outcome, although patients with no response to sunitinib presented higher bFGF levels than the ones with a temporary clinical benefit or a better response [71]. These data are consistent with previous clinical findings demonstrating that bFGF is responsible for sunitinib resistance, indicating the necessity of targeting both VEGF and bFGF pathways simultaneously [72,73]. Patients under anti-VEGF therapy can still present beneficial outcome by a multi-kinase inhibitor, such as sorafenib. When sunitinib-resistant patients were treated with sorafenib the overall survival was improved, revealing both the importance of the proper timing and order of each targeted approach [72,86].

3.8. Breast Cancer

In breast cancer cells, the role of VEGF is indispensable for the initial tumor growth, but bFGF upregulation can compensate for the VEGF downregulation at later stages. This was elegantly demonstrated by Tet-regulated VEGF expression in the T-47D breast cancer cells. VEGF downregulation was detrimental for tumor inoculation or early tumor growth, however, upon VEGF suppression at later stages, bFGF expression was upregulated without affecting tumor growth. bFGF was not detectable in tumors of the same size overexpressing VEGF during the entire experimental period [75].

Tumor growth directly depends on the tumor microenvironment, and obesity, as a systemic condition associated with hypoxic adipose tissues, affects the tumor microenvironment, regulating tumor growth and outcome of anti-cancer therapeutic approaches. It was recently shown that the plasma concentration of bFGF is higher in obese breast cancer patients. Adipose tissue size inversely correlated with vascular density and bFGF overexpression was particularly abundant in adipose-rich tissues based on the immunohistochemical observation of human breast tumor samples from obese patients. Additionally, obesity has been inversely correlated with the response to anti-VEGF treatment. Similarly, the baseline bFGF levels were higher in untreated obese compared to untreated lean mice and anti-VEGF treatment increased them further. bFGF overexpression was identified in the adipocyte-rich tumor periphery and in activated cancer-associated fibroblasts, which is consistent with bFGF localization in adipocyte-rich human breast cancer. In two syngeneic breast cancer tumor models, it was demonstrated that tumors were less vascularized and more hypoxic and anti-VEGF therapy was less potent in reducing vessel density in obese, compared to lean mice. FGF receptor blockade with AZD4547, a pan-FGFR inhibitor, improved tumor responsiveness to anti-VEGF treatment in obese mice, not in lean mice, but showed toxicity [74]. Instead, metformin, a safe and popular anti-diabetic drug, previously shown to reduce cellular bFGF expression and with anti-cancer effect in obese settings [87,88], reduced vessel density and re-sensitized to anti-VEGF therapy in obese mice. Mechanistically, metformin treatment reduced bFGF mRNA and protein expression and inhibited bFGF downstream signaling pathways, such as AKT, S6, ERK and STAT3 [74].

3.9. Cervical Carcinoma

The role of pericytes is equally important to the one of endothelial cells in angiogenesis, as they provide survival signaling to endothelial cells and play an important functional role in mediating blood flow and endothelial cell permeability [89]. Similarly, in tumors, the inhibition of VEGF signaling leads to the reduction in immature (without pericyte coverage) tumor microvasculature with an increase in the percentage of vessels with pericyte coverage (mature vessels) [89,90]. The PDGF/PDGFR signaling is the predominant mediator of pericyte migration and proliferation [91]. bFGF shared the same expression pattern with the PDGF receptor in stromal fibroblasts in a genetically engineered model of cervical carcinogenesis and their expression was increased in cancer-associated fibroblasts (CAFs), but not in other cell types. Moreover, bFGF was demonstrated to be a downstream effector in PDGF signaling, as its expression was decreased upon treatment with the selective PDGFR inhibitor imatinib in cervical carcinoma [76]. Therefore, bFGF plays a key regulating role in PDGF-induced angiogenesis and in acquired resistance induced by VEGF-targeted therapy [76,89].

3.10. Prostate Cancer

Prostate cancer is considered one of the resistant cancers to anti-angiogenic treatments and one of the reported reasons is the involvement of the FGF-FGFR family in transformation and angiogenesis [10]. VEGF overexpression and microvessel density have been associated with tumor growth, poor prognosis and increased metastatic potential. In phase II clinical trials of castration-resistant advanced prostate cancer, anti-angiogenic therapy improved relapsed-free survival and led to disease stabilization, whereas in phase III trials, no significant outcome was identified in terms of overall survival. Instead, anti-angiogenic treatment caused increased toxicity and greater incidence of treatment-related death [92]. The trials included anti-VEGF antibodies, such as bevacizumab, decoy receptors, such as aflibercept, as well as tyrosine kinase inhibitors, such as sunitinib (targeting VEGFR2, PDGFRβ, c-Kit and RET) [77,92,93]. bFGF, along with interleukin-6 (IL-6) are known contributors of androgen ablation, chemotherapy resistance and metastatic dissemination of prostate cancer cells. It was further shown that bFGF triggers IL-6 and prostate-specific membrane antigen (PSMA) expression, markers of chronic inflammation and prostate cancer prostate cancer progression in advanced stages [94]. Finally, the high FGFR expression levels of the prostate cancer cells have been taken into consideration for the design of tumor cell- and tumor endothelial cell-specific liposomes for improved doxorubicin delivery, with promising results [95,96].

4. Mechanisms of bFGF Release or Upregulation with Angiogenic Potential

The origin and release mechanisms of the bFGF pool driving angiogenesis has been well-reported (Figure 1). Most of the endothelial-synthesized bFGF remains cell-associated, however, a portion of the bFGF pool is sequestered in the subendothelial extracellular matrix (ECM) for deposit [97]. Like the other FGFs, bFGF is a heparin-binding molecule, bound to heparan sulfate, which constitutes more than 90% of the subendothelial ECM glycosaminoglycan side chains and serves as a sink to concentrate and stabilize bFGF, protecting it from degradative enzymes. The endothelial cells also synthesize secreted or cell membrane- and extracellular matrix (ECM)-associated heparan sulfate proteoglycans (HSPGs) [98,99]. HSPGs and heparan sulfate protect bFGF from thermal denaturation and proteolytic degradation and further modulate bFGF activity. Moreover, bFGF binding to HSPGs serves as a reservoir, from which FGFs can be released in response to specific triggering events [98,100,101]. The interaction of bFGF with HSPG modulates FGF activity by increasing its receptor binding affinity with the establishment of stable growth factor-receptor complexes and the facilitation of FGFR dimerization with subsequent activation [102,103]. Release from the ECM storage takes place after injury, mild perturbation of endothelial cells or release of proteases, further stimulating the autocrine proliferation of adjacent endothelial cells and leading to angiogenesis [97,104]. One of the drivers of diseases characterized by aberrant angiogenesis, such as choroidal neovascularization, is the increased ECM cleavage and subsequent release of bFGF [105]. During tumor-induced angiogenesis, the release of bFGF is partly regulated by the activity of tumor-derived heparan sulfate-degrading enzymes, which release bFGF in the capillary basement membrane [104,106].

One group of these enzymes are the MMPs, a family of soluble and membrane-anchored proteolytic enzymes which can degrade components of ECM. It is well-established that MMPs are important regulators of angiogenesis, as they break down matrix components and thus clear the path for migrating ECs during angiogenesis. Additionally, MMPs can also switch on angiogenesis by liberating matrix-bound bFGF [105,107,108]. MMP-2 expression has been correlated with the de novo formation of small capillaries in tumors. Bevacizumab treatment led to increased expression and enzymatic activity of MMP-2 and MMP-9, common metalloproteinases associated with neovascularization of tumors, in glioblastoma cells both in vitro and in vivo [64,109]. Bevacizumab treatment also resulted in the upregulation of bFGF and of the MMP inhibitors TIMP-1 and TIMP-2, as a potential response to MMP upregulation in U87 and NSC23 glioblastoma cells, suggesting that tumors can overcome anti-VEGF treatment via the release of bFGFs from ECM with the help of MMPs, supporting an autocrine pattern of bFGF signal transduction that results in neovascularization [64].

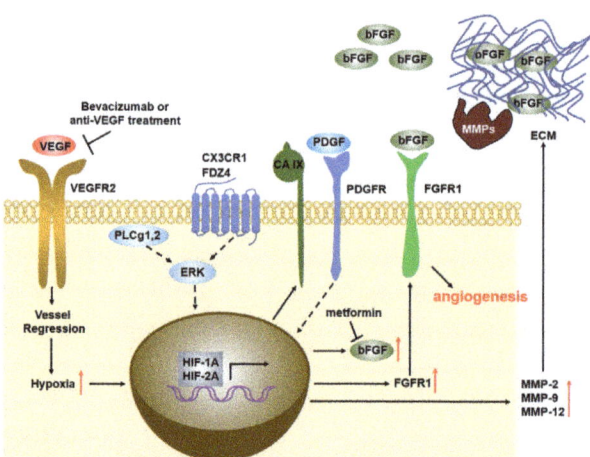

Figure 1. Pathways of bFGF-induced compensation upon anti-VEGF treatment. Bevacizumab or anti-VEGF treatment leads to vascular regression, inducing hypoxia in the surrounding tissues. Hypoxia drives the expression of carbonic anhydrase IX and activates HIF-1A and HIF-2A, increasing bFGF levels. Metformin treatment blocks bFGF mRNA and protein levels. A similar increase in bFGF levels is achieved upon anti-VEGF treatment in cancer cells, via the upregulation of PLCg1,2, FDZ4 and CX3CL1 (ligand of CX3CR1) with a subsequent ERK activation. PDGFR activation in smooth muscle cells leads to FGFR1 expression. HIF-2A activation induces the expression of MMP-2, -9 and -12, releasing bFGF molecules via extracellular matrix (ECM) degradation.

There is an intimate, but well-characterized crosstalk between bFGF and the different members of the VEGF family during angiogenesis, lymphangiogenesis and vasculogenesis. Among the members of the VEGF family, VEGF-A/VEGFR2 appears to play a major role in blood vessel angiogenesis and VEGF-C and VEGF-D are involved in lymphangiogenesis by interacting with VEGFR3 [38,110,111]. Previous studies have the reported synergistic and complementary activity of bFGF with VEGF and PDGF-BB [112–114]. bFGF upregulates PDGFR expression to increase the responsiveness to PDGF-BB in endothelial cells and PDGF-BB-treated vascular smooth muscle cells may contribute to the increased responsiveness to bFGF by upregulating FGFR1 expression [114,115]. In turn, bFGF can also contribute to the increased expression of other proangiogenic factors, highlighting the complex compensatory mechanisms that regulate angiogenic processes and contributing to resistance upon anti-VEGF treatment [11,69].

The most common and widely accepted mechanism of bFGF upregulation upon VEGF inhibition is related to the induction of tumor hypoxia. Antiangiogenic therapy in different tumor types induces the elevation of hypoxia markers HIF-1A, HIF-2A and CA IX, followed by increased bFGF expression [10,64]. In bevacizumab-resistant HNSCCs, bFGF upregulation was mediated by ERK, which was induced due to higher expression of upstream activator genes including phospholipase C (PLCg2), frizzled receptor-4 (FDZ4), chemokine C-X3-C motif (CX3CL1), and chemokine C-C motif ligand 5 (CCL5). This was confirmed by the decreased activation of ERK and the corresponding decrease in bFGF levels upon the downregulation of each of these genes [30].

5. Targeting Anti-VEGF Resistance: Combinatorial Therapies

As bFGF is a prominent factor in anti-VEGF therapy resistance, experimental evidence suggests that targeting bFGF in addition to VEGF may provide synergistic outcome and prove beneficial for the treatment of angiogenesis-related diseases, including cancer. Different chemical structures and mechanisms of action of several bFGF inhibitors have been described (Figure 2). One soluble pattern recognition receptor long-pentraxin-3 (PTX3), which binds with bFGF with high affinity and specificity, has been shown to antagonize bFGF activity. This interaction leads to the inhibition of the angiogenic activity of bFGF, as it can no longer bind with FGFRs, ultimately blocking bFGF-mediated tumor angiogenesis and growth. PTX3 has a unique N-terminal extension which has been identified as a bFGF binding domain. PTX3-derived synthetic peptides have shown significant anti-angiogenic activity in vitro and in vivo, with potential implications in cancer therapy [57].

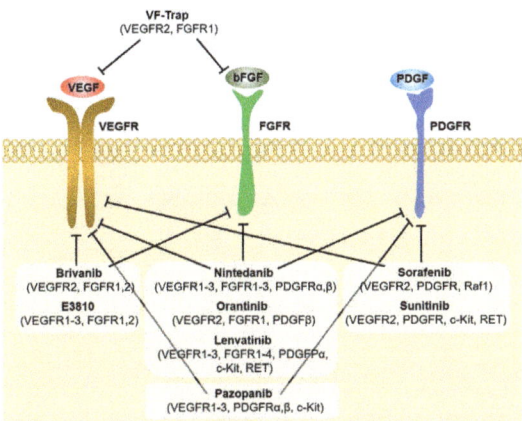

Figure 2. Inhibitors targeting combination of angiogenesis pathways either blocking ligand interaction (VF-Trap) or downstream signaling pathways. The parenthesis below each inhibitor highlight the growth factor receptors targeted by each inhibitor.

Growing evidence suggests that the dual inhibition of VEGFR and FGFR in preclinical models can overcome anti-VEGF therapy resistance. Pancreatic islet carcinogenesis was one of the first models where the FGF family of ligands was identified to be among the primary resistance mechanisms [69,89]. Treatment with an anti-VEGFR2-blocking monoclonal antibody decreased the vascular density after 10 days in the RIP-Tag2 mouse model of islet cell carcinogenesis. However, an angiogenic rebound in tumors at 4 weeks of treatment was noted, which was associated with an increase in bFGF expression. The concomitant blockade of VEGF signaling with the VEGFR2-blocking monoclonal antibody and FGF signaling by adenovirus-delivered soluble form of FGFR2 (FGF-trap) significantly reduced tumor burden and vessel density compared to the anti-VEGFR2 alone [69]. Co-targeting VEGF by bevacizumab and FGFRs by the small molecule inhibitor PD173074 abrogated tumor growth in the bevacizumab-resistant HNSCC xenograft model by inhibiting tumor angiogenesis [30].

A novel chimeric decoy receptor VF-Trap fusion protein that binds both VEGF and bFGF was developed by Li et al. to simultaneously block activity of both VEGF and bFGF pathways and achieve an additive anti-tumor effect. In vitro, VF-trap blocked VEGF- and bFGF-induced vascular endothelial cell proliferation and migration, while in vivo, combined VEGF and bFGF sequestration resulted in a significant inhibition of renal and lung xenograft tumor growth compared to the single VEGF inhibition [116].

The efforts for the combined blockade of VEGF and FGF pathways have led to the development of tyrosine kinase inhibitors, which unlike the antibodies, target the downstream signaling pathways of VEGF, FGF and other growth factors, with brivanib and E-3810 being characteristic examples. Brivanib is a tyrosine kinase inhibitor that targets VEGFR2, FGFR1 and FGFR2 [117,118]. In preclinical studies, brivanib administration demonstrated encouraging results in different cancer models, but it mostly led to tumor inhibition rather than tumor regression and its efficacy depended on endogenous bFGF expression [117,118]. In the clinical setting, brivanib in combination with standard chemotherapy and monoclonal antibodies demonstrated moderate and manageable side effects and provided encouraging results for hepatocellular carcinoma and colorectal cancer, increasing progression-free survival [118]. Unfortunately, in a phase III clinical trial for unresectable hepatocellular carcinoma, brivanib in combination with chemotherapy failed to improve overall survival. In fact, this multinational study was terminated earlier, when two other phase III studies with brivanib on advanced HCC patients failed to meet their overall survival objectives [119]. Interestingly, in a recent case report, brivanib demonstrated excellent antitumor efficacy for an HCC patient as second-line therapy, bringing up the possibility of a better clinical outcome of brivanib after HCC resection, with long-term treatment and the delayed onset of administration. Specifically, brivanib was administered as a monotherapy to a patient who had developed lung metastases one year after HCC resection, and after sorafenib treatment for three months failed to hinder disease progression. A period of 2.5 months after brivanib treatment, lung metastases decreased or disappeared and lymph node metastases decreased, a trend that continued at later evaluations. The total duration of brivanib treatment was 11 months due to grade 2 thrombocytopenia, but with tolerable side effects, and 4 months after the end of treatment the patient remained in good condition without signs of deterioration. This could suggest that brivanib may be more effective with long-term treatment in a delayed-onset fashion [120]. In a meta-analysis, the efficacy of brivanib in combination with cetuximab and chemotherapy was found to be better than the efficacy of the combination of cetuximab with chemotherapy or sorafenib with chemotherapy, although it presented toxicity. The superiority of this combination could be explained due the simultaneous inhibition of VEGF-induced angiogenesis in the endothelial cells with the EGFR signaling blockade in the tumor cells [121]. E3810 is a tyrosine kinase inhibitor that targets VEGFR1, VEGFR2, VEGFR3, FGFR1 and FGFR2 and retrieves responses in tumors that are not responsive to other small inhibitors, such as sunitinib. In preclinical studies, E3810 showed tumor regression and significantly delayed tumor growth, although tumors resumed their growth when treatment was suspended [122].

Sorafenib and sunitinib are other prominent members of this group, that have been tested, shown to increase progression free survival in a variety of cancers and are FDA-approved. Sorafenib targets VEGFR2, PDGFRβ and Raf1 kinase activity and sunitinib targets VEGFR2, PDGFRβ, c-Kit and RET [20]. Even these, however, have not increased the overall survival significantly [16,20]. Similarly, nintedanib, a tyrosine kinase inhibitor that blocks the VEGF, FGF and PDGF pathways has been approved for non-small-cell lung cancer and recently for idiopathic pulmonary fibrosis [123–125], while pazopanib, a small molecule multi-kinase inhibitor that blocks VEGF, FGF, PDGF pathways and c-Kit and has been approved for advanced soft-tissue sarcoma and renal cell carcinoma [126]. Orantinib (SU6668) is another small molecule inhibitor that binds and inhibits the phosphorylation of VEGFR2, FGFR1 and PDGFR-β, thus blocking the signal transduction of the corresponding ligands. In vivo, it inhibited the growth of glioma, melanoma, lung, colon, ovarian and epidermoid tumor xenografts and suppressed tumor angiogenesis, by inhibiting tumor endothelial cell survival directly (apoptosis of endothelial and tumor cells) or via inhibition of pericyte coverage [127,128]. In a phase III clinical trial of hepatocellular carcinoma, orantinib increased the time to progression but did not improve the overall survival [129]. Lenvatinib, a VEGFR1-3, FGFR1-4, PDGFRα, c-Kit and RET inhibitor, is one of the six approved systemic therapies for hepatocellular carcinoma, the most common form of liver cancer [130,131].

In terms of tumor vessel normalization, the information existing regarding the efficacy of these inhibitors is still limited. It was demonstrated that the effect of lenvatinib with anti-PD-1 treatment was superior to the outcome of single sorafenib or FGFR treatment and improved anti-cancer activity. This was due to inhibition of immunosuppressive effects and the induction of vessel normalization, opening up the potential of combined anti-angiogenic treatments and tumor vascular normalization for immunotherapy [70].

6. Conclusions and Perspectives

The last few decades, intense scientific efforts on anti-angiogenic therapies have provided beneficial outcome in the clinical setting for some diseases, while they are still far from the desired therapeutic outcome in others, including cancer. Contrary to the classical notion of vascular regression, the main goal of current anti-angiogenic treatments is tumor vascular normalization and maturity, which provides increased tumor access to chemotherapeutic drugs and higher efficacy of cancer immunotherapy. The variety of cytokines and growth factors, the complexity of their signaling pathways and the interplay and compensation among them have hindered the generation of potent therapies. Targeting several growth factors with combinatory therapies, downstream signaling adaptors, where different growth factor pathways converge, important endothelial functions, such as metabolism, and the induction of vascular normalization remain promising areas that drive the common efforts towards novel anti-angiogenic therapies and cancer treatment. The compensatory mechanisms triggered upon anti-angiogenic monotherapies have driven the establishment of the current multitargeting anti-angiogenic inhibitors in the clinical practice. The identification and potent inhibition of downstream kinases and key signaling molecules where many angiogenic pathways converge could overcome current issues driven by the diversity of angiogenic ligands and receptors and should be the focus of future research. Moreover, the combination of current or future broad-spectrum anti-angiogenic inhibitors with immunotherapy in different cancers bears high potential to significantly advance the outcome of anticancer treatments and provides a promising field for clinical research.

Author Contributions: Conceptualization, F.T.Z. and C.M.M.; writing—review and editing, F.T.Z., M.S.S. and C.M.M.; funding acquisition: C.M.M. All authors have read and agreed to the published version of the manuscript.

Funding: This work was funded in part by the National Institutes of Health Grant (NCI) R15CA231339 and Texas Tech University Health Sciences Center (TTUHSC) School of Pharmacy Office of the Sciences grant. The funders had no role in study design, decision to write, or preparation of the manuscript.

Conflicts of Interest: The authors declare no competing interests.

References

1. Folkman, J. Angiogenesis: An organizing principle for drug discovery? *Nat. Rev. Drug Discov.* **2007**, *6*, 273–286. [CrossRef]
2. Carmeliet, P. Mechanisms of angiogenesis and arteriogenesis. *Nat. Med.* **2000**, *6*, 389–395. [CrossRef]
3. Augustin, H.G. Commentary on Folkman: "How Is Blood Vessel Growth Regulated in Normal and Neoplastic Tissue?". *Cancer Res.* **2016**, *76*, 2854–2856. [CrossRef] [PubMed]
4. Senger, D.R.; Galli, S.J.; Dvorak, A.M.; A Perruzzi, C.; Harvey, V.S.; Dvorak, H.F. Tumor cells secrete a vascular permeability factor that promotes accumulation of ascites fluid. *Science* **1983**, *219*, 983–985. [CrossRef]
5. Leung, D.W.; Cachianes, G.; Kuang, W.J.; Goeddel, D.V.; Ferrara, N. Vascular endothelial growth factor is a secreted angiogenic mitogen. *Science* **1989**, *246*, 1306–1309. [CrossRef]
6. Plouet, J.; Schilling, J.; Gospodarowicz, D. Isolation and characterization of a newly identified endothelial cell mitogen produced by AtT-20 cells. *EMBO J.* **1989**, *8*, 3801–3806. [CrossRef] [PubMed]
7. Rosenthal, R.A.; Megyesi, J.F.; Henzel, W.J.; Ferrara, N.; Folkman, J. Conditioned medium from mouse sarcoma 180 cells contains vascular endothelial growth factor. *Growth Factors* **1990**, *4*, 53–59. [CrossRef] [PubMed]
8. Shing, Y.; Folkman, J.; Sullivan, R.; Butterfield, C.; Murray, J.; Klagsbrun, M. Heparin affinity: Purification of a tumor-derived capillary endothelial cell growth factor. *Science* **1984**, *223*, 1296–1299. [CrossRef]
9. Ye, W. The Complexity of Translating Anti-angiogenesis Therapy from Basic Science to the Clinic. *Dev. Cell* **2016**, *37*, 114–125. [CrossRef] [PubMed]
10. Jayson, G.C.; Kerbel, R.; Ellis, L.M.; Harris, A.L. Antiangiogenic therapy in oncology: Current status and future directions. *Lancet* **2016**, *388*, 518–529. [CrossRef]
11. Choi, H.J.; Armaiz Pena, G.N.; Pradeep, S.; Cho, M.S.; Coleman, R.L.; Sood, A.K. Anti-vascular therapies in ovarian cancer: Moving beyond anti-VEGF approaches. *Cancer Metastasis Rev.* **2015**, *34*, 19–40. [CrossRef] [PubMed]
12. Jain, R.K. Tumor angiogenesis and accessibility: Role of vascular endothelial growth factor. *Semin. Oncol.* **2002**, *29*, 3–9. [CrossRef]
13. Ferrara, N.; Hillan, K.J.; Novotny, W. Bevacizumab (Avastin), a humanized anti-VEGF monoclonal antibody for cancer therapy. *Biochem. Biophys. Res. Commun.* **2005**, *333*, 328–335. [CrossRef] [PubMed]

14. Tuma, R.S. Success of bevacizumab trials raises questions for future studies. *J. Natl. Cancer Inst.* **2005**, *97*, 950–951. [CrossRef]
15. Mitamura, T.; Gourley, C.; Sood, A.K. Prediction of anti-angiogenesis escape. *Gynecol. Oncol.* **2016**, *141*, 80–85. [CrossRef] [PubMed]
16. Mitchell, D.C.; Bryan, B.A. Anti-angiogenic therapy: Adapting strategies to overcome resistant tumors. *J. Cell. Biochem.* **2010**, *111*, 543–553. [CrossRef] [PubMed]
17. Cao, D.; Zheng, Y.; Xu, H.; Ge, W.; Xu, X. Bevacizumab improves survival in metastatic colorectal cancer patients with primary tumor resection: A meta-analysis. *Sci. Rep.* **2019**, *9*, 20326. [CrossRef]
18. Cohen, M.H.; Gootenberg, J.; Keegan, P.; Pazdur, R. FDA drug approval summary: Bevacizumab (Avastin) plus Carboplatin and Paclitaxel as first-line treatment of advanced/metastatic recurrent nonsquamous non-small cell lung cancer. *Oncologist* **2007**, *12*, 713–718. [CrossRef]
19. Abdalla, A.M.E.; Xiao, L.; Ullah, M.W.; Yu, M.; Ouyang, C.; Yang, G. Current Challenges of Cancer Anti-angiogenic Therapy and the Promise of Nanotherapeutics. *Theranostics* **2018**, *8*, 533–548. [CrossRef]
20. Wong, P.P.; Bodrug, N.; Hodivala-Dilke, K.M. Exploring Novel Methods for Modulating Tumor Blood Vessels in Cancer Treatment. *Curr. Biol.* **2016**, *26*, R1161–R1166. [CrossRef]
21. Bergers, G.; Hanahan, D. Modes of resistance to anti-angiogenic therapy. *Nat. Rev. Cancer* **2008**, *8*, 592–603. [CrossRef]
22. Beenken, A.; Mohammadi, M. The FGF family: Biology, pathophysiology and therapy. *Nat. Rev. Drug Discov.* **2009**, *8*, 235–253. [CrossRef]
23. Johnson, D.E.; Williams, L.T. Structural and functional diversity in the FGF receptor multigene family. *Adv. Cancer Res.* **1993**, *60*, 1–41. [CrossRef]
24. Bae, J.H.; Schlessinger, J. Asymmetric tyrosine kinase arrangements in activation or autophosphorylation of receptor tyrosine kinases. *Mol. Cells* **2010**, *29*, 443–448. [CrossRef]
25. Fantl, W.J.; Escobedo, J.A.; Martin, G.A.; Turck, C.W.; del Rosario, M.; McCormick, F.; Williams, L.T. Distinct phosphotyrosines on a growth factor receptor bind to specific molecules that mediate different signaling pathways. *Cell* **1992**, *69*, 413–423. [CrossRef]
26. Yang, X.; Liaw, L.; Prudovsky, I.; Brooks, P.C.; Vary, C.; Oxburgh, L.; Friesel, R. Fibroblast growth factor signaling in the vasculature. *Curr. Atheroscler. Rep.* **2015**, *17*, 509. [CrossRef]
27. De Moerlooze, L.; Spencer-Dene, B.; Revest, J.M.; Hajihosseini, M.; Rosewell, I.; Dickson, C. An important role for the IIIb isoform of fibroblast growth factor receptor 2 (FGFR2) in mesenchymal-epithelial signalling during mouse organogenesis. *Development* **2000**, *127*, 483–492.
28. Kimelman, D.; Kirschner, M. Synergistic induction of mesoderm by FGF and TGF-beta and the identification of an mRNA coding for FGF in the early Xenopus embryo. *Cell* **1987**, *51*, 869–877. [CrossRef]
29. Ardizzone, A.; Scuderi, S.A.; Giuffrida, D.; Colarossi, C.; Puglisi, C.; Campolo, M.; Cuzzocrea, S.; Esposito, E.; Paterniti, I. Role of Fibroblast Growth Factors Receptors (FGFRs) in Brain Tumors, Focus on Astrocytoma and Glioblastoma. *Cancers* **2020**, *12*, 3825. [CrossRef]
30. Gyanchandani, R.; Ortega Alves, M.V.; Myers, J.N.; Kim, S. A proangiogenic signature is revealed in FGF-mediated bevacizumab-resistant head and neck squamous cell carcinoma. *Mol. Cancer Res.* **2013**, *11*, 1585–1596. [CrossRef]
31. Zhu, D.L.; Tuo, X.M.; Rong, Y.; Zhang, K.; Guo, Y. Fibroblast growth factor receptor signaling as therapeutic targets in female reproductive system cancers. *J. Cancer* **2020**, *11*, 7264–7275. [CrossRef]
32. Navid, S.; Fan, C.; P, O.F.-V.; Generali, D.; Li, Y. The Fibroblast Growth Factor Receptors in Breast Cancer: From Oncogenesis to Better Treatments. *Int. J. Mol. Sci.* **2020**, *21*, 2011. [CrossRef]
33. Cunningham, D.L.; Sweet, S.M.; Cooper, H.J.; Heath, J.K. Differential phosphoproteomics of fibroblast growth factor signaling: Identification of Src family kinase-mediated phosphorylation events. *J. Proteome Res.* **2010**, *9*, 2317–2328. [CrossRef]
34. Redington, A.E.; Roche, W.R.; Madden, J.; Frew, A.J.; Djukanovic, R.; Holgate, S.T.; Howarth, P.H. Basic fibroblast growth factor in asthma: Measurement in bronchoalveolar lavage fluid basally and following allergen challenge. *J. Allergy Clin. Immunol.* **2001**, *107*, 384–387. [CrossRef] [PubMed]
35. Kanazawa, S.; Tsunoda, T.; Onuma, E.; Majima, T.; Kagiyama, M.; Kikuchi, K. VEGF, basic-FGF, and TGF-beta in Crohn's disease and ulcerative colitis: A novel mechanism of chronic intestinal inflammation. *Am. J. Gastroenterol.* **2001**, *96*, 822–828. [CrossRef] [PubMed]
36. Yamashita, A.; Yonemitsu, Y.; Okano, S.; Nakagawa, K.; Nakashima, Y.; Irisa, T.; Iwamoto, Y.; Nagai, Y.; Hasegawa, M.; Sueishi, K. Fibroblast growth factor-2 determines severity of joint disease in adjuvant-induced arthritis in rats. *J. Immunol.* **2002**, *168*, 450–457. [CrossRef]
37. Gibran, N.S.; Isik, F.F.; Heimbach, D.M.; Gordon, D. Basic fibroblast growth factor in the early human burn wound. *J. Surg. Res.* **1994**, *56*, 226–234. [CrossRef]
38. Presta, M.; Dell'Era, P.; Mitola, S.; Moroni, E.; Ronca, R.; Rusnati, M. Fibroblast growth factor/fibroblast growth factor receptor system in angiogenesis. *Cytokine Growth Factor Rev.* **2005**, *16*, 159–178. [CrossRef]
39. Ornitz, D.M.; Itoh, N. The Fibroblast Growth Factor signaling pathway. *Wiley Interdiscip. Rev. Dev. Biol.* **2015**, *4*, 215–266. [CrossRef] [PubMed]
40. Belov, A.A.; Mohammadi, M. Molecular mechanisms of fibroblast growth factor signaling in physiology and pathology. *Cold Spring Harb. Perspect. Biol.* **2013**, *5*. [CrossRef]

41. Antoine, M.; Wirz, W.; Tag, C.G.; Mavituna, M.; Emans, N.; Korff, T.; Stoldt, V.; Gressner, A.M.; Kiefer, P. Expression pattern of fibroblast growth factors (FGFs), their receptors and antagonists in primary endothelial cells and vascular smooth muscle cells. *Growth Factors* **2005**, *23*, 87–95. [CrossRef] [PubMed]
42. Presta, M.; Tiberio, L.; Rusnati, M.; Dell'Era, P.; Ragnotti, G. Basic fibroblast growth factor requires a long-lasting activation of protein kinase C to induce cell proliferation in transformed fetal bovine aortic endothelial cells. *Cell Regul.* **1991**, *2*, 719–726. [CrossRef]
43. Shono, T.; Kanetake, H.; Kanda, S. The role of mitogen-activated protein kinase activation within focal adhesions in chemotaxis toward FGF-2 by murine brain capillary endothelial cells. *Exp. Cell Res.* **2001**, *264*, 275–283. [CrossRef]
44. Nicoli, S.; De Sena, G.; Presta, M. Fibroblast growth factor 2-induced angiogenesis in zebrafish: The zebrafish yolk membrane (ZFYM) angiogenesis assay. *J. Cell. Mol. Med.* **2009**, *13*, 2061–2068. [CrossRef]
45. Gualandris, A.; Rusnati, M.; Belleri, M.; Nelli, E.E.; Bastaki, M.; Molinari-Tosatti, M.P.; Bonardi, F.; Parolini, S.; Albini, A.; Morbidelli, L.; et al. Basic fibroblast growth factor overexpression in endothelial cells: An autocrine mechanism for angiogenesis and angioproliferative diseases. *Cell Growth Differ.* **1996**, *7*, 147–160.
46. Sajib, S.; Zahra, F.T.; Lionakis, M.S.; German, N.A.; Mikelis, C.M. Mechanisms of angiogenesis in microbe-regulated inflammatory and neoplastic conditions. *Angiogenesis* **2018**, *21*, 1–14. [CrossRef]
47. Jackson, J.R.; Seed, M.P.; Kircher, C.H.; Willoughby, D.A.; Winkler, J.D. The codependence of angiogenesis and chronic inflammation. *FASEB J.* **1997**, *11*, 457–465. [CrossRef]
48. Carmeliet, P.; Jain, R.K. Angiogenesis in cancer and other diseases. *Nature* **2000**, *407*, 249–257. [CrossRef] [PubMed]
49. Schmid, M.C.; Varner, J.A. Myeloid cell trafficking and tumor angiogenesis. *Cancer Lett.* **2007**, *250*, 1–8. [CrossRef] [PubMed]
50. Presta, M.; Andres, G.; Leali, D.; Dell'Era, P.; Ronca, R. Inflammatory cells and chemokines sustain FGF2-induced angiogenesis. *Eur. Cytokine Netw.* **2009**, *20*, 39–50. [CrossRef] [PubMed]
51. Andres, G.; Leali, D.; Mitola, S.; Coltrini, D.; Camozzi, M.; Corsini, M.; Belleri, M.; Hirsch, E.; Schwendener, R.A.; Christofori, G.; et al. A pro-inflammatory signature mediates FGF2-induced angiogenesis. *J. Cell. Mol. Med.* **2009**, *13*, 2083–2108. [CrossRef] [PubMed]
52. ZhuGe, D.L.; Javaid, H.M.A.; Sahar, N.E.; Zhao, Y.Z.; Huh, J.Y. Fibroblast growth factor 2 exacerbates inflammation in adipocytes through NLRP3 inflammasome activation. *Arch. Pharm. Res.* **2020**, *43*, 1311–1324. [CrossRef] [PubMed]
53. Im, J.H.; Buzzelli, J.N.; Jones, K.; Franchini, F.; Gordon-Weeks, A.; Markelc, B.; Chen, J.; Kim, J.; Cao, Y.; Muschel, R.J. FGF2 alters macrophage polarization, tumour immunity and growth and can be targeted during radiotherapy. *Nat. Commun.* **2020**, *11*, 4064. [CrossRef] [PubMed]
54. Fujii, T.; Kuwano, H. Regulation of the expression balance of angiopoietin-1 and angiopoietin-2 by Shh and FGF-2. *In Vitro Cell. Dev. Biol. Anim.* **2010**, *46*, 487–491. [CrossRef]
55. Pepper, M.S.; Ferrara, N.; Orci, L.; Montesano, R. Potent synergism between vascular endothelial growth factor and basic fibroblast growth factor in the induction of angiogenesis in vitro. *Biochem. Biophys. Res. Commun.* **1992**, *189*, 824–831. [CrossRef]
56. Akwii, R.G.; Sajib, M.S.; Zahra, F.T.; Mikelis, C.M. Role of Angiopoietin-2 in Vascular Physiology and Pathophysiology. *Cells* **2019**, *8*, 471. [CrossRef]
57. Alessi, P.; Leali, D.; Camozzi, M.; Cantelmo, A.; Albini, A.; Presta, M. Anti-FGF2 approaches as a strategy to compensate resistance to anti-VEGF therapy: Long-pentraxin 3 as a novel antiangiogenic FGF2-antagonist. *Eur. Cytokine Netw.* **2009**, *20*, 225–234. [CrossRef]
58. Presta, L.G.; Chen, H.; O'Connor, S.J.; Chisholm, V.; Meng, Y.G.; Krummen, L.; Winkler, M.; Ferrara, N. Humanization of an anti-vascular endothelial growth factor monoclonal antibody for the therapy of solid tumors and other disorders. *Cancer Res.* **1997**, *57*, 4593–4599.
59. Willett, C.G.; Boucher, Y.; di Tomaso, E.; Duda, D.G.; Munn, L.L.; Tong, R.T.; Chung, D.C.; Sahani, D.V.; Kalva, S.P.; Kozin, S.V.; et al. Direct evidence that the VEGF-specific antibody bevacizumab has antivascular effects in human rectal cancer. *Nat. Med.* **2004**, *10*, 145–147. [CrossRef] [PubMed]
60. Garcia, J.; Hurwitz, H.I.; Sandler, A.B.; Miles, D.; Coleman, R.L.; Deurloo, R.; Chinot, O.L. Bevacizumab (Avastin(R)) in cancer treatment: A review of 15 years of clinical experience and future outlook. *Cancer Treat. Rev.* **2020**, *86*, 102017. [CrossRef]
61. Mitchell, P.; Liew, G.; Gopinath, B.; Wong, T.Y. Age-related macular degeneration. *Lancet* **2018**, *392*, 1147–1159. [CrossRef]
62. Furuta, T.; Nakada, M.; Misaki, K.; Sato, Y.; Hayashi, Y.; Nakanuma, Y.; Hamada, J. Molecular analysis of a recurrent glioblastoma treated with bevacizumab. *Brain Tumor Pathol.* **2014**, *31*, 32–39. [CrossRef] [PubMed]
63. Okamoto, S.; Nitta, M.; Maruyama, T.; Sawada, T.; Komori, T.; Okada, Y.; Muragaki, Y. Bevacizumab changes vascular structure and modulates the expression of angiogenic factors in recurrent malignant gliomas. *Brain Tumor Pathol.* **2016**, *33*, 129–136. [CrossRef]
64. Lucio-Eterovic, A.K.; Piao, Y.; de Groot, J.F. Mediators of glioblastoma resistance and invasion during antivascular endothelial growth factor therapy. *Clin. Cancer Res.* **2009**, *15*, 4589–4599. [CrossRef]
65. Navas, T.; Kinders, R.J.; Lawrence, S.M.; Ferry-Galow, K.V.; Borgel, S.; Hollingshead, M.G.; Srivastava, A.K.; Alcoser, S.Y.; Makhlouf, H.R.; Chuaqui, R.; et al. Clinical Evolution of Epithelial-Mesenchymal Transition in Human Carcinomas. *Cancer Res.* **2020**, *80*, 304–318. [CrossRef]
66. Yamashita-Kashima, Y.; Fujimoto-Ouchi, K.; Yorozu, K.; Kurasawa, M.; Yanagisawa, M.; Yasuno, H.; Mori, K. Biomarkers for antitumor activity of bevacizumab in gastric cancer models. *BMC Cancer* **2012**, *12*, 37. [CrossRef] [PubMed]

67. Kopetz, S.; Hoff, P.M.; Morris, J.S.; Wolff, R.A.; Eng, C.; Glover, K.Y.; Adinin, R.; Overman, M.J.; Valero, V.; Wen, S.; et al. Phase II trial of infusional fluorouracil, irinotecan, and bevacizumab for metastatic colorectal cancer: Efficacy and circulating angiogenic biomarkers associated with therapeutic resistance. *J. Clin. Oncol.* **2010**, *28*, 453–459. [CrossRef] [PubMed]
68. Zhao, M.; Yu, Z.; Li, Z.; Tang, J.; Lai, X.; Liu, L. Expression of angiogenic growth factors VEGF, bFGF and ANG1 in colon cancer after bevacizumab treatment in vitro: A potential self-regulating mechanism. *Oncol. Rep.* **2017**, *37*, 601–607. [CrossRef] [PubMed]
69. Casanovas, O.; Hicklin, D.J.; Bergers, G.; Hanahan, D. Drug resistance by evasion of antiangiogenic targeting of VEGF signaling in late-stage pancreatic islet tumors. *Cancer Cell* **2005**, *8*, 299–309. [CrossRef]
70. Deng, H.; Kan, A.; Lyu, N.; Mu, L.; Han, Y.; Liu, L.; Zhang, Y.; Duan, Y.; Liao, S.; Li, S.; et al. Dual Vascular Endothelial Growth Factor Receptor and Fibroblast Growth Factor Receptor Inhibition Elicits Antitumor Immunity and Enhances Programmed Cell Death-1 Checkpoint Blockade in Hepatocellular Carcinoma. *Liver Cancer* **2020**, *9*, 338–357. [CrossRef]
71. Porta, C.; Paglino, C.; Imarisio, I.; Ganini, C.; Sacchi, L.; Quaglini, S.; Giunta, V.; de Amici, M. Changes in circulating pro-angiogenic cytokines, other than VEGF, before progression to sunitinib therapy in advanced renal cell carcinoma patients. *Oncology* **2013**, *84*, 115–122. [CrossRef]
72. Schmidinger, M. Third-line dovitinib in metastatic renal cell carcinoma. *Lancet Oncol.* **2014**, *15*, 245–246. [CrossRef]
73. Welti, J.C.; Gourlaouen, M.; Powles, T.; Kudahetti, S.C.; Wilson, P.; Berney, D.M.; Reynolds, A.R. Fibroblast growth factor 2 regulates endothelial cell sensitivity to sunitinib. *Oncogene* **2011**, *30*, 1183–1193. [CrossRef] [PubMed]
74. Incio, J.; Ligibel, J.A.; McManus, D.T.; Suboj, P.; Jung, K.; Kawaguchi, K.; Pinter, M.; Babykutty, S.; Chin, S.M.; Vardam, T.D.; et al. Obesity promotes resistance to anti-VEGF therapy in breast cancer by up-regulating IL-6 and potentially FGF-2. *Sci. Transl. Med.* **2018**, *10*, eaag0945. [CrossRef]
75. Yoshiji, H.; Harris, S.R.; Thorgeirsson, U.P. Vascular endothelial growth factor is essential for initial but not continued in vivo growth of human breast carcinoma cells. *Cancer Res.* **1997**, *57*, 3924–3928.
76. Pietras, K.; Pahler, J.; Bergers, G.; Hanahan, D. Functions of paracrine PDGF signaling in the proangiogenic tumor stroma revealed by pharmacological targeting. *PLoS Med.* **2008**, *5*, e19. [CrossRef]
77. Michaelson, M.D.; Oudard, S.; Ou, Y.-C.; Sengeløv, L.; Saad, F.; Houede, N.; Ostler, P.; Stenzl, A.; Daugaard, G.; Jones, R.; et al. Randomized, Placebo-Controlled, Phase III Trial of Sunitinib Plus Prednisone Versus Prednisone Alone in Progressive, Metastatic, Castration-Resistant Prostate Cancer. *J. Clin. Oncol.* **2014**, *32*, 76–82. [CrossRef] [PubMed]
78. Motoo, N.Y.; Hayashi, Y.; Shimizu, A.; Ura, M.; Nishikawa, R. Safety and effectiveness of bevacizumab in Japanese patients with malignant glioma: A post-marketing surveillance study. *Jpn. J. Clin. Oncol.* **2019**, *49*, 1016–1023. [CrossRef] [PubMed]
79. Astolfi, A.; Pantaleo, M.A.; Indio, V.; Urbini, M.; Nannini, M. The Emerging Role of the FGF/FGFR Pathway in Gastrointestinal Stromal Tumor. *Int. J. Mol. Sci.* **2020**, *21*, 3313. [CrossRef]
80. Galzie, Z.; Fernig, D.G.; Smith, J.A.; Poston, G.J.; Kinsella, A.R. Invasion of human colorectal carcinoma cells is promoted by endogenous basic fibroblast growth factor. *Int. J. Cancer* **1997**, *71*, 390–395. [CrossRef]
81. Forner, A.; Reig, M.; Bruix, J. Hepatocellular carcinoma. *Lancet* **2018**, *391*, 1301–1314. [CrossRef]
82. Kudo, M. Lenvatinib May Drastically Change the Treatment Landscape of Hepatocellular Carcinoma. *Liver Cancer* **2018**, *7*, 1–19. [CrossRef] [PubMed]
83. Aziz, S.A.; Sznol, J.; Adeniran, A.; Colberg, J.W.; Camp, R.L.; Kluger, H.M. Vascularity of primary and metastatic renal cell carcinoma specimens. *J. Transl. Med.* **2013**, *11*, 15. [CrossRef] [PubMed]
84. Welsh, S.J.; Fife, K. Pazopanib for the treatment of renal cell carcinoma. *Futur. Oncol.* **2015**, *11*, 1169–1179. [CrossRef]
85. Rini, B.I.; Atkins, M.B. Resistance to targeted therapy in renal-cell carcinoma. *Lancet Oncol.* **2009**, *10*, 992–1000. [CrossRef]
86. Fischer, S.; Gillessen, S.; Rothermundt, C. Sequence of treatment in locally advanced and metastatic renal cell carcinoma. *Transl. Androl. Urol.* **2015**, *4*, 310–325.
87. Wang, X.F.; Zhang, J.-Y.; Li, L.; Zhao, X.-Y. Beneficial effects of metformin on primary cardiomyocytes via activation of adenosine monophosphate-activated protein kinase. *Chin. Med. J.* **2011**, *124*, 1876–1884.
88. Muti, P.; Berrino, F.; Krogh, V.; Villarini, A.; Barba, M.; Strano, S.; Blandino, G. Metformin, diet and breast cancer: An avenue for chemoprevention. *Cell Cycle* **2009**, *8*, 2661. [CrossRef] [PubMed]
89. Ellis, L.M.; Hicklin, D.J. Pathways Mediating Resistance to Vascular Endothelial Growth Factor–Targeted Therapy. *Clin. Cancer Res.* **2008**, *14*, 6371–6375. [CrossRef] [PubMed]
90. Benjamin, L.E.; Golijanin, D.; Itin, A.; Pode, D.; Keshet, E. Selective ablation of immature blood vessels in established human tumors follows vascular endothelial growth factor withdrawal. *J. Clin. Investig.* **1999**, *103*, 159–165. [CrossRef]
91. Reinmuth, N.; Liu, W.; Jung, Y.D.; Ahmad, S.A.; Shaheen, R.M.; Fan, F.; Bucana, C.D.; McMahon, G.; Gallick, G.E.; Ellis, L.M. Induction of VEGF in perivascular cells defines a potential paracrine mechanism for endothelial cell survival. *FASEB J.* **2001**, *15*, 1239–1241. [CrossRef]
92. Melegh, Z.; Oltean, S. Targeting Angiogenesis in Prostate Cancer. *Int. J. Mol. Sci.* **2019**, *20*, 2676. [CrossRef]
93. Tannock, I.F.; Fizazi, K.; Ivanov, S.; Karlsson, C.T.; Fléchon, A.; Skoneczna, I.; Orlandi, F.; Gravis, G.; Matveev, V.; Bavbek, S.; et al. Aflibercept versus placebo in combination with docetaxel and prednisone for treatment of men with metastatic castration-resistant prostate cancer (VENICE): A phase 3, double-blind randomised trial. *Lancet Oncol.* **2013**, *14*, 760–768. [CrossRef]
94. Ben Jemaa, A.; Sallami, S.; Ramarli, D.; Colombatti, M.; Oueslati, R. The Proinflammatory Cytokine, IL-6, and its Interference with bFGF Signaling and PSMA in Prostate Cancer Cells. *Inflammation* **2012**, *36*, 643–650. [CrossRef] [PubMed]

95. Chen, X.; Wang, X.; Wang, Y.; Yang, L.; Hu, J.; Xiao, W.; Fu, A.; Cai, L.; Li, X.; Ye, X. Improved tumor-targeting drug delivery and therapeutic efficacy by cationic liposome modified with truncated bFGF peptide. *J. Control. Release* **2010**, *145*, 17–25. [CrossRef] [PubMed]
96. Sarkar, C.; Goswami, S.; Basu, S.; Chakroborty, D. Angiogenesis Inhibition in Prostate Cancer: An Update. *Cancers* **2020**, *12*, 2382. [CrossRef] [PubMed]
97. Vlodavsky, I.; Folkman, J.; Sullivan, R.; Fridman, R.; Ishai-Michaeli, R.; Sasse, J.; Klagsbrun, M. Endothelial cell-derived basic fibroblast growth factor: Synthesis and deposition into subendothelial extracellular matrix. *Proc. Natl. Acad. Sci. USA* **1987**, *84*, 2292–2296. [CrossRef] [PubMed]
98. Saksela, O.; Moscatelli, D.; Sommer, A.; Rifkin, D.B. Endothelial cell-derived heparan sulfate binds basic fibroblast growth factor and protects it from proteolytic degradation. *J. Cell Biol.* **1988**, *107*, 743–751. [CrossRef]
99. Saksela, O.; Rifkin, D.B. Release of basic fibroblast growth factor-heparan sulfate complexes from endothelial cells by plasminogen activator-mediated proteolytic activity. *J. Cell Biol.* **1990**, *110*, 767–775. [CrossRef]
100. Pineda-Lucena, A.; Nunez De Castro, I.; Lozano, R.M.; Munoz-Willery, I.; Zazo, M.; Gimenez-Gallego, G. Effect of low pH and heparin on the structure of acidic fibroblast growth factor. *Eur. J. Biochem.* **1994**, *222*, 425–431. [CrossRef]
101. Vlodavsky, I.; Fuks, Z.; Ishai-Michaeli, R.; Bashkin, P.; Levi, E.; Korner, G.; Bar-Shavit, R.; Klagsbrun, M. Extracellular matrix-resident basic fibroblast growth factor: Implication for the control of angiogenesis. *J. Cell. Biochem.* **1991**, *45*, 167–176. [CrossRef] [PubMed]
102. Nugent, M.A.; Edelman, E.R. Kinetics of basic fibroblast growth factor binding to its receptor and heparan sulfate proteoglycan: A mechanism for cooperativity. *Biochemistry* **1992**, *31*, 8876–8883. [CrossRef]
103. Roghani, M.; Mansukhani, A.; Dell'Era, P.; Bellosta, P.; Basilico, C.; Rifkin, D.B.; Moscatelli, D. Heparin increases the affinity of basic fibroblast growth factor for its receptor but is not required for binding. *J. Biol. Chem.* **1994**, *269*, 3976–3984. [CrossRef]
104. Kramer, R.H.; Vogel, K.G.; Nicolson, G.L. Solubilization and degradation of subendothelial matrix glycoproteins and proteoglycans by metastatic tumor cells. *J. Biol. Chem.* **1982**, *257*, 2678–2686. [CrossRef]
105. Qi, J.H.; Bell, B.; Singh, R.; Batoki, J.; Wolk, A.; Cutler, A.; Prayson, N.; Ali, M.; Stoehr, H.; Anand-Apte, B. Sorsby Fundus Dystrophy Mutation in Tissue Inhibitor of Metalloproteinase 3 (TIMP3) promotes Choroidal Neovascularization via a Fibroblast Growth Factor-dependent Mechanism. *Sci. Rep.* **2019**, *9*, 17429. [CrossRef]
106. Vlodavsky, I.; Fuks, Z.; Bar-Ner, M.; Ariav, Y.; Schirrmacher, V. Lymphoma cell-mediated degradation of sulfated proteoglycans in the subendothelial extracellular matrix: Relationship to tumor cell metastasis. *Cancer Res.* **1983**, *43*, 2704–2711. [PubMed]
107. Gonzalez-Avila, G.; Sommer, B.; Garcia-Hernandez, A.A.; Ramos, C. Matrix Metalloproteinases' Role in Tumor Microenvironment. *Adv. Exp. Med. Biol.* **2020**, *1245*, 97–131. [CrossRef]
108. Wang, X.; Khalil, R.A. Matrix Metalloproteinases, Vascular Remodeling, and Vascular Disease. *Adv. Pharmacol.* **2018**, *81*, 241–330. [CrossRef] [PubMed]
109. Djonov, V.; Cresto, N.; Aebersold, D.M.; Burri, P.H.; Altermatt, H.J.; Hristic, M.; Berclaz, G.; Ziemiecki, A.; Andres, A.C. Tumor cell specific expression of MMP-2 correlates with tumor vascularisation in breast cancer. *Int. J. Oncol.* **2002**, *21*, 25–30. [CrossRef]
110. Shibuya, M. Vascular endothelial growth factor and its receptor system: Physiological functions in angiogenesis and pathological roles in various diseases. *J. Biochem.* **2013**, *153*, 13–19. [CrossRef] [PubMed]
111. Melincovici, C.S.; Bosca, A.B.; Susman, S.; Marginean, M.; Mihu, C.; Istrate, M.; Moldovan, I.M.; Roman, A.L.; Mihu, C.M. Vascular endothelial growth factor (VEGF)-key factor in normal and pathological angiogenesis. *Rom. J. Morphol. Embryol.* **2018**, *59*, 455–467. [PubMed]
112. Compagni, A.; Wilgenbus, P.; Impagnatiello, M.A.; Cotten, M.; Christofori, G. Fibroblast growth factors are required for efficient tumor angiogenesis. *Cancer Res.* **2000**, *60*, 7163–7169.
113. Giavazzi, R.; Sennino, B.; Coltrini, D.; Garofalo, A.; Dossi, R.; Ronca, R.; Tosatti, M.P.; Presta, M. Distinct role of fibroblast growth factor-2 and vascular endothelial growth factor on tumor growth and angiogenesis. *Am. J. Pathol.* **2003**, *162*, 1913–1926. [CrossRef]
114. Nissen, L.J.; Cao, R.; Hedlund, E.M.; Wang, Z.; Zhao, X.; Wetterskog, D.; Funa, K.; Brakenhielm, E.; Cao, Y. Angiogenic factors FGF2 and PDGF-BB synergistically promote murine tumor neovascularization and metastasis. *J. Clin. Investig.* **2007**, *117*, 2766–2777. [CrossRef] [PubMed]
115. Lieu, C.; Heymach, J.; Overman, M.; Tran, H.; Kopetz, S. Beyond VEGF: Inhibition of the fibroblast growth factor pathway and antiangiogenesis. *Clin. Cancer Res.* **2011**, *17*, 6130–6139. [CrossRef]
116. Li, D.; Xie, K.; Zhang, L.; Yao, X.; Li, H.; Xu, Q.; Wang, X.; Jiang, J.; Fang, J. Dual blockade of vascular endothelial growth factor (VEGF) and basic fibroblast growth factor (FGF-2) exhibits potent anti-angiogenic effects. *Cancer Lett.* **2016**, *377*, 164–173. [CrossRef] [PubMed]
117. Bhide, R.S.; Lombardo, L.J.; Hunt, J.T.; Cai, Z.W.; Barrish, J.C.; Galbraith, S.; Sr, J.R.; Mortillo, S.; Wautlet, B.S.; Krishnan, B.; et al. The antiangiogenic activity in xenograft models of brivanib, a dual inhibitor of vascular endothelial growth factor receptor-2 and fibroblast growth factor receptor-1 kinases. *Mol. Cancer Ther.* **2010**, *9*, 369–378. [CrossRef] [PubMed]
118. Dempke, W.C.; Zippel, R. Brivanib, a novel dual VEGF-R2/bFGF-R inhibitor. *Anticancer Res.* **2010**, *30*, 4477–4483.
119. Kudo, M.; Han, G.; Finn, R.S.; Poon, R.T.; Blanc, J.F.; Yan, L.; Yang, J.; Lu, L.; Tak, W.Y.; Yu, X.; et al. Brivanib as adjuvant therapy to transarterial chemoembolization in patients with hepatocellular carcinoma: A randomized phase III trial. *Hepatology* **2014**, *60*, 1697–1707. [CrossRef]

120. Zhu, H.; Zhang, C.; Yang, X.; Yi, C. Treatment with Brivanib alaninate as a second-line monotherapy after Sorafenib failure in hepatocellular carcinoma: A case report. *Medicine* **2019**, *98*, e14823. [CrossRef]
121. Ba-Sang, D.Z.; Long, Z.W.; Teng, H.; Zhao, X.P.; Qiu, J.; Li, M.S. A network meta-analysis on the efficacy of sixteen targeted drugs in combination with chemotherapy for treatment of advanced/metastatic colorectal cancer. *Oncotarget* **2016**, *7*, 84468–84479. [CrossRef] [PubMed]
122. Bello, E.; Colella, G.; Scarlato, V.; Oliva, P.; Berndt, A.; Valbusa, G.; Serra, S.C.; D'Incalci, M.; Cavalletti, E.; Giavazzi, R.; et al. E-3810 is a potent dual inhibitor of VEGFR and FGFR that exerts antitumor activity in multiple preclinical models. *Cancer Res.* **2011**, *71*, 1396–1405. [CrossRef]
123. Caglevic, C.; Grassi, M.; Raez, L.; Listi, A.; Giallombardo, M.; Bustamante, E.; Gil-Bazo, I.; Rolfo, C. Nintedanib in non-small cell lung cancer: From preclinical to approval. *Ther. Adv. Respir. Dis.* **2015**, *9*, 164–172. [CrossRef] [PubMed]
124. Roskoski, R., Jr. Properties of FDA-approved small molecule protein kinase inhibitors: A 2020 update. *Pharmacol. Res.* **2020**, *152*, 104609. [CrossRef] [PubMed]
125. Hilberg, F.; Roth, G.J.; Krssak, M.; Kautschitsch, S.; Sommergruber, W.; Tontsch-Grunt, U.; Garin-Chesa, P.; Bader, G.; Zoephel, A.; Quant, J.; et al. BIBF 1120: Triple angiokinase inhibitor with sustained receptor blockade and good antitumor efficacy. *Cancer Res.* **2008**, *68*, 4774–4782. [CrossRef] [PubMed]
126. Miyamoto, S.; Kakutani, S.; Sato, Y.; Hanashi, A.; Kinoshita, Y.; Ishikawa, A. Drug review: Pazopanib. *Jpn. J. Clin. Oncol.* **2018**, *48*, 503–513. [CrossRef] [PubMed]
127. Laird, A.D.; Vajkoczy, P.; Shawver, L.K.; Thurnher, A.; Liang, C.; Mohammadi, M.; Schlessinger, J.; Ullrich, A.; Hubbard, S.R.; Blake, R.A.; et al. SU6668 is a potent antiangiogenic and antitumor agent that induces regression of established tumors. *Cancer Res.* **2000**, *60*, 4152–4160.
128. Shaheen, R.M.; Tseng, W.W.; Davis, D.W.; Liu, W.; Reinmuth, N.; Vellagas, R.; Wieczorek, A.A.; Ogura, Y.; McConkey, D.J.; Drazan, K.E.; et al. Tyrosine kinase inhibition of multiple angiogenic growth factor receptors improves survival in mice bearing colon cancer liver metastases by inhibition of endothelial cell survival mechanisms. *Cancer Res.* **2001**, *61*, 1464–1468. [PubMed]
129. Hidaka, H.; Izumi, N.; Aramaki, T.; Ikeda, M.; Inaba, Y.; Imanaka, K.; Okusaka, T.; Kanazawa, S.; Kaneko, S.; Kora, S.; et al. Subgroup analysis of efficacy and safety of orantinib in combination with TACE in Japanese HCC patients in a randomized phase III trial (ORIENTAL). *Med. Oncol.* **2019**, *36*, 52. [CrossRef] [PubMed]
130. Llovet, J.M.; Kelley, R.K.; Villanueva, A.; Singal, A.G.; Pikarsky, E.; Roayaie, S.; Lencioni, R.; Koike, K.; Zucman-Rossi, J.; Finn, R.S. Hepatocellular carcinoma. *Nat. Rev. Dis. Primers* **2021**, *7*, 6. [CrossRef] [PubMed]
131. Kudo, M.; Finn, R.S.; Qin, S.; Han, K.H.; Ikeda, K.; Piscaglia, F.; Baron, A.; Park, J.W.; Han, G.; Jassem, J.; et al. Lenvatinib versus sorafenib in first-line treatment of patients with unresectable hepatocellular carcinoma: A randomised phase 3 non-inferiority trial. *Lancet* **2018**, *391*, 1163–1173. [CrossRef]

Review

Angiogenesis in the Normal Adrenal Fetal Cortex and Adrenocortical Tumors

Sofia S. Pereira [1,2,3,4], Sofia Oliveira [1,5], Mariana P. Monteiro [1,2] and Duarte Pignatelli [3,4,6,7,*]

1. Department of Anatomy, Instituto de Ciências Biomédicas Abel Salazar, University of Porto, 4050-313 Porto, Portugal; sspereira@icbas.up.pt (S.S.P.); up201909110@fc.up.pt (S.O.); mpmonteiro@icbas.up.pt (M.P.M.)
2. Clinical and Experimental Endocrinology, Multidisciplinary Unit for Biomedical Research (UMIB), 4050-313 Porto, Portugal
3. Instituto de Investigação e Inovação em Saúde (I3S), Universidade do Porto, 4200-135 Porto, Portugal
4. Institute of Molecular Pathology and Immunology, University of Porto (IPATIMUP), 4200-135 Porto, Portugal
5. Faculty of Sciences, University of Porto, 4169-007 Porto, Portugal
6. Department of Endocrinology, Hospital S. João, 4200-319 Porto, Portugal
7. Department of Biomedicine, Faculty of Medicine, University of Porto, 4200-319 Porto, Portugal
* Correspondence: dpignatelli@ipatimup.pt; Tel.: +351-91-288-0313

Citation: Pereira, S.S.; Oliveira, S.; Monteiro, M.P.; Pignatelli, D. Angiogenesis in the Normal Adrenal Fetal Cortex and Adrenocortical Tumors. *Cancers* **2021**, *13*, 1030. https://doi.org/10.3390/cancers13051030

Academic Editor: Domenico Ribatti

Received: 31 December 2020
Accepted: 24 February 2021
Published: 1 March 2021

Publisher's Note: MDPI stays neutral with regard to jurisdictional claims in published maps and institutional affiliations.

Copyright: © 2021 by the authors. Licensee MDPI, Basel, Switzerland. This article is an open access article distributed under the terms and conditions of the Creative Commons Attribution (CC BY) license (https://creativecommons.org/licenses/by/4.0/).

Simple Summary: Pharmacological angiogenesis modulation was robustly demonstrated to be a powerful clinical resource in oncotherapy. Adrenocortical carcinomas (ACC) often have a poor prognosis for which therapeutic options are limited. Understanding the mechanisms that regulate adrenocortical angiogenesis both under physiological conditions and in ACC could provide important clues on how these processes could be modulated for clinical purposes. This report summarizes the current knowledge on adrenal cortex angiogenesis regulation in physiological conditions and ACC. Embryonic adrenal angiogenesis is regulated by VEGF and Ang-Tie signaling pathways. VEGF angiogenic pathway was initially considered a promising therapeutic target for improving ACC prognosis. However, every single VEGF pathway-targeting clinical trial in ACC so far conducted yielded disappointing results. In contrast, the potential of Ang-Tie pathway-targeting in ACC is yet to be explored. Therefore, further investigation on the role and efficacy of modulating both Ang-Tie and VEGF pathways in ACC is still an unmet need.

Abstract: Angiogenesis plays an important role in several physiological and pathological processes. Pharmacological angiogenesis modulation has been robustly demonstrated to achieve clinical benefits in several cancers. Adrenocortical carcinomas (ACC) are rare tumors that often have a poor prognosis. In addition, therapeutic options for ACC are limited. Understanding the mechanisms that regulate adrenocortical angiogenesis along the embryonic development and in ACC could provide important clues on how these processes could be pharmacologically modulated for ACC treatment. In this report, we performed an integrative review on adrenal cortex angiogenesis regulation in physiological conditions and ACC. During embryonic development, adrenal angiogenesis is regulated by both VEGF and Ang-Tie signaling pathways. In ACC, early research efforts were focused on VEGF signaling and this pathway was identified as a good prognostic factor and thus a promising therapeutic target. However, every clinical trial so far conducted in ACC using VEGF pathway-targeting drugs, alone or in combination, yielded disappointing results. In contrast, although the Ang-Tie pathway has been pointed out as an important regulator of fetal adrenocortical angiogenesis, its role is yet to be explored in ACC. In the future, further research on the role and efficacy of modulating both Ang-Tie and VEGF pathways in ACC is needed.

Keywords: angiogenesis; adrenal fetal cortex; adrenocortical tumors; adrenocortical carcinoma; anti-angiogenic drugs

1. Introduction

Angiogenesis is a dynamic process during which new blood vessels are formed derived from pre-existing vasculature. Angiogenesis is an extensively studied process in tumors and a well-recognized hallmark of cancer [1]. Angiogenesis was previously studied in adrenocortical carcinomas (ACC), although the relative rarity of these tumors represents a limitation to conduct extensive clinical and molecular characterization studies. This review aims to bring together all the available data on angiogenesis regulation during the adrenocortical development and in ACC, which could be potentially useful to identify future research avenues to achieve advances in ACC clinical management and disease prognosis. Data source and study selection approach is described in the Supplementary File S1.

2. Angiogenesis Regulation

Angiogenesis plays a central role in several physiological (e.g., fetal development and wound healing) and pathological processes (e.g., vascular overgrowth for tumor expansion and metastasis) [2–4]. Angiogenesis, either in normal or tumor tissues, usually occurs via one or more of the following mechanisms:

(1) Sprouting angiogenesis, one the most well characterized mechanism leading to angiogenesis, relies on endothelial cells function specification into either tip or stalk cells. Tip cells are derived from the parent vessel, degrade the basement membrane, extend large filopodia which can sense angiogenic factor gradients, such as vascular endothelial growth factor (VEGF), and migrate along the chemotactic paths. In contrast, stalk cells proliferate behind tip cells to form the sprout body, start the process of lumen formation, and connect with neighboring vessels [5–7].

(2) Intussusceptive angiogenesis is a process that consists in the splitting of pre-existing vessels into two new vessels. It starts with the formation of transluminal tissue pillars through the invagination of opposing capillary endothelial cells into the vascular lumen, creating a zone of contact. Commonly, intussusceptive and sprouting angiogenesis are complementary mechanisms [5,8].

(3) Recruitment of endothelial progenitor cells and vasculogenesis, a process through which endothelial progenitor cells are recruited in response to several growth factors, cytokines and/or hypoxia-inducible factors. Endothelial progenitor cells differentiate into mature endothelial cells and are incorporated into the angiogenic sprout, thus contributing to new blood vessel formation [4,9].

(4) Vasculogenic mimicry: malignant tumor cells form de novo vessel-like structures without endothelial cells. The newly formed channels mimic the embryonic vascular network pattern, being able to provide enough blood supply to the tumor tissue [10,11].

Multiple signaling pathways regulate blood vessel growth and maintenance. Among these, VEGF and Ang-Tie pathways are particularly important and have been the focus of multiple studies, especially in the context of cancer [12]. VEGF receptor and Tie ligands are widely distributed and were shown to play a coordinated role in endothelial cell proliferation and vessel wall assembly in normal and pathological conditions.

2.1. VEGF Pathway in Angiogenesis Regulation

In mammals, the VEGF system mainly includes five secreted ligands (VEGF-A, VEGF-B, VEGF-C, VEGF-D and placental growth factor) and three primary tyrosine kinase receptors (VEGF-R1, VEGF-R2, VEGF-R3) [13]. The VEGF system also includes the cell-surface proteins, heparan sulfate proteoglycans and neuropilin-1 and -2, which operate as VEGF coreceptors [14,15].

VEGFR-1 and VEGFR-2 are expressed in vascular endothelial cells, while VEGF-R3 seems to be prominently expressed in lymphatic endothelial cells [16]. VEGF ligands have different affinities for one of the three VEGF-R. As tyrosine kinases receptors, upon dimerization by a VEFG ligand, the VEGF-Rs auto-phosphorylate, a phenomenon which in turn activates downstream signaling pathways including mitogen-activated protein kinase (MAPK) pathway, the phosphatidylinositol-3 kinase (PI3K-AKT) pathway, and the phospholipase-C-γ pathway. Those pathways drive various intracellular effects in endothelial cells, such as migration, proliferation, and cell survival. The activation of phospholipase-C-γ pathway via VEGF-A-VEGFR-2 binding was reported to be a key signal for endothelial proliferation [17].

In 2001, a new VEGF was identified, the endocrine-gland-derived VEGF (EG-VEGF). This ligand does not show any structural homology to the VEGF family, but displays several biological similarities to VEGF ligands, including hypoxic regulation and ability to induce fenestration in target cells. Moreover, EG-VEGF expression is restricted to steroidogenic tissues (adrenal, ovary, testis and placenta) and its effects seem to be restricted to endothelial cells derived from these organs [18].

2.2. Ang-Tie Pathway in Angiogenesis Regulation

Ang-Tie signaling pathway regulates vascular permeability and remodeling during tumor angiogenesis and metastasis. Ang/Tie signaling seems to complement the VEGF signaling pathway by controlling later stages of angiogenesis and by being involved in vascular maturation (Figure 1) [19].

The angiopoietin family includes two type 1 transmembrane protein receptors: Tie1 and Tie2 and four ligands: Ang1, Ang2, Ang3 and Ang4. Ang1 and Ang2 have been identified as the main ligands for Tie receptors, while the Ang3 and Ang4 biological function is still poorly characterized [20–22].

Ang1 binds and activates Tie2 resulting in Tie2 internalization and ligand release. Then it leads to Tie2 tyrosine residues phosphorylation that in turn recruits adaptor proteins and ignites PI3K/Akt and MAPK signaling pathways, promoting pro-survival, anti-permeability, and anti-inflammatory effects on endothelial cells [23]. Tie2 is not required for the endothelial cells' differentiation but is rather reported as necessary for cell maintenance [24].

Ang2 that shares approx. 60% amino acid homology with Ang1, binds to Tie2 with a similar affinity as Ang1. Ang2 seems to block the Ang1-induced Tie2 phosphorylation. Ang2 is upregulated during tumor angiogenesis and so was considered as a potential antiangiogenic target. However, recent studies found Ang2 to have a dual function, acting as a Tie2 antagonist in the presence of Ang1 or acting as a Tie2 agonist in the absence of Ang1 [25]. Different studies reported that it is unlikely that Tie2 can act differently when binding to Ang1 and Ang2, since both angiopoietins interact with Tie2 in a structurally similar manner and pointed out that other still unidentified mechanisms were likely to be involved [26]. One of the proposed mechanisms involve the Tie1 receptor [27].

Contrary to Tie2, the Tie1 has been less well characterized. Tie1 is considered an orphan receptor and is mainly expressed at vascular bifurcations and branching points, with no yet identified in vivo ligand [28]. It is well known, however, that Tie1 has an important role in vascular development, since its inactivation causes late embryonic lethality and vasculature maturation failure [29,30]. Recent studies proposed that Tie1 forms a complex with Tie2 on the endothelial cell surface and acts as a Tie2 inhibitor [27]. Cells expressing both receptors are responsive to chemotactic signals and able to promote vessel branching and sprouting that is required for angiogenesis. On the other hand, Tie1 is absent in stable and quiescent mature vessels [27].

A mechanistic study indicated Tie1 as being responsible for angiopoietin's differential function. In mature vessels, as Tie1 is absent, Tie2 can be activated by either Ang1 or Ang2, to promote vessel stability. On active angiogenesis sites, Tie1 and Tie2 form a complex and Ang2 fails to activate Tie2, allowing vessel branching to be promoted. On the other hand, Ang1 is able to dissociate Tie2 from the Tie1-Tie2 complex, activating Tie2 and thus enhancing vascular stability [27,31].

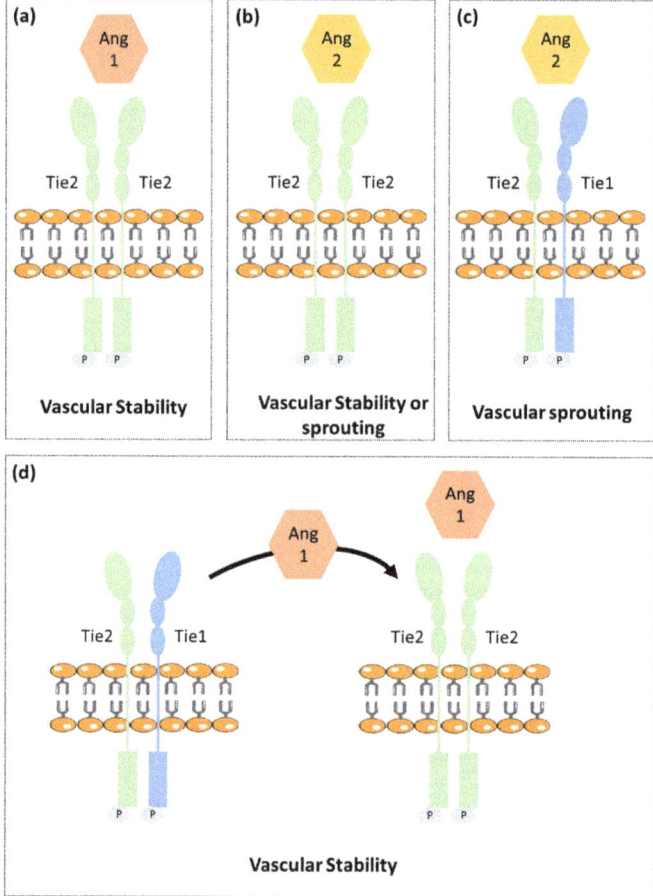

Figure 1. Schematic illustration of the vessel stability regulation by Ang/Tie signaling. (**a**) Ang1 binds to Tie2 promoting vascular stability. (**b**) Ang2 has a dual function acting as a Tie2 agonist or antagonist to promote vascular stability or sprouting, respectively. (**c**) Tie1 forms a complex with Tie2. Upon Ang2 stimulation, Tie1 and Tie2 remain associated and Ang2 induces vascular sprouting. (**d**) Ang1 stimulation promotes Tie2 clustering leading to vascular stability. This schematic representation includes the most consensual theories; however, this pathway is not yet fully understood. Besides that, this figure is a schematic representation that is not intended to translate the real chemical conformation of the proteins.

3. Angiogenesis in Normal Adrenal Cortex

3.1. Fetal Adrenal Cortex

Human fetal adrenal (HFA) plays a critical role in fetal maturation and perinatal survival. HFA steroid hormones regulate intrauterine homeostasis and appropriate fetal organ systems maturation [32,33].

Contrary to the adult adrenal cortex that includes three distinct zones: glomerulosa, fasciculata and reticularis; the HFA is primarily composed of two single distinct zones: outer zone or definitive zone and inner zone or fetal zone [33,34]. The definitive zone comprises a narrow band of small cells that exhibit typical characteristics of cells in proliferative state. Definitive zone does not produce steroids until the third trimester. However, as gestation advances, definitive zone cells start to accumulate lipids and resemble steroidogenic active cells. The fetal zone is the largest adrenal cortex zone and consists of large cells that exhibit features characteristic of steroid-secreting cells [33–37]. In ultrastructural studies, a third zone in between definitive zone and fetal zone, named transitional zone, has been described. The transitional zone is composed by cells with intermediate characteristics, but capable to synthetize cortisol and so cells can be considered analogous to fasciculata layer cells of mature adrenal cortex [33,38–41].

Due to the HFA critical role in fetal maturation, the early and extensive vasculature development that occurs in this gland, is not only necessary but also particularly important. Angiogenesis is not only required for HFA growth and maturation, but it is also necessary for the influx of steroid precursors and trophic factors into the gland to enable mature steroids synthesis and secretion into circulation. Indeed, the fetal adrenal gland is one of the most highly vascularized organs in the human fetus [41].

Previous studies have reported that VEGF-A, FGF-2, Ang1, Ang2, and Tie2 are expressed in HFA since midgestation and to have a putative role in adrenal gland angiogenesis [34,42,43].

Ang2 expression in HFA is markedly higher when compared to the mature adrenal gland, whereas Ang1 and Tie2 expression seem to be similar in both fetal and adult adrenals. Thus, supporting higher angiogenesis activity and vascular instability in developing adrenal glands [42].

Ang2, FGF-2 and VEGF-A expression are mainly expressed in the gland periphery suggesting that the HFA periphery is the primary site of angiogenesis, in parallel to cell proliferation [42,43]. Further supporting this hypothesis, a dense network of irregular capillaries was also observed at the HFA periphery [44].

On the contrary, Ang1 is mainly expressed in the fetal zone, suggesting that the inner adrenal zone presents a greater vessel maturity. Tie2, was exclusively identified to be present in endothelial cells throughout the gland [42,43].

Adrenocorticotropic hormone (ACTH), the main regulator of HFA growth and function, also seems to be implicated in angiogenesis control. In vitro studies found that ACTH upregulates VEGF-A, FGF-2 and Ang2 in the HFA, therefore controlling angiogenesis while simultaneously exerting growth and secretion stimulatory actions [42,43,45,46].

The steroidogenic factor 1 has a critical role in adrenal development, steroidogenesis, and also in gonadal differentiation [47]. In addition, steroidogenic factor 1 also seems to be implicated in HFA angiogenesis regulation by direct interaction and activation of the Ang2 gene promoter. Furthermore, the authors demonstrated that steroidogenic factor 1 and Ang2 are strongly co-expressed in HFA periphery in early stages of development [48].

Overall, these findings support that the adrenal gland growth, steroidogenesis and blood vessel formation, are synchronized phenomena [42,43,45,46].

3.2. Adult Adrenal Cortex

The adrenal gland is one of the most vascularized organs in adult mammalian organisms. Its developed intrinsic vasculature is required for an efficient secretion of steroid hormones into the systemic blood flow. The adrenal gland is supplied by three different arterial branches derived the abdominal aorta: inferior phrenic artery, middle adrenal artery and renal artery. The arterial blood enters in the adrenal gland and flows centripetally through the adrenal cortex into the adrenal medulla [49,50].

Previous studies have found that adrenocortical cells highly express VEGF-A and EG-VEGF—a VEGF specific of steroidogenic organs, both having been pointed out as important molecules for maintenance of the dense and fenestrated vasculature of the adrenal cortex. This expression also seems to be regulated by ACTH [51–55].

In addition, the vasculature of the adrenal cortex seems to be coordinated with the mass of the adrenal cortex, since it suffers fluctuations decreasing or increasing along regression or expansion of the adrenal cortex, respectively [52].

4. Angiogenesis in Adrenocortical Tumors

Adrenocortical tumors (ACT) are common adrenal tumors affecting 3% to 10% of the human population [56]. The majority of ACT are benign non-functioning adrenocortical adenomas (ACA), while malignant ACC are rare with an incidence of 0.7 to 2 per million per year [56]. ACC most often have a poor prognosis and are frequently already metastasized when first diagnosed. ACC pathogenesis is still largely unclear, which results in a lack of biomarkers available for diagnosis and in limited treatment options [57,58].

The status of the VEGF pathway in adrenocortical tumors has been already addressed in multiple studies (Table 1).

Table 1. VEGF pathway findings in adrenocortical tumors.

	Patient Group Comparisons	Results
VEGF	Patients with ACT vs. Healthy individuals	↑ VEGF serum levels in patients with ACT [59,60]
	Aldosterone secreting ACA vs. Non-functioning ACA	↑ VEGF tumor expression in aldosterone producing ACA [61]
	Cortisol secreting ACA vs. Aldosterone secreting ACA	↑ VEGF serum levels patients with cortisol secreting ACA [60]
	ACC vs. Normal adrenal glands	↑ VEGF expression in ACC [61,62]
	ACC vs. ACA	↑ VEGF serum levels in ACC ↑ VEGF tumor expression in ACC [59,61,63,64]
	Patients with recurrent ACC vs. Patients with non-recurrent ACC	↑ VEGF serum levels in recurrent ACC ↑ VEGF tumor expression in recurrent ACC [60,63]
	Localized ACC vs. Invasive ACC	No difference in VEGF tumor expression [63]
VEGF-R2	ACC vs. Normal adrenal glands	↑ VEGF-R2 tumor expression in ACC [62]
	ACC vs. ACA	↑ VEGF-R2 tumor expression in ACC [64]

ACA—Adrenocortical Adenomas; ACC—Adrenocortical carcinomas; ACT—Adrenocortical tumors; VEGF—Vascular endothelial growth factor; VEGFR—Vascular endothelial growth factor receptor; ↑—Increased protein levels or expression.

Patients with ACT were found to present higher VEGF serum levels as compared to healthy controls [59,60]. In addition, Kolomecki et al. demonstrated that VEGF serum levels were significantly higher in patients with non-functioning malignant tumors than in patients with non-functioning ACA. Noteworthy, VEGF serum levels in patients with ACC were shown to decrease after tumor surgical resection and increase in patients who experienced tumor recurrence [59]. de Fraipont et al. found that cytosolic VEGF-A concentrations were higher in ACC when compared to ACA, although not being significantly different when localized and more invasive ACC were compared [63]. Nevertheless, cytosolic VEGF-A concentrations were higher in recurrent as compared to non-recurrent ACC after primary tumor resection [63].

Tumor VEGF expression was also found to be higher in ACC as compared to normal adrenal glands and ACA [61,62,64]. VEGF receptor 2 tumor expression was also found to be higher in ACC when compared with ACA and normal adrenal glands [62,64].

Bernini et al., however, found that tumor VEGF expression was not directly related with vascular density, which was lower in ACC as compared to ACA and normal adrenal tissue. The fact that a higher VEGF expression was not shown to be associated with increased vascular density in ACC, was somehow unsurprising since a high vascular density already characterizes normal adrenal cortex tissue. What surprised researchers was that despite ACC lower vascular density, patients still had a very short survival time [61].

Other studies reported that although no differences in vascular density were noticed when ACC, ACA and normal adrenal glands were compared, blood vessels perimeter and area were higher in ACC when compared to ACA [65,66]. In addition, endothelial cell proliferation was higher in ACC [66].

On an opposed direction, another group reported vascular density to be higher in malignant ACT as compared to benign ACT [67]. Another study observed that in their series VEGF expression was positively correlated with vessel density [64]. Pereira et al. also reported ACC to present a higher vascular density, but only when compared to cortisol secreting ACA [68]. This could, however, be derived from cortisol anti-angiogenic effects [69]. There is additional evidence supporting that adrenocortical angiogenic status could be tightly related to the tumor's hormonal functionality. Bernini et al. found that VEGF tumor expression was higher in aldosterone secreting ACA as compared to non-functioning ACA and normal adrenal glands [61]. In addition, in another study patients with cortisol-secreting ACA were found to have higher circulating VEGF levels than patients with aldosterone secreting adenomas [60].

The discovery of EG-VEGF, a steroidogenic organ specific VEGF, brought some enthusiasm to the scientific community as a potential explanation to the contradictory angiogenic patterns in ACTs as well as a potential target for ACC treatment. Heck et al. characterized the expression of EG-VEGF and its receptors [prokineticin receptor 1 (PKR1) and 2 (PKR2)] in a large number of ACC, ACA and normal adrenal glands. In this study, EG-VEGF and both receptors PKR1 and PKR2 were found to be present in the majority of ACT. Moreover, the nuclear protein expression of either EG-VEGF or PKR1 or both in ACC was reported to be associated with higher mortality, suggesting that these could be used as prognostic markers for overall patient survival [53].

New prognostic and diagnostic markers are needed to improve ACC clinical practice. As described in this section, the usefulness of angiogenic factors for ACC diagnosis and/or prognosis was already investigated. From those, VEGF was the one with more consistent and replicable results, being increased in ACC when compared with ACA [59,61,63,64], in particular in the recurrent malignant tumors [60,63]. However, due to the rarity of ACC, the number of patients included in each study is small. So, in the future, to validate this result, multi-center studies are needed to increase the samples/participants' number and to uniformize the methodological approach to analyze the VEGF tumors expression in ACT. Stratified analysis according to tumors functionality are needed since in previous studies, it showed to influence VEGF levels.

5. Anti-Angiogenic Agents' Efficacy in Adrenocortical Carcinomas Treatment

The demonstration that patients with ACC had high VEGF circulating levels and tumor expression, along with the recent evidence on the efficacy of anti-VEGF drugs for other types of neoplasia treatment, such as, advanced colorectal cancer [70], opened the promising perspective of using this drug class agents for ACC treatment as well.

The first report using an anti-angiogenic agent for the treatment of patients with ACC was released in 2010 (Table 2). Ten patients with advanced ACC, refractory to several cytotoxic chemotherapies, were treated with the monoclonal anti-VEGF antibody bevacizumab in combination with capecitabine, an adrenolytic agent. The results were disappointing since the disease progressed in all patients [71].

A phase II clinical trial using sorafenib in combination with paclitaxel was conducted in ten patients with advanced ACC after treatment with mitotane plus one or two chemotherapy lines. Sorafenib is a tyrosine kinase inhibitor (TKI) drug that inhibits several receptors, such as VEGFR2, VEGFR3, platelet-derived growth factor receptor (PDFGR) and RAF-1, a key enzyme in the MAPK-ERK signaling pathways. The sorafenib plus paclitaxel drug combination was demonstrated to be ineffective in patients with ACC, as progressive disease was observed in nine consecutive patients leading to clinical trial interruption 2 months after initiation [72].

In another trial 35 patients with ACC refractory to mitotane and cytotoxic chemotherapies were treated with the TKI sunitinib, a drug that inhibits multiple receptors, such as VEGFR1 and VEGFR2, c-KIT, Fms-like tyrosine kinase 3, and PDGFR. Six of the thirty-five patients in that trial died of progressive disease. Of the remaining twenty-nine patients, five patients had stable disease, and 23 patients had progressive disease on first evaluation (12 weeks). Three of the five patients with stable disease on first evaluation, had disease progression later. In addition, authors reported that concomitant mitotane had a negative impact on treatment outcome, by lowering sunitinib blood levels. Therefore, sunitinib only demonstrated to have a modest efficacy in the treatment of patients with advanced ACC, while the efficacy in patients without mitotane exposure needs to be further assessed [73].

In a phase II clinical trial, axitinib, a potent VEGFR-1, VEGFR-2, and VEGFR-3 selective inhibitor, was administrated to thirteen patients with metastatic ACC previously treated with at least one chemotherapy regimen, with or without mitotane. No patient in trial achieved a partial or complete response, and only eight patients experienced stable disease for more than 3 months [74].

Thalidomide is an immunomodulatory agent with anti-angiogenic properties by targeting TNF-α, ILs, VEGF, bFGF. The effectiveness of thalidomide was investigated in a trial that included twenty-seven patients with advanced ACC refractory to mitotane and other systemic drug treatments. Twenty-five of the twenty-seven patients experienced clinical or radiological disease progression at the time of first evaluation. So, thalidomide also, only showed to be marginally effective in patients with refractory advanced ACC [75].

Lenvatinib is another TKI drug that inhibits multiple receptors including VEGFR-1, VEGFR-2 and VEGFR-3, FGFRs, PDGFR-α, KIT and RET. The efficacy of lenvatinib in combination with pembrolizumab, an immune checkpoint inhibitor, was also investigated in eight patients with ACC and progressive and/or metastatic disease after receiving previous treatment interventions. None of the eight patients had to discontinue the treatment in result of toxicity. One patient had stable disease, lasting for 8 months, two patients had a partial response while receiving therapy and five patients developed progressive disease, so lenvatinib plus pembrolizumab combined therapy was demonstrated to achieve positive responses in a subset of patients without significant toxicity [76]. Phase II clinical trials with larger patient cohorts are still needed to confirm these conclusions.

The clinical efficacy and safety of the TKI cabozantinib was investigated in a retrospective cohort study in sixteen patients with advanced ACC after other treatments having failed. Cabozantinib is a multi-inhibitor of c-MET, VEGFR-2, AXL, and RET. At first evaluation, two patients had partial response and six had stable disease. At four months evaluation, half of patients were alive and progression free [77]. Although these results were not brilliant, they were superior to the ones previously reported for other anti-angiogenic agents.

Although previous studies using anti-angiogenic therapies did not show encouraging results in patients with ACC, there are several registered clinical trials using VEGF-R inhibitors ongoing or due to be initiated. Two phase II clinical trials designed to test the efficacy of cabozantinib in patients with advanced ACC (NCT03370718 and NCT03612232) are currently recruiting, and a phase II clinical trial to test the efficacy of the VEGFR-2 inhibitor apatinib plus camrelizumab an immune checkpoint inhibitor, is registered (NCT04318730).

Table 2. Clinical studies using anti-angiogenic drugs for the treatment of patients with adrenocortical carcinomas.

Anti-Angiogenic Drug	Mechanism of Action	Study Type	Patient Population	Results	Ref.
Bevacizumab (+capecitabine)	Monoclonal anti-VEGF antibody	Observational retrospective cohort study	Patients with refractory ACC ($n = 10$)	PFS: 59 days OS: 124 days	[71]
Thalidomide	Immunomodulatory agent that targets TNF-α, ILs, VEGF, bFGF	Observational retrospective cohort study	Patients with refractory ACC ($n = 27$)	PFS: 11.2 weeks (4.4–22.8 weeks) OS: 36.4 weeks (5.1–111.1 weeks)	[75]
Lenvatinib (+pembrolizumab)	Multi-TKI that inhibits VEGFR-1, VEGFR-2 and VEGFR-3, FGFRs, PDGFR-α, KIT, RET	Observational retrospective cohort study	Patients with recurrent and/or metastatic ACC ($n = 8$)	PFS: 5.5 months OS: NA	[76]
Cabozatinib	TKI that targets VEGFR-2 and c-Met	Observational retrospective cohort study	Patients with refractory metastatic ACC ($n = 16$)	PFS: 16.2 weeks (2.8–61 weeks) OS: 56 weeks (5.6–83.1 weeks)	[77]
Sorafenib (+paclitaxel)	Multi-TKI inhibitor that VEGFR-2 VEGFR-3, PDGFR and RAF-1	Phase II, single-arm, open label clinical trial	Patients with refractory metastatic ACC ($n = 10$)	Trial interrupted due disease progression in all enrolled patients	[72]
Sunitinib	Multi-TKI that inhibits VEGFR-1 and VEGFR-2, c-KIT, FLT3 and PDGFR	Phase II, single-arm, open label clinical trial	Patients with advanced ACC after mitotane or others cytotoxic drugs ($n = 35$)	PFS: 2.8 months (5.6–11.2 months) OS: 5.4 months (14.0–35.5 months)	[73]
Axitinib	Selective inhibitor of VEGFR-1, VEGFR-2 and VEGFR-3	Phase II, single-arm, open label clinical trial	Patients with metastatic ACC previously treated with at least one chemotherapy regimen ($n = 13$)	PFS: 5.48 months (1.8–10.92 months) OS: 13.7 months	[74]

FPS and OS median and ranges were included in the table, when available in the original manuscript. ACC—Adrenocortical carcinoma; bFGF—basic fibroblast growth factor; FGFRs—fibroblast growth factor receptors; FLT3—FMS-like tyrosine kinase 3; Ils—interleukins; NA—not available; OS—Overall Survival; PFS—Progression-free survival; TKI—Tyrosine kinase inhibitor; TNF-α—Tumor necrosis factor alpha; VEGF—Vascular endothelial growth factor; VEGFR—Vascular endothelial growth factor receptor; PDGFR—platelet-derived growth factor receptor.

It is important to highlight that although the VEGF pathway is one of the most important angiogenic pathways, this is not the only pathway involved in angiogenesis regulation. As far as we are aware, no clinical trial has tested the effect of drugs targeting the Ang-Tie pathway in patients with ACC.

In 2014, one single patient with ACC was enrolled in a phase 1b clinical trial designed to test the efficacy of trebananib, a dual Ang1 and Ang2 inhibitor plus VEGFR inhibitors (bevacizumab or motesanib) in various solid tumors. Stable disease was achieved in the patient with ACC. However, since this clinical trial included patients with different solid tumors no further data specifically related to this patient is available [78].

Further investigation of molecules involved in Ang-Tie pathway in ACC tissues is needed in order to understand whether this pathway has a role in the pathophysiology of this type of tumor. This knowledge is needed to provide a rationale for conducting clinical trials targeting both Ang-Tie and VEGF pathways in patients with ACC.

There are no doubts that angiogenesis is an important process in ACC progression and the rationale for the use of anti-angiogenic drugs in ACC treatment is unquestionable.

However, so far clinical trials to test the efficacy of these drugs were conducted in patients with advanced/metastatic ACC that precluded the possibilities of success, since even if the angiogenic capacity of the tumor is decreased, the disease is already in an uncontrolled stage. In contrast, the efficacy of anti-angiogenic drugs in non-advanced ACC is unknown yet. Therefore, ex-vivo studies could be useful to assess these drugs efficacy as compared to mitotane, which is the only drug licensed for ACC treatment.

In addition, whenever possible, conducting studies to evaluate whether there is a correlation between drug efficacy and the tumor molecular profile, would provide additional insights on the mechanisms responsible for successful or unsuccessful treatment and support future clinical trials.

6. Conclusions

Angiogenesis is well-known to be required for cancer cell expansion and considered an important hallmark of cancer [1]. Adrenal glands have a very dense vascular network that is necessary to support their hormonal secretion functions. Therefore, it may not be surprising that this normal adrenal gland high vascular density is unlikely to be further increased in the context of adrenocortical neoplasia.

Although several molecules within the VEGF pathway were identified as prognostic markers and promising targets for ACC treatment, the cohort studies and clinical trials so far concluded yielded disappointing results. The most promising results were observed for cabozantinib, a multi-TKI inhibitor, which induced stable or partial response in half of the patients with advanced ACC [77]. Given this effect and its overall safety with a tolerable side effect profile, two clinical trials to assess the efficacy of this drug are now recruiting (NCT03370718 and NCT03612232). Both trials require the discontinuation of mitotane and exclude patients with mitotane levels higher than 2 mg/L. Future clinical trials should take in consideration prior mitotane exposure on the study design, since previous or concomitant mitotane use may influence the investigational drug treatment outcomes due to its impact on drug metabolism [73,79].

Another registered clinical trial will assess the efficacy of the anti-angiogenic drug Apatinib combined with an immunomodulatory agent (PD-1 inhibitor: camrelizumab) in ACC (NCT04318730), a drug that was previously tested alone and was demonstrated to induce a stable disease or an objective response in 52% of the patients [80]. Since this drug combination elicited impressive clinical results in many solid tumors [81–83], there is a great expectation for this clinical trial outcomes.

Furthermore, the Ang-Tie pathway which is known to have an important role on fetal adrenal gland angiogenesis should also receive attention [42,43]. The role of Ang-Tie pathway in adrenocortical tumors has not yet been investigated. Future studies and clinical trials investigating the role of Ang-Tie pathway in adrenocortical tumors and the efficacy of targeting Ang-Tie or both Ang-Tie and VEGF pathways in ACC treatment are needed.

Supplementary Materials: The following are available online at https://www.mdpi.com/2072-6694/13/5/1030/s1. Supplementary File S1: Data Source and Study Selection.

Author Contributions: Conceptualization, S.S.P. and D.P.; writing—original draft preparation, S.S.P. and S.O.; writing—review and editing, M.P.M. and D.P. All authors have read and agreed to the published version of the manuscript.

Funding: This research was funded by the Foundation for Science and Technology (FCT)-Portugal (PTDC/MEC-ONC/31384/2017) and by the Associação dos Amigos do Serviço de Endocrinologia do Hospital de São João. Unit for Multidisciplinary Research in Biomedicine (UMIB) is also funded by FCT (UIDB/00215/2020 and UIDP/00215/2020).

Conflicts of Interest: The authors declare no conflict of interest.

References

1. Hanahan, D.; Weinberg, R.A. Hallmarks of cancer: The next generation. *Cell* **2011**, *144*, 646–674. [CrossRef]
2. Tonnesen, M.G.; Feng, X.; Clark, R.A. Angiogenesis in wound healing. *J. Investig. Dermatol. Symp. Proc.* **2000**, *5*, 40–46. [CrossRef]

3. Chung, A.S.; Ferrara, N. Developmental and Pathological Angiogenesis. *Annu. Rev. Cell Dev. Biol.* **2011**, *27*, 563–584. [CrossRef]
4. Zuazo-Gaztelu, I.; Casanovas, O. Unraveling the Role of Angiogenesis in Cancer Ecosystems. *Front. Oncol.* **2018**, *8*, 248. [CrossRef]
5. Mentzer, S.J.; Konerding, M.A. Intussusceptive angiogenesis: Expansion and remodeling of microvascular networks. *Angiogenesis* **2014**, *17*, 499–509. [CrossRef] [PubMed]
6. Lugano, R.; Ramachandran, M.; Dimberg, A. Tumor angiogenesis: Causes, consequences, challenges and opportunities. *Cell. Mol. Life Sci.* **2019**. [CrossRef] [PubMed]
7. Duran, C.L.; Howell, D.W.; Dave, J.M.; Smith, R.L.; Torrie, M.E.; Essner, J.J.; Bayless, K.J. Molecular Regulation of Sprouting Angiogenesis. *Compr. Physiol.* **2017**, *8*, 153–235. [CrossRef]
8. Makanya, A.N.; Hlushchuk, R.; Djonov, V.G. Intussusceptive angiogenesis and its role in vascular morphogenesis, patterning, and remodeling. *Angiogenesis* **2009**, *12*, 113. [CrossRef]
9. Kolte, D.; McClung, J.A.; Aronow, W.S. Vasculogenesis and angiogenesis. In *Translational Research in Coronary Artery Disease*; Elsevier: Amsterdam, The Netherlands, 2016; pp. 49–65.
10. Ge, H.; Luo, H. Overview of advances in vasculogenic mimicry—A potential target for tumor therapy. *Cancer Manag. Res.* **2018**, *10*, 2429–2437. [CrossRef] [PubMed]
11. Fernández-Cortés, M.; Delgado-Bellido, D.; Oliver, F.J. Vasculogenic Mimicry: Become an Endothelial Cell "But Not So Much". *Front. Oncol.* **2019**, *9*. [CrossRef] [PubMed]
12. Saharinen, P.; Eklund, L.; Pulkki, K.; Bono, P.; Alitalo, K. VEGF and angiopoietin signaling in tumor angiogenesis and metastasis. *Trends Mol. Med.* **2011**, *17*, 347–362. [CrossRef] [PubMed]
13. Takahashi, H.; Shibuya, M. The vascular endothelial growth factor (VEGF)/VEGF receptor system and its role under physiological and pathological conditions. *Clin. Sci.* **2005**, *109*, 227–241. [CrossRef]
14. Guo, H.-F.; Vander Kooi, C.W. Neuropilin Functions as an Essential Cell Surface Receptor. *J. Biol. Chem.* **2015**, *290*, 29120–29126. [CrossRef]
15. Chiodelli, P.; Mitola, S.; Ravelli, C.; Oreste, P.; Rusnati, M.; Presta, M. Heparan sulfate proteoglycans mediate the angiogenic activity of the vascular endothelial growth factor receptor-2 agonist gremlin. *Arterioscler. Thromb. Vasc. Biol.* **2011**, *31*, e116–e127. [CrossRef]
16. Cross, M.J.; Dixelius, J.; Matsumoto, T.; Claesson-Welsh, L. VEGF-receptor signal transduction. *Trends Biochem. Sci.* **2003**, *28*, 488–494. [CrossRef]
17. Shibuya, M. Vascular Endothelial Growth Factor (VEGF) and Its Receptor (VEGFR) Signaling in Angiogenesis: A Crucial Target for Anti- and Pro-Angiogenic Therapies. *Genes Cancer* **2011**, *2*, 1097–1105. [CrossRef]
18. LeCouter, J.; Kowalski, J.; Foster, J.; Hass, P.; Zhang, Z.; Dillard-Telm, L.; Frantz, G.; Rangell, L.; DeGuzman, L.; Keller, G.A.; et al. Identification of an angiogenic mitogen selective for endocrine gland endothelium. *Nature* **2001**, *412*, 877–884. [CrossRef] [PubMed]
19. Augustin, H.G.; Koh, G.Y.; Thurston, G.; Alitalo, K. Control of vascular morphogenesis and homeostasis through the angiopoietin-Tie system. *Nat. Rev. Mol. Cell Biol.* **2009**, *10*, 165–177. [CrossRef] [PubMed]
20. Lee, H.J.; Cho, C.H.; Hwang, S.J.; Choi, H.H.; Kim, K.T.; Ahn, S.Y.; Kim, J.H.; Oh, J.L.; Lee, G.M.; Koh, G.Y. Biological characterization of angiopoietin-3 and angiopoietin. *FASEB J* **2004**, *18*, 1200–1208. [CrossRef]
21. Eklund, L.; Saharinen, P. Angiopoietin signaling in the vasculature. *Exp. Cell Res.* **2013**, *319*, 1271–1280. [CrossRef]
22. Fagiani, E.; Christofori, G. Angiopoietins in angiogenesis. *Cancer Lett.* **2013**, *328*, 18–26. [CrossRef]
23. Davis, S.; Aldrich, T.H.; Jones, P.F.; Acheson, A.; Compton, D.L.; Jain, V.; Ryan, T.E.; Bruno, J.; Radziejewski, C.; Maisonpierre, P.C.; et al. Isolation of Angiopoietin-1, a Ligand for the TIE2 Receptor, by Secretion-Trap Expression Cloning. *Cell* **1996**, *87*, 1161–1169. [CrossRef]
24. Dumont, D.J.; Gradwohl, G.; Fong, G.H.; Puri, M.C.; Gertsenstein, M.; Auerbach, A.; Breitman, M.L. Dominant-negative and targeted null mutations in the endothelial receptor tyrosine kinase, tek, reveal a critical role in vasculogenesis of the embryo. *Genes Dev.* **1994**, *8*, 1897–1909. [CrossRef] [PubMed]
25. Yuan, H.T.; Khankin, E.V.; Karumanchi, S.A.; Parikh, S.M. Angiopoietin 2 is a partial agonist/antagonist of Tie2 signaling in the endothelium. *Mol. Cell. Biol.* **2009**, *29*, 2011–2022. [CrossRef] [PubMed]
26. Barton, W.A.; Tzvetkova-Robev, D.; Miranda, E.P.; Kolev, M.V.; Rajashankar, K.R.; Himanen, J.P.; Nikolov, D.B. Crystal structures of the Tie2 receptor ectodomain and the angiopoietin-2-Tie2 complex. *Nat. Struct. Mol. Biol.* **2006**, *13*, 524–532. [CrossRef]
27. Seegar, T.C.M.; Eller, B.; Tzvetkova-Robev, D.; Kolev, M.V.; Henderson, S.C.; Nikolov, D.B.; Barton, W.A. Tie1-Tie2 interactions mediate functional differences between angiopoietin ligands. *Mol. Cell* **2010**, *37*, 643–655. [CrossRef] [PubMed]
28. Porat, R.M.; Grunewald, M.; Globerman, A.; Itin, A.; Barshtein, G.; Alhonen, L.; Alitalo, K.; Keshet, E. Specific Induction of *tie1* Promoter by Disturbed Flow in Atherosclerosis-Prone Vascular Niches and Flow-Obstructing Pathologies. *Circ. Res.* **2004**, *94*, 394–401. [CrossRef] [PubMed]
29. Puri, M.C.; Rossant, J.; Alitalo, K.; Bernstein, A.; Partanen, J. The receptor tyrosine kinase TIE is required for integrity and survival of vascular endothelial cells. *EMBO J.* **1995**, *14*, 5884–5891. [CrossRef] [PubMed]
30. La Porta, S.; Roth, L.; Singhal, M.; Mogler, C.; Spegg, C.; Schieb, B.; Qu, X.; Adams, R.H.; Baldwin, H.S.; Savant, S.; et al. Endothelial Tie1-mediated angiogenesis and vascular abnormalization promote tumor progression and metastasis. *J. Clin. Investig.* **2018**, *128*, 834–845. [CrossRef]

31. Zhang, Y.; Kontos, C.D.; Annex, B.H.; Popel, A.S. Angiopoietin-Tie Signaling Pathway in Endothelial Cells: A Computational Model. *iScience* **2019**, *20*, 497–511. [CrossRef]
32. Liggins, G.C. Adrenocortical-related maturational events in the fetus. *Am. J. Obstet. Gynecol.* **1976**, *126*, 931–941. [CrossRef]
33. Mesiano, S.; Jaffe, R.B. Developmental and functional biology of the primate fetal adrenal cortex. *Endocr. Rev.* **1997**, *18*, 378–403. [CrossRef] [PubMed]
34. Ishimoto, H.; Jaffe, R.B. Development and function of the human fetal adrenal cortex: A key component in the feto-placental unit. *Endocr. Rev.* **2011**, *32*, 317–355. [CrossRef]
35. Lanman, J.T. The fetal zone of the adrenal gland: Its developmental course, comparative anatomy, and possible physiologic functions. *Medicine* **1953**, *32*, 389–430. [CrossRef] [PubMed]
36. Johannisson, E. The foetal adrenal cortex in the human. Its ultrastructure at different stages of development and in different functional states. *Acta Endocrinol.* **1968**, *58*, S7–S107. [CrossRef]
37. Spencer, S.J.; Mesiano, S.; Lee, J.Y.; Jaffe, R.B. Proliferation and apoptosis in the human adrenal cortex during the fetal and perinatal periods: Implications for growth and remodeling. *J. Clin. Endocrinol. Metab.* **1999**, *84*, 1110–1115. [CrossRef] [PubMed]
38. McNutt, N.S.; Jones, A.L. Observations on the ultrastructure of cytodifferentiation in the human fetal adrenal cortex. *Lab. Investig. J. Tech. Methods Pathol.* **1970**, *22*, 513–527.
39. Sucheston, M.E.; Cannon, M.S. Development of zonular patterns in the human adrenal gland. *J. Morphol.* **1968**, *126*, 477–491. [CrossRef]
40. Mesiano, S.; Coulter, C.L.; Jaffe, R.B. Localization of cytochrome P450 cholesterol side-chain cleavage, cytochrome P450 17 alpha-hydroxylase/17, 20-lyase, and 3 beta-hydroxysteroid dehydrogenase isomerase steroidogenic enzymes in human and rhesus monkey fetal adrenal glands: reappraisal of functional zonation. *J. Clin. Endocrinol. Metab.* **1993**, *77*, 1184–1189. [CrossRef] [PubMed]
41. McClellan, M.C.; Brenner, R.M. Development of the fetal adrenals in nonhuman primates: Electron microscopy. In *Fetal Endocrinology*; Novy, M.J., Resko, J.A., Eds.; Academic Press: Cambridge, MA, USA, 1981; Volume 1, pp. 383–403. [CrossRef]
42. Ishimoto, H.; Ginzinger, D.G.; Jaffe, R.B. Adrenocorticotropin preferentially up-regulates angiopoietin 2 in the human fetal adrenal gland: Implications for coordinated adrenal organ growth and angiogenesis. *J. Clin. Endocrinol. Metab.* **2006**, *91*, 1909–1915. [CrossRef]
43. Ishimoto, H.; Minegishi, K.; Higuchi, T.; Furuya, M.; Asai, S.; Kim, S.H.; Tanaka, M.; Yoshimura, Y.; Jaffe, R.B. The periphery of the human fetal adrenal gland is a site of angiogenesis: Zonal differential expression and regulation of angiogenic factors. *J. Clin. Endocrinol. Metab.* **2008**, *93*, 2402–2408. [CrossRef]
44. Pitynski, K.; Litwin, J.A.; Nowogrodzka-Zagorska, M.; Miodonski, A.J. Vascular architecture of the human fetal adrenal gland: a SEM study of corrosion casts. *Ann. Anat. Anat. Anz Off. Organ Anat. Ges.* **1996**, *178*, 215–222. [CrossRef]
45. Mesiano, S.; Mellon, S.H.; Gospodarowicz, D.; Di Blasio, A.M.; Jaffe, R.B. Basic fibroblast growth factor expression is regulated by corticotropin in the human fetal adrenal: A model for adrenal growth regulation. *Proc. Natl. Acad. Sci. USA* **1991**, *88*, 5428–5432. [CrossRef] [PubMed]
46. Shifren, J.L.; Mesiano, S.; Taylor, R.N.; Ferrara, N.; Jaffe, R.B. Corticotropin regulates vascular endothelial growth factor expression in human fetal adrenal cortical cells. *J. Clin. Endocrinol. Metab.* **1998**, *83*, 1342–1347. [CrossRef] [PubMed]
47. Ozisik, G.; Achermann, J.C.; Meeks, J.J.; Jameson, J.L. SF1 in the development of the adrenal gland and gonads. *Horm. Res.* **2003**, *59* (Suppl. 1), 94–98. [CrossRef]
48. Ferraz-de-Souza, B.; Lin, L.; Shah, S.; Jina, N.; Hubank, M.; Dattani, M.T.; Achermann, J.C. ChIP-on-chip analysis reveals angiopoietin 2 (Ang2, ANGPT2) as a novel target of steroidogenic factor-1 (SF-1, NR5A1) in the human adrenal gland. *FASEB J.* **2010**, *25*, 1166–1175. [CrossRef]
49. Sapirstein, L.A.; Goldman, H. Adrenal blood flow in the albino rat. *Am. J. Physiol* **1959**, *196*, 159–162. [CrossRef]
50. Bassett, J.R.; West, S.H. Vascularization of the adrenal cortex: Its possible involvement in the regulation of steroid hormone release. *Microsc. Res. Tech.* **1997**, *36*, 546–557. [CrossRef]
51. Thomas, M.; Keramidas, M.; Monchaux, E.; Feige, J.J. Role of adrenocorticotropic hormone in the development and maintenance of the adrenal cortical vasculature. *Microsc. Res. Tech.* **2003**, *61*, 247–251. [CrossRef] [PubMed]
52. Vittet, D.; Ciais, D.; Keramidas, M.; De Fraipont, F.; Feige, J.J. Paracrine control of the adult adrenal cortex vasculature by vascular endothelial growth factor. *Endocr. Res.* **2000**, *26*, 843–852. [CrossRef] [PubMed]
53. Heck, D.; Wortmann, S.; Kraus, L.; Ronchi, C.L.; Sinnott, R.O.; Fassnacht, M.; Sbiera, S. Role of Endocrine Gland-Derived Vascular Endothelial Growth Factor (EG-VEGF) and Its Receptors in Adrenocortical Tumors. *Horm. Cancer* **2015**, *6*, 225–236. [CrossRef] [PubMed]
54. Senger, D.R. Vascular endothelial growth factor: Much more than an angiogenesis factor. *Mol. Biol. Cell* **2010**, *21*, 377–379. [CrossRef]
55. Gomez-Sanchez, C.E. Regulation of Adrenal Arterial Tone by Adrenocorticotropin: The Plot Thickens. *Endocrinology* **2007**, *148*, 3566–3568. [CrossRef] [PubMed]
56. Else, T.; Kim, A.C.; Sabolch, A.; Raymond, V.M.; Kandathil, A.; Caoili, E.M.; Jolly, S.; Miller, B.S.; Giordano, T.J.; Hammer, G.D. Adrenocortical Carcinoma. *Endocr. Rev.* **2014**, *35*, 282–326. [CrossRef]
57. Libé, R. Adrenocortical carcinoma (ACC): Diagnosis, prognosis, and treatment. *Front. Cell Dev. Biol.* **2015**, *3*, 45. [CrossRef]

58. Fassnacht, M.; Arlt, W.; Bancos, I.; Dralle, H.; Newell-Price, J.; Sahdev, A.; Tabarin, A.; Terzolo, M.; Tsagarakis, S.; Dekkers, O.M. Management of adrenal incidentalomas: European Society of Endocrinology Clinical Practice Guideline in collaboration with the European Network for the Study of Adrenal Tumors. *Eur. J. Endocrinol.* **2016**, *175*, G1–G34. [CrossRef]
59. Kolomecki, K.; Stepien, H.; Bartos, M.; Kuzdak, K. Usefulness of VEGF, MMP-2, MMP-3 and TIMP-2 serum level evaluation in patients with adrenal tumours. *Endocr. Regul.* **2001**, *35*, 9–16.
60. Zacharieva, S.; Atanassova, I.; Orbetzova, M.; Nachev, E.; Kalinov, K.; Kirilov, G.; Shigarminova, R.; Ivanova, R.; Dashev, G. Circulating vascular endothelial growth factor and active renin concentrations and prostaglandin E2 urinary excretion in patients with adrenal tumours. *Eur. J. Endocrinol.* **2004**, *150*, 345–349. [CrossRef]
61. Bernini, G.P.; Moretti, A.; Bonadio, A.G.; Menicagli, M.; Viacava, P.; Naccarato, A.G.; Iacconi, P.; Miccoli, P.; Salvetti, A. Angiogenesis in Human Normal and Pathologic Adrenal Cortex. *J. Clin. Endocrinol. Metab.* **2002**, *87*, 4961–4965. [CrossRef] [PubMed]
62. Kroiss, M.; Reuss, M.; Kühner, D.; Johanssen, S.; Beyer, M.; Zink, M.; Hartmann, M.F.; Dhir, V.; Wudy, S.A.; Arlt, W.; et al. Sunitinib Inhibits Cell Proliferation and Alters Steroidogenesis by Down-Regulation of HSD3B2 in Adrenocortical Carcinoma Cells. *Front. Endocrinol.* **2011**, *2*, 27. [CrossRef]
63. de Fraipont, F.; El Atifi, M.; Gicquel, C.; Bertagna, X.; Chambaz, E.M.; Feige, J.J. Expression of the Angiogenesis Markers Vascular Endothelial Growth Factor-A, Thrombospondin-1, and Platelet-Derived Endothelial Cell Growth Factor in Human Sporadic Adrenocortical Tumors: Correlation with Genotypic Alter. *J. Clin. Endocrinol. Metab.* **2000**, *85*, 4734–4741. [CrossRef]
64. Xu, Y.Z.; Zhu, Y.; Shen, Z.J.; Sheng, J.Y.; He, H.C.; Ma, G.; Qi, Y.C.; Zhao, J.P.; Wu, Y.X.; Rui, W.B.; et al. Significance of heparanase-1 and vascular endothelial growth factor in adrenocortical carcinoma angiogenesis: Potential for therapy. *Endocrine* **2011**, *40*, 445–451. [CrossRef]
65. Diaz-Cano, S.J.; De Miguel, M.; Blanes, A.; Galera, H.; Wolfe, H.J. Contribution of the microvessel network to the clonal and kinetic profiles of adrenal cortical proliferative lesions. *Hum. Pathol.* **2001**, *32*, 1232–1239. [CrossRef] [PubMed]
66. Sasano, H.; Ohashi, Y.; Suzuki, T.; Nagura, H. Vascularity in human adrenal cortex. *Mod. Pathol.* **1998**, *11*, 329–333. [PubMed]
67. Zhu, Y.; Xu, Y.; Chen, D.; Zhang, C.; Rui, W.; Zhao, J.; Zhu, Q.; Wu, Y.; Shen, Z.; Wang, W.; et al. Expression of STAT3 and IGF2 in adrenocortical carcinoma and its relationship with angiogenesis. *Clin. Transl. Oncol.* **2014**, *16*, 644–649. [CrossRef]
68. Pereira, S.S.; Costa, M.M.; Guerreiro, S.G.; Monteiro, M.P.; Pignatelli, D. Angiogenesis and Lymphangiogenesis in the Adrenocortical Tumors. *Pathol. Oncol. Res.* **2018**, *24*, 689–693. [CrossRef] [PubMed]
69. Logie, J.J.; Ali, S.; Marshall, K.M.; Heck, M.M.S.; Walker, B.R.; Hadoke, P.W.F. Glucocorticoid-mediated inhibition of angiogenic changes in human endothelial cells is not caused by reductions in cell proliferation or migration. *PLoS ONE* **2010**, *5*, e14476. [CrossRef]
70. Hurwitz, H.; Fehrenbacher, L.; Novotny, W.; Cartwright, T.; Hainsworth, J.; Heim, W.; Berlin, J.; Baron, A.; Griffing, S.; Holmgren, E.; et al. Bevacizumab plus Irinotecan, Fluorouracil, and Leucovorin for Metastatic Colorectal Cancer. *N. Engl. J. Med.* **2004**, *350*, 2335–2342. [CrossRef]
71. Wortmann, S.; Quinkler, M.; Ritter, C.; Kroiss, M.; Johanssen, S.; Hahner, S.; Allolio, B.; Fassnacht, M. Bevacizumab plus capecitabine as a salvage therapy in advanced adrenocortical carcinoma. *Eur. J. Endocrinol.* **2010**, *162*, 349. [CrossRef]
72. Berruti, A.; Sperone, P.; Ferrero, A.; Germano, A.; Ardito, A.; Priola, A.M.; Francia, S.D.; Volante, M.; Daffara, F.; Generali, D.; et al. Phase II study of weekly paclitaxel and sorafenib as second/third-line therapy in patients with adrenocortical carcinoma. *Eur. J. Endocrinol.* **2012**, *166*, 451. [CrossRef]
73. Kroiss, M.; Quinkler, M.; Johanssen, S.; van Erp, N.P.; Lankheet, N.; Pöllinger, A.; Laubner, K.; Strasburger, C.J.; Hahner, S.; Müller, H.-H.; et al. Sunitinib in Refractory Adrenocortical Carcinoma: A Phase II, Single-Arm, Open-Label Trial. *J. Clin. Endocrinol. Metab.* **2012**, *97*, 3495–3503. [CrossRef]
74. O'Sullivan, C.; Edgerly, M.; Velarde, M.; Wilkerson, J.; Venkatesan, A.M.; Pittaluga, S.; Yang, S.X.; Nguyen, D.; Balasubramaniam, S.; Fojo, T. The VEGF inhibitor axitinib has limited effectiveness as a therapy for adrenocortical cancer. *J. Clin. Endocrinol. Metab.* **2014**, *99*, 1291–1297. [CrossRef] [PubMed]
75. Kroiss, M.; Deutschbein, T.; Schlötelburg, W.; Ronchi, C.L.; Hescot, S.; Körbl, D.; Megerle, F.; Beuschlein, F.; Neu, B.; Quinkler, M.; et al. Treatment of Refractory Adrenocortical Carcinoma with Thalidomide: Analysis of 27 Patients from the European Network for the Study of Adrenal Tumours Registry. *Exp. Clin. Endocrinol. Diabetes* **2019**, *127*, 578–584. [CrossRef] [PubMed]
76. Bedrose, S.; Miller, K.C.; Altameemi, L.; Ali, M.S.; Nassar, S.; Garg, N.; Daher, M.; Eaton, K.D.; Yorio, J.T.; Daniel, D.B.; et al. Combined lenvatinib and pembrolizumab as salvage therapy in advanced adrenal cortical carcinoma. *J. Immunother. Cancer* **2020**, *8*, e001009. [CrossRef] [PubMed]
77. Kroiss, M.; Megerle, F.; Kurlbaum, M.; Zimmermann, S.; Wendler, J.; Jimenez, C.; Lapa, C.; Quinkler, M.; Scherf-Clavel, O.; Habra, M.A.; et al. Objective Response and Prolonged Disease Control of Advanced Adrenocortical Carcinoma with Cabozantinib. *J. Clin. Endocrinol. Metab.* **2020**, *105*. [CrossRef] [PubMed]
78. Hong, D.S.; Kurzrock, R.; Mulay, M.; Rasmussen, E.; Wu, B.M.; Bass, M.B.; Zhong, Z.D.; Friberg, G.; Rosen, L.S. A phase 1b, open-label study of trebananib plus bevacizumab or motesanib in patients with solid tumours. *Oncotarget* **2014**, *5*, 11154–11167. [CrossRef]
79. Kroiss, M.; Quinkler, M.; Lutz, W.K.; Allolio, B.; Fassnacht, M. Drug interactions with mitotane by induction of CYP3A4 metabolism in the clinical management of adrenocortical carcinoma. *Clin. Endocrinol.* **2011**, *75*, 585–591. [CrossRef]

80. Raj, N.; Zheng, Y.; Kelly, V.; Katz, S.S.; Chou, J.; Do, R.K.G.; Capanu, M.; Zamarin, D.; Saltz, L.B.; Ariyan, C.E.; et al. PD-1 Blockade in Advanced Adrenocortical Carcinoma. *J. Clin. Oncol.* **2020**, *38*, 71–80. [CrossRef]
81. Qin, S.; Ren, Z.; Meng, Z.; Chen, Z.; Chai, X.; Xiong, J.; Bai, Y.; Yang, L.; Zhu, H.; Fang, W.; et al. Camrelizumab in patients with previously treated advanced hepatocellular carcinoma: A multicentre, open-label, parallel-group, randomised, phase 2 trial. *Lancet Oncol.* **2020**, *21*, 571–580. [CrossRef]
82. Zhou, C.; Chen, G.; Huang, Y.; Zhou, J.; Lin, L.; Feng, J.; Wang, Z.; Shu, Y.; Shi, J.; Hu, Y.; et al. Camrelizumab plus carboplatin and pemetrexed versus chemotherapy alone in chemotherapy-naive patients with advanced non-squamous non-small-cell lung cancer (CameL): A randomised, open-label, multicentre, phase 3 trial. *Lancet Respir. Med.* **2020**. [CrossRef]
83. Zhang, L.; Yang, Y.; Chen, X.; Li, J.; Pan, J.; He, X.; Lin, L.; Shi, Y.; Feng, W.; Xiong, J.; et al. 912MO A single-arm, open-label, multicenter phase II study of camrelizumab in patients with recurrent or metastatic (R/M) nasopharyngeal carcinoma (NPC) who had progressed on ≥2 lines of chemotherapy: CAPTAIN study. *Ann. Oncol.* **2020**, *31*, S659. [CrossRef]

Review

Targeting E-selectin to Tackle Cancer Using Uproleselan

Barbara Muz [1], Anas Abdelghafer [1,2], Matea Markovic [1,2], Jessica Yavner [1], Anupama Melam [1], Noha Nabil Salama [2,3] and Abdel Kareem Azab [1,*]

1. Department of Radiation Oncology, Cancer Biology Division, Washington University in St. Louis School of Medicine, St. Louis, MO 63108, USA; bmuz@wustl.edu (B.M.); Anas.Abdelghafer@stlcop.edu (A.A.); Matea.Markovic@stlcop.edu (M.M.); jyavner@wustl.edu (J.Y.); anupama.melam@wustl.edu (A.M.)
2. Department of Pharmaceutical and Administrative Sciences, St. Louis College of Pharmacy, University of Health Sciences and Pharmacy in St. Louis, St. Louis, MO 63110, USA; Noha.Salama@stlcop.edu
3. Department of Pharmaceutics and Industrial Pharmacy, Faculty of Pharmacy, Cairo University, Cairo 11562, Egypt
* Correspondence: kareem.azab@wustl.edu; Tel.: +1-(314)-362-9254; Fax: +1-(314)-362-9790

Simple Summary: This review focuses on eradicating cancer by targeting a surface protein expressed on the endothelium—E-selectin—with a novel drug, uproleselan (GMI-1271). Blocking E-selectin in the tumor microenvironment acts on multiple levels; uproleselan was shown (i) to inhibit cancer cell tethering, rolling and extravasating, i.e., cancer dissemination, (ii) to reduce adhesion and lose stem cell-like properties, (iii) to mobilize cancer cells to circulation where they are more susceptible to chemotherapy, which altogether contributes (iv) to overcome drug resistance. Uproleselan has been tested effective in leukemia, myeloma, pancreatic, colon and breast cancer cells, all of which can be found in the bone marrow as a primary or as a metastatic tumor site. In addition, uproleselan has a good safety profile in patients. It improves the efficacy of chemotherapy, reduces side effects such as neutropenia, intestinal mucositis and infections, and extends overall survival.

Abstract: E-selectin is a vascular adhesion molecule expressed mainly on endothelium, and its primary role is to facilitate leukocyte cell trafficking by recognizing ligand surface proteins. E-selectin gained a new role since it was demonstrated to be involved in cancer cell trafficking, stem-like properties and therapy resistance. Therefore, being expressed in the tumor microenvironment, E-selectin can potentially be used to eradicate cancer. Uproleselan (also known as GMI-1271), a specific E-selectin antagonist, has been tested on leukemia, myeloma, pancreatic, colon and breast cancer cells, most of which involve the bone marrow as a primary or as a metastatic tumor site. This novel therapy disrupts the tumor microenvironment by affecting the two main steps of metastasis—extravasation and adhesion—thus blocking E-selectin reduces tumor dissemination. Additionally, uproleselan mobilized cancer cells from the protective vascular niche into the circulation, making them more susceptible to chemotherapy. Several preclinical and clinical studies summarized herein demonstrate that uproleselan has favorable safety and pharmacokinetics and is a tumor microenvironment-disrupting agent that improves the efficacy of chemotherapy, reduces side effects such as neutropenia, intestinal mucositis and infections, and extends overall survival. This review highlights the critical contribution of E-selectin and its specific antagonist, uproleselan, in the regulation of cancer growth, dissemination, and drug resistance in the context of the bone marrow microenvironment.

Keywords: selectins; E-selectin; uproleselan; cancer

1. Introduction

E-selectin, a vascular adhesion molecule, plays a pivotal role in cell trafficking in both physiological and pathophysiological conditions. It is involved in extravasation, homing, adhesion, proliferation, stemness/cell dormancy, and drug resistance of leukocytes, hematopoietic stem cells (HSCs), and cancer cells. E-selectin is a potentially promising

target for several therapeutic and medical imaging applications due to its overexpression in tissues affected by inflammation, infection, or malignancy.

E-selectin plays an important role in the interaction of cancer cells with the bone marrow (BM) microvasculature; hence, impeding these interactions not only blocks cell tethering, rolling and extravasating but also mobilizes cancer cells to circulation where they are more susceptible to chemotherapy [1]. One of the current methods used to eradicate cancer cells is to cause programmed cell death, also known as anoikis, which occurs in adhesion-dependent cells when they are forced to detach from the environment and the surrounding extracellular matrix (ECM) or are prevented from homing into a new protective BM niche [2]. A number of newly developed drugs aim to cause this programmed cell death through the disruption of the tumor microenvironment (TME) in order to target the cell–cell and cell–ECM interactions by targeting E-selectin on the supporting cells such as endothelial cells [3,4]. Moreover, there is increasing evidence showing that immune cell accumulation in the tumor as a response to chemotherapy contributes to tumor survival, less efficacious therapy, and adverse clinical events [5]. Therefore, another strategy to the improved therapeutic effect of chemotherapy is by blocking E-selectin-mediated infiltration of immune cells into tumors, as demonstrated in the breast cancer model [6].

This review focuses on the novel glycomimetic E-selectin antagonist, uproleselan (GMI-1271; GMI-1687), as an adjuvant cancer therapy. Preclinical studies demonstrated that uproleselan disrupts the interaction between the BM microenvironment and cancer cells, including leukemia, myeloma, colon, prostate, pancreatic and breast cancer cells. The results of blocking E-selectin with uproleselan were determined in vitro—where it reduced adhesion, chemotaxis, trans-endothelial migration and stroma-induced drug resistance and in animal models—where it induced stem and cancer cell mobilization from the BM to circulation and resensitized cancer cells to chemotherapies. Supporting evidence demonstrates that combination treatment with uproleselan reduced multiple myeloma (MM) resistance to carfilzomib and lenalidomide, as well as acute myeloid leukemia (AML) to cytarabine, and enhanced their therapeutic effects demonstrated by reduced tumor growth and prolonged mice survival. Moreover, uproleselan has been successfully used in clinical trials to treat patients with AML and demonstrated improved efficacy of chemotherapy and reduction of side effects such as neutropenia and infections. The trials on MM are undergoing. Based on the promising preclinical and clinical findings, targeting E-selectin has clear potential as an adjuvant cancer therapy.

2. The Role of E-Selectin in Cancer Pathophysiology

2.1. Inhibition of Selectins as a Therapeutic Strategy in Cancer

There are 3 types of selectins with a distinctive tissue expression—E-(endothelium), L-(leukocytes), and P-(platelets) selectin [1]. Selectins are major cell-surface adhesion molecules that serve as biologic brakes, rapidly decelerating leukocytes as they tether and roll on the endothelium [1,7]. Despite the low binding affinity between selectins and their ligands, it is crucial for the leukocytes to reach their destination. Following the rolling step, chemoattractant-activated leukocytes further increase the affinity to the integrins (such as VLA4) [8–12], then squeeze between the endothelial cells (ECs) and extravasate into specific tissues [13]. However, not only leukocytes cell trafficking is regulated by selectins, but also cancer cell adhesion, chemotaxis, stemness of the HSCs and cancer (stem) cells, and the response to anticancer drugs.

Disrupting the interaction between tumor cells and the endothelium and the TME affects cancer cell dissemination and sensitization to therapy—this can be achieved by blocking selectins [14–23]. The blockade of selectins allows for the mobilization of cancer cells, causing anoikis, which further increases their sensitivity and thus enhances the efficacy of chemotherapy. It appears that each selectin plays a major role and/or has been investigated in certain cancer models. For instance, E-selectin has been shown to be a key receptor in leukemia [24,25], myeloma [23], as well as in solid tumors such as pancreatic [26–28], prostate [29,30], colon [31,32] and breast [33–35] cancer cells. L-selectin has been shown to be a key receptor for chronic lymphocytic leukemia [20,36]. Moreover, P-selectin has been shown to play an important role in myeloma [15,18,22,23].

2.2. The Role of E-selectin in Cancer Progression

E-selectin, also known as CD62E, is constitutively expressed on vascular endothelium, and in BM stromal cells [23,25]. Moreover, E-selectin is upregulated in microvasculature in the presence of tumors that commonly metastasize to the bone marrow. There is a number of E-selectin ligands that are expressed on migrating cancer cells (Table 1) including E-selectin ligand (ESL-1) [37], L-selectin (CD62L) [38], P-selectin glycoprotein ligand-1 (PSGL-1, CD162) [15,18,22,39–42], CD43 [43,44], homing cell adhesion molecule 1 (HCAM1; CD44) [35,42,45,46], death receptor 3 (DR-3) [31,32] and cutaneous lymphocyte-associated antigen (CLA) [45,46].

Table 1. E-selectin ligands expressed in cancer.

E-Selectin Ligand	Full Name	Expression in Cancer	References
ESL-1	E-selectin ligand	Prostate	[47,49]
PSGL-1	P-selectin glycoprotein ligand-1; CD162	MM AML	[15,18,22] [40,42]
L-selectin	CD62L	CLL	[36]
CD43	Leukosialin, sialophorin, galactoglycoprotein	ALL DLBCL CLL Lung	[43] [48] [50] [51]
CD44	Homing cell adhesion molecule 1 (HCAM1)	AML Breast	[42,52] [35]
DR-3	Death receptor 3	Colon	[31,32]
CLA	Cutaneous lymphocyte-associated antigen	AML MM	[17] [47,49,53]

Abbreviations: MM, multiple myeloma; AML, acute myeloid leukemia; CLL, chronic lymphocytic leukemia; ALL, acute lymphocytic leukemia; DLBCL, diffuse large B-cell lymphoma.

Frequently, overexpression of functional cancer surface proteins serves as a biomarker for cancer progression and patients' response to treatment. For instance, recent evidence suggested that CLA can play such a role in AML [17] and MM [45]. Chien et al. examined CLA expression in almost 90 AML patient samples from the peripheral blood and the BM and found a 4-fold higher expression for relapsed/refractory patients than for newly diagnosed AML patients [17]. These results were in line with increased CLA expression in cancerous plasma cells from relapsed/refractory patients compared to newly diagnosed MM patients [15,47,48]. Moreover, it was shown that CLA was increased in hypoxic MM cells, indicating the progression of MM to more advanced stages. In the mouse model, CLA^{high} MM cells were more aggressive, metastasized faster facilitating tumorigenesis, and contributed to bortezomib-mediated resistance in vivo that was reversed by blocking E-selectin [45,46]. It was also demonstrated that MM cell rolling on E-selectin in vitro was proportional to CLA levels [45]. Furthermore, circulating tumor cells were more CLA positive in relapsed MM patients than in the one isolated from the BM [45], indicating more invasive and metastatic cancer cells. These results imply that CLA undergoes dynamic changes with cancer growth and metastasis, its expression was unfavorable and correlated with worse prognosis and thus could be a potential biomarker of tumor progression and a prognostic factor of drug resistance development.

2.3. Signaling Pathways Regulated by E-Selectin

Some of the signaling pathways involved in E-selectin-mediated cancer functions were shown to include p38 and extracellular signal-regulated kinases (ERK)/mitogen-activated protein kinases (MAPK) ERK/MAPK), phosphatidylinositol 3-kinases (PI3K) and nuclear factor kappa-light-chain-enhancer of activated B cells (NF-kB), Wnt and Hedgehog (Table 2). The p38 and ERK MAPK pathways were shown to be involved in the migratory capabilities of colon cancer [31,32]. Esposito et al. demonstrated that the Wnt pathway is induced in breast cancer cell metastasis to the bone through activation of mesenchymal-epithelial transition (MET) and induction of stemness at the new metastatic site [54].

Table 2. E-selectin-mediated signaling pathways.

E-Selectin-Mediated Function	Signaling Pathway	Role	Reference
Cell Trafficking and metastasis	p38 ERK/MAPK	Pro-migratory	[31] [32]
Adhesion and tumor growth	NF-kB and PI3K ERK/AKT Wnt	Pro-survival Antiapoptotic	[32] [42,55,56] [57]
Stemness and self-renewal	Wnt Hedgehog	Maintaining stemness	[54,57]
Drug resistance	ERK/AKT NF-kB	Chemoresistance Pro-survival	[32,42,55,56]

A mechanism of E-selectin-mediated tumor adhesion and proliferation was demonstrated to be regulated by pro-survival NF-kB and ERK signaling pathways [32,42,55,56]. Porquet et al. demonstrated that DR-3 overexpressed on HT29 and SW620 colon cancer cells interact with E-selectin, activates the antiapoptotic PI3K/NF-kB pathways, thus protects cancer cells from apoptosis [32]. Following the inhibition of PI3K and AKT pathways concurrently, the colon carcinoma cell apoptosis was increased as demonstrated by cleaved caspase-8 and caspase-3, as well as DNA fragmentation assay [32].

E-selectin is also considered a self-renewal regulator [53] by activating the cancer stemness [54,57]. Bone-homing cancer cells, especially hematological malignancies, are "hiding" in the protective and discrete E-selectin+ BM milieu that facilitates dormancy and stemness in that niche. E-selectin slows down cell division promoted by direct activation of the pro-stemness Wnt [54,57] and Hedgehog pathways (as shown in AML blasts and leukemia stem cells) [57], and pro-survival NF-kB signaling pathway [42,55,56]. It was shown that E-selectin contributes to chemotherapy resistance through cancer pro-survival (ERK/AKT), NF-kB and antiapoptotic pathways [32,42,55,56].

There is growing evidence showing that E-selectin is involved in several aspects of cancer pathophysiology:

2.3.1. Cell Trafficking and Metastasis

It has been shown that cancer cells, especially hematological malignancies, use a similar system of cell trafficking to leukocytes [12,13,15,25,58]. E-selectin is involved in cancer cell trafficking and metastasis through regulating homing and engraftment [23,33,40,59,60]. Metastatic dissemination is initiated and tightly regulated by the interactions between activated E-selectin and their counter-ligands [7,29,30,40,61]. In addition, it was shown that soluble E-selectin (which sheds from the activated endothelium) contributes to CD44-expressing breast cancer cells migration and shear-resistant adhesion, facilitating leukocytes and cancer cells homing to tissues [35]. Therefore, hindering cancer cell migratory abilities by blocking E-selectin and/or their ligands is believed to hamper cancer cell extravasation and formation of new metastatic lesions in distant organs, all of which also has been scrutinized by specifically targeting E-selectin [7,15,19,22,29,62,63]. This interaction, however, may be tumor-specific since in vivo E-selectin knockout studies demonstrated that lung metastasis is not affected by the genetic deletion of E-selectin [54,55].

2.3.2. Adhesion and Tumor Growth

E-selectin is a major vascular adhesion molecule [1,19]. E-selectin overexpression in cancer contributes to tumor growth due to adhesion-mediated pro-survival and antiapoptotic pathways supporting cancer proliferation. This is induced by the interaction between cancer cells with the BM microenvironment, with selectin-expressing ECs and stromal cells [15,22,32]. E-selectin is involved in cancer adhesion and adhesion-dependent cancer survival and proliferation [56]. Blocking selectins with monoclonal antibodies, by silencing the gene or by using the pan-selectin inhibitor (Rivipansel, GMI-1070), inhibited tumor adhesion dynamics and adhesion-mediated proliferation.

Additionally, Morita et al. demonstrated that E-selectin in breast cancer vasculature promotes immune cell accumulation, which facilitates tumor growth [6]. Thus, blocking E-selectin with aptamer (ESTA) significantly decreased CD45+ immune cell tumor homing in doxorubicin-treated mice, causing inhibition of tumor growth and lung metastasis. These results imply that tumor growth can be indirectly controlled by immune cell homing to the tumor through E-selectin regulation. Moreover, soluble E-selectin in the serum was described to facilitate circulating CD44-expressing cancer cells and immune cells homing to tissues, thus contributing to tumor metastasis and growth [35].

2.3.3. Stemness and Self-Renewal

E-selectin is involved in HSC and cancer stemness and dormancy [25,55,57,64]. E-selectin is vital to hematopoiesis in terms of its ability to maintain steady-state expression in the BM vasculature and to retain HSCs proliferation [25]. Interestingly, the absence or blockade of E-selectin resulted in an increased proportion of quiescent HSCs, enhanced HSC survival by promoting chemoresistance [25]. Therefore, due to higher HSCs recuperation and lower BM toxicity through accelerated blood neutrophil recovery, mice with E-selectin $-/-$ were able to survive chemotherapy 2–6-fold better than the control group; after treating both groups with antimetabolite cytotoxic 5-fluorouracil (5-FU), E-selectin $-/-$ mice survived over 140 days while wild-type mice only survived about 48 days [57,65].

BM is hijacked by cancer cells explicitly metastasizing to the bone and utilizing this E-selectin-rich environment to become quiescent and stem-cell-like [54]. On top of that, it is facilitated by physoxia (low physiological oxygenation) present in the BM, which in the presence of the growing and expanding tumor drops, even more, contributing to hypoxic conditions mediating further stemness and drug resistance [58,64]. The presence of cancer cells in the BM is unnatural and contributes to a stressful and inflammatory environment, topped by the overexpression of E-selectin in the microvasculature [23,57,65]. Since chemotherapeutics mainly kill rapidly dividing cells, BM acts as a shield for the dormant cancer stem cells protecting them from killing [25,55,59].

2.3.4. Drug Resistance

E-selectin is also involved in cancer drug resistance [15,16,66,67]. It was shown in MM and leukemia that cancer cells are protected from cytotoxic drugs due to the cell interaction with the BM vasculature inducing pro-survival signals, thus promoting cancer progression [59]. It was shown that leukemic cells with a stronger ability to bind E-selectin were 12-fold more resistant to chemotherapy in the AML mouse model [16]. In addition, gaining adhesion properties by cancer cells in suspension due to cytotoxic drug exposure, upregulated ligands and/or receptors and, as a result, conveyed drug resistance [60]. For instance, cancer cells (such as MM) overexpressing the E-selectin ligand, such as CLA, were more aggressive and more resistant to proteasome inhibitors, including bortezomib [23,45]. In addition, inhibition of these interactions using the pan-selectin inhibitor Rivipansel (GMI-1070) reversed the adhesion-mediated drug resistance induced by ECs and BM stroma in preclinical models through the sensitization of MM cells to bortezomib, which improved survival of the MM-bearing mice [15]. However, poor pharmacokinetics and a short half-life of Rivipansel requires administration of high concentration, making it inconvenient for patients. Therefore, there is a need for a selectin-specific inhibitor with better pharmacokinetics. With growing evidence demonstrating that E-selectin is involved in tumor progression and recurrence through regulation of metastasis, adhesion, stemness and drug resistance—specifically targeting E-selectin became of high interest and high importance.

3. Uproleselan in Cancer Therapy

The field of glycobiology in cancer has emerged since anomalous glycosylation patterns, and sialic acids and sialic acid-containing glycoconjugates associated with tumors became attractive targets for anticancer therapies. The idea started with the investigation of an enzyme called sialyltransferase (ST3 Gal-6) that mainly functions to generate E-selectin ligands. E-selectin recognizes sialylated carbohydrates/fucosylated glycoprotein ligands such as ESL-1, PSGL-1, CD44 and CLA, among others (Table 1), that are expressed on circulating leukocytes and overexpressed on cancer cells. This research provided a rationale to target E-selectin in cancer using a novel glycomimetic E-selectin antagonist, GMI-1271 (later named uproleselan), to overcome cancer spread and chemoresistance summarized in Table 3 [33].

Table 3. Role of uproleselan in preclinical cancer models.

Role of Uproleselan in Cancer	Results	Reference
Metastasis	Prevented MM dissemination Inhibited pancreatic ductal adenocarcinoma to the lymph nodes, as well as to the liver, lung and diaphragm in combination with gemcitabine Inhibited breast cancer metastasis to the bone marrow	[23] [27] [54]
Adhesion	Decreased the adhesion of cancer cells to stromal and endothelial cells in vitro Reduced adhesion of CML leukemic stem cells to E-selectin in the vascular niche	[23,45] [53]
Mobilization	Enhanced mobilization of cancer cells out of the bone marrow into the circulation Mobilized myeloma and leukemic cells from the marrow into the peripheral blood after a single injection Activated the tumor-reactive and tumor-specific marrow infiltrating lymphocytes	[62] [23,45] [68]
Cancer stem-cell like	Inhibited cancer (stem) cell quiescence and induced cell maturation Resensitized leukemic stem cell to chemotherapy in AML-bearing mice	[57,64,65,69] [16,70]
Chemotherapy sensitization	Improved CML killing in combination with imatinib Sensitized AML in combination with daunorubicin (DNR) and cytarabine (AraC) in different mouse models (syngeneic, xenogeneic and patient blasts) Overcame MM drug resistance and improved the efficacy to proteasome inhibitors (bortezomib and carfilzomib) and IMiDs (lenalidomide)	[53] [16,62] [23,47,49,53]
Reducing adverse events	Reduced bone marrow toxicity including neutropenia, protected and increased percentile of HSCs, enhanced neutrophilic recovery, reduced small intestine mucositis by decreasing the number of infiltrating inflammatory macrophages	[16,65]

3.1. Uproleselan—Chemical Structure and Properties

Uproleselan (synonym GMI-1271; chemical abstracts service (CAS) registry number: 1914993-95-5) is a small molecule glycomimetic rationally designed based on the bioactive conformation of sialyl Lea/x. Further, it is a potent and specific antagonist of E-selectin. In the target-based drug classification (PubChem.ncbi.nlm.nih.gov), uproleselan is considered a drug targeting (i) cell surface molecule and ligand, (ii) a cell adhesion molecule, and (iii) a selectin.

Uproleselan (Kd = 0.46 µM) mainly inhibits E-selectin (IC50 = 1.75 µM), but also weakly inhibits L-selectin (IC50 = 2.9 µM) and P-selectin (>10 µM). Uproleselan's chemical formula $C_{60}H_{108}N_3NaO_{27}$ (molecular weight of 1325.70679 g/mol) is demonstrated in Figure 1.

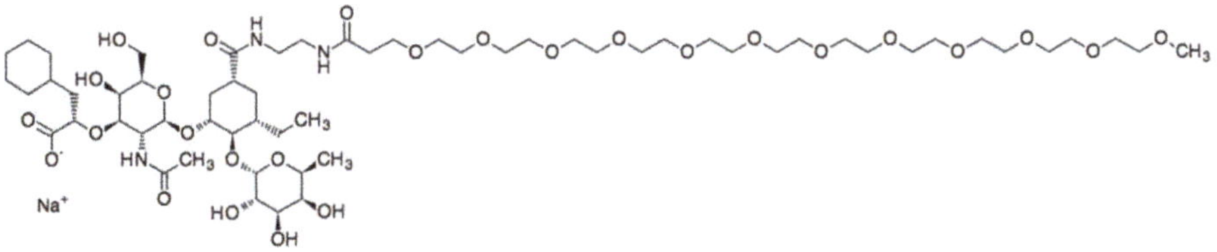

Figure 1. Chemical structure of uproleselan.

3.2. The Role of Uproleselan in Cancer Therapy

3.2.1. Uproleselan Inhibits Metastasis

Extravasation (egress) followed by the homing of circulating cancer cells is a crucial step of metastasis. It has been shown that uproleselan offers a promising treatment in preventing metastasis through blocking E-selectin, both in vitro (trans-endothelial migration assay) and in vivo in myeloma, pancreatic and breast cancer models [23,27,57]. The anti-homing properties of uproleselan in cancer were confirmed by inhibiting cellular interactions at every stage of cancer cell trafficking, including cancer cell retention in the blood after treating MM mouse endothelium and at the same time blocking MM cell homing to the BM and spreading the disease [23]. This specific antagonist was also shown in combination with gemcitabine to significantly reduce the frequency of metastasis of pancreatic ductal adenocarcinoma to the lymph nodes, as well as to the liver, lung and diaphragm, but did not alter primary tumor size [27]. Esposito et al. neatly demonstrated that breast cancer bone metastasis was facilitated via the E-selectin-enriched bone vascular niche, which induced MET and was inhibited by uproleselan [54].

3.2.2. Uproleselan Decreases Adhesion and Activates Cancer Cell Mobilization

E-selectin performs as a gatekeeper for cancer (stem) cells from leaving or entering the BM. We and others have shown that uproleselan decreased the adhesion of cancer cells to stromal and endothelial cells in vitro [23,49,65]. In addition, static adhesion and dynamic rolling of cancer cells to E-selectin were proportional to E-selectin ligand levels and were inhibited using uproleselan [45]. As a result, blocking E-selectin activity causes de-adhesion and releases cells into the peripheral blood.

First, it was revealed by Winkler et al. that the absence of E-selectin in mice (Esel−/−) improved mobilization of HSCs, especially after granulocyte colony-stimulating factor (G-CSF) administration, which increased E-selectin expression at HSC vascular niche [25]. Then, following uproleselan administration into mice decreased cell adhesion and acted as a mobilizing agent [61].

Uproleselan combined with G-CSF enhanced mobilization of cancer cells out of the BM into the circulation much more than G-CSF alone [62]. In addition, it was shown that the E-selectin antagonist mobilized myeloma and leukemia cells from the marrow into the peripheral blood gradually within 60 min following a single injection; these cancer cells persisted in the circulation for up to 24 h and reached a ~10-fold increase at 48 h post-injection [23,45]. In comparison, a well-known CXCR4 inhibitor (AMD3100, plerixafor) rapidly mobilized tumor cells by 11-fold by 1 h, which returned to baseline within 24 h [63,71,72]. Interestingly, the most efficient cell mobilization was achieved by dual inhibition of E-selectin, and CXCR4 (using GMI-1359) mobilized leukemic cells by ~16-fold at 8 h post-injection and remained elevated even at 72 h.

One of the strategies to kill cancer cells is to expose them to systemically administered chemotherapy by anoikis [12,15,23]. Inducing de-adhesion and mobilization of cancer cells to the circulation and simultaneously not letting them home back to the marrow is achieved by the novel strategy via blocking E-selectin [22,45]. These findings demonstrate that blocking E-selectin with uproleselan mobilizes cancer cells over a long period of time, sustains the presence of tumor cells in circulation, inhibits their reentry into the BM, and thereby provides a longer window to target these cells in the circulation, sensitizing them to chemotherapy thus longer exposure to chemotherapy, and as a result significant reduction of the tumor burden [23,45].

Amongst mobilizing HSCs and cancer cells into the circulation, it was also shown that disrupting the TME with uproleselan activated the tumor-reactive and tumor-specific marrow infiltrating lymphocytes (MILs) [68]. CT26-immune mice were treated for three days with saline, G-CSF (0.125 mg/kg) or uproleselan (40 mg/kg) followed by determination of the phenotype and functional CD8+ T cells in the BM and peripheral blood 12 h after the last injection [68]. Treatment of mice with uproleselan, but not with G-CSF, led to an approximate 3–4-fold increase in naïve T cells (CD8+CD62L+CD44−) and central memory T cells (CD8+CD62L+CD44+) in peripheral blood and correlated with increased interferon gamma (IFNγ) ex vivo in response to treatment [68].

3.2.3. Uproleselan Causes Maturity of Cancer Stem Cells

E-selectin was shown to be a pivotal regulator in the BM in switching between stemness/quiescence and activation/maturation of HSCs [25]. Disrupting the protective interaction between cancer cells and supportive E-selectin with uproleselan caused inhibition of quiescence through the downregulation of Wnt activity [57], and increased cell cycle and thus the maturity of cancer (stem) cells [57,65,69]. Barbier et al. demonstrated that AML-bearing mice treated with uproleselan along with chemotherapy survived longer due to chemo-sensitization of the regenerating leukemic stem cells [16].

3.2.4. Uproleselan Resensitizes Cancer Cells to Therapies in Pre-Clinical Models

The main strength of utilizing multifactorial uproleselan involves its combination with other therapies. Very frequently, a single drug is not enough to successfully battle cancer and prevent tumor recurrence. Further, a plethora of evidence shows that administration of uproleselan in combination with different chemotherapies overcomes drug resistance and/or improves the efficacy of standard chemotherapy through the re-sensitization to therapies in multiple cancer models.

It was reported that in a xenotransplantation CML model, murine recipients of human CML-initiating cells treated with uproleselan and imatinib, the tyrosine kinase inhibitor, which is a standard of care in CML, further decreased the engraftment of cancer cells by decreasing their interaction time with the BM endothelium, compared to imatinib alone [53]. Uproleselan reduced adhesion of BCR-ABL1+ leukemic stem cells (LSCs) to E-selectin in the vascular niche, increased the cell cycle with simultaneous overexpression of the transcriptional regulator and protooncogene SCL/TAL1, as well as decreased CD44 expression in vitro and in vivo, thus improved eradication by imatinib [53].

Similarly, uproleselan used jointly with daunorubicin (DNR) and cytarabine (AraC) significantly improved the killing of AML cells in multiple different AML mouse models (syngeneic, xenogeneic and patient blasts), improving mice survival. Sensitization of LSCs (CD34+CD38−CD123+) to AraC chemotherapy was demonstrated by a single administration of AraC into the wild-type and Esel−/− mice, which showed that the absence of E-selectin improved LSC killing. Therefore, blocking E-selectin with uproleselan combined with AraC reversed E-selectin-induced chemoresistance to AraC and significantly decreased the number of LSCs in the femur by 95% over AraC alone [25,47,64,69].

Furthermore, it was also shown in MM mouse models (xenograft and syngeneic TGM1 disseminated model) that tumors become resistant to chemotherapies due to hypoxia, adhesion to cellular and non-cellular components of the BM, and stemness [56,58,60,66,67,73,74]. Uproleselan overcame drug resistance and improved the efficacy of chemotherapies, such as proteasome inhibitors (bortezomib and carfilzomib) and IMiDs such as lenalidomide [23,47,49]. The interaction between MM cells and the TME was disrupted with uproleselan through decreasing E-selectin-mediated adhesion, stroma-induced drug resistance, chemotaxis and stemness of MM cells, which sensitized them to therapy in vitro. Additionally, uproleselan inhibited the dissemination process and therefore extended the exposure of MM cells to chemotherapies, which resulted in delayed tumor growth and prolonged mice survival [23,47,49].

One of the explanations of the improved drug efficacy that results from blocking E-selectin in the TME is the fact that uproleselan mobilizes cancer cells (AML and MM) out of the BM, causing anoikis and thus making them more susceptible to chemotherapy [4,23,69,75]. Another possible mechanism of reduced mortality of mice treated with uproleselan in combination with chemotherapy involves reducing some of the adverse effects such as BM toxicity, including neutropenia through an increased percentile of HSCs, which survived each round of treatment that facilitates blood and BM recovery and enhanced neutrophilic recovery [61,65]. Moreover, small intestine mucositis (inflammation and sloughing of the mucous membranes lining of the digestive tract) was also reduced through a reduction in the number of infiltrating inflammatory macrophages (F4/80+Ly-6C+), which normally exacerbate mucosal damage [65]. These events resulted in slowed down weight loss and eventually improved mouse survival.

The preclinical studies performed on uproleselan have revealed that blocking E-selectin affects interactions between the cancer cells and the tumor microenvironment, and thus is a valid and vital therapeutic strategy. This strategy relies on hampering the homing of already circulating tumor cells to a new metastatic niche and/or causing anoikis, a cancer death through cell de-adhesion and mobilization to the circulation, overcoming stemness and drug resistance, and further sensitizing cancer cell to chemotherapies (Figure 2).

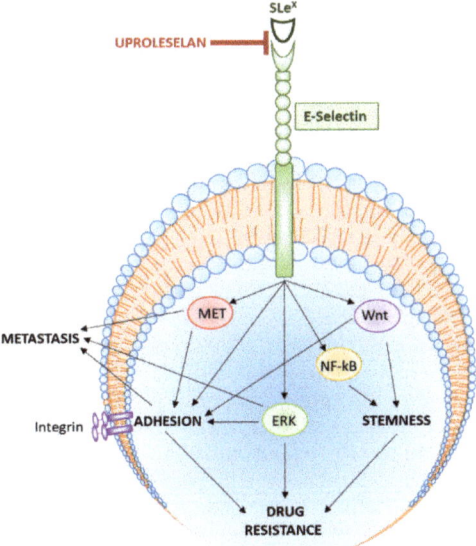

Figure 2. Mechanism of action of uproleselan.

4. Clinical Trials Using Uproleselan

4.1. Pharmacokinetics of Uproleselan

In phase I clinical trials (ClinicalTrials.gov Identifiers: NCT02168595; NCT03606447; NCT02271113), uproleselan was administered intravenously and showed favorable pharmacokinetic (PK) profiles in single doses up to 10 mg/kg in one study and up to 40 mg/kg in the second study [14,76]. Uproleselan exhibited a dose-dependent plasma levels maximum plasma-concentration (Cmax) and area under the plasma concentration time curve (AUC). Approximately two-thirds of uproleselan was excreted unchanged in the urine, and renal clearance (CLr) was on average 86 mL/min, being less than the estimated creatinine clearance (CrCl), suggesting tubular reabsorption with the apparent half-life ($T_{1/2}$) averaging at 2.3 h [76].

4.2. Safety of Uproleselan (Phase I and Phase I/II Trials)

In a phase I single-dose escalation, a double-blind, placebo-controlled trial in healthy subjects was conducted to evaluate the safety, tolerability, toxicity, adverse events and PK profile of uproleselan. The drug was administered intravenously at concentrations 2 mg/kg, 5 mg/kg, 20 mg/kg and 40 mg/kg (ClinicalTrials.gov Identifiers: NCT03606447; NCT02271113). The most common adverse events were fatigue, headache, infusion site adverse events, oropharyngeal pain, presyncope and rash; however, all the above were reported as mild and unrelated to uproleselan [14,77].

In phase I/II clinical trial (ClinicalTrials.gov Identifier: NCT02306291), uproleselan was investigated in combination with chemotherapy in a total of 66 AML patients (including 44 relapses and 22 refractory patients). Increasing doses of uproleselan from 5 to 20 mg/kg for eight days were administered to patients in combination with mitoxantrone, etoposide, and cytarabine chemotherapy and monitored for adverse effects. The most frequent grade 3/4 adverse events included sepsis (18%), gastrointestinal (11%), and cardiac effects (9%). The recommended phase 2 dose (RP2 D) of 10 mg/kg was then administered to a total of 54 patients providing an optimal exposure with mucositis (2%) as the most frequent grade 3/4 adverse event. These results imply that uproleselan safety in AML is favorable even when administered with chemotherapy also in elderly AML patients [69,75].

Moreover, another phase I clinical trial (ClinicalTrials.gov Identifier: NCT02811822) for dose-escalation of uproleselan in conjunction with chemotherapy (i.e., proteasomal inhibitors such as bortezomib and carfilzomib) is currently being performed in 10 MM patients.

4.3. Clinical Efficacy

In a phase I/II clinical trial (ClinicalTrials.gov Identifier: NCT02306291), uproleselan was administered to 66 AML patients in combination with chemotherapy. In phase I, patients were given increasing doses of uproleselan (5–20 mg/kg) 24 h prior, every 12 h during, and 48 h post-chemotherapy, including mitoxantrone, etoposide, and cytarabine. Overall, the 60-day mortality rate of all AML patients in the study was 9%, the median overall survival was 9.2 months (95% CI, 3–12.6), and event-free survival was 12.6 months (95% CI, 9.9–NA) [78]. Currently, uproleselan is being tested in AML patients in combination with chemotherapy and compared to chemotherapy alone in phase III randomized, double-blind trial in the U.S., Australia, and Europe (ClinicalTrials.gov Identifier: NCT03616470).

In addition, in a phase I/II trial (ClinicalTrials.gov Identifier: NCT02744833), uproleselan was used as an antithrombotic treatment and tested in two patients with venous thromboembolism disease [77,79]. The symptoms of thrombosis improved, the clotting was resolved, and the associated inflammation was decreased in these patients with isolated calf-level deep vein thrombosis, with positive biological effect and improved biomarkers of coagulation, cell adhesion, and leukocyte/platelet activation [77,79].

5. Conclusions

Tumor cells engage specific cellular BM stroma and microvasculature and non-cellular components for tumor propagation and outgrowth, facilitating extravasation, stemness and acquisition of drug resistance. The body of work reviewed herein demonstrates that targeting E-selectin is a potential and promising adjuvant therapy to successfully disrupt the tumor microenvironment and thus kill the cancer cells more efficiently as well as reduce side effects. Recent progress in glycobiology, as well as in investigating the E-selectin—ligand-binding mechanisms, has rendered the use of specific antagonists such as uproleselan, which binds E-selectin more specifically and effectively. Uproleselan has been assessed as an adjuvant therapeutic treatment in several cancers, especially the ones relying on the protective BM milieu (leukemia and myeloma) and solid tumors metastasizing to the bone (breast, prostate, and colon cancer). This summary clearly suggests that E-selectin is a valid target, considering that it reduced cancer metastasis, detached them from a safe E-selectin-rich BM niche, mobilized cancer cells to the circulation, and overcame drug resistance by sensitizing cancer cells to chemotherapies. Both in vitro and in vivo studies supported the use of uproleselan by demonstrating significant extended mice survival rate with alleviated adverse effects such as intestinal mucositis in AML and MM-bearing mice treated with uproleselan administered together with different chemotherapies. Furthermore, completed clinical trials in AML patients using uproleselan showed high remission rates and improved overall survival with favorable safety of the drug.

Nevertheless, based on E-selectin's important role and expression at the site of inflammation and infection, the main concern regarding targeting selectively and specifically E-selectin in the cancer tissue, and not in other inflammation sites, remains a challenge that needs to be addressed in future research.

Author Contributions: B.M., A.A., M.M. wrote the manuscript; A.M. prepared the Schematic Figure 2; J.Y. edited the manuscript for English grammar and spelling, N.N.S. and A.K.A. supervised and revised the manuscript. All authors have read and agreed to the published version of the manuscript.

Funding: This study was supported by the Paula C. and Rodger O. Riney Blood Cancer Research Initiative Fund, the National Institutes of Health (NIH) grant U54CA199092, as well as Internal grant funding through the Faculty Research Incentive Funds by the University of Health Sciences and Pharmacy in St. Louis.

Conflicts of Interest: Dr. Azab received research support from Glycomimetics and is the founder and owner of Targeted Therapeutics LLC and Cellatrix LLC; however, these have no contribution to this study. Other authors state no conflicts of interest.

References

1. Bevilacqua, M.P.; Nelson, R.M. Selectins. *J. Clin. Investig.* **1993**, *91*, 379–387. [CrossRef]
2. Sai, B.; Xiang, J. Disseminated tumour cells in bone marrow are the source of cancer relapse after therapy. *J. Cell. Mol. Med.* **2018**, *22*, 5776–5786. [CrossRef]
3. Pals, S.T.; Kersten, M.J.; Spaargaren, M. Targeting cell adhesion and homing as strategy to cure Waldenström's macroglobulinemia. *Best Pr. Res. Clin. Haematol.* **2016**, *29*, 161–168. [CrossRef] [PubMed]
4. Paoli, P.; Giannoni, E.; Chiarugi, P. Anoikis molecular pathways and its role in cancer progression. *Biochim. et Biophys. Acta (BBA) Bioenerg.* **2013**, *1833*, 3481–3498. [CrossRef] [PubMed]
5. García-Martínez, E.; Luengo-Gil, G.; Benito, A.C.; González-Billalabeitia, E.; Conesa, M.A.V.; García, T.G.; García-Garre, E.; Vicente, V.; De La Peña, F.A. Tumor-infiltrating immune cell profiles and their change after neoadjuvant chemotherapy predict response and prognosis of breast cancer. *Breast Cancer Res.* **2014**, *16*, 1–17. [CrossRef] [PubMed]
6. Morita, Y.; Leslie, M.; Kameyama, H.; Lokesh, G.L.R.; Ichimura, N.; Davis, R.; Hills, N.; Hasan, N.; Zhang, R.; Kondo, Y.; et al. Functional Blockade of E-Selectin in Tumor-Associated Vessels Enhances Anti-Tumor Effect of Doxorubicin in Breast Cancer. *Cancers* **2020**, *12*, 725. [CrossRef] [PubMed]
7. Borsig, L. Selectins Facilitate Carcinoma Metastasis and Heparin Can Prevent Them. *Physiology* **2004**, *19*, 16–21. [CrossRef] [PubMed]
8. Kucia, M.; Jankowski, K.; Reca, R.; Wysoczynski, M.; Bandura, L.; Allendorf, D.J.; Zhang, J.; Ratajczak, J.; Ratajczak, M.Z. CXCR4–SDF-1 Signalling, Locomotion, Chemotaxis and Adhesion. *J. Mol. Histol.* **2003**, *35*, 233–245. [CrossRef]

9. Roccaro, A.M.; Mishima, Y.; Sacco, A.; Moschetta, M.; Tai, Y.-T.; Shi, J.; Zhang, Y.; Reagan, M.R.; Huynh, D.; Kawano, Y.; et al. CXCR4 Regulates Extra-Medullary Myeloma through Epithelial-Mesenchymal-Transition-like Transcriptional Activation. *Cell Rep.* **2015**, *12*, 622–635. [CrossRef]
10. Sipkins, D.A.; Wei, X.; Wu, J.W.; Runnels, J.M.; Côté, D.; Means, T.K.; Luster, A.D.; Scadden, D.T.; Lin, C.P. In vivo imaging of specialized bone marrow endothelial microdomains for tumour engraftment. *Nat. Cell Biol.* **2005**, *435*, 969–973. [CrossRef]
11. Azab, A.K.; Azab, F.; Blotta, S.; Pitsillides, C.M.; Thompson, B.; Runnels, J.M.; Roccaro, A.M.; Ngo, H.T.; Melhem, M.R.; Sacco, A.; et al. RhoA and Rac1 GTPases play major and differential roles in stromal cell–derived factor-1–induced cell adhesion and chemotaxis in multiple myeloma. *Blood* **2009**, *114*, 619–629. [CrossRef] [PubMed]
12. Azab, A.K.; Runnels, J.M.; Pitsillides, C.; Moreau, A.-S.; Azab, F.; Leleu, X.; Jia, X.; Wright, R.; Ospina, B.; Carlson, A.L.; et al. CXCR4 inhibitor AMD3100 disrupts the interaction of multiple myeloma cells with the bone marrow microenvironment and enhances their sensitivity to therapy. *Blood* **2009**, *113*, 4341–4351. [CrossRef] [PubMed]
13. Butcher, E.C.; Picker, L.J. Lymphocyte Homing and Homeostasis. *Science* **1996**, *272*, 60–67. [CrossRef]
14. Angelini, D.E.; Devata, S.; Hawley, A.E.; Blackburn, S.A.; Grewal, S.; Hemmer, M.V.; Flanner, H.; Kramer, W.; E Parker, W.; Li, Y.-L.; et al. E-Selectin Antagonist GMI-1271 Shows a Favorable Safety, PK and Bleeding Profile in Phase I Studies of Healthy Volunteers. *Blood* **2016**, *128*, 3826. [CrossRef]
15. Azab, A.K.; Quang, P.; Azab, F.; Pitsillides, C.; Thompson, B.; Chonghaile, T.N.; Patton, J.T.; Maiso, P.; Monrose, V.; Sacco, A.; et al. P-selectin glycoprotein ligand regulates the interaction of multiple myeloma cells with the bone marrow microenvironment. *Blood* **2012**, *119*, 1468–1478. [CrossRef]
16. Barbier, V.; Erbani, J.; Fiveash, C.; Davies, J.M.; Tay, J.; Tallack, M.R.; Lowe, J.; Magnani, J.L.; Pattabiraman, D.R.; Perkins, A.C.; et al. Endothelial E-selectin inhibition improves acute myeloid leukaemia therapy by disrupting vascular niche-mediated chemoresistance. *Nat. Commun.* **2020**, *11*, 1–15. [CrossRef]
17. Chien, S.S.; Dai, J.; Magnani, J.L.; Sekizaki, T.S.; E Fogler, W.; Thackray, H.M.; Gardner, K.; Estey, E.H.; Becker, P.S. E-Selectin Ligand Expression By Leukemic Blasts Is Associated with Prognosis in Patients with AML. *Blood* **2018**, *132*, 1513. [CrossRef]
18. Federico, C.; Alhallak, K.; Sun, J.; Duncan, K.; Azab, F.; Sudlow, G.P.; De La Puente, P.; Muz, B.; Kapoor, V.; Zhang, L.; et al. Tumor microenvironment-targeted nanoparticles loaded with bortezomib and ROCK inhibitor improve efficacy in multiple myeloma. *Nat. Commun.* **2020**, *11*, 1–13. [CrossRef]
19. Jubeli, E.; Moine, L.; Vergnaud, J.; Barratt, G. E-selectin as a target for drug delivery and molecular imaging. *J. Control. Release* **2012**, *158*, 194–206. [CrossRef]
20. Kimura, A.; Kawaishi, K.; Sasaki, A.; Hyodo, H.; Oguma, N. L-Selectin Expression in CD34 Positive Cells in Chronic Myeloid Leukemia. *Leuk. Lymphoma* **1998**, *28*, 399–404. [CrossRef]
21. Laird, C.; Hassanein, W.; O'Neill, N.A.; French, B.M.; Cheng, X.; Fogler, W.; Magnani, J.L.; Parsell, D.; Cimeno, A.; Phelps, C.J.; et al. P- and E-selectin receptor antagonism prevents human leukocyte adhesion to activated porcine endothelial monolayers and attenuates porcine endothelial damage. *Xenotransplantation* **2018**, *25*, e12381. [CrossRef] [PubMed]
22. Muz, B.; Azab, F.; De La Puente, P.; Rollins, S.; Alvarez, R.; Kawar, Z.; Azab, A.K. Inhibition of P-Selectin and PSGL-1 Using Humanized Monoclonal Antibodies Increases the Sensitivity of Multiple Myeloma Cells to Bortezomib. *BioMed Res. Int.* **2015**, *2015*, 1–8. [CrossRef] [PubMed]
23. Muz, B.; Azab, F.; Fiala, M.; King, J.; Kohnen, D.; Fogler, W.E.; Smith, T.; Magnani, J.L.; Vij, R.; Azab, A.K. Inhibition of E-Selectin (GMI-1271) or E-selectin together with CXCR4 (GMI-1359) re-sensitizes multiple myeloma to therapy. *Blood Cancer J.* **2019**, *9*, 1–6. [CrossRef] [PubMed]
24. Azab, A.K.; Weisberg, E.; Sahin, I.; Liu, F.; Awwad, R.; Azab, F.; Liu, Q.; Griffin, J.D.; Ghobrial, I.M. The influence of hypoxia on CML trafficking through modulation of CXCR4 and E-cadherin expression. *Leukemia* **2012**, *27*, 961–964. [CrossRef]
25. Winkler, I.G.; Barbier, V.; Nowlan, B.; Jacobsen, R.N.; E Forristal, C.; Patton, J.T.; Magnani, J.L.; Lévesque, J.-P. Vascular niche E-selectin regulates hematopoietic stem cell dormancy, self renewal and chemoresistance. *Nat. Med.* **2012**, *18*, 1651–1657. [CrossRef]
26. Shea, D.J.; Li, Y.W.; Stebe, K.J.; Konstantopoulos, K. E-selectin-mediated rolling facilitates pancreatic cancer cell adhesion to hyaluronic acid. *FASEB J.* **2017**, *31*, 5078–5086. [CrossRef]
27. Steele, M.M.; Radhakrishnan, P.; Magnani, J.L.; Hollingsworth, M.A. A Small Molecule Glycomimetic Antagonist of E-selectin (GMI-1271) Prevents Pancreatic Tumor Metastasis and Offers Improved Efficacy of Chemotherapy. Available online: https://cancerres.aacrjournals.org/content/74/19_Supplement/4503 (accessed on 1 January 2020).
28. Takada, A.; Ohmori, K.; Yoneda, T.; Tsuyuoka, K.; Hasegawa, A.; Kiso, M.; Kannagi, R. Contribution of carbohydrate antigens sialyl Lewis A and sialyl Lewis X to adhesion of human cancer cells to vascular endothelium. *Cancer Res.* **1993**, *53*, 354–361.
29. Festuccia, C.; Mancini, A.; Gravina, G.L.; Colapietro, A.; Vetuschi, A.; Pompili, S.; Ventura, L.; Monache, S.D.; Iorio, R.; Del Fattore, A.; et al. Dual CXCR4 and E-Selectin Inhibitor, GMI-1359, Shows Anti-Bone Metastatic Effects and Synergizes with Docetaxel in Prostate Cancer Cell Intraosseous Growth. *Cells* **2019**, *9*, 32. [CrossRef]
30. Dimitroff, C.J.; Lechpammer, M.; Long-Woodward, D.; Kutok, J.L. Rolling of Human Bone-Metastatic Prostate Tumor Cells on Human Bone Marrow Endothelium under Shear Flow Is Mediated by E-Selectin. *Cancer Res.* **2004**, *64*, 5261–5269. [CrossRef]
31. Gout, S.; Morin, C.; Houle, F.; Huot, J. Death Receptor-3, a New E-Selectin Counter-Receptor that Confers Migration and Survival Advantages to Colon Carcinoma Cells by Triggering p38 and ERK MAPK Activation. *Cancer Res.* **2006**, *66*, 9117–9124. [CrossRef]

32. Porquet, N.; Poirier, A.; Houle, F.; Pin, A.-L.; Gout, S.; Tremblay, P.-L.; Paquet, E.R.; Klinck, R.; Auger, F.A.; Huot, J. Survival advantages conferred to colon cancer cells by E-selectin-induced activation of the PI3K-NFκB survival axis downstream of Death receptor-3. *BMC Cancer* **2011**, *11*, 285. [CrossRef] [PubMed]
33. Price, T.T.; Sipkins, D.A. E-Selectin and SDF-1 regulate metastatic trafficking of breast cancer cells within the bone. *Mol. Cell. Oncol.* **2016**, *4*, e1214771. [CrossRef] [PubMed]
34. Price, T.T.; Burness, M.L.; Sivan, A.; Warner, M.J.; Cheng, R.; Lee, C.H.; Olivere, L.; Comatas, K.; Magnani, J.; Lyerly, H.K.; et al. Dormant breast cancer micrometastases reside in specific bone marrow niches that regulate their transit to and from bone. *Sci. Transl. Med.* **2016**, *8*, 340ra73. [CrossRef] [PubMed]
35. Kang, S.-A.; Blache, C.A.; Bajana, S.; Hasan, N.; Kamal, M.; Morita, Y.; Gupta, V.; Tsolmon, B.; Suh, K.S.; Gorenstein, D.G.; et al. The effect of soluble E-selectin on tumor progression and metastasis. *BMC Cancer* **2016**, *16*, 331. [CrossRef]
36. Debreceni, I.B.; Szász, R.; Kónya, Z.; Erdődi, F.; Kiss, F.; Kappelmayer, J. L-Selectin Expression is Influenced by Phosphatase Activity in Chronic Lymphocytic Leukemia. *Cytom. Part B Clin. Cytom.* **2019**, *96*, 149–157. [CrossRef]
37. Steegmaler, M.; Levinovitz, A.; Isenmann, S.; Borges, E.; Lenter, M.; Kocher, H.P.; Kleuser, B.; Vestweber, D. The E-selectin-ligand ESL-1 is a variant of a receptor for fibroblast growth factor. *Nat. Cell Biol.* **1995**, *373*, 615–620. [CrossRef]
38. Picker, L.J.; Warnock, R.; Burns, A.R.; Doerschuk, C.M.; Berg, E.L.; Butchert, E.C. The neutrophil selectin LECAM-1 presents carbohydrate ligands to the vascular selectins ELAM-1 and GMP-140. *Cell* **1991**, *66*, 921–933. [CrossRef]
39. Videira, P.A.; Silva, M.; Martin, K.C.; Sackstein, R. Ligation of the CD44 Glycoform HCELL on Culture-Expanded Human Monocyte-Derived Dendritic Cells Programs Transendothelial Migration. *J. Immunol.* **2018**, *201*, 1030–1043. [CrossRef]
40. Krause, D.S.; Lazarides, K.; Lewis, J.B.; Von Andrian, U.H.; Van Etten, R.A. Selectins and their ligands are required for homing and engraftment of BCR-ABL1+ leukemic stem cells in the bone marrow niche. *Blood* **2014**, *123*, 1361–1371. [CrossRef]
41. Moore, K.L.; Eaton, S.F.; F Lyons, D.; Lichenstein, H.S.; Cummings, R.D.; McEver, R.P. The P-selectin glycoprotein ligand from human neutrophils displays sialylated, fucosylated, O-linked poly-N-acetyllactosamine. *J. Biol. Chem.* **1994**, *269*, 23318–23327.
42. Winkler, I.G.; Erbani, J.M.; Barbier, V.; Davies, J.M.; Tay, J.; Fiveash, C.E.; Lowe, J.; Tallack, M.; Magnani, J.L.; Levesque, J.-P. Vascular E-Selectin Mediates Chemo-Resistance in Acute Myeloid Leukemia Initiating Cells Via Canonical Receptors PSGL-1 (CD162) and Hcell (CD44) and AKT Signaling. Available online: http://glycomimetics.com/wp-content/uploads/2018/11/ASH2017_Winkler-talk-12dec2017_for-GMI_with-text2-1.pdf (accessed on 1 January 2020).
43. Nonomura, C.; Kikuchi, J.; Kiyokawa, N.; Ozaki, H.; Mitsunaga, K.; Ando, H.; Kanamori, A.; Kannagi, R.; Fujimoto, J.; Muroi, K.; et al. CD43, but not P-Selectin Glycoprotein Ligand-1, Functions as an E-Selectin Counter-Receptor in Human Pre-B–Cell Leukemia NALL-1. *Cancer Res.* **2008**, *68*, 790–799. [CrossRef] [PubMed]
44. Matsumoto, M.; Atarashi, K.; Umemoto, E.; Furukawa, Y.; Shigeta, A.; Miyasaka, M.; Hirata, T. CD43 Functions as a Ligand for E-Selectin on Activated T Cells. *J. Immunol.* **2005**, *175*, 8042–8050. [CrossRef] [PubMed]
45. Natoni, A.; Smith, T.A.G.; Keane, N.; McEllistrim, C.; Connolly, C.; Jha, A.; Andrulis, M.; Ellert, E.; Raab, M.S.; Glavey, S.V.; et al. E-selectin ligands recognised by HECA452 induce drug resistance in myeloma, which is overcome by the E-selectin antagonist, GMI-1271. *Leukemia* **2017**, *31*, 2642–2651. [CrossRef] [PubMed]
46. Natoni, A.; Smith, T.A.; Keane, N.; Locatelli-Hoops, S.C.; Oliva, I.; Fogler, W.E.; Magnani, J.L.; O'Dwyer, M. E-Selectin Ligand Expression Increases with Progression of Myeloma and Induces Drug Resistance in a Murine Transplant Model, Which Is Overcome By the Glycomimetic E-Selectin Antagonist, GMI-1271. *Blood* **2015**, *126*, 1805. [CrossRef]
47. Yasmin-Karim, S.; King, M.R.; Messing, E.M.; Lee, Y.-F. E-selectin ligand-1 controls circulating prostate cancer cell rolling/adhesion and metastasis. *Oncotarget* **2014**, *5*, 12097–12110. [CrossRef]
48. Xiaobo, M.; Zhong, Y.-P.; Zheng, Y.; Jiang, J.; Wang, Y.-P. Coexpression of CD 5 and CD 43 predicts worse prognosis in diffuse large B-cell lymphoma. *Cancer Med.* **2018**, *7*, 4284–4295. [CrossRef]
49. Dimitroff, C.J.; Descheny, L.; Trujillo, N.; Kim, R.; Nguyen, V.; Huang, W.; Pienta, K.J.; Kutok, J.L.; Rubin, M.A. Identification of Leukocyte E-Selectin Ligands, P-Selectin Glycoprotein Ligand-1 and E-Selectin Ligand-1, on Human Metastatic Prostate Tumor Cells. *Cancer Res.* **2005**, *65*, 5750–5760. [CrossRef]
50. Sorigue, M.; Juncà, J.; Sarrate, E.; Grau, J. Expression of CD43 in chronic lymphoproliferative leukemias. *Cytom. Part B Clin. Cytom.* **2017**, *94*, 136–142. [CrossRef]
51. Fu, Q.; Cash, S.E.; Andersen, J.J.; Kennedy, C.R.; Oldenburg, D.G.; Zander, V.B.; Foley, G.R.; Shelley, C.S. CD43 in the nucleus and cytoplasm of lung cancer is a potential therapeutic target. *Int. J. Cancer* **2012**, *132*, 1761–1770. [CrossRef]
52. Krause, D.S.; Lazarides, K.; Von Andrian, U.H.; A Van Etten, R. Requirement for CD44 in homing and engraftment of BCR-ABL–expressing leukemic stem cells. *Nat. Med.* **2006**, *12*, 1175–1180. [CrossRef]
53. Godavarthy, P.S.; Kumar, R.; Herkt, S.C.; Pereira, R.S.; Hayduk, N.; Weissenberger, E.S.; Aggoune, D.; Manavski, Y.; Lucas, T.; Pan, K.-T.; et al. The vascular bone marrow niche influences outcome in chronic myeloid leukemia via the E-selectin - SCL/TAL1 - CD44 axis. *Haematologica* **2019**, *105*, 136–147. [CrossRef] [PubMed]
54. Esposito, M.; Mondal, N.; Greco, T.M.; Wei, Y.; Spadazzi, C.; Lin, S.-C.; Zheng, H.; Cheung, C.; Magnani, J.L.; Lin, S.-H.; et al. Bone vascular niche E-selectin induces mesenchymal–epithelial transition and Wnt activation in cancer cells to promote bone metastasis. *Nat. Cell Biol.* **2019**, *21*, 627–639. [CrossRef] [PubMed]
55. Läubli, H.; Borsig, L. Selectins as Mediators of Lung Metastasis. *Cancer Microenviron.* **2010**, *3*, 97–105. [CrossRef]
56. Meads, M.B.; Gatenby, R.A.; Dalton, W.S. Environment-mediated drug resistance: A major contributor to minimal residual disease. *Nat. Rev. Cancer* **2009**, *9*, 665–674. [CrossRef] [PubMed]

57. Chien, S.; Haq, S.U.; Pawlus, P.M.; Moon, P.R.T.; Estey, M.E.H.; Appelbaum, F.R.; Othus, M.; Magnani, J.L.; Becker, P.S. Adhesion Of Acute Myeloid Leukemia Blasts To E-Selectin In The Vascular Niche Enhances Their Survival By Mechanisms Such As Wnt Activation. *Blood* **2013**, *122*, 61. [CrossRef]
58. Muz, B.; De La Puente, P.; Azab, F.; Luderer, M.; Azab, A.K. Hypoxia promotes stem cell-like phenotype in multiple myeloma cells. *Blood Cancer J.* **2014**, *4*, e262. [CrossRef]
59. Martínez-Moreno, M.; Leiva, M.; Aguilera-Montilla, N.; Sevilla-Movilla, S.; De Val, S.I.; Arellano-Sánchez, N.; Gutiérrez, N.C.; Maldonado, R.; Martínez-López, J.; Buño, I.; et al. In vivo adhesion of malignant B cells to bone marrow microvasculature is regulated by α4β1 cytoplasmic-binding proteins. *Leukemia* **2016**, *30*, 861–872. [CrossRef]
60. Damiano, J.S.; Cress, A.E.; Hazlehurst, L.A.; Shtil, A.A.; Dalton, W.S. Cell adhesion mediated drug resistance (CAM-DR): Role of integrins and resistance to apoptosis in human myeloma cell lines. *Blood* **1999**, *93*, 1658–1667. [CrossRef]
61. Winkler, I.G.; Barbier, V.; Perkins, A.C.; Magnani, J.L.; Levesque, J.-P. Mobilisation of Reconstituting HSC Is Boosted By Synergy Between G-CSF and E-Selectin Antagonist GMI-1271. *Blood* **2014**, *124*, 317. [CrossRef]
62. Chien, S.; Zhao, X.; Brown, M.; Saxena, A.; Patton, J.T.; Magnani, J.L.; Becker, P.S. A Novel Small Molecule E-Selectin Inhibitor GMI-1271 Blocks Adhesion of AML Blasts to E-Selectin and Mobilizes Blood Cells in Nodscid IL2Rgc-/- Mice Engrafted with Human AML. *Blood* **2012**, *120*. [CrossRef]
63. Zhang, W.; Zhang, Q.; Mu, H.; Battula, V.L.; Patel, N.; Schober, W.; Han, X.; Fogler, W.E.; Magnani, J.L.; Andreeff, M. Dual E-Selectin/CXCR4 Antagonist GMI-1359 Exerts Efficient Anti-Leukemia Effects in a FLT3 ITD Mutated Acute Myeloid Leukemia Patient-Derived Xenograft Murine Model. *Blood* **2016**, *128*, 3519. [CrossRef]
64. Muz, B.; De La Puente, P.; Azab, F.; Azab, A.K. The role of hypoxia in cancer progression, angiogenesis, metastasis, and resistance to therapy. *Hypoxia* **2015**, *3*, 83–92. [CrossRef] [PubMed]
65. Winkler, I.G.; Barbier, V.; Nutt, H.; Hasnain, S.Z.; Levesque, J.P.; Magnani, J.L.; McGuckin, M.A. Administration of E-selectin Antagonist GMI-1271 Improves Survival to High-Dose Chemotherapy by Alleviating Mucositis and Accelerating Neutrophil Recovery. Available online: http://glycomimetics.com/wp-content/uploads/2018/11/Ingrid-mucositis-poster-layout-29nov13-5pm-Compatibility-Mode.pdf (accessed on 1 January 2020).
66. Manier, S.; Sacco, A.; Leleu, X.; Ghobrial, I.M.; Roccaro, A.M. Bone Marrow Microenvironment in Multiple Myeloma Progression. *J. Biomed. Biotechnol.* **2012**, *2012*, 1–5. [CrossRef]
67. Muz, B.; De La Puente, P.; Azab, F.; Luderer, M.; Azab, A.K. The Role of Hypoxia and Exploitation of the Hypoxic Environment in Hematologic Malignancies. *Mol. Cancer Res.* **2014**, *12*, 1347–1354. [CrossRef] [PubMed]
68. Fogler, W.E.; Smith, T.A.; King, R.K.; Magnani, J.L. Abstract 1757: Mobilization of tumor-primed, marrow-infiltrating lymphocytes into peripheral blood with inhibitors of E-selectin or E-selectin and CXCR4. *Immunology* **2018**, *78*, 1757. [CrossRef]
69. DeAngelo, D.J.; Jonas, B.A.; Becker, P.S.; O'Dwyer, M.; Advani, A.S.; Marlton, P.; Magnani, J.L.; Thackray, H.M.; Liesveld, J. GMI-1271, a novel E-selectin antagonist, combined with induction chemotherapy in elderly patients with untreated AML. *J. Clin. Oncol.* **2017**, *35*, 2560. [CrossRef]
70. Winkler, I.G.; Barbier, V.; Pattabiraman, D.R.; Gonda, T.J.; Magnani, J.L.; Levesque, J. Vascular Niche E-Selectin Protects Acute Myeloid Leukemia Stem Cells from Chemotherapy. *Blood* **2014**, *124*. [CrossRef]
71. Fogler, W.E.; Flanner, H.; Wolfgang, C.; Smith, J.A.; Thackray, H.M.; Magnani, J.L. Administration of the Dual E-Selectin/CXCR4 Antagonist, GMI-1359, Results in a Unique Profile of Tumor Mobilization from the Bone Marrow and Facilitation of Chemotherapy in a Murine Model of FLT3 ITD AML. *Blood* **2016**, *128*, 2826. [CrossRef]
72. Zhang, W.; Fogler, W.E.; Magnani, J.L.; Andreeff, M. The Dual E-Selectin/CXCR4 Inhibitor, GMI-1359, Enhances Efficacy of Anti-Leukemia Chemotherapy in FLT3-ITD Mutated Acute Myeloid Leukemia. *Blood* **2015**, *126*, 3790. [CrossRef]
73. Azab, A.K.; Hu, J.; Quang, P.; Azab, F.; Pitsillides, C.; Awwad, R.; Thompson, B.; Maiso, P.; Sun, J.D.; Hart, C.P.; et al. Hypoxia promotes dissemination of multiple myeloma through acquisition of epithelial to mesenchymal transition-like features. *Blood* **2012**, *119*, 5782–5794. [CrossRef]
74. Fei, M.; Hang, Q.; Hou, S.; He, S.; Ruan, C. Adhesion to fibronectin induces p27Kip1 nuclear accumulation through down-regulation of Jab1 and contributes to cell adhesion-mediated drug resistance (CAM-DR) in RPMI 8226 cells. *Mol. Cell. Biochem.* **2013**, *386*, 177–187. [CrossRef] [PubMed]
75. DeAngelo, D.J.; Jonas, B.A.; Liesveld, J.L.; Bixby, D.L.; Advani, A.S.; Marlton, P.; O'Dwyer, M.E.; Magnani, J.L.; Thackray, H.M.; Becker, P.S. GMI-1271 improves efficacy and safety of chemotherapy in R/R and newly diagnosed older patients with AML: Results of a Phase 1/2 study. *Blood* **2017**, *130*. [CrossRef]
76. Devata, S.; Sood, S.L.; Hemmer, M.M.V.; Flanner, M.H.; Kramer, W.; Nietubicz, C.; Hawley, A.; E Angelini, D.; Myers, D.D.D.; Blackburn, S.; et al. First in Human Phase 1 Single Dose Escalation Studies of the E-Selectin Antagonist GMI-1271 Show a Favorable Safety, Pharmacokinetic, and Biomarker Profile. *Blood* **2015**, *126*, 1004. [CrossRef]
77. Devata, S.; Angelini, D.E.; Rn, M.S.B.; Hawley, A.; Myers, D.D.; Schaefer, J.K.; Ma, M.H.; Magnani, J.L.; Thackray, H.M.; Wakefield, T.W.; et al. Use of GMI-1271, an E-selectin antagonist, in healthy subjects and in 2 patients with calf vein thrombosis. *Res. Pr. Thromb. Haemost.* **2020**, *4*, 193–204. [CrossRef] [PubMed]

78. DeAngelo, D.J.; Erba, H.; Jonas, B.A.; O'Dwyer, M.E.; Marlton, P.; Huls, G.; Liesveld, J.L.; Cooper, B.; Bhatnagar, B.; Armstrong, M.; et al. Trials in Progress: A phase 3 trial to evaluate the efficacy of Uproleselan (GMI-1271) with chemotherapy in patients with relapsed/refractory acute myeloid leukemia. *J. Clin. Oncol.* **2019**, *37*. [CrossRef]
79. Myers, J.D.; Lester, P.; Adili, R.; Hawley, A.; Durham, L.; Dunivant, V.; Reynolds, G.; Crego, K.; Zimmerman, Z.; Sood, S.; et al. A new way to treat proximal deep venous thrombosis using E-selectin inhibition. *J. Vasc. Surg. Venous Lymphat. Disord.* **2020**, *8*, 268–278. [CrossRef] [PubMed]

Review

Potential Roles of Tumor Cell- and Stroma Cell-Derived Small Extracellular Vesicles in Promoting a Pro-Angiogenic Tumor Microenvironment

Nils Ludwig [1], Dominique S. Rubenich [2], Łukasz Zaręba [3], Jacek Siewiera [4], Josquin Pieper [5], Elizandra Braganhol [2], Torsten E. Reichert [1] and Mirosław J. Szczepański [3,*]

1. Department of Oral and Maxillofacial Surgery, University Hospital Regensburg, 93053 Regensburg, Germany; nils.ludwig@ukr.de (N.L.); torsten.reichert@ukr.de (T.E.R.)
2. Programa de Pós-Graduação em Biociências, Universidade Federal de Ciências da Saúde de Porto Alegre (UFCSPA), Porto Alegre 90050-170, Brazil; dominiquesr@ufcspa.edu.br (D.S.R.); ebraganhol@ufcspa.edu.br (E.B.)
3. Department of Biochemistry, Medical University of Warsaw, 02-091 Warsaw, Poland; s068791@student.wum.edu.pl
4. Department of Hyperbaric Medicine, Military Institute of Medicine, 04-141 Warsaw, Poland; jsiewiera@wim.mil.pl
5. Department of Oral and Maxillofacial Surgery, University Hospital Bochum, 44892 Bochum, Germany; josquin.pieper@kk-bochum.de
* Correspondence: mszczepanski@wum.edu.pl

Received: 7 November 2020; Accepted: 30 November 2020; Published: 2 December 2020

Simple Summary: In this review, we focus on the distinct functions of tumor-cell-derived small extracellular vesicles in promotion of angiogenesis and describe their potential as a therapeutic target for anti-angiogenic therapies. Also, we focus on extracellular vesicles derived from non-cancer cells and their potential role in stimulating a pro-angiogenic tumor microenvironment. The article describes the biogenesis of small extracellular vesicles and refers to their proteomic cargo components that play a role in promoting angiogenesis. Moreover, we explain how small extracellular vesicles derived from tumors and non-cancer cells can interact with recipient cells and alter their functions. We particularly focus on phenotypical and functional changes in endothelial cells, macrophages, and neutrophils that result in proangiogenic signaling.

Abstract: Extracellular vesicles (EVs) are produced and released by all cells and are present in all body fluids. They exist in a variety of sizes, however, small extracellular vesicles (sEVs), the EV subset with a size range from 30 to 150 nm, are of current interest. They are characterized by a distinct biogenesis and complex cargo composition, which reflects the cytosolic contents and cell-surface molecules of the parent cells. This cargo consists of proteins, nucleic acids, and lipids and is competent in inducing signaling cascades in recipient cells after surface interactions or in initiating the generation of a functional protein by delivering nucleic acids. Based on these characteristics, sEVs are now considered as important mediators of intercellular communication. One hallmark of sEVs is the promotion of angiogenesis. It was shown that sEVs interact with endothelial cells (ECs) and promote an angiogenic phenotype, ultimately leading to increased vascularization of solid tumors and disease progression. It was also shown that sEVs reprogram cells in the tumor microenvironment (TME) and act in a functionally cooperative fashion to promote angiogenesis by a paracrine mechanism involving the differential expression and secretion of angiogenic factors from other cell types. In this review, we will focus on the distinct functions of tumor-cell-derived sEVs (TEX) in promotion of angiogenesis and describe their potential as a therapeutic target for anti-angiogenic therapies. Also, we will focus on non-cancer stroma-cell-derived small extracellular vesicles and their potential role in stimulating a pro-angiogenic TME.

Keywords: tumor-derived exosomes; small extracellular vesicles; angiogenesis; cancer; tumor microenvironment

1. Introduction

Small extracellular vesicles (sEVs) are a fraction of the extracellular vesicles produced by all cell types, including tumor cells [1,2]. They are described as small membranous vesicles with diameters ranging from 30 to 150 nm, that carry selected proteins, lipids, nucleic acids, and glycoconjugates. Growing evidence suggests that sEVs are present in all body fluids such as plasma, cerebrospinal fluid, urine, and saliva [1,3]. They are produced during the inward invagination of the endosome which results in the formation of small vesicles, called intraluminal bodies (ILVs), encapsulated in a larger multivesicular body (MVB). After the maturation and transport of the MVB to the peripheral areas of the cell, it fuses with the cell membrane and releases sEVs into the extracellular space [4]. sEVs were considered to function as cellular bins, but research in recent years has brought evidence that they play a crucial role in intercellular communication under physiological and pathological conditions [5]. In particular, sEVs produced by tumor cells, thus tumor-derived sEVs (TEX), gained a lot of interest due to their role in tumor growth, metastasis, immune escape, and angiogenesis [6,7]. Growing evidence suggests that cancer cells release larger quantities of sEVs compared to non-cancer cells and, therefore, TEX are enriched in the tumor microenvironment (TME) and plasma of cancer patients [8]. This enhanced release of TEX was shown to be associated with cellular stress, such as hypoxia, acidic pH, and many other triggers present in the TME [9–11]. Due to their large amount in biofluids, sEVs were proposed as useful biomarkers for non-invasive cancer monitoring in the context of a liquid biopsy [2]. Among the above-mentioned functions, the stimulation of tumor angiogenesis appears to be one hallmark of TEX and it is crucial for tumor persistence and progression [12]. It was shown that TEX carry a plethora of angiogenic factors and molecules, which can contribute to the formation of new blood vessels [6]. In this review, we focus on the role of TEX in promoting a pro-angiogenic TME. Furthermore, we discuss stroma non-cancer cell-derived sEVs and their contribution to the promotion of angiogenesis with special regard to endothelial cells, macrophages, and neutrophils.

2. Biogenesis of sEVs

The biogenesis of sEVs begins with the formation of an early endosome (EE) by inward invagination of the plasma membrane [5]. Subsequently, the EE matures into a late endosome (LE), which begins to produce intraluminal vesicles (ILVs) by using an endosomal sorting complex required for transport (ESCRT) [13,14]. ILVs originate from inward budding of the endosomal membrane, which encloses fragments of cytosolic content and transmembrane and peripheral proteins into smaller vesicles located in the endosome. A newly-formatted endosome enriched in luminar ILVs is considered a multivesicular body (MVB) [15]. Finally, mature MVBs can fuse with the lysosome membrane and become involved in the degradation pathway or fuse with the cell membrane, which results in sEV release into the extracellular fluid (ECF) [15].

The formation of ILVs and their sorting process in MVBs is defined by a highly complicated mechanism, which is precisely regulated by the ESCRT complex. This complex consists of four proteins—ESCRT-0, ESCRT-I, ESCRT-II, and ESCRT-III [16,17]. ILV formation begins with sorting and sequestrating of ubiquitinated proteins by ESCRT-0 in specific sections of the MVB membrane [18]. Afterwards, ESCRT-I and ESCRT-II connect with ESCRT-0 and assemble a complex with high avidity for the ubiquitinated proteins. Subsequently, ESCRT-III joins to the rest of the complex and induces membrane deformation, thereby commencing the inward invagination and ILV formation [19]. Finally, newly-originated ILVs are released into the MVB lumen. Now, the ubiquitinated ILV cargo is destined for lysosomal degradation unless deubiquitylating enzymes (DUBs) change this fate.

TSG101 which is an ESCRT-machinery component when ubiquitinated with ISG15 (ISGylated) can trigger the MVB aggregation with lysosomes and its protein degradation [20]. Interestingly, in some conditions deubiquitylation of TSG101 may induce lysosomal trafficking [16]. Moreover, ATPase VPS4 dissociates the ESCRT protein complex from the membrane and enables its recycling [21–23].

Mature MVBs can directly fuse with the lysosome and deliver its content for a degradation or target the cell membrane and release ILVs, from what are now called sEVs, into the ECF. Rab27a and Rab27b are important regulators of MVB transport to the peripheral areas of the cell and promote its docking with the cell membrane, ultimately releasing sEVs into the ECF [24]. sEVs are secreted by virtually all types of cells, however, cancer cells show outstanding activity in sEV production when compared to non-cancer cells [5]. The underlying mechanisms of sEV production in different cell types, with special regards to potential discriminations between cancer and non-cancer cells, are still to be determined. Uncovering the precise molecular interactions and pathways of sEV production is part of ongoing research. The biogenesis of sEVs is illustrated in Figure 1.

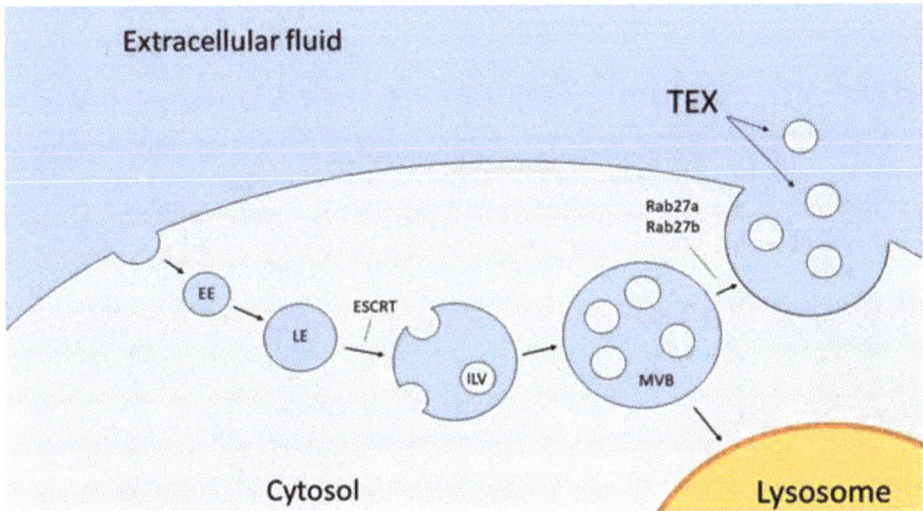

Figure 1. Biogenesis and secretion of small extracellular vesicles. The EE is created by inward invagination of the cell membrane. Afterwards, the EEs maturate into LEs. The ESCRT molecular machinery is involved in producing ILVs leading to the formation of MVBs, which contain many ILVs in its lumen. MVBs undergo one of two main pathways, since they (1) fuse with a lysosome or (2) degrade or fuse with the cell membrane and release small extracellular vesicles (sEVs). Abbreviations: EE, early endosome; LE, late endosome; ESCRT, endosomal sorting complex required for transport; ILV, intraluminal vesicle; MVB, multivesicular body; and TEX, tumor-derived sEVs.

3. Proteomic Cargo of TEX

TEX play a key role in mediating intercellular communication by transporting proteins, lipids, nucleic acids, and many other molecules involved in signaling processes from cancer cells to recipient cells [25–27]. TEX are enriched in a variety of factors, such as proteases, enzymes, growth factors, and cytokines, which are transported in the lumen of TEX and are delivered to different recipient cells in the TME, facilitating tumor growth and expansion [28]. Moreover, TEX carry cargo components on their surface such as receptors/ligands, adherent molecules, or tetraspanins (e.g., CD63, CD9, and CD81) [29]. The molecular cargo composition of TEX reflects the contents of the donor cell [30], however, recent evidence suggests that some cargo components of TEX are a result of a special sorting mechanism rather than an exact reflection of donor cell composition [31]. Furthermore, the sphingosine 1-phosphate receptor was suggested to play a role in distributing molecules to ILVs [32].

As mentioned above, the cargo transported by TEX can be distinguished between intraluminal and surface-bound molecules. It was demonstrated, that TEX carry immunosuppressive ligands on their surface such as programmed cell death protein 1 (PD-1), Fas, and TNF-related apoptosis-inducing ligand (TRAIL) [28]. Moreover, the membrane of TEX is enriched in adherent molecules such as intercellular adhesion molecule (ICAM), epithelial cell adhesion molecule (EpCAM) and CD44. Furthermore, the membrane-associated cargo of TEX may contain transmembrane receptors such as chemokine receptor CXC type 4 (CXCR-4), c-MET, heat shock proteins, and above-mentioned tetraspanins, which are commonly used as sEV markers [1,29]. The lumen of TEX incorporates transport proteins such as programmed cell death 6-interacting protein (ALIX), Rab proteins, dynamin, and lysosome associated membrane proteins (LAMPs) and signaling molecules including mitogen-activated protein kinase (MAPK), Rho, extracellular signal-regulated protein kinases 1 and 2 (ERK1/2), Wnt, and even cytoskeletal proteins such as actin and tubulin, as indicated in Figure 2 [1]. The cargo components of sEVs and their biological effects in relation to their cellular origins are summarized in Table 1.

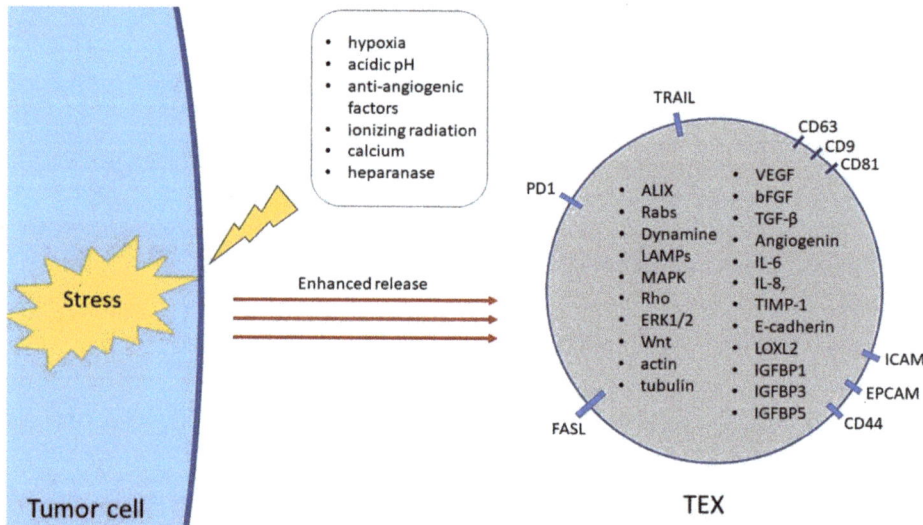

Figure 2. Release of TEX and their molecular cargo composition. Stress conditions such as hypoxia, acidic pH, anti-angiogenic factors, ionizing radiation, increased Ca^{2+} concentration, or heparanase activity are considered to stimulate the release of TEX. The protein cargo of TEX can be distinguished between luminal and membrane-bound molecules. The membrane-associated factors may include CD63, CD9, CD81, intercellular adhesion molecule (ICAM), epithelial adhesion molecule (EpCAM), CD44, FasL, programmed cell death protein 1 (PD-1), TNF-related apoptosis-inducing ligand (TRAIL), while TEX encapsulate proteins such as programmed cell death 6-interacting protein (ALIX), Rab proteins, dynamine, lysosome associated membrane proteins (LAMPs); signaling pathway components such as mitogen-activated protein kinase (MAPK), Rho, extracellular signal-regulated protein kinases 1 and 2 (ERK1/2), Wnt; cytoskeletal proteins such as actin and tubulin; and proangiogenic factors such as vascular endothelial growth factor (VEGF), basic fibroblast growth factor (bFGF), transforming growth factor-β (TGF-β).

Table 1. Comparison of articles from our literature search with emphasis on source of TEX or sEVs, their cargo, and biological effects. Abbreviations: IL: Interleukin; TRAIL: TNF-related apoptosis-inducing ligand; PD-L1: Programmed death-ligand 1; HSC70: Heat shock 70 kDa protein 8; PDGF: Platelet-derived growth factor; VEGF: Vascular endothelial growth factor; Vash1: Vasohibin-1; Angpt1: Angiopoietin 1; Flk1: Fetal liver kinase 1; uPA: Urokinase-type plasminogen activator, MMP-9: Matrix metallopeptidase 9; IGFBP3: insulin like growth factor binding protein 3; HGF: Hepatocyte growth factor; EPHB2: ephrin type B receptor 2; TGF-β: Transforming growth factor-β; STAT3: Signal transducer and activator of transcription 3.

Reference	Cancer Type/Source of sEVs	TEX Cargo	Tumor-Promoting Biological Effects
Skog et al. [27]	Glioblastoma	Angiogenin, IL-6, IL-8	Angiogenesis
Sharma et al. [7]	Melanoma	FasL, TRAIL, PD-L1	Immunosuppression
Ludwig et al. [25]	Head and neck squamous-cell carcinoma cell line UMSCC47	CD39/CD73, adenosine	M2 macrophage Polarization and enhanced secretion of angiogenic factors
Umezu et al. [26]	Leukemia cells (K562)	miR-92a	Enhanced endothelial cell migration and tube formation
Kucharzewska et al. [30]	Glioma cells	Matrix metalloproteinases, IL-8, PDGFs, caveolin 1, and lysyl oxidase	Activation of vascular cells
Ko et al. [33]	ES2, HCT116, and 786-0 cell lines	Heparin-bound VEGF on the surface of sEVs	Endothelial cells migration and tube formation
Xue et al. [34]	adipose Mesenchymal stem cells	Vash1, Angpt1 and Flk1	enhancement of Angiogenesis through the PKA-signaling pathway
Ludwig et al. [35]	Head and neck squamous-cell carcinoma cell lines (PCI-13, UMSCC47)	uPA, MMP-9, coagulation factor III, thrombospondin-1, uPA, IGFBP-3, endostatin	Reprogramming of HUVECs
Thompson et al. [36]	Bacterial heparinase-III-treated CAG, ARH-77, MDA-MB-231	Syndecan-1, VEGF, and HGF	Enhanced endothelial cell invasion
Zeng et al. [37]	Hepatocellular carcinoma	VEGF	Tumor vasculogenesis despite anti-angiogenic therapy
Sato et al. [38]	Head and neck squamous cell carcinoma	EPHB2	Promotion of angiogenesis
Carrasco-Ramirez et al. [39]	Melanoma	Podoplanin	Modulation of lymphatic vessel formation
Hong et al. [40]	Colorectal cancer	Cell-cycle–related mRNAs	Proliferation of endothelial cells
Lang et al. [41]	Glioma cells	Long non-coding RNA CCAT2	Promotion of angiogenesis and inhibition of endothelial cell apoptosis
Van Balkom et al. [42]	Human microvascular endothelial cell line (HMEC-1)	miR-214	Prevention from cell cycle arrest and, thus, stimulation of blood vessel formation
Azambuja et al. [43]	Reprogrammed macrophages	Arginase-1	Glioblastoma progression
Webber et al. [44]	Prostate, bladder, colorectal and breast cancer cell lines	TGF-β	Differentiation of fibroblasts to myofibroblasts
Hong et al. [45]	Acute myeloid leukemia	TGF-β	Immunosuppression
Zheng et al. [46]	Tumor-associated macrophages	Apolipoprotein-E	Migration of gastric cancer cells
Shi et al. [47]	Gastric cancer cells	STAT3	PD-L1 expression on neutrophils to suppress T-cell-mediated immunity

Different studies demonstrate that TEX are involved in inducing or promoting the process of new blood vessel formation during all stages of tumor development [12,27,48]. The analysis of TEX showed that cargo components also include factors that are essential for angiogenic pathways. Among these factors are pro-angiogenic growth factors such as basic fibroblast growth factor (bFGF), vascular endothelial growth factor (VEGF), and transforming growth factor-β (TGF-β), which were shown to have a strong impact on angiogenesis [6]. Moreover, TEX carry many other proteins that play a role in angiogenic processes including angiogenin, interleukin-6 (IL-6), IL-8, TIMP metallopeptidase inhibitor 1 (TIMP-1), and E-cadherin [27,49]. The clinical relevance of the pro-angiogenic cargo of TEX was supported in a recent paper showing that TEX carry a specific isoform of VEGF on their surface. Interestingly, this isoform preferentially localizes on the surface of TEX through its high affinity for heparin and has a profoundly-increased half-life compared to soluble VEGF. Additionally, this TEX-associated VEGF is not neutralized by bevacizumab and high levels were associated with disease progression in bevacizumab-treated cancer patients [33]. These results indicate that TEX may increase the resistance to anti-angiogenic therapies or may even interfere with these therapies.

The release of TEX is highly dependent on environmental conditions. Various signals have been described to influence the release of sEVs by tumor cells. One of the well-established triggers for the enhanced release of TEX, which is also a common characteristic of the TME, is hypoxia. It was shown that hypoxic conditions induce the release of TEX in breast cancer, bladder cancer, prostate cancer, head and neck cancer, and many other malignant entities [11,34,35,50]. An acidic environment is considered to be another important trigger for TEX release and is provided by increased anaerobic metabolism of tumor cells. Moreover, it was shown that higher levels of acidity and TEX release are strictly related to and strongly contribute to malignant tumor phenotypes [51]. Furthermore, TEX release may be induced by anti-angiogenic factors, ionizing radiation, increased Ca^{2+} concentration, or active forms of heparanase, as illustrated in Figure 2 [6,36,37,52–54]. It was recently shown that the cargo composition of TEX can be altered depending on the environmental factors which trigger the release of TEX. In particular, anti-cancer therapies were reported to promote the release of TEX, which are enriched in immunosuppressive or pro-angiogenic cargo components and ultimately weaken response to therapy or promote metastasis [55]. TEX generated under hypoxic conditions were shown to be especially enriched in pro-angiogenic factors, bearing a greater potential to contribute to new blood vessel formation [30]. Several pro-angiogenic cargo components of TEX were found to be regulated by hypoxia, including TGF-β, VEGF, lysyl oxidase homolog 2 (LOXL2), IL-8, insulin like growth factor binding protein 1 (IGFBP1), IGFBP3, and IGFBP5 [12].

4. Interactions of TEX with Recipient Cells

TEX are considered to be an important contributor to intercellular communication mechanisms and facilitate the tumor/stroma crosstalk. It was shown that this crosstalk allows the reprogramming of stroma cells and ultimately shapes a tumor-promoting microenvironment [28]. This microenvironment is characterized by the presence of pro-angiogenic factors, which are either secreted by the tumor cells or other cells present in the TME. TEX might carry a fraction of these pro-angiogenic factors and might also be an important regulator for the release of pro-angiogenic factors by other cells in the TME. Therefore, the interactions of TEX with recipient cells are of major importance to understanding the role of TEX in angiogenesis.

Several mechanisms have been described concerning the TEX/recipient cell interaction. Most data focusses on the internalization of TEX by recipient cells which can lead to the generation of a functional protein by delivery of nucleic acids [27,56]. The detailed mechanism for TEX uptake varies depending on the cargo of TEX and the recipient cell, as well as microenvironmental conditions [56]. The most-described pathways for the internalization of TEX are endocytosis, macropinocytosis, phagocytosis, and membrane fusion. Interestingly, most cell types are able to utilize several of these pathways to internalize TEX and blocking of one pathway might enhance the uptake of TEX by another pathway. Therefore, the uptake mechanisms need to be carefully studied and investigated for each cell type. For endothelial cells (ECs) it was shown, that TEX are internalized rapidly within 4 h of co-incubation and the most dominant internalization pathway was described to be endocytosis [35]. In other cell types, for example macrophages, phagocytosis is the most dominant uptake mechanism and macrophages are even more efficient in internalizing TEX compared to ECs [57]. However, it was reported that macrophages internalize TEX not only via phagocytosis, but also via clathrin- and caveolin-dependent endocytosis and macropinocytosis [56]. T cells were shown to internalize only minimal quantities of TEX, indicating that other forms of interaction are responsible for the TEX-mediated effects on this cell type [58].

Surface-mediated receptor–ligand interactions are another form of interaction between TEX and recipient cells. TEX carry ligands on their surface, which can bind to receptors expressed by the recipient cells and initiate a signaling cascade. This was shown for multiple cell types, including ECs. Reported signaling pathways are, among others, the notch pathway [59], the adenosine pathway [25], ephrin pathway [38], and the E-cadherin pathway [49].

TEX also carry functionally-active enzymes on their surface which can produce factors that stimulate surrounding cells in a paracrine fashion. This was shown for the ectonucleotidases CD39 and CD73, which generate adenosine and stimulate EC growth depending on adenosine A_{2B} receptor signaling [25].

5. Effects of TEX on Endothelial Cells

Most reports which focus on the pro-angiogenic effects of TEX describe the interactions of TEX and ECs. Interestingly, it was shown for multiple types of cancer, that TEX induce a pro-angiogenic phenotype in ECs and stimulate their proliferation, migration, and tube formation [12]. These functional alterations of ECs were described to be mediated via different pathways. It was demonstrated that TEX carry ephrin type B receptor 2 (EPHB2) and that EPHB2 promotes angiogenesis by ephrin-B reverse signaling, inducing STAT3 phosphorylation. Accordingly, a STAT3 inhibitor was presented as a strategy to inhibit TEX-induced angiogenesis [38]. Another study demonstrated that the underlying mechanism for the pro-angiogenic effects of TEX is adenosine A2B receptor signaling [25]. The transmembrane glycoprotein podoplanin was also suggested to play a crucial role in mediating effects of TEX on ECs [39]. Besides that, the delivery of nucleic acids to ECs by TEX is considered to promote pro-angiogenic functions of ECs [27]. In particular, miRNAs in the lumen of TEX were described to be involved in reprogramming ECs. Most frequently, the following miRNAs appeared in the literature and seem to play an important role in angiogenic processes—miR-21, miR-23a, miR-30b, miR-126a, and miR-210 [12]. However, other RNA classes were also described in the same context. TEX derived from colorectal cancer cells are enriched with mRNAs such as CDK8, ERH, and RAD21, which are mainly related to the cell cycle [40]. Also, TEX-associated long noncoding RNAs (lncRNAs) were reported to promote angiogenesis. The transfer of long intergenic noncoding RNA CCAT2 to ECs by TEX enhanced in vitro and in vivo angiogenesis and upregulated VEGFA and TGF-β levels on ECs [41]. It seems that the pro-angiogenic effects of TEX on ECs are orchestrated via multiple pathways which probably converge, therefore, making TEX efficacious promotors of angiogenesis.

Besides these direct effects of TEX on ECs, it was shown that reprogrammed ECs secrete several potent growth factors and cytokines and stimulate pericyte PI3K/AKT signaling activation and migration [30]. Additionally, it was demonstrated that ECs also release sEVs themselves which can impact the process of new blood vessel formation [60]. These stroma-cell-derived small extracellular vesicles (EC-derived sEVs) in human plasma carried vascular cell adhesion molecule-1 (VCAM-1), endothelial nitric oxide synthase, von Willebrand factor (vWF), platelet derived growth factor BB (PDGF-BB), large neutral amino acid transporter (LAT-1), angiopoietin 1 and 2, glucose transporter 1 (GLUT-1), and lysyl oxidase homolog 2 (LOXL-2) [61]. The cargo of EC-derived sEVs also consists of nucleic acids, such as miR-214, which can be transferred to other ECs and stimulate their migration and angiogenesis. sEVs from miR-214-depleted ECs fail to stimulate these processes [42]. However, only limited data are available regarding EC-derived sEVs and their role in tumor angiogenesis or tumor progression. Future studies are necessary to analyze the detailed cargo composition of sEVs derived from tumor-associated ECs as well as studying their functions in the TME.

6. Effects of TEX on Macrophages

Tumor-associated macrophages (TAMs) constitute a plastic and heterogeneous cell population of the TME that can account for up to 50% of some solid neoplasms. Most often, TAMs support disease progression and resistance to therapy, however, TAMs can also mediate antineoplastic effects, especially in response to pharmacological agents that boost their phagocytic and oxidative functions [62]. The phenotype of TAMs is heterogenous and translates into distinct functions in the TME. TAMs can be categorized into two subsets, classically-activated (M1) and alternatively-activated (M2) macrophages based on their phenotype and distinct functional abilities. In the TME, the M2 phenotype is considered to be the dominant macrophage population and is responsible for tumor progression and associated with poor prognosis [43,63]. The polarization towards M1 or M2 relies on the presence of different stimuli. Macrophages exposed to cytokines like IL-12, TNF, or IFNγ, microbe-associated molecular patterns (MAMPs) such as bacterial lipopolysaccharide (LPS), or other toll-like receptor (TLR) agonists acquire an M1 state. Conversely, IL-4, IL-5, IL-10, IL-13, CSF1, TFG-β, and PGE2 promote macrophage polarization towards an M2 state [62]. Some of these stimuli, such as TGF-β, were shown to be part of the TEX cargo [44,45], therefore, it was suggested, that TEX are involved in polarizing macrophages. Glioblastoma-derived TEX were shown to polarize macrophages towards the M2 phenotype, whereas melanoma and head- and neck-cancer-derived TEX were shown to induce a mixed M1/M2 phenotype of macrophages [25,43,57]. These results indicate that TEX are indeed involved in macrophage polarization, however, the direction of this polarization is dependent on the composition of TEX. It was also demonstrated that TEX are able to stimulate the recruitment of macrophages [64] and injection of TEX-containing plugs into mice promoted a massive infiltration of M2 macrophages [25]. The reprogramming of macrophages by TEX was also connected to the formation of a pre-metastatic niche, which is considered to be one of the hallmarks of TEX functions [64].

One important effect of TAMs is the provision of trophic and nutritional support for cancer cells or other cells in the TME, ultimately promoting disease progression. This also includes the stimulation of angiogenesis and, therefore, the crosstalk between TAMs and ECs. M2-like TAMs were described as key effectors of stimulating angiogenesis, since they produce diverse pro-angiogenic factors, like TGF-β, VEGF, PDGF, and angiogenic chemokines, as indicated in Figure 3 [65]. Macrophage infiltration in tumors is generally associated with high vascular density and M2-like TAMs predominantly localize in hypoxic tumor areas [66]. As indicated above, TEX are able to reprogram macrophages and, therefore, might promote the pro-angiogenic effects of TAMs. Recent work supports this concept and shows that TEX not only directly interact with ECs to stimulate new blood vessel formation, but also interact in an indirect way by reprogramming macrophages towards a pro-angiogenic phenotype [67]. In head and neck cancer it was demonstrated that TEX stimulate the release of pro-angiogenic factors by macrophages and therefore stimulate ECs in an indirect way. Especially angiopoietin-1 and 2, IL-8, MMP-9, serpin E1, and TIMP-1 were reported to be released by macrophages in larger quantities after co-incubation with TEX [25]. This reprogramming of macrophages towards a pro-angiogenic M2 phenotype by TEX can be either induced by proteins such as TGF-β and nucleic acids such as miRNAs, or other TEX-associated factors such as adenosine [25,68].

It is also important to mention that reprogrammed TAMs again release sEVs with tumor-promoting functions. Recent studies reported a functional role for miRNA-containing sEVs derived from M2-like macrophages in regulating migration and invasion of colorectal cancer cells [69]. Another study demonstrated the sEV-mediated transfer of functional apolipoprotein E from TAMs to tumor cells, resulting in an enhanced migration of gastric cancer cells [46]. Arginase-1(+) sEVs derived from TAMs promoted tumor cell migration and proliferation in glioblastoma and were considered as a distributor of proteins with pro-tumor functions in the TME [43]. Although no evidence exists so far, it is likely that TAM-derived sEVs also carry a pro-angiogenic cargo, which reflects the cytosolic contents and cell-surface molecules associated with the M2-like phenotype. Analog to TEX, the sEVs derived from TAMs might also be involved in stimulating ECs and inducing or promoting angiogenesis. However, future studies are necessary to confirm this hypothesis.

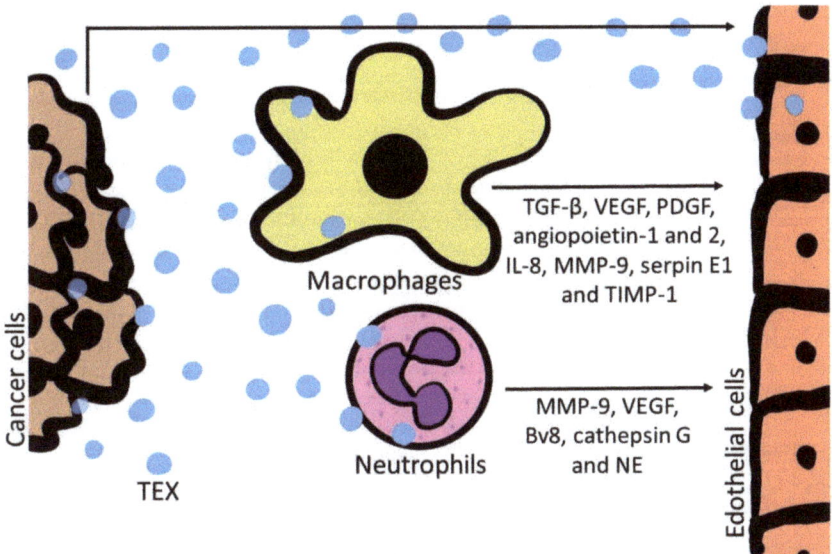

Figure 3. A schematic visualizing the reprogramming of endothelial cells by TEX. TEX carry a variety of pro-angiogenic factors, which are either surface-bound or encapsulated in their lumen. TEX interact directly with endothelial cells or reprogram other cells in the tumor microenvironment such as macrophages and neutrophils and promote the release of pro-angiogenic factors to stimulate a pro-angiogenic phenotype in endothelial cells.

7. Effects of TEX on Neutrophils

As described above, the M2-like phenotype of macrophages contributes to tumor progression by induction and stimulation of angiogenesis through the secretion of pro-angiogenic factors [70]. Analog to macrophages, it was reported that neutrophils can play a similar role, however, much less is currently known about their pro-tumor effects. It is commonly observed, that patients with solid tumors show elevated levels of neutrophils compared to healthy individuals [71]. Neutrophils are the most abundant type of granulocytes and form part of the polymorphonuclear cell family (PMNs). They are an essential part of the innate immune system, with being one of the first responders of inflammatory cells to migrate towards the site of inflammation or tissue damage, acting against a wide variety of pathogens [72]. Defining the exact roles of neutrophils is still ongoing research, however, they are considered to participate in the immune response through the regulation and recruitment of other immune cells, such as monocytes/macrophages and dendritic cells [73], as well as through modulation of the interaction between B and T cells [72]. Analogous to macrophages, neutrophils show heterogenous phenotypes which allow them to be proficient in distinct functions [74]. Compared to M1 and M2 macrophages, neutrophils have an activation spectrum that has not yet been clarified, however, they can be grouped according to the same logic—N1 (pro-inflammatory polarization) and N2 (anti-inflammatory polarization) [72]. The N2 phenotype is acquired by the presence of TGF-β, favoring the infiltration of neutrophils with high expression of CXCR-4, vascular endothelial growth factor A (VEGF-A), and metalloprotease 9 (MMP-9) [75]. Neutrophils are efficacious producers of pro-angiogenic factors and especially cells with the N2 phenotype contribute to angiogenesis and tissue invasion, ultimately promoting tumor progression [76–78]. Although it is still ongoing research to define the role of TEX in neutrophil polarization, TEX carry factors that are considered important for a polarization towards the N2-like phenotype. These factors include TGF-β, MMP-9, and arginase-1 [35,43,44]. Evidence for functional reprogramming of neutrophils by TEX was reported in a recent paper showing that gastric-cancer-cell-derived sEVs enhanced PD-L1 expression of neutrophils, thereby suppressing the activity of T cells [47]. The concept that TEX interact with neutrophils and promote their activation is also supported by co-incubation studies leading to both pro-inflammatory [79] and anti-inflammatory profiles [80,81] depending on the source and cargo components of TEX. One effector for the promotion of pro- or anti-inflammatory responses by neutrophils is the release of granules [72]. It should be noted that the neutrophil secondary granules are enriched in MMP-9, which is a decisive component of angiogenic pathways [75,78,82]. Interestingly, neutrophils are the only cell type that is capable of releasing MMP-9 without being attached to its physiological inhibitor TIMP-1. Therefore, the secretion of highly-active MMP-9 by neutrophils can have a major impact on angiogenic sites [83]. Additionally to MMP-9, neutrophils release other potent growth factors to promote angiogenesis, such as VEGF or Bv8, as indicated in Figure 3 [75,77,82,84,85]. It was observed that neutrophils generate heterogenous microparticles, which are capable of altering epithelial gene expression as well as cell proliferation in HUVECs [86].

Another activity of neutrophils in the TME is the formation of extracellular traps (NETs), which can be stimulated by tumor cells [87]. NETs consists of nuclear contents, combined with cytosolic proteins and granules [83]. Among these components, cathepsin G, elastase (NE), and MMP-9 are most frequently described in the literature due to their functional relevance. It was demonstrated that the release of NE and cathepsin G activates platelets, which in turn activate the coagulation cascade through inactivation of the anticoagulant protein factor pathway inhibitor [88,89]. Thus, the formation of NETs is essential for accelerating the process of thrombus formation in solid tumors. A study with sEVs derived from breast cancer cells highlighted that TEX are able to induce the release of NETs and accelerate thrombi formation. These are significant findings since the crosstalk between TEX and neutrophils might play a major role in the establishment of cancer-associated thrombosis [90].

To conclude, it is well established that neutrophils are involved in promoting angiogenesis and interacting with ECs in the TME. It has also been shown that TEX play a substantial role in activating neutrophils and promoting their functional activity. However, the link between angiogenesis and neutrophils, which were reprogrammed by TEX, is still ongoing research. Future studies are necessary to focus on the detailed mechanisms of the interaction between neutrophils and TEX, with special regards to their pro-angiogenic functions.

8. Conclusions

Research in recent years had a strong focus on TEX and uncovered multiple effects of sEVs in the TME. One hallmark of TEX is the induction and promotion of angiogenesis, which is probably orchestrated by several signaling pathways, depending on the cargo composition of TEX and the recipient cells, as well as environmental factors. Interestingly, TEX do not only directly communicate with ECs and reprogram them to an angiogenic phenotype, they also interact with other cells in the TME that, in response, contribute to tumor angiogenesis. Defining these direct and indirect pathways to determine whether pharmacologic treatments, chemo- or radiotherapy, or exposure to compounds may influence the amount and functionality of TEX is an area of substantial interest. Blocking TEX-mediated effects may be a promising strategy to overcome therapy resistance to anti-angiogenic therapies or reduce tumor vascularization to ultimately ameliorate disease progression.

Author Contributions: Conceptualization, N.L. and M.J.S.; writing—original draft preparation, N.L., D.S.R., Ł.Z., J.S., J.P., E.B., and T.E.R.; writing—review and editing, N.L. and M.J.S.; supervision, M.J.S.; funding acquisition, M.J.S. All authors have read and agreed to the published version of the manuscript.

Funding: This work was partially funded from the National Science Centre, Poland UMO-2017/26/M/NZ5/00877# to Mirosław J. Szczepański.

Conflicts of Interest: The authors declare no conflict of interest.

References

1. Whiteside, T.L. The effect of tumor-derived exosomes on immune regulation and cancer immunotherapy. *Futur. Oncol.* **2017**, *13*, 2583–2592. [CrossRef]
2. Whiteside, T.L. The potential of tumor-derived exosomes for noninvasive cancer monitoring. *Expert Rev. Mol. Diagn.* **2015**, *15*, 1293–1310. [CrossRef]
3. Franzen, C.A.; Blackwell, R.H.; Foreman, K.E.; Kuo, P.C.; Flanigan, R.C.; Gupta, G.N. Urinary Exosomes: The Potential for Biomarker Utility, Intercellular Signaling and Therapeutics in Urological Malignancy. *J. Urol.* **2016**, *195*, 1331–1339. [CrossRef]
4. Whiteside, T.L. Exosomes carrying immunoinhibitory proteins and their role in cancer. *Clin. Exp. Immunol.* **2017**, *189*, 259–267. [CrossRef]
5. Whiteside, T.L. Tumor-Derived Exosomes and Their Role in Cancer Progression. *Int. Rev. Cytol.* **2016**, *74*, 103–141. [CrossRef]
6. Głuszko, A.; Szczepański, M.J.; Ludwig, N.; Mirza, S.M.; Olejarz, W. Exosomes in Cancer: Circulating Immune-Related Biomarkers. *BioMed Res. Int.* **2019**, *2019*, 1–9. [CrossRef]
7. Sharma, P.; Diergaarde, B.; Ferrone, S.; Kirkwood, J.M.; Whiteside, T.L. Melanoma cell-derived exosomes in plasma of melanoma patients suppress functions of immune effector cells. *Sci. Rep.* **2020**, *10*, 1–11. [CrossRef]
8. Ludwig, N.; Gillespie, D.G.; Reichert, T.E.; Jackson, E.K.; Whiteside, T.L. Purine Metabolites in Tumor-Derived Exosomes May Facilitate Immune Escape of Head and Neck Squamous Cell Carcinoma. *Cancers* **2020**, *12*, 1602. [CrossRef]
9. Ludwig, N.; Yerneni, S.S.; Menshikova, E.V.; Gillespie, D.G.; Jackson, E.K.; Whiteside, T.L. Simultaneous Inhibition of Glycolysis and Oxidative Phosphorylation Triggers a Multi-Fold Increase in Secretion of Exosomes: Possible Role of 2′3′-cAMP. *Sci. Rep.* **2020**, *10*, 1–12. [CrossRef]
10. Chen, T.; Guo, J.; Yang, M.; Zhu, X.; Cao, X. Chemokine-Containing Exosomes Are Released from Heat-Stressed Tumor Cells via Lipid Raft-Dependent Pathway and Act as Efficient Tumor Vaccine. *J. Immunol.* **2011**, *186*, 2219–2228. [CrossRef]
11. King, H.W.; Michael, M.Z.; Gleadle, J.M. Hypoxic enhancement of exosome release by breast cancer cells. *BMC Cancer* **2012**, *12*, 421. [CrossRef] [PubMed]
12. Ludwig, N.; Whiteside, T.L. Potential roles of tumor-derived exosomes in angiogenesis. *Expert Opin. Ther. Targets* **2018**, *22*, 409–417. [CrossRef] [PubMed]
13. Huotari, J.; Helenius, A. Endosome maturation. *EMBO J.* **2011**, *30*, 3481–3500. [CrossRef]
14. Hessvik, N.P.; Llorente, A. Current knowledge on exosome biogenesis and release. *Cell. Mol. Life Sci.* **2018**, *75*, 193–208. [CrossRef]
15. Urbanelli, L.; Magini, A.; Buratta, S.; Brozzi, A.; Sagini, K.; Polchi, A.; Tancini, B.; Emiliani, C. Signaling Pathways in Exosomes Biogenesis, Secretion and Fate. *Genes* **2013**, *4*, 152–170. [CrossRef]
16. Henne, W.M.; Buchkovich, N.J.; Emr, S.D. The ESCRT Pathway. *Dev. Cell* **2011**, *21*, 77–91. [CrossRef]
17. Henne, W.M.; Stenmark, H.; Emr, S.D. Molecular Mechanisms of the Membrane Sculpting ESCRT Pathway. *Cold Spring Harb. Perspect. Biol.* **2013**, *5*, a016766. [CrossRef]
18. Zhang, Y.; Liu, Y.; Liu, H.; Tang, W.H. Exosomes: Biogenesis, biologic function and clinical potential. *Cell Biosci.* **2019**, *9*, 1–18. [CrossRef]
19. McGough, I.J.; Vincent, J.-P. Exosomes in developmental signalling. *Development* **2016**, *143*, 2482–2493. [CrossRef]
20. Villarroya-Beltri, C.; Baixauli, F.; Mittelbrunn, M.; Delgado, I.F.; Torralba, D.; Moreno-Gonzalo, O.; Baldanta, S.; Enrich, C.; Guerra, S.; Sánchez-Madrid, F. ISGylation controls exosome secretion by promoting lysosomal degradation of MVB proteins. *Nat. Commun.* **2016**, *7*, 13588. [CrossRef]
21. Yeates, E.F.A.; Tesco, G. The Endosome-associated Deubiquitinating Enzyme USP8 Regulates BACE1 Enzyme Ubiquitination and Degradation. *J. Biol. Chem.* **2016**, *291*, 15753–15766. [CrossRef]
22. Van Niel, G.; Porto-Carreiro, I.; Simoes, S.; Raposo, G. Exosomes: A Common Pathway for a Specialized Function. *J. Biochem.* **2006**, *140*, 13–21. [CrossRef]
23. Mashouri, L.; Yousefi, H.; Aref, A.R.; Ahadi, A.M.; Molaei, F.; Alahari, S.K. Exosomes: Composition, biogenesis, and mechanisms in cancer metastasis and drug resistance. *Mol. Cancer* **2019**, *18*, 1–14. [CrossRef] [PubMed]
24. Ostrowski, M.; Carmo, N.B.; Krumeich, S.; Fanget, I.; Raposo, G.; Savina, A.; Moita, C.F.; Schauer, K.; Hume, A.N.; Freitas, R.P.; et al. Rab27a and Rab27b control different steps of the exosome secretion pathway. *Nat. Cell Biol.* **2010**, *12*, 19–30. [CrossRef]

25. Ludwig, N.; Yerneni, S.S.; Azambuja, J.H.; Gillespie, D.G.; Menshikova, E.V.; Jackson, E.K.; Whiteside, T.L. Tumor-derived exosomes promote angiogenesis via adenosine A2B receptor signaling. *Angiogenesis* **2020**, *23*, 599–610. [CrossRef]
26. Umezu, T.; Ohyashiki, K.; Kuroda, M.I.; Ohyashiki, J.H. Leukemia cell to endothelial cell communication via exosomal miRNAs. *Oncogene* **2013**, *32*, 2747–2755. [CrossRef]
27. Skog, J.; Wurdinger, T.; Van Rijn, S.; Meijer, D.H.; Gainche, L.; Curry, W.T.; Carter, B.S.; Krichevsky, A.M.; Breakefield, X.O. Glioblastoma microvesicles transport RNA and proteins that promote tumour growth and provide diagnostic biomarkers. *Nat. Cell Biol.* **2008**, *10*, 1470–1476. [CrossRef]
28. Whiteside, T.L. Exosome and mesenchymal stem cell cross-talk in the tumor microenvironment. *Semin. Immunol.* **2018**, *35*, 69–79. [CrossRef]
29. Cui, S.; Cheng, Z.; Qin, W.; Jiang, L. Exosomes as a liquid biopsy for lung cancer. *Lung Cancer* **2018**, *116*, 46–54. [CrossRef]
30. Kucharzewska, P.; Christianson, H.C.; Welch, J.E.; Svensson, K.J.; Fredlund, E.; Ringnér, M.; Mörgelin, M.; Bourseau-Guilmain, E.; Bengzon, J.; Belting, M. Exosomes reflect the hypoxic status of glioma cells and mediate hypoxia-dependent activation of vascular cells during tumor development. *Proc. Natl. Acad. Sci. USA* **2013**, *110*, 7312–7317. [CrossRef]
31. Minciacchi, V.R.; Freeman, M.R.; Di Vizio, D. Extracellular Vesicles in Cancer: Exosomes, Microvesicles and the Emerging Role of Large Oncosomes. *Semin. Cell Dev. Biol.* **2015**, *40*, 41–51. [CrossRef]
32. Tang, M.K.S.; Yue, P.Y.K.; Ip, P.P.; Huang, R.-L.; Lai, H.-C.; Cheung, A.N.Y.; Tse, K.Y.; Ngan, H.Y.S.; Wong, A.S.T. Soluble E-cadherin promotes tumor angiogenesis and localizes to exosome surface. *Nat. Commun.* **2018**, *9*, 1–15. [CrossRef]
33. Ko, S.Y.; Lee, W.; Kenny, H.A.; Dang, L.H.; Ellis, L.M.; Jonasch, E.; Lengyel, E.; Naora, H. Cancer-derived small extracellular vesicles promote angiogenesis by heparin-bound, bevacizumab-insensitive VEGF, independent of vesicle uptake. *Commun. Biol.* **2019**, *2*, 1–17. [CrossRef]
34. Deep, G.; Panigrahi, G.K. Hypoxia-Induced Signaling Promotes Prostate Cancer Progression: Exosomes Role as Messenger of Hypoxic Response in Tumor Microenvironment. *Crit. Rev. Oncog.* **2015**, *20*, 419–434. [CrossRef]
35. Jelonek, K.; Widlak, P.; Pietrowska, M. The Influence of Ionizing Radiation on Exosome Composition, Secretion and Intercellular Communication. *Protein Pept. Lett.* **2016**, *23*, 656–663. [CrossRef]
36. Thompson, C.A.; Purushothaman, A.; Ramani, V.C.; Vlodavsky, I.; Sanderson, R.D. Heparanase Regulates Secretion, Composition, and Function of Tumor Cell-derived Exosomes. *J. Biol. Chem.* **2013**, *288*, 10093–10099. [CrossRef]
37. Sheldon, H.; Heikamp, E.; Turley, H.; Dragovic, R.; Thomas, P.; Oon, C.E.; Leek, R.; Edelmann, M.; Kessler, B.; Sainson, R.C.A.; et al. New mechanism for Notch signaling to endothelium at a distance by Delta-like 4 incorporation into exosomes. *Blood* **2010**, *116*, 2385–2394. [CrossRef]
38. Sato, S.; Vasaikar, S.; Eskaros, A.; Kim, Y.; Lewis, J.S.; Zhang, B.; Zijlstra, A.; Weaver, A.M. EPHB2 carried on small extracellular vesicles induces tumor angiogenesis via activation of ephrin reverse signaling. *JCI Insight* **2019**, *4*, e132447. [CrossRef]
39. Carrasco-Ramírez, P.; Greening, D.W.; Andrés, G.; Gopal, S.K.; Martín-Villar, E.; Renart, J.; Simpson, R.J.; Quintanilla, M. Podoplanin is a component of extracellular vesicles that reprograms cell-derived exosomal proteins and modulates lymphatic vessel formation. *Oncotarget* **2016**, *7*, 16070–16089. [CrossRef]
40. Hong, B.S.; Cho, J.-H.; Kim, H.; Choi, E.-J.; Rho, S.; Kim, J.; Kim, J.H.; Choi, D.; Kim, Y.-K.; Hwang, D.; et al. Colorectal cancer cell-derived microvesicles are enriched in cell cycle-related mRNAs that promote proliferation of endothelial cells. *BMC Genom.* **2009**, *10*, 556. [CrossRef]
41. Ludwig, N.; Lotze, M.T. A treatise on endothelial biology and exosomes: Homage to Theresa Maria Listowska Whiteside. *HNO* **2020**, *68*, 71–79. [CrossRef]
42. Rhee, I. Diverse macrophages polarization in tumor microenvironment. *Arch. Pharmacal Res.* **2016**, *39*, 1588–1596. [CrossRef]
43. Azambuja, J.H.; Ludwig, N.; Yerneni, S.S.; Braganhol, E.; Whiteside, T.L. Arginase-1+ Exosomes from Reprogrammed Macrophages Promote Glioblastoma Progression. *Int. J. Mol. Sci.* **2020**, *21*, 3990. [CrossRef]
44. Webber, J.; Steadman, R.; Mason, M.D.; Tabi, Z.; Clayton, A. Cancer Exosomes Trigger Fibroblast to Myofibroblast Differentiation. *Cancer Res.* **2010**, *70*, 9621–9630. [CrossRef]
45. Lan, J.; Sun, L.; Xu, F.; Liu, L.; Hu, F.; Song, D.; Hou, Z.; Wu, W.; Luo, X.; Wang, J.; et al. M2 Macrophage-Derived Exosomes Promote Cell Migration and Invasion in Colon Cancer. *Cancer Res.* **2019**, *79*, 146–158. [CrossRef]

46. Gonzalez-Avila, G.; Sommer, B.; García-Hernández, A.A.; Ramos, C. Matrix Metalloproteinases' Role in Tumor Microenvironment. *Adv. Exp. Med. Biol.* **2020**, 97–131. [CrossRef]
47. Kajimoto, T.; Okada, T.; Miya, S.; Zhang, L.; Nakamura, S.-I. Ongoing activation of sphingosine 1-phosphate receptors mediates maturation of exosomal multivesicular endosomes. *Nat. Commun.* **2013**, *4*, 2712. [CrossRef]
48. Olejarz, W.; Kubiak-Tomaszewska, G.; Chrzanowska, A.; Lorenc, T. Exosomes in Angiogenesis and Anti-angiogenic Therapy in Cancers. *Int. J. Mol. Sci.* **2020**, *21*, 5840. [CrossRef]
49. Xue, C.; Shen, Y.; Li, X.; Li, B.; Zhao, S.; Gu, J.; Chen, Y.; Ma, B.; Wei, J.; Han, Q.; et al. Exosomes Derived from Hypoxia-Treated Human Adipose Mesenchymal Stem Cells Enhance Angiogenesis Through the PKA Signaling Pathway. *Stem Cells Dev.* **2018**, *27*, 456–465. [CrossRef]
50. Ludwig, N.; Yerneni, S.S.; Razzo, B.M.; Whiteside, T.L. Exosomes from HNSCC Promote Angiogenesis through Reprogramming of Endothelial Cells. *Mol. Cancer Res.* **2018**, *16*, 1798–1808. [CrossRef]
51. Logozzi, M.; Spugnini, E.; Mizzoni, D.; Di Raimo, R.; Fais, S. Extracellular acidity and increased exosome release as key phenotypes of malignant tumors. *Cancer Metastasis Rev.* **2019**, *38*, 93–101. [CrossRef]
52. Zeng, Y.; Yao, X.; Liu, X.; He, X.; Li, L.; Liu, X.; Yan, Z.; Wu, J.; Fu, B.M. Anti-angiogenesis triggers exosomes release from endothelial cells to promote tumor vasculogenesis. *J. Extracell. Vesicles* **2019**, *8*, 1629865. [CrossRef]
53. Messenger, S.W.; Woo, S.S.; Sun, Z.; Martin, T.F. A Ca2+-stimulated exosome release pathway in cancer cells is regulated by Munc13-4. *J. Cell Biol.* **2018**, *217*, 2877–2890. [CrossRef]
54. Savina, A.; Furlán, M.; Vidal, M.; Colombo, M.I.; Gamel-Didelon, K.; Kunz, L.; Föhr, K.J.; Gratzl, M.; Mayerhofer, A. Exosome Release Is Regulated by a Calcium-dependent Mechanism in K562 Cells. *J. Biol. Chem.* **2003**, *278*, 20083–20090. [CrossRef]
55. Keklikoglou, I.; Cianciaruso, C.; Güç, E.; Squadrito, M.L.; Spring, L.M.; Tazzyman, S.; Lambein, L.; Poissonnier, A.; Ferraro, G.B.; Baer, C.; et al. Chemotherapy elicits pro-metastatic extracellular vesicles in breast cancer models. *Nat. Cell Biol.* **2019**, *21*, 190–202. [CrossRef]
56. Mulcahy, L.A.; Pink, R.C.; Carter, D.R.F. Routes and mechanisms of extracellular vesicle uptake. *J. Extracell. Vesicles* **2014**, *3*, 24641. [CrossRef]
57. Azambuja, J.H.; Ludwig, N.; Yerneni, S.; Rao, A.; Braganhol, E.; Whiteside, T.L. Molecular profiles and immunomodulatory activities of glioblastoma-derived exosomes. *Neuro-Oncol. Adv.* **2020**, *2*, vdaa056. [CrossRef]
58. Muller, L.; Simms, P.; Hong, C.-S.; Nishimura, M.I.; Jackson, E.K.; Watkins, S.C.; Whiteside, T.L. Human tumor-derived exosomes (TEX) regulate Treg functions via cell surface signaling rather than uptake mechanisms. *OncoImmunology* **2017**, *6*, e1261243. [CrossRef]
59. Lang, H.-L.; Hu, G.-W.; Zhang, B.; Kuang, W.; Chen, Y.; Wu, L.; Xu, G.-H. Glioma cells enhance angiogenesis and inhibit endothelial cell apoptosis through the release of exosomes that contain long non-coding RNA CCAT2. *Oncol. Rep.* **2017**, *38*, 785–798. [CrossRef]
60. Van Balkom, B.W.M.; Eisele, A.S.; Pegtel, D.M.; Bervoets, S.; Verhaar, M.C. Quantitative and qualitative analysis of small RNAs in human endothelial cells and exosomes provides insights into localized RNA processing, degradation and sorting. *J. Extracell. Vesicles* **2015**, *4*, 26760. [CrossRef]
61. Van Balkom, B.W.M.; De Jong, O.G.; Smits, M.; Brummelman, J.; Ouden, K.D.; De Bree, P.M.; Van Eijndhoven, M.A.J.; Pegtel, D.M.; Stoorvogel, W.; Würdinger, T.; et al. Endothelial cells require miR-214 to secrete exosomes that suppress senescence and induce angiogenesis in human and mouse endothelial cells. *Blood* **2013**, *121*, 3997–4006. [CrossRef]
62. Vitale, I.; Manic, G.; Coussens, L.M.; Kroemer, G.; Garg, A.D. Macrophages and Metabolism in the Tumor Microenvironment. *Cell Metab.* **2019**, *30*, 36–50. [CrossRef]
63. Hong, C.-S.; Sharma, P.; Yerneni, S.S.; Simms, P.; Jackson, E.K.; Whiteside, T.L.; Boyiadzis, M. Circulating exosomes carrying an immunosuppressive cargo interfere with cellular immunotherapy in acute myeloid leukemia. *Sci. Rep.* **2017**, *7*, 1–10. [CrossRef]
64. Costa-Silva, B.; Aiello, N.M.; Ocean, A.J.; Singh, S.; Zhang, H.; Thakur, B.K.; Becker, A.; Hoshino, A.; Mark, M.T.; Molina, H.; et al. Pancreatic cancer exosomes initiate pre-metastatic niche formation in the liver. *Nat. Cell Biol.* **2015**, *17*, 816–826. [CrossRef]
65. Goswami, K.K.; Ghosh, T.; Ghosh, S.; Sarkar, M.; Bose, A.; Baral, R. Tumor promoting role of anti-tumor macrophages in tumor microenvironment. *Cell. Immunol.* **2017**, *316*, 1–10. [CrossRef]
66. Kim, J.; Bae, J.-S. Tumor-Associated Macrophages and Neutrophils in Tumor Microenvironment. *Mediat. Inflamm.* **2016**, *2016*, 1–11. [CrossRef]

67. Ludwig, N.; Jackson, E.K.; Whiteside, T.L. Role of exosome-associated adenosine in promoting angiogenesis. *Vessel. Plus* **2020**, *4*. [CrossRef]
68. Moradi-Chaleshtori, M.; Hashemi, S.M.; Soudi, S.; Bandehpour, M.; Mohammadi-Yeganeh, S. Tumor-derived exosomal microRNAs and proteins as modulators of macrophage function. *J. Cell. Physiol.* **2019**, *234*, 7970–7982. [CrossRef]
69. Zheng, P.; Luo, Q.; Wang, W.; Li, J.; Wang, T.; Wang, P.; Chen, L.; Zhang, P.; Chen, H.; Liu, Y.; et al. Tumor-associated macrophages-derived exosomes promote the migration of gastric cancer cells by transfer of functional Apolipoprotein E. *Cell Death Dis.* **2018**, *9*, 1–14. [CrossRef]
70. Fiani, M.L.; Barreca, V.; Sargiacomo, M.; Ferrantelli, F.; Manfredi, F.; Federico, M. Exploiting Manipulated Small Extracellular Vesicles to Subvert Immunosuppression at the Tumor Microenvironment Through Mannose Receptor/CD206 Targeting. *Int. J. Mol. Sci.* **2020**, *21*, 6318. [CrossRef]
71. Caldeira, P.C.; Vieira, É.L.M.; Sousa, A.A.; Teixeira, A.L.; Aguiar, M.C.F. Immunophenotype of neutrophils in oral squamous cell carcinoma patients. *J. Oral Pathol. Med.* **2017**, *46*, 703–709. [CrossRef]
72. Giese, M.A.; Hind, L.E.; Huttenlocher, A. Neutrophil plasticity in the tumor microenvironment. *Blood* **2019**, *133*, 2159–2167. [CrossRef]
73. Mantovani, A.; Cassatella, M.A.; Costantini, C.; Jaillon, S. Neutrophils in the activation and regulation of innate and adaptive immunity. *Nat. Rev. Immunol.* **2011**, *11*, 519–531. [CrossRef]
74. Lecot, P.; Sarabi, M.; Abrantes, M.P.; Mussard, J.; Koenderman, L.; Caux, C.; Bendriss-Vermare, N.; Michallet, M.-C. Neutrophil Heterogeneity in Cancer: From Biology to Therapies. *Front. Immunol.* **2019**, *10*, 2155. [CrossRef]
75. Mollinedo, F. Neutrophil Degranulation, Plasticity, and Cancer Metastasis. *Trends Immunol.* **2019**, *40*, 228–242. [CrossRef]
76. Aldabbous, L.; Abdul-Salam, V.; McKinnon, T.; Duluc, L.; Pepke-Zaba, J.; Southwood, M.; Ainscough, A.J.; Hadinnapola, C.; Wilkins, M.R.; Toshner, M.; et al. Neutrophil Extracellular Traps Promote Angiogenesis. *Arter. Thromb. Vasc. Biol.* **2016**, *36*, 2078–2087. [CrossRef]
77. Cassatella, M.A.; Östberg, N.K.; Tamassia, N.; Soehnlein, O. Biological Roles of Neutrophil-Derived Granule Proteins and Cytokines. *Trends Immunol.* **2019**, *40*, 648–664. [CrossRef]
78. Shi, Y.; Zhang, J.; Mao, Z.; Jiang, H.; Liu, W.; Shi, H.; Ji, R.; Xu, W.; Qian, H.; Zhang, X. Extracellular Vesicles from Gastric Cancer Cells Induce PD-L1 Expression on Neutrophils to Suppress T-Cell Immunity. *Front. Oncol.* **2020**, *10*, 629. [CrossRef]
79. Jiang, M.; Fang, H.; Shao, S.; Dang, E.; Zhang, J.; Qiao, P.; Yang, A.; Wang, G. Keratinocyte exosomes activate neutrophils and enhance skin inflammation in psoriasis. *FASEB J.* **2019**, *33*, 13241–13253. [CrossRef]
80. Shang, A.; Gu, C.; Zhou, C.; Yang, Y.; Chen, C.; Zeng, B.; Wu, J.; Lu, W.; Wang, W.; Sun, Z.; et al. Exosomal KRAS mutation promotes the formation of tumor-associated neutrophil extracellular traps and causes deterioration of colorectal cancer by inducing IL-8 expression. *Cell Commun. Signal.* **2020**, *18*, 52. [CrossRef]
81. Mahmoudi, M.; Taghavi-Farahabadi, M.; Namaki, S.; Baghaei, K.; Rayzan, E.; Rezaei, N.; Hashemi, S.M. Exosomes derived from mesenchymal stem cells improved function and survival of neutrophils from severe congenital neutropenia patients in vitro. *Hum. Immunol.* **2019**, *80*, 990–998. [CrossRef] [PubMed]
82. Borregaard, N.; Sørensen, O.E.; Theilgaard-Mönch, K. Neutrophil granules: A library of innate immunity proteins. *Trends Immunol.* **2007**, *28*, 340–345. [CrossRef]
83. Wang, J. Neutrophils in tissue injury and repair. *Cell Tissue Res.* **2018**, *371*, 531–539. [CrossRef] [PubMed]
84. Liang, W.; Ferrara, N. The Complex Role of Neutrophils in Tumor Angiogenesis and Metastasis. *Cancer Immunol. Res.* **2016**, *4*, 83–91. [CrossRef]
85. Varricchi, G.; Loffredo, S.; Galdiero, M.R.; Marone, G.; Cristinziano, L.; Granata, F.; Marone, G. Innate effector cells in angiogenesis and lymphangiogenesis. *Curr. Opin. Immunol.* **2018**, *53*, 152–160. [CrossRef]
86. Dalli, J.; Montero-Melendez, T.; Norling, L.V.; Yin, X.; Hinds, C.; Haskard, D.; Mayr, M.; Perretti, M. Heterogeneity in Neutrophil Microparticles Reveals Distinct Proteome and Functional Properties. *Mol. Cell. Proteom.* **2013**, *12*, 2205–2219. [CrossRef]
87. Cools-Lartigue, J.; Spicer, J.; Najmeh, S.; Ferri, L. Neutrophil extracellular traps in cancer progression. *Cell. Mol. Life Sci.* **2014**, *71*, 4179–4194. [CrossRef]
88. Zarbock, A.; Ley, K. Protein tyrosine kinases in neutrophil activation and recruitment. *Arch. Biochem. Biophys.* **2011**, *510*, 112–119. [CrossRef]

89. Engelmann, B.; Massberg, S. Thrombosis as an intravascular effector of innate immunity. *Nat. Rev. Immunol.* **2013**, *13*, 34–45. [CrossRef]
90. Leal, A.C.; Mizurini, D.M.; Gomes, T.; Rochael, N.C.; Saraiva, E.M.; Dias, M.S.; Werneck, C.C.; Sielski, M.S.; Vicente, C.P.; Monteiro, R.Q. Tumor-Derived Exosomes Induce the Formation of Neutrophil Extracellular Traps: Implications for The Establishment of Cancer-Associated Thrombosis. *Sci. Rep.* **2017**, *7*, 1–12. [CrossRef]

Publisher's Note: MDPI stays neutral with regard to jurisdictional claims in published maps and institutional affiliations.

© 2020 by the authors. Licensee MDPI, Basel, Switzerland. This article is an open access article distributed under the terms and conditions of the Creative Commons Attribution (CC BY) license (http://creativecommons.org/licenses/by/4.0/).

Review

Cancer-Associated Angiogenesis: The Endothelial Cell as a Checkpoint for Immunological Patrolling

Antonio Giovanni Solimando [1,2,*], Simona De Summa [3], Angelo Vacca [1] and Domenico Ribatti [4,*]

1. Department of Biomedical Sciences and Human Oncology, Section of Internal Medicine 'G. Baccelli', University of Bari Medical School, 70124 Bari, Italy; angelo.vacca@uniba.it
2. Istituto di Ricovero e Cura a Carattere Scientifico-IRCCS Istituto Tumori "Giovanni Paolo II" of Bari, 70124 Bari, Italy
3. Molecular Diagnostics and Pharmacogenetics Unit, IRCCS Istituto Tumori Giovanni Paolo II, 70124 Bari, Italy; desumma.simona@gmail.com
4. Department of Basic Medical Sciences, Neurosciences, and Sensory Organs, University of Bari Medical School, 70124 Bari, Italy
* Correspondence: antonio.solimando@uniba.it (A.G.S.); domenico.ribatti@uniba.it (D.R.); Tel.: +39-3395626475 (A.G.S.); +39-080-5478326 (D.R.)

Received: 25 October 2020; Accepted: 12 November 2020; Published: 15 November 2020

Simple Summary: A clinical decision and study design investigating the level and extent of angiogenesis modulation aimed at vascular normalization without rendering tissues hypoxic is key and represents an unmet medical need. Specifically, determining the active concentration and optimal times of the administration of antiangiogenetic drugs is crucial to inhibit the growth of any microscopic residual tumor after surgical resection and in the pre-malignant and smolder neoplastic state. This review uncovers the pre-clinical translational insights crucial to overcome the caveats faced so far while employing anti-angiogenesis. This literature revision also explores how abnormalities in the tumor endothelium harm the crosstalk with an effective immune cell response, envisioning a novel combination with other anti-cancer drugs and immunomodulatory agents. These insights hold vast potential to both repress tumorigenesis and unleash an effective immune response.

Abstract: Cancer-associated neo vessels' formation acts as a gatekeeper that orchestrates the entrance and egress of patrolling immune cells within the tumor milieu. This is achieved, in part, via the directed chemokines' expression and cell adhesion molecules on the endothelial cell surface that attract and retain circulating leukocytes. The crosstalk between adaptive immune cells and the cancer endothelium is thus essential for tumor immune surveillance and the success of immune-based therapies that harness immune cells to kill tumor cells. This review will focus on the biology of the endothelium and will explore the vascular-specific molecular mediators that control the recruitment, retention, and trafficking of immune cells that are essential for effective antitumor immunity. The literature revision will also explore how abnormalities in the tumor endothelium impair crosstalk with adaptive immune cells and how targeting these abnormalities can improve the success of immune-based therapies for different malignancies, with a particular focus on the paradigmatic example represented by multiple myeloma. We also generated and provide two original bio-informatic analyses, in order to sketch the physiopathology underlying the endothelial–neoplastic interactions in an easier manner, feeding into a vicious cycle propagating disease progression and highlighting novel pathways that might be exploited therapeutically.

Keywords: tumor angiogenesis; endothelium; microenvironment; multiple myeloma; immunotherapy; anti-angiogenesis; adhesion molecules; immune-checkpoint inhibitor

1. Introduction

The interface between malignant cells and neighboring vessels, both recently sprouted during angiogenesis, or resident ones, is one of the pivotal physiological events tangled in the expansion of neoplastic cells and their dissemination [1]. Cancer vessels' formation is deemed as the result of an angiogenic switch driven by both genetic and epigenetic mechanisms that hijack the tumor trajectory through a full blown self-sustaining entity able to interact with the surrounding niche [2]. The newly formed tumor blood vessels have specific characteristics that allow discrimination from resting blood vessels [3]. They are characterized by rapid proliferation, increased permeability, and disorganized architecture [4]. Initially thought to be a must for the growth and progression of tumors, the formation of new vessels was regarded as one of the hallmarks of both solid [5,6] and hematological malignancies [7–9]. However, this has turned out not to be the case, as tumors have been uncovered to also be able to grow without neo-angiogenesis, mainly by co-opting pre-existing vessels, but also through vascular mimicry [10]. Since its discovery by Dr. Judah Folkman, tumor angiogenesis has been proposed as a target for novel tumor therapies [11]. However, the success in the clinic of anti-angiogenic compounds has been limited in contrast to many preclinical positive results obtained in animal models [12]. This is partly determined by heterogeneous vascular and immunological pattern dependencies fueling the boundary between the cancer cells and the endothelium counterpart [13,14].

Solid tumor is made up of a plethora of cell types rather than just a homogeneous mass of cancer cell, such as cancer associated fibroblasts, an heterogenous immune cell infiltrate, and the individual cells that form the blood and lymphatic vessels [15]. The biology of the individual cells that form the tumor vasculature is central to many processes in the tumor microenvironment, providing oxygen and nutrients, forming conduits for metastases, and directly signaling into nearby cancer cells or other stromal cells [16,17]. The niche is also important during the crosstalk with immune cells and the endothelium has been uncovered to be a gatekeeper, representing the first cell type that immune cells contact as they are exiting the circulation into the tumor, but also as they leave the tumor back into circulation [18]. The endothelial cells can thus act as a director in many ways in this process of tumor immune surveillance by its ability to interact directly with immune cells and malignant cells.

The ability to develop an angiogenetic response is a property common to all tissues. Tumor angiogenesis has historically been uncovered to be one of the key hallmarks of cancer [19]. Nonetheless, one of the main problems in comparing the different clinical studies that have used antiangiogenetic therapies is the lack of reliable markers for the assessment of the antiangiogenetic activity and efficacy of the drugs used [20]. Moreover, a tumor response to these drugs, in the form of reduction of tumor mass alone, may not be an appropriate index of the effectiveness of the treatment, owing to the cytostatic nature of the treatment and the potential contribution of the vasculature in promoting tumor immunosuppression [21]. This seems to be related to the chaotic and disorganized nature of the tumor vasculature, but also to a plethora of ancillary mechanisms [22]. Furthermore, the ability of an antiangiogenetic drug to induce a prolonged stabilization of the disease and an increase in survival should be considered more significant in the assessment of the response to antiangiogenetic therapies [23]. Here, we recapitulate the available data from a translational standpoint and support the picture we draw of the pathophysiological dysregulated endothelial–neoplastic interactions with two bio-informatic interrogations that show, on the one hand, a vicious cycle of disease progression and, on the other hand, pinpoint pathways of potential therapeutic interest.

2. Antitumor Immunity Impairment: Role of Structural and Functional Abnormalities

Despite its essential role, tumor vasculature is structurally and functionally aberrant, with intercellular junctions and extracellular matrix attachments may not form normally in tumors, leading to impaired monolayer formation and barrier function. Completely chaotic loss of tight junctions between adjacent individual cells in the overlapping endothelium, with odd sprouts being cast across the lumen of tumour vasculature, would be an impediment to proper tumor immune surveillance. These abnormalities also occur at the levels of the vasculature in the individual cells directly interacting with the extracellular matrix (ECM) and the pericytes that typically wrap around the outside of the vessel, providing support and stability; nonetheless, in the tumor microenvironment, pericytes are sparse and they are loosely attached to the surface of the tumour vessels, directly contributing to some of the vascular dysfunction [24]. Consistent with preclinical models, patient tumour vessels are disorganized and half of the vessels do not seem to support blood flow at all; alternatively, blood could be detected pooling and flowing in the opposite direction and the vessel diameters have been uncovered to be atypical; a lower wall shear stress can influence the delivery of drugs and immunotherapy along with impaired cancer immune surveillance due to disorganization in the tumor vessels [25]. Thus, there are many aspects of the cancer niche that make it inhospitable to infiltrating immune cells, thus inspiring several strategies aimed to target different aspects of the tumor microenvironment with the goal of improving both the quantity and the quality of infiltrating immune cells [21]. Defining tumors based on the quantity and the quality of immune cell infiltrates allowed to dissect cancer milieu with abundant immune cells, namely inflamed and cold malignancies, as well as immune cells able to enter the tumor microenvironment despite being suppressed [26]. The cancer endothelium can thus be considered a gatekeeper for leukocyte entry and egress from solid and hematological cancers, triggering a cascade that implicates the leukocyte capture by the vessel wall as well as their rolling along the activated surface, and eventually immune cells arrest; next, in order to spread, the patrolling leukocytes ultimately pass through the endothelial boundary via paracellular routes between two adjacent endothelial cells, also being prone to infiltrate via transcellular route, directly through the endothelial cells cytoplasm [27]. Chemokines and integrins play a pivotal role in extracellular matrix (ECM) degradation. In more detail, integrins as heterodimeric molecules constituted by alpha and beta subunits on the cell surface bind to microenvironmental structures via fibronectin and laminin, while activating degradation pathways such as matrix-metallo-proteinases (MMPs) and urokinase-type plasminogen activator (uPA) [28,29]. Moreover, cell-to-cell and cell–ECM interactions are also mediated by adhesion molecules in both solid [30] and hematological malignancies [31,32]. Modifications in adhesion molecules have been related to invasiveness [33,34], angiogenesis [35,36], and druggable targeting [37–39]. Furthermore, the tumor vasculature restricts the infiltration of adaptive immune cells [40,41]. Thus, modifying the tumor vasculature can result in improved immune therapeutic outcome [42]. Consequently, a modern technique such as single cell RNA sequencing has been used to identify diverse subpopulations of tumor-associated endothelial cells [43]. It is conceivable to envision gene expression patterns and individual cells found throughout solid and hematological malignancies and a high grade of modulation in genes implicated in homing, trafficking, and retention of anti-tumor immune cells, corroborating at single-cell level that tumor cells are actively suppressing those pathways important for anti-tumor immunity [43,44].

3. Improving Immune–Vascular Crosstalk for Cancer Immunotherapy

The cancer immunotherapy has revolutionized the way we treat neoplastic patients in the last years. Since the first Food and Drug Administration (FDA) and European Medicines Agency (EMA) approval of the immune checkpoint inhibitor (ICI) ipilimumab for melanoma, which targets the anti CTLA-4 checkpoint, an explosion of approval of different ICIs that target a PD1 or programmed death-ligand 1 (PDL1) for a wide range of cancer indications has been observed [45,46]. The ICIs have provided significant clinical benefit including improvement in overall survival for some of the most aggressive and often lethal cancers [47]; however, despite the promising results, the overall objective response rate gained by ICIs as a monotherapy remains suboptimal, ranging between 20 to 30%, and overall survival and toxicity profile still need to be improved [48–50]. One strategy applied to accomplish higher clinical response is to generate more effective antitumor shrinkage by combining multiple checkpoints [51]. Nonetheless, the toxicity profile is higher [49,52]. Therefore, there is a growing interest aimed to identify alternative strategies to improve the clinical outcome and antitumor response of ICIs, without significantly increasing the risk of toxicities. In the frame of this thinking, cancer immunotherapy points towards a multifaceted profiling and, given the basic pathophysiology underlying cancer immune surveillance evasion modalities, multiple strategies, besides ICIs-based ones, are aimed at targeting immunosuppressant metabolites [53]. The T cells can shape tumor blood vessels and cancer endothelium and prevent the recruitment and infiltration of the effector immune cells while remodeling the ECM, further inhibiting the migration and infiltration of functional patrolling immune cells [54]. Tumor vasculature actively contributes to the immune suppression, as tumor vessels are highly abnormal and functionally impaired, determining a significant degree of hypoxia, acidosis, and necrosis within the tumor [55]. These pathophysiological mechanisms can lead to the production of immunosuppressive molecules such as small ions, lactate, and reactive oxygen species, all of which work to suppress effective T cytotoxic cell function; at the same time, the production of chemokines and cytokines fosters the differentiation and the activation of immunosuppressive cells such as myeloid derived stem cells (MDSCs) and M2, like tumor macrophages, that also act to inhibit the activities of cytotoxic T cells [56]. Conversely, on the vessel, these mechanisms also downregulate multiple adhesion molecules that are essential for the rolling, adhesion, and transmigration of T cells to enter the cancer milieu [57–59], creating a highly immunosuppressive microenvironment, dominated by immune suppressive signals and largely devoid effector T cells. Contrariwise, normalizing tumor vasculature improves T cell infiltration, boosting the immune reaction and halting the immune suppressing environment to a more immune activating phenotype and working in synergy with the cancer immunotherapy.

Anti-vascular endothelial growth factor receptor (anti-VEGFR) pioneered the attempts to normalize tumor vasculature and restore its function, as indicated by tissue perfusion and decreasing intratumoral hypoxia, and fostered further investigations aimed at shaping the intratumoral immune cell phenotype in parallel with vascular normalization [23], while polarizing macrophages throughout and M1 gene-expression phenotype, paralleling an increase in adaptive immune cells' infiltration in the setting of this antiangiogenic treatment [23,60]. Vascular endothelial growth factor (VEGF) and inflammatory molecules are not merely key proangiogenic elements, but are also immune modulators, which boost vascular formation and cooperate in creating permissive environment in most lethal malignancies, and lead to poor drug response [61–63] and survival [19,64]. Remarkably, evidence obtained from pre-clinical and clinical breast cancer models points toward a link between favorable prognostic-related angiogenesis genes and T cell signaling, effective immune cell infiltration that is also pericyte-dependent [65]. In more detail, pericytes seem to be crucial for recruiting immune cells into the tumor niche and orchestrating an immune–vascular crosstalk involving CD4/CD8 T cells and pericytes. Furthermore, to efficiently unleash immune effector cells, Tian et al. uncovered tumor vascular normalization synergism and ICIs (either anti-PD1 or anti-CTLA4 antibodies) to be operative and parallel CD4 T cell activation [65]. Collectively, the interplay between T cells and tumor vasculature primes a CD4 T cell activation and the interferon gamma (IFNγ) production, associated with the normalization of tumor vessels and consequent hypoxia attenuation, reduced intra tumor immunoparesis and further recruitment of bystanders' immune infiltrates, leading to an even enhanced angiogenesis homeostasis. Contrariwise, pericytes or CD4 T cells elimination and major histocompatibility complex (MHC)II inactivation boosted cancer hypoxia, immunosuppression, and metastatic potential [54,65]. Compelling additional evidence corroborated the existence of close interactions between the tumor endothelium and immune effectors cells with therapeutic implications for ICIs treatment in a colorectal cancer model in an interferon gamma (IFNγ)-dependent fashion [66]. In the frame of this thinking, Zheng et al. highlight the importance of IFNγ receptor signaling in host cell populations for both immune response and vascular tumor homeostasis. Thus, a boosting feedback loop of immune reprogramming and tumor vascular regularization shapes the immunoparetic cancer, frequently rich in immunosuppressive cells and dysfunctional effector T lymphocytes being potentially druggable by ICIs, which can in turn stimulate the regularization of blood vessels and ultimately facilitate the infiltration of effector T cells and improve their function, further halting the immune permissive cancer niche [56].

4. Multiple Myeloma (MM) as a Paradigm for Endothelial Gatekeeper Function within the Neoplastic Niche: In Silico Functional Enrichment Study Identifies Prognostic Relevant Gene Profiles in MM Bone Marrow Derived Endothelial Cells

Numerous cell types can be mobilized from the bone marrow and directed to the sites of new vessel formation, where they strengthen the proangiogenic effects [1]. Among them, there are non-hematopoietic bone marrow populations, CD45-, called endothelial progenitor cells (EPCs) [67]. Unlike perivascular cells, which function with paracrine mechanisms by secreting VEGF, endothelial progenitors are incorporated into the wall of nascent vessels, where they differentiate into mature endothelial cells. Being VEGFR-1 positive, they bind VEGF and other proangiogenic factors produced by cancer cells [68]. EPCs facilitate vasculogenesis and are deemed a novel target, particularly at the pre-malignant phase of neoplastic process and in the smoldering stage of disease, fostering the "angiogenic switch". Moreover, during neoplastic dissemination, EPCs stimulate the shift from subclinical to macroscopic secondary lesions [69]. Hematological cancers represent a paradigmatic condition in which EPCs-mediated priming of cancer angiogenesis takes place, given the close cross talk with the neoplastic clone, and the putative shared ontogeny. Thus, the description of neoplastic-infiltrating EPCs in hematological malignancies may shed more light on a more precise anti-angiogenic strategy, with the advantage of tipping the balance of critical phases of disease progression [70]. Multiple myeloma represents a poster child condition in this regard, being characterized by a multistep natural history, as well as by variable pre-neoplastic stages preceding full-blown disease [70,71].

Multiple myeloma (MM) is a clonal proliferation of malignant plasma cells (PCs) accumulating and disseminating in the bone marrow (BM) with ensuing induction of focal skeletal lesions and osteoporosis driving myeloma bone disease, anemia, renal insufficiency, hypercalcemia [72], higher infection rates [73–75], and secondary life-threatening complications [76–78]. MM represents an ideal model of colonization and interaction of tumor cells in the bone microenvironment [79–81], where the immune-milieu [82,83] and aberrant angiogenesis shape a permissive ecosystem, supporting disease progression via a plethora of autocrine [84,85] and paracrine loops [86,87].

Recently, we demonstrated that bone marrow endothelial cells from both newly diagnosed (NDMM) and relapsed-refractory multiple myeloma (RRMM) patients feed into a vicious cycle orchestrated by aberrant adhesion molecules on the bone marrow endothelial cells and plasma cell surface and correlate with poor clinical prognosis [31,35,88]. Based on this evidence and several pieces of data [89,90], increased adhesion molecules levels have been uncovered to contribute to more aggressive phenotype [29,91]. Direct contact of endothelial cells and endothelial progenitors with MM plasma cells would enhance adhesion molecules levels [92,93]. In silico analysis has been performed on dataset GSE28331 (https://www.ncbi.nlm.nih.gov/geo/query/acc.cgi?acc=GSE28331) [93]. Raw data were RMA normalized, using "affy" package (1.56.0) [94]. The method limma [95] was used to detect differentially expressed genes. The results were considered as statistically significant when adjusted p-value < 0.05. K-means and hierarchical clustering were executed using "Factoextra" (1.0.5) [96], "dendextend" (1.9.0) [97], "colorspace" (1.3-2) [98], and "ggplot2" (2.2.1) [99].

To characterize the adhesion molecules-related angiogenic switch in more detail and to corroborate available at gene-expression level in a broader spectrum of disease phenotype, we interrogated different independent public datasets. Given that mobilization of endothelial precursors cells (EPCs) occurs at the early stages of MM progression [70], preceding MM progression, we selected the GSE28331 data collection.

Next, determining whether MM EPCs could be distinguished from MM-cells according to the natural grouping of their gene expression profiles, we analyzed publically available data from 20 EPC and 12 MM-cell samples (GSE28331). The analyses clearly split the MM-cells and EPCs into two branches (heatmap, Figure 1A), according to the expression values of the top 100 different regulated genes. Over-expression of angiogenic genes in EPCs deemed statistically significant and relevant for pro-angiogenic biological processes increased expression of angiogenic genes in EPCs deemed statistically significant and relevant for pro-angiogenic biological processes (Figure 1A,B). Based on these different expression patterns, we performed an enrichment pathway and functional annotation analysis (Figure 1B). These in silico unpublished data together with the previously described autocrine loop pinpoint that the cell adhesion molecules have noteworthy qualities; they can be involved in the homophilic network on two opposing cell types; moreover, adhesion molecules are shed as soluble isoforms being able to bind to cell-bound isoforms, which in turn even enhances its binding capacity (Figure 2). What develops is a vicious cycle of neoplastic MM cells expressing and shedding adhesion molecules, increasing membrane-bound expression on the endothelium and boosting angiogenesis. In turn, increasing numbers of activated vessels can increasingly bind cancer cells, which promptly catch enhanced space within the neoplastic milieu for contact-mediated interactions [35,100] (Figure 2).

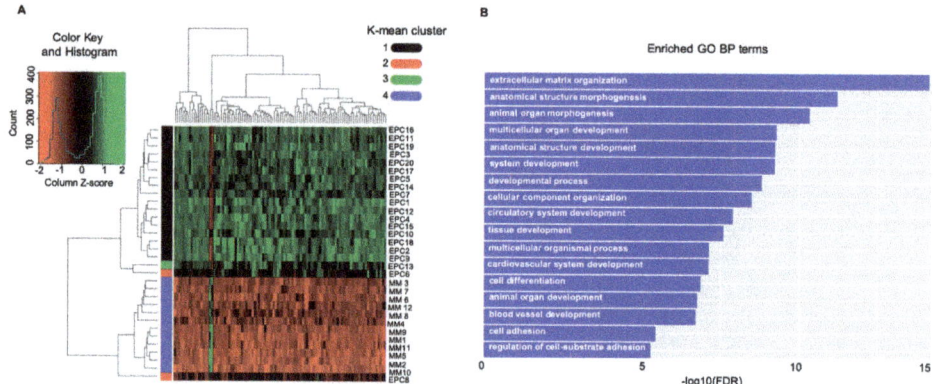

Figure 1. In silico data interrogation points towards a significant crosstalk between the neoplastic-cells and the vasculature counterpart: adhesion-system boosts multiple myeloma (MM)-related angiogenesis in the bone marrow microenvironment. (**A**) Heatmap, showing expression value of the top 100 deregulated genes, includes a dendrogram with two major branches; one containing MM-cells and one EPC sample, and the other grouping the leftover EPCs. (**B**) GO functional enrichment results showed that genes are involved in several biological processes. Cell adhesion and angiogenesis were significantly enriched in the gene network analysis. EPCs: endothelial precursor; GO: gene ontology; BP: biological process.

The K-means clustering from the above-mentioned GSE28331 dataset (Figure 3A) showed highly ranked enriched biological processes including blood vessel formation, cell adhesion, and developmental processes; the network analysis highlighted a significant enrichment for focal adhesion and matrix-receptor interaction Kyoto Encyclopedia of Genes and Genomes (KEGG) pathways (Figure 3B).

Consequently, using several pre-clinical models [30,88,102], blocking the adhesion system seems to halt blood vessel formation, reduce adhesion-mediated networks, and weaken neoplastic disease progression. These therapeutic effects of interfering with the adhesion system were observed in translational animal models, not in patients and, therefore, must be interpreted with caution. Nevertheless, these pieces of evidence may be a warning of a pivotal druggable targets of MM and, more generally, microenvironment addicted malignancies that might be investigated therapeutically.

The dysregulated endothelial–neoplastic interactions sketched by our bio-informatic investigations show, on the one hand, a vicious cycle of disease progression and, on the other hand, point out pathways of potential therapeutic interest. These gene expression profiles were observed in one model of disease, and thus must be interpreted with caution and need further validation on a broad spectrum of malignancies. Nonetheless, solid and hematological malignancies share common mechanisms involving the cross talk between the cancer endothelium and the immune microenvironment, as summarized in Table 1.

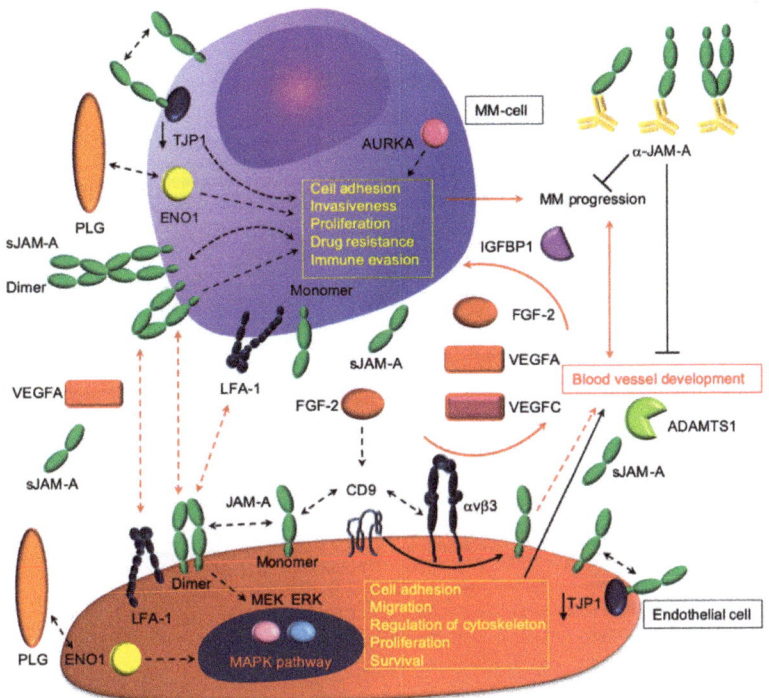

Figure 2. Proposed paradigmatic model of how junctional adhesion molecule-A (JAM-A) plays a pivotal role in angiogenesis, disease progression, and aggressive phenotype. As proof of concept, JAM-A localizes at endothelial tight junctions, in association with the alphaVβ3 integrin. Besides being expressed by MM-cells, JAM-A orchestrates MM angiogenesis: upon stimulation with fibroblast growth factor-2 (FGF-2), the JAM-A-alphaVβ3 complex can dissociate and localizes diffusely along the cell membrane, where it can drive signaling processes, leading to the activation of extracellular signal-regulated MAPK, which leads to angiogenesis and cytoskeleton rearrangement. Trans- homo/heterophilic JAM-A interactions: angiogenesis appears prevalent in MM, as indicated by the results presented in Figure 1 and in [31,35,101]. JAM-A binds heterotypically with lymphocyte function-associated antigen-1 (LFA-1), thus promoting potential interactions of MM-cells and endothelial cells with immune cells. These intricate interactions between ligands and receptors within the MM milieu appear to enhance a pro-survival and immunosuppressive environment, where angiogenesis, immune response, and intrinsic tumor cell resistance depend on each other. ADAMTS: A disintegrin and metalloproteinase with thrombospondin motifs 1; AURKA: Aurora kinase A; CD9: CD9 molecule; ENO1: Enolase 1; FGF-2: fibroblast growth factor-2; LFA-1: lymphocyte function-associated antigen 1; MAPK: mitogen-activated protein kinase; PLG: plasminogen; TJP1: tight junction protein-1; αVβ3: integrin alpha V beta 3; VEGFA: vascular endothelial growth factor A. See the results and [35] for additional details.

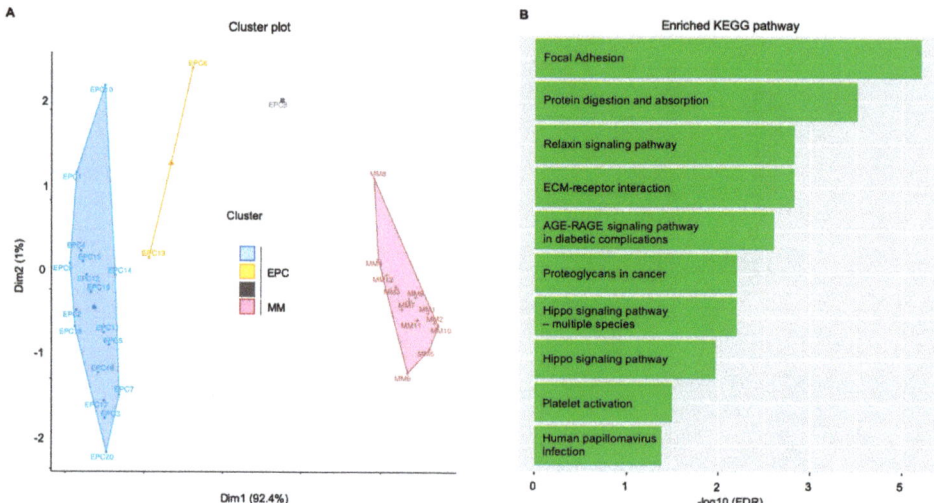

Figure 3. In silico validation confirmed the pivotal role of focal adhesion in sustaining the MM clone. Endothelial cells and MM gene expression supported the bioinformatic findings. (**A**) K-mean clustering results represented as distribution of samples in clusters. (**B**) Gene network functional enrichment: histogram representation of significantly enriched KEGG pathways. Overall, focal adhesion and extracellular matrix (ECM)–receptor interaction confirmed the in vitro and ex vivo evidences. Dim: dimension. KEGG: Kyoto Encyclopedia of Genes and Genomes; AGE: advanced glycation end products; EPC: endothelial progenitor cells; FDR: false discovery rate; MM: multiple myeloma plasma cells; RAGE: receptor for advanced glycation endproducts.

Table 1. Endothelial cells function as a gatekeeper for immunological patrolling in solid and hematological malignancies: synthetic overview of the molecular actors. PD-L1, programmed death-ligand 1; IFNγ, interferon gamma; FGF-2, fibroblast growth factor-2; ICAM, intercellular adhesion molecule; VCAM, vascular cell-adhesion molecule; JAM, junctional adhesion molecule; NEU1, epidermal growth factor like domain 7; VEGF, vascular endothelial growth factor; HLA-E, human leukocyte antigen E; ENO-1, Enolase 1; CCL/CXCL, chemokine ligand; TNF, tumour necrosis factor; NO, nitric oxide; TIM3, T-cell immunoglobulin and mucin domain 3; IDO1, indoleamine 2,3-dioxygenase 1; LFA1, lymphocyte function-associated antigen 1; VLA4, very late antigen 4; VE-cadherin, vascular endothelial cadherin; PECAM1, platelet/endothelial-cell adhesion molecule 1; ESAM, endothelial cell-selective adhesion molecule.

Proangiogenic Molecules *	Soluble Factors *	Immune Checkpoints	Major Histocompatibility Complex (MHC)	Adhesion Molecules *
FGF2 Modulate selective up- and down-regulation [56,103,104] of adhesion molecules (ICAM [30,105], VCAM [106], JAMs [35,107,108])	*Chemokines (CCL2/18, CXCL10/11, CXCL4)* Deregulated chemokines, halting immune effector surveillance and attracting immune tolerogenic cells [27,43,105,109,110]	PD-L1/2 Cancer endothelium, also express immune checkpoints: a cross talk between aberrant vasculature, immune, and cancer cells creates an immune permissive tumor milieu [58,101,111–114]	*MHC I* Often overexpressed within the tumor niche, where the cancer associated endothelium is characterized by a lack of co-stimulatory molecules (B7.1–and B7.2) [58,115–117]	*Selectin-mediated leukocyte rolling* E-selectin/P-selectin Orchestrate leukocyte recruitment. [79,118,119]

Table 1. Cont.

Proangiogenic Molecules *	Soluble Factors *	Immune Checkpoints	Major Histocompatibility Complex (MHC)	Adhesion Molecules *
NEU1 Induces a decreased adhesion molecule expression and boosts angiogenesis via NOTCH pathway [120–122]	*Cytokines (IFNγ,TNFα)* unresponsiveness and anergy along with PD-L1 overexpression [112,123–127]	**ENO-1** acts both as a glycolytic enzyme and a plasminogen receptor expressed on the cell surface of tumor cells. Surface ENO1 plays a crucial role in cancer metabolism, tumor invasion and immune suppression in the cancer immune-microenvironment [35,128]	**MHC II** Can be decreased on tumor infiltrating vessels, thus contributing to an immune tolerogenic niche [117]	*Integrin-mediated leukocyte rolling:* ICAM1 binds to LFA1 (αLβ2 integrin); VCAM1 binds to VLA4 (α4β1 integrin) Multiple functions [27] ** Function
VEGF-A/C Vascular endothelial growth factors induce cell phenotype changes, recruiting immune suppressive cells [58,62,129,130]	*NO* Directly and indirectly affect effective immune response by altering leukocyte infiltration and suppressing CD8+ T cells [131–134]	**IDO1 and TIM3** Immune regulatory checkpoints overexpress in cancer endothelial cells upon cytokines stimulation (i.e., IFNγ) able to induce T cells programmed cell death and cell cycle arrest, respectively [114]	**HLA-E** CD8+T cells infiltration in ovarian cancer correlated with improved better survival when HLA-E expression is decreased [135]	*VE-cadherin and intracellular membrane compartments,* containing PECAM1, JAMs, ESAM, ICAM2, and CD99 promote paracellular migration [27,136–138] #

* Molecules with demonstrated immunological function influencing microenvironment patrolling are summarized. ** Endothelial cells express selectins and integrins, the most important leukocyte adhesion cascade tumour-associated endothelial cells, express lower levels of cell adhesion molecules, promoting endothelial anergy and reducing the ability of effector T cells to infiltrate tumours. # Aberrant expression of cancer associated vessels surface proteins contribute to the hypoxia and acidosis, which, in turn, enhance adhesion molecules' expression, recruiting immune-suppressive cells and conversely excluding effector T cells, by downregulating key integrin and selectins. Adhesion molecules are pivotal in gatekeeping function of endothelial-mediated transmigration.

5. Measuring T Cell Exit from Tumors: How Do Lymphatic Vessels Shape the Intratumoral Repertoire

The lymphatic vasculature is a hierarchical network of vessels found within nearly all peripheral tissues. The main function of lymphatic vessels is to unidirectionally transport interstitial fluids proteins and leukocytes from tissue periphery to the draining lymph nodes structure [139]. The organization of lymphatic vessels is uniquely designed to carry out transport functions in tissues and allow leukocyte egress in order to target solid and hematological malignancies [58,140]. The lymphatic system has been explored by several methods, such as in vivo approaches aiming to quantify leukocyte egress upon the uptake of a labeled tracer and microparticle injection [141]. Specifically, pre-labeled cells are injected into the skin, as the most convenient site, and then labeled cells are detected in the draining lymph node, while quantifying the number of migrating lymphocytes as a readout for the amount of egress occurring [141]. Alternatively, interstitial adoptive transfer and intravital microscopy served as lymphatic vasculature investigating tools [142]. In more detail, intravital microscopy allows to actually visualize and track the movements of pre-labeled leukocytes within tissues as well perceive them entering into pre-labeled lymphatics [142]. Moreover, photoconvertible mice using cell type-specific expression of photoconvertible fluorescent protein Kik Green-Red offered a novel strategy to T cell egress quantification in vivo [143,144]. Remarkably, tumor egressed immune cells are transcriptionally distinct from intratumoral T cells [145] and the CD8 T cells seem not to express markers of exhaustion [146]. Of note, a physical barrier to egress enhances adoptive T cell therapy efficacy in preclinical models [130]. On top on this, T cell egress from tumors can represent a potential mechanism of immune escape. Nonetheless, the limitations to all these methods are represented by biases in selecting cell types prone to be evaluated for the egress, while the application of a tracer only allows to track cells that can uptake that tracer, namely phagocytic cells. Conversely, by adoptive transfer models, the major caveat is the labeling, limiting the assay to two or three cell types at a time [147]. Even using intravital microscopy, a limited number of labeled leukocytes can be tracked at a single time, and it can also be time-consuming and low throughput for tracking immune cell egress in vivo [147]. Thus, by elucidating the mechanisms that govern egress, it is not only possible to gain a significantly better understanding of how the immune landscape of a tumor is formed, but also to manipulate egress mechanisms in a therapeutically beneficial way [148]. Nonetheless, the translational value of the available finding is still debated and standard histological analysis or flow cytometry profiling of intratumoral leukocyte pools does not really provide any information regarding leukocyte trafficking dynamics [149].

To overcome these caveats, promising new avenues have recently been optimized to study the fate of tumor infiltrating immune cell populations, cancer metastasis, migration patterns of alloreactive T cells, or the dynamics and plasticity of immune cell subsets in different scenarios such as infection, inflammation, and immunotolerance using the in vivo photoconvertible fluorescence protein "kaede" transgenic mice [143]. The unique property of kaede protein is that it is influenced by violet light pulse exposition. This state-of-the-art method uncovered lymph nodes to be heavily infiltrated with myeloid cells, predominantly inflammatory monocytes and macrophages [150]. However, some lymphocytes in these tumors are also present and the egressing population seems mostly represented by CD4 and CD8 T cells [150]. Collectively, the available shreds of evidence point toward a vicious cycle between the lymph nodal endothelium and the patrolling immune cells, implying that egressed and retained T cells differ substantially. An acquisition of markers associated with T cell exhaustion in cells that are retained within the tumor, indicated by high expression of PD1 Tim3 and CD 39, characterizes lymphocytes that are also unable to produce effector cytokines such as interferon gamma and TNF alpha. Contrariwise, T cells that have passed through the tumor and exited to the draining lymph nodes are not expressing markers of exhaustion and retain their ability to produce effector cytokines [151]. It might be advantageous to keep these tumor-specific T cells within the tumor for a much longer period of time, potentially improving their function. Despite that direct translation of subclasses based on the vascular phenotype into clinical decision-making is yet to be achieved, these findings may also point towards a potential Achilles' heel of multiple cancer that might be exploited therapeutically.

6. Boosting Cancer Immunotherapy Using Anti-Angiogenics: Therapeutic Windows and Challenges Offered by the Visualization and Reprogramming of the Tumor Milieu

Across the timeline of the development of various imaging techniques, both clinical and preclinical models greatly contributed to the imaging of tumor vasculature and microenvironment [152–156]. The translational value of imaging tumor blood vessels allowed to identify the abnormal microvasculature, visualizing the shape and diameter, the vessel wall, the abnormal branching, and even the blood flow, characterizing the level of heterogeneity in vivo [157]. Based on these pieces of evidence, cancer vasculature appears to be functionally abnormal [158], corroborating previous findings regarding abnormal blood flow as a consequence of aberrant vessel formation [158,159]. While comparing with normal vessels in the cancer tissue, there is a lack of correlation between the size of the vessels' diameter and red blood cells velocity [158,159]. Remarkably, the next generation of experimental immunodiagnostics in cancer model also provided imaging understandings regarding immune cell trafficking in tumour vessels, namely monocytes, interacting with the vessel wall [155] and leading to patrolling immune cells' recruitment. From the above mentioned standpoint, the traditional anti-angiogenesis can deeply affect anti-tumour immunity, as full doses of drugs shrink the tumor, leading to cancer hypoxia and priming immune suppressive cells' infiltration [160]. A wise use of therapeutic strategies halting the cancer angiogenesis must thus take into account the abnormal metabolic microenvironment characterizing a heterogeneous oxygenation [161,162]. Assessing oxygenation in the different layers of tumour pinpoint that there is a progressive increase of nutrient and oxygen levels across the inner depth [162], thus fueling genomic instability [163], the cancer progression (PD) [16], the switch to anaerobic metabolism [164], as well as the epithelial–mesenchymal transition, metastases [165], and the induction of cancer "stem cell" phenotype [166]. Hypoxia is a hallmark of cancer, inducing many abnormalities with prognostic consequences linked to defects in apoptosis and autophagy [167,168] and the resistance to radio-chemotherapy [169–171] and immunotherapy [13,172,173] likewise hamper the cancer aggressive phenotype acquisition, while shaping a pro-angiogenic, inflamed, and immunosuppressive neoplastic ecosystem [154,174,175]. Consequently, it is necessary either to target many different actors on the scene within the neoplastic niche or attempt to homogenize the cancer heterogeneity [161].

Hypoxia as a Key Factor for Angiogenesis and Immune Equilibrium

Sufficient oxygen pressure is required for our organs to function properly. Conversely, insufficient oxygen supply is a prominent feature in various pathological processes, including tumor development and metastasis [176,177]. Hypoxic malignant cells are more prone to increase their genetic instability [178], while decreasing the immune response. Moreover, insufficient oxygen supply influences ECM remodeling and stiffness [179], further halting the susceptibility to chemotherapy and radiation therapy [180]. Notably, the enhanced angiogenesis is deemed to counteract the neoplastic metabolic and energetic need, but also shapes the tumor microenvironment and boosts the malignant cells faculty to gain immunosuppression, fueling the cancer progression [181].

The association of cell signaling driving cellular adaptation to hypoxia prompted the investigation on targets that might halt the proliferation of hypoxic tumors if halted. The three pivotal oxygen-dependent molecular mechanism during metabolic adaptation rely on hypoxia inducible factors (HIF members), unfolded protein response (UPR) [182], and mammalian target of rapamycin (mTOR) [8]. Specific targeting of hypoxia in cancer therapy has been extensively investigated and trials exploiting hypoxia-dependent druggable signaling are ongoing [181,183].

Nonetheless, normalizing the tumor vasculature with the judicious use of antiangiogenics can revert this process, directing intervening in oxygen delivery [184–186].

As proangiogenic factors typically predominate, tumour perfusion and oxygenation are usually impaired; the re-establishment of physiologic equilibrium aims to vasogenic edema and interstitial pressure reduction, while enhancing the drug delivery and indirectly reducing neoplastic cells shedding and invasiveness [160,187]. A paradigmatic example of in vivo modelling of judicious use of anti-angiogenic treatment has been pioneered by Winkler et al. using anti-VEGFR2 targeting in glioblastoma multiforme (GBM), able to increase pericyte coverage in mature vessels [188], and further corroborated in other tumour types [153,189]. Because of this improvement in the vessels' structure, functional consequences such as radiation and anti-angiogenic synergism occur during the vessel "normalization window" [188]. The pericyte recruitment parallels angiopoietin-1 (Ang1) and angiopoietin-2 (Ang2) crosstalk. Ang1 promotes vessel maturation and survival through Tie-2 receptor phosphorylation and via the PI3K-AKT-mediated signaling pathway. The development that follows after the formation of immature vessels is mainly due to Ang1 and ephrin B2. Conversely, Ang2 is a context-dependent molecule that counterbalances Ang1 [190,191]. Thus, the Ang1/Ang2 ratio might correlate with vascular normalization [188]. Notably, Ang2 overexpression decreases the prognostic advantage gained by anti-VEGFR strategies [192], uncovering Ang2 to be a rate-determining step for anti-VEGFR treatment. In fact, the dual anti-Ang2/VEGFRs therapy has been shown to enhance the length of the window of vessel normalization in vivo, thus achieving survival improvement and tumour burden reduction upon dual VEGFR2-Ang2 inhibition [193]. These treatment effects of simultaneous VEGFR and Ang2 halting were observed in preclinical models, not in patients and, therefore, must be translated with carefulness. Nonetheless, reprogramming of tumour milieu for immunotherapeutical purposes seems to be conceivable because of the plethora of pathophysiological effects played by VEGF on the immune innate and adaptive compartment [160,194], by enhancing the recruitment and proliferation of cancer tolerogenic Tregs cells [195] and tumour associate macrophages (TAM). Both actors nurse the milieu, making it tolerogenic, and feed into auto-paracrine myeloid-derived suppressor cells (MDSCs) [196] via VEGF and break cytotoxic T lymphocytes' (CTLs) effector functions [58,197]. Collectively, the abnormal cancer vasculature contributes to immunosuppression in the niche [194,198–200] and enhances the shedding of systemic factors hijacking the anti-cancer response [22,198,201].

Current advances in tumour immunotherapy consent to proficiently unleash immune effector cells [202]. What ensues is an immune-supportive skewing, also originating from the vascular normalization [23,203]. Typically, the highest anti-angiogenic doses have been employed at the maximum tolerated doses until PD. Nonetheless, the dosage is key because increasing the amount and a sustained extent of anti-angiogenic therapy are themselves associated with cancer hypoxia and, eventually, PD [23,198,204]. The insights regarding the window of normalized perfusion from vascular normalization depend on the dose and potency of the antiangiogenic therapy. Precisely, the degree of neo-vessels normalization in localized and disseminated cancers is liable determined by the dose of anti-angiogenic compounds and the amount of the angiogenic stimulus in the given neoplasia [204]. Disproportionate perfusion reduction can boost oxidative stress and dissemination potential, while halting the immune infiltrate [205,206]. Therefore, as the stage of normalized cancer oxygen delivery after tailored anti-angiogenic treatment is transitory, the choice of the proper timing matching the vascular normalization "window", the tailored dose of anti-angiogenic treatment, as well as the most effective immune-modulatory agent appear critical. An elevated concentration and therapy extent of anti-VEGF therapy are associated with decreased cancer oxygen supply and elevated hypoxia [23]. Notably, pre-clinical models uncovered a lower concentration of anti-angiogenic agents to be correlated with sustained vascular normalization [22,198], as low as one-quarter of the conventional dose. Clinical studies corroborated these findings, demonstrating that a decreased dose of anti-VEGF (<3.6 mg/kg, weekly) combined with cytoreduction resulted in improved survival over a high dose (5 mg/kg, week) in subjects suffering from glioblastoma [207,208]. Many attempts have been proposed to unbridle an effective immune response while breaking the vicious cycle between abnormal angiogenesis and immune patrolling actors in aggressive and refractory malignancies [209–211] Collectively, the combination of angiogenesis and immunity targeting has been studied a lot in pre-clinical as well as clinical settings, some of them showing promising results [160,212,213]. Overall, the knowledge on the abnormal vasculature and microenvironment provides the backbone for normalization of tumour vasculature strategy, with the judicious use of antiangiogenics and niche reprogramming with the goal of immunotherapy improvement [161].

7. Conclusions

Critical mechanisms fostering blood and lymphatic vessels' formation and facilitating immunosuppression throughout tumor growth and progression have been uncovered. Cancer cells grow and progress through a persistent crosstalk with the neighboring milieu. Next generation techniques sketch at a high resolution such that the new vessels' formation and immune paresis regularly occur to fuel this vicious cycle. Consequently, state-of-the-art therapeutic strategies merging anti-angiogenic and immune-directed treatments appear to hold promise to shape the neoplastic ecosystem and boost the therapeutic efficacy.

Author Contributions: Conceptualization, D.R., A.G.S., and A.V.; methodology, A.G.S. and S.D.S.; software, S.D.S. and A.G.S.; validation, D.R. and A.V.; formal analysis, A.G.S. and S.D.S.; investigation, A.G.S. and D.R.; resources, A.G.S., D.R., and A.V.; data curation, A.G.S. and S.D.S.; writing—original draft preparation, A.G.S., D.R., and A.V.; writing—review and editing, D.R. and A.V; visualization, A.G.S. and S.D.S.; supervision, D.R. and A.V.; project administration, D.R. and A.G.S.; funding acquisition, D.R., A.G.S., and A.V. All authors have read and agreed to the published version of the manuscript.

Funding: This research was funded by GLOBALDOC Project—CUP H96J17000160002 approved with A.D. n. 9 on 18 January 2017, from Puglia Region, financed under the Action Plan for Cohesion approved with Commission decision C (2016) 1417 of 3 March 2016 to A.G.S., as well as the Apulian Regional project: medicina di precisione to A.G.S.

Acknowledgments: We thank Antonella Argentiero and Matteo Claudio Da Già for technical support and valuable discussions. We also thank Mary Victoria Pragnell and Maria Gemma Pizzolla for linguistic editing.

Conflicts of Interest: The authors declare no conflict of interest.

Abbreviations

CCL	Chemokine ligand
CXCL	Chemokine ligand
ENO1	Enolase1
ESAM	Endothelial cell-selective adhesion molecule
FGF2	Fibroblast growth factor 2
ICAM	Intercellular adhesion molecule
IDO1	Indoleamine 2,3-Dioxygenase 1
HLA-E	Human leukocyte antigen E
IFNγ	Interferon gamma
JAMs	Junctional adhesion molecules
LFA1	Lymphocyte function-associated antigen 1 (also known as αLβ2-integrin)
MHC	Major histocompatibility complex
NEU1	Epidermal growth factor like domain 7 (Egfl7)
NO	Nitric oxide
PD-L1/2	Programmed death-ligand 1/2
PECAM1	Platelet/endothelial-cell adhesion molecule 1
TIM3	T-cell immunoglobulin and mucin domain 3
TNFα	Tumour necrosis factor alpha
VCAM	Vascular cell-adhesion molecule
VE-cadherin	Vascular endothelial cadherin
VEGF	Vascular endothelial growth factor
VLA4	Very late antigen 4 (also known as α4β1-integrin)

References

1. Kerbel, R.S. Tumor angiogenesis. *N. Engl. J. Med.* **2008**, *358*, 2039–2049. [CrossRef] [PubMed]
2. Baeriswyl, V.; Christofori, G. The angiogenic switch in carcinogenesis. *Semin. Cancer Biol.* **2009**, *19*, 329–337. [CrossRef] [PubMed]
3. Ribatti, D.; Nico, B.; Crivellato, E.; Roccaro, A.M.; Vacca, A. The history of the angiogenic switch concept. *Leukemia* **2007**, *21*, 44–52. [CrossRef] [PubMed]
4. Ribatti, D.; Vacca, A.; Presta, M. The discovery of angiogenic factors. *Gen. Pharmacol. Vasc. Syst.* **2000**, *35*, 227–231. [CrossRef]
5. Vermeulen, P.B.; Gasparini, G.; Fox, S.B.; Toi, M.; Martin, L.; Mcculloch, P.; Pezzella, F.; Viale, G.; Weidner, N.; Harris, A.L.; et al. Quantification of angiogenesis in solid human tumours: An international consensus on the methodology and criteria of evaluation. *Eur. J. Cancer* **1996**, *32*, 2474–2484. [CrossRef]
6. Hasan, J.; Byers, R.; Jayson, G.C. Intra-tumoural microvessel density in human solid tumours. *Br. J. Cancer* **2002**, *86*, 1566–1577. [CrossRef]
7. Leone, P.; Buonavoglia, A.; Fasano, R.; Solimando, A.G.; De Re, V.; Cicco, S.; Vacca, A.; Racanelli, V. Insights into the Regulation of Tumor Angiogenesis by Micro-RNAs. *J. Clin. Med.* **2019**, *8*, 2030. [CrossRef]
8. Lamanuzzi, A.; Saltarella, I.; Desantis, V.; Frassanito, M.A.; Leone, P.; Racanelli, V.; Nico, B.; Ribatti, D.; Ditonno, P.; Prete, M.; et al. Inhibition of mTOR complex 2 restrains tumor angiogenesis in multiple myeloma. *Oncotarget* **2018**, *9*, 20563–20577. [CrossRef]
9. Li, W.W.; Hutnik, M.; Gehr, G. Antiangiogenesis in haematological malignancies. *Br. J. Haematol.* **2008**, *143*, 622–631. [CrossRef]

10. Hendrix, M.J.C.; Seftor, E.A.; Seftor, R.E.B.; Chao, J.-T.; Chien, D.-S.; Chu, Y.-W. Tumor cell vascular mimicry: Novel targeting opportunity in melanoma. *Pharmacol. Ther.* **2016**, *159*, 83–92. [CrossRef]
11. Folkman, J. Angiogenesis. In *Biology of Endothelial Cells*; Jaffe, E.A., Ed.; Developments in Cardiovascular Medicine; Springer: Boston, MA, USA, 1984; Volume 27, pp. 412–428. ISBN 978-1-4612-9786-4.
12. Moserle, L.; Jimenez-Valerio, G.; Casanovas, O. Antiangiogenic Therapies: Going beyond Their Limits. *Cancer Discov.* **2014**, *4*, 31–41. [CrossRef] [PubMed]
13. Argentiero, A.; Solimando, A.G.; Krebs, M.; Leone, P.; Susca, N.; Brunetti, O.; Racanelli, V.; Vacca, A.; Silvestris, N. Anti-angiogenesis and Immunotherapy: Novel Paradigms to Envision Tailored Approaches in Renal Cell-Carcinoma. *J. Clin. Med.* **2020**, *9*, 1594. [CrossRef] [PubMed]
14. Sakariassen, P.O.; Prestegarden, L.; Wang, J.; Skaftnesmo, K.-O.; Mahesparan, R.; Molthoff, C.; Sminia, P.; Sundlisaeter, E.; Misra, A.; Tysnes, B.B.; et al. Angiogenesis-independent tumor growth mediated by stem-like cancer cells. *Proc. Natl. Acad. Sci. USA* **2006**, *103*, 16466–16471. [CrossRef] [PubMed]
15. Junttila, M.R.; de Sauvage, F.J. Influence of tumour micro-environment heterogeneity on therapeutic response. *Nature* **2013**, *501*, 346–354. [CrossRef] [PubMed]
16. Muz, B.; de la Puente, P.; Azab, F.; Azab, A.K. The role of hypoxia in cancer progression, angiogenesis, metastasis, and resistance to therapy. *Hypoxia Auckl. N. Z.* **2015**, *3*, 83–92. [CrossRef] [PubMed]
17. Sozzani, S.; Rusnati, M.; Riboldi, E.; Mitola, S.; Presta, M. Dendritic cell-endothelial cell cross-talk in angiogenesis. *Trends Immunol.* **2007**, *28*, 385–392. [CrossRef] [PubMed]
18. Zuazo-Gaztelu, I.; Casanovas, O. Unraveling the Role of Angiogenesis in Cancer Ecosystems. *Front. Oncol.* **2018**, *8*, 248. [CrossRef]
19. Hanahan, D.; Weinberg, R.A. Hallmarks of Cancer: The Next Generation. *Cell* **2011**, *144*, 646–674. [CrossRef]
20. Abdalla, A.M.E.; Xiao, L.; Ullah, M.W.; Yu, M.; Ouyang, C.; Yang, G. Current Challenges of Cancer Anti-angiogenic Therapy and the Promise of Nanotherapeutics. *Theranostics* **2018**, *8*, 533–548. [CrossRef]
21. Joyce, J.A.; Fearon, D.T. T cell exclusion, immune privilege, and the tumor microenvironment. *Science* **2015**, *348*, 74–80. [CrossRef]
22. Huang, Y.; Goel, S.; Duda, D.G.; Fukumura, D.; Jain, R.K. Vascular normalization as an emerging strategy to enhance cancer immunotherapy. *Cancer Res.* **2013**, *73*, 2943–2948. [CrossRef] [PubMed]
23. Huang, Y.; Yuan, J.; Righi, E.; Kamoun, W.S.; Ancukiewicz, M.; Nezivar, J.; Santosuosso, M.; Martin, J.D.; Martin, M.R.; Vianello, F.; et al. Vascular normalizing doses of antiangiogenic treatment reprogram the immunosuppressive tumor microenvironment and enhance immunotherapy. *Proc. Natl. Acad. Sci. USA* **2012**, *109*, 17561–17566. [CrossRef] [PubMed]
24. Morikawa, S.; Baluk, P.; Kaidoh, T.; Haskell, A.; Jain, R.K.; McDonald, D.M. Abnormalities in pericytes on blood vessels and endothelial sprouts in tumors. *Am. J. Pathol.* **2002**, *160*, 985–1000. [CrossRef]
25. Fisher, D.T.; Muhitch, J.B.; Kim, M.; Doyen, K.C.; Bogner, P.N.; Evans, S.S.; Skitzki, J.J. Intraoperative intravital microscopy permits the study of human tumour vessels. *Nat. Commun.* **2016**, *7*, 10684. [CrossRef] [PubMed]
26. Galon, J.; Bruni, D. Approaches to treat immune hot, altered and cold tumours with combination immunotherapies. *Nat. Rev. Drug Discov.* **2019**, *18*, 197–218. [CrossRef]
27. Ley, K.; Laudanna, C.; Cybulsky, M.I.; Nourshargh, S. Getting to the site of inflammation: The leukocyte adhesion cascade updated. *Nat. Rev. Immunol.* **2007**, *7*, 678–689. [CrossRef]
28. Lu, P.; Takai, K.; Weaver, V.M.; Werb, Z. Extracellular matrix degradation and remodeling in development and disease. *Cold Spring Harb. Perspect. Biol.* **2011**, *3*. [CrossRef]
29. Da Vià, M.C.; Solimando, A.G.; Garitano-Trojaola, A.; Barrio, S.; Munawar, U.; Strifler, S.; Haertle, L.; Rhodes, N.; Teufel, E.; Vogt, C.; et al. CIC Mutation as a Molecular Mechanism of Acquired Resistance to Combined BRAF-MEK Inhibition in Extramedullary Multiple Myeloma with Central Nervous System Involvement. *Oncologist* **2019**. [CrossRef]
30. Harjunpää, H.; Llort Asens, M.; Guenther, C.; Fagerholm, S.C. Cell Adhesion Molecules and Their Roles and Regulation in the Immune and Tumor Microenvironment. *Front. Immunol.* **2019**, *10*, 1078. [CrossRef]
31. Solimando, A.G.; Brandl, A.; Mattenheimer, K.; Graf, C.; Ritz, M.; Ruckdeschel, A.; Stühmer, T.; Mokhtari, Z.; Rudelius, M.; Dotterweich, J.; et al. JAM-A as a prognostic factor and new therapeutic target in multiple myeloma. *Leukemia* **2018**, *32*, 736–743. [CrossRef]
32. Rudelius, M.; Rosenfeldt, M.T.; Leich, E.; Rauert-Wunderlich, H.; Solimando, A.G.; Beilhack, A.; Ott, G.; Rosenwald, A. Inhibition of focal adhesion kinase overcomes resistance of mantle cell lymphoma to ibrutinib in the bone marrow microenvironment. *Haematologica* **2018**, *103*, 116–125. [CrossRef] [PubMed]

33. Anderson, A.R.A. A hybrid mathematical model of solid tumour invasion: The importance of cell adhesion. *Math. Med. Biol. J. IMA* **2005**, *22*, 163–186. [CrossRef] [PubMed]
34. Argentiero, A.; Calabrese, A.; Solimando, A.G.; Notaristefano, A.; Panarelli, M.M.G.; Brunetti, O. Bone metastasis as primary presentation of pancreatic ductal adenocarcinoma: A case report and literature review. *Clin. Case Rep.* **2019**, *7*, 1972–1976. [CrossRef] [PubMed]
35. Solimando, A.G.; Da Vià, M.C.; Leone, P.; Borrelli, P.; Croci, G.A.; Tabares, P.; Brandl, A.; Di Lernia, G.; Bianchi, F.P.; Tafuri, S.; et al. Halting the vicious cycle within the multiple myeloma ecosystem: Blocking JAM-A on bone marrow endothelial cells restores the angiogenic homeostasis and suppresses tumor progression. *Haematologica* **2020**. [CrossRef]
36. Moghadam, A.R.; Patrad, E.; Tafsiri, E.; Peng, W.; Fangman, B.; Pluard, T.J.; Accurso, A.; Salacz, M.; Shah, K.; Ricke, B.; et al. Ral signaling pathway in health and cancer. *Cancer Med.* **2017**, *6*, 2998–3013. [CrossRef]
37. Solimando, A.G.; Ribatti, D.; Vacca, A.; Einsele, H. Targeting B-cell non Hodgkin lymphoma: New and old tricks. *Leuk. Res.* **2016**, *42*, 93–104. [CrossRef]
38. Farahani, E.; Patra, H.K.; Jangamreddy, J.R.; Rashedi, I.; Kawalec, M.; Rao Pariti, R.K.; Batakis, P.; Wiechec, E. Cell adhesion molecules and their relation to (cancer) cell stemness. *Carcinogenesis* **2014**, *35*, 747–759. [CrossRef]
39. Seibold, M.; Stühmer, T.; Kremer, N.; Mottok, A.; Scholz, C.-J.; Schlosser, A.; Leich, E.; Holzgrabe, U.; Brünnert, D.; Barrio, S.; et al. RAL GTPases mediate multiple myeloma cell survival and are activated independently of oncogenic RAS. *Haematologica* **2019**. [CrossRef]
40. Wu, N.Z.; Klitzman, B.; Dodge, R.; Dewhirst, M.W. Diminished leukocyte-endothelium interaction in tumor microvessels. *Cancer Res.* **1992**, *52*, 4265–4268.
41. Ganss, R.; Hanahan, D. Tumor microenvironment can restrict the effectiveness of activated antitumor lymphocytes. *Cancer Res.* **1998**, *58*, 4673–4681.
42. Hamzah, J.; Jugold, M.; Kiessling, F.; Rigby, P.; Manzur, M.; Marti, H.H.; Rabie, T.; Kaden, S.; Gröne, H.-J.; Hämmerling, G.J.; et al. Vascular normalization in Rgs5-deficient tumours promotes immune destruction. *Nature* **2008**, *453*, 410–414. [CrossRef] [PubMed]
43. Lambrechts, D.; Wauters, E.; Boeckx, B.; Aibar, S.; Nittner, D.; Burton, O.; Bassez, A.; Decaluwé, H.; Pircher, A.; Van den Eynde, K.; et al. Phenotype molding of stromal cells in the lung tumor microenvironment. *Nat. Med.* **2018**, *24*, 1277–1289. [CrossRef] [PubMed]
44. Zavidij, O.; Haradhvala, N.J.; Mouhieddine, T.H.; Sklavenitis-Pistofidis, R.; Cai, S.; Reidy, M.; Rahmat, M.; Flaifel, A.; Ferland, B.; Su, N.K.; et al. Single-cell RNA sequencing reveals compromised immune microenvironment in precursor stages of multiple myeloma. *Nat. Cancer* **2020**, *1*, 493–506. [CrossRef]
45. Couzin-Frankel, J. Cancer Immunotherapy. *Science* **2013**, *342*, 1432–1433. [CrossRef]
46. Robert, C. A decade of immune-checkpoint inhibitors in cancer therapy. *Nat. Commun.* **2020**, *11*, 3801. [CrossRef]
47. Siu, L.L.; Ivy, S.P.; Dixon, E.L.; Gravell, A.E.; Reeves, S.A.; Rosner, G.L. Challenges and Opportunities in Adapting Clinical Trial Design for Immunotherapies. *Clin. Cancer Res. Off. J. Am. Assoc. Cancer Res.* **2017**, *23*, 4950–4958. [CrossRef]
48. Nishino, M.; Ramaiya, N.H.; Hatabu, H.; Hodi, F.S. Monitoring immune-checkpoint blockade: Response evaluation and biomarker development. *Nat. Rev. Clin. Oncol.* **2017**, *14*, 655–668. [CrossRef]
49. Solimando, A.G.; Crudele, L.; Leone, P.; Argentiero, A.; Guarascio, M.; Silvestris, N.; Vacca, A.; Racanelli, V. Immune Checkpoint Inhibitor-Related Myositis: From Biology to Bedside. *Int. J. Mol. Sci.* **2020**, *21*, 3054. [CrossRef]
50. Martins, F.; Sofiya, L.; Sykiotis, G.P.; Lamine, F.; Maillard, M.; Fraga, M.; Shabafrouz, K.; Ribi, C.; Cairoli, A.; Guex-Crosier, Y.; et al. Adverse effects of immune-checkpoint inhibitors: Epidemiology, management and surveillance. *Nat. Rev. Clin. Oncol.* **2019**, *16*, 563–580. [CrossRef]
51. Larkin, J.; Chiarion-Sileni, V.; Gonzalez, R.; Grob, J.-J.; Rutkowski, P.; Lao, C.D.; Cowey, C.L.; Schadendorf, D.; Wagstaff, J.; Dummer, R.; et al. Five-Year Survival with Combined Nivolumab and Ipilimumab in Advanced Melanoma. *N. Engl. J. Med.* **2019**, *381*, 1535–1546. [CrossRef]
52. Motzer, R.J.; Tannir, N.M.; McDermott, D.F.; Arén Frontera, O.; Melichar, B.; Choueiri, T.K.; Plimack, E.R.; Barthélémy, P.; Porta, C.; George, S.; et al. Nivolumab plus Ipilimumab versus Sunitinib in Advanced Renal-Cell Carcinoma. *N. Engl. J. Med.* **2018**, *378*, 1277–1290. [CrossRef] [PubMed]
53. Martin, J.D.; Cabral, H.; Stylianopoulos, T.; Jain, R.K. Improving cancer immunotherapy using nanomedicines: Progress, opportunities and challenges. *Nat. Rev. Clin. Oncol.* **2020**, *17*, 251–266. [CrossRef] [PubMed]
54. Khan, K.A.; Kerbel, R.S. Improving immunotherapy outcomes with anti-angiogenic treatments and vice versa. *Nat. Rev. Clin. Oncol.* **2018**, *15*, 310–324. [CrossRef] [PubMed]

55. Balamurugan, K. HIF-1 at the crossroads of hypoxia, inflammation, and cancer. *Int. J. Cancer* **2016**, *138*, 1058–1066. [CrossRef] [PubMed]
56. Huang, Y.; Kim, B.Y.S.; Chan, C.K.; Hahn, S.M.; Weissman, I.L.; Jiang, W. Improving immune-vascular crosstalk for cancer immunotherapy. *Nat. Rev. Immunol.* **2018**, *18*, 195–203. [CrossRef]
57. Liu, Y.; Shaw, S.K.; Ma, S.; Yang, L.; Luscinskas, F.W.; Parkos, C.A. Regulation of Leukocyte Transmigration: Cell Surface Interactions and Signaling Events. *J. Immunol.* **2004**, *172*, 7–13. [CrossRef] [PubMed]
58. Leone, P.; Di Lernia, G.; Solimando, A.G.; Cicco, S.; Saltarella, I.; Lamanuzzi, A.; Ria, R.; Frassanito, M.A.; Ponzoni, M.; Ditonno, P.; et al. Bone marrow endothelial cells sustain a tumor-specific CD8+ T cell subset with suppressive function in myeloma patients. *Oncoimmunology* **2019**, *8*, e1486949. [CrossRef]
59. Chae, Y.K.; Choi, W.M.; Bae, W.H.; Anker, J.; Davis, A.A.; Agte, S.; Iams, W.T.; Cruz, M.; Matsangou, M.; Giles, F.J. Overexpression of adhesion molecules and barrier molecules is associated with differential infiltration of immune cells in non-small cell lung cancer. *Sci. Rep.* **2018**, *8*, 1023. [CrossRef]
60. Orecchioni, M.; Ghosheh, Y.; Pramod, A.B.; Ley, K. Macrophage Polarization: Different Gene Signatures in M1(LPS+) vs. Classically and M2(LPS–) vs. Alternatively Activated Macrophages. *Front. Immunol.* **2019**, *10*, 1084. [CrossRef]
61. Porcelli, L.; Iacobazzi, R.M.; Di Fonte, R.; Serratì, S.; Intini, A.; Solimando, A.G.; Brunetti, O.; Calabrese, A.; Leonetti, F.; Azzariti, A.; et al. CAFs and TGF-β Signaling Activation by Mast Cells Contribute to Resistance to Gemcitabine/Nabpaclitaxel in Pancreatic Cancer. *Cancers* **2019**, *11*, 330. [CrossRef]
62. Albini, A.; Bruno, A.; Noonan, D.M.; Mortara, L. Contribution to Tumor Angiogenesis from Innate Immune Cells Within the Tumor Microenvironment: Implications for Immunotherapy. *Front. Immunol.* **2018**, *9*, 527. [CrossRef] [PubMed]
63. Quail, D.F.; Joyce, J.A. Microenvironmental regulation of tumor progression and metastasis. *Nat. Med.* **2013**, *19*, 1423–1437. [CrossRef] [PubMed]
64. Argentiero, A.; De Summa, S.; Di Fonte, R.; Iacobazzi, R.M.; Porcelli, L.; Da Vià, M.; Brunetti, O.; Azzariti, A.; Silvestris, N.; Solimando, A.G. Gene Expression Comparison between the Lymph Node-Positive and -Negative Reveals a Peculiar Immune Microenvironment Signature and a Theranostic Role for WNT Targeting in Pancreatic Ductal Adenocarcinoma: A Pilot Study. *Cancers* **2019**, *11*, 942. [CrossRef] [PubMed]
65. Tian, L.; Goldstein, A.; Wang, H.; Ching Lo, H.; Sun Kim, I.; Welte, T.; Sheng, K.; Dobrolecki, L.E.; Zhang, X.; Putluri, N.; et al. Mutual regulation of tumour vessel normalization and immunostimulatory reprogramming. *Nature* **2017**, *544*, 250–254. [CrossRef] [PubMed]
66. Zheng, X.; Fang, Z.; Liu, X.; Deng, S.; Zhou, P.; Wang, X.; Zhang, C.; Yin, R.; Hu, H.; Chen, X.; et al. Increased vessel perfusion predicts the efficacy of immune checkpoint blockade. *J. Clin. Investig.* **2018**, *128*, 2104–2115. [CrossRef]
67. Hill, J.M.; Zalos, G.; Halcox, J.P.J.; Schenke, W.H.; Waclawiw, M.A.; Quyyumi, A.A.; Finkel, T. Circulating endothelial progenitor cells, vascular function, and cardiovascular risk. *N. Engl. J. Med.* **2003**, *348*, 593–600. [CrossRef]
68. Goon, P.K.Y.; Lip, G.Y.H.; Boos, C.J.; Stonelake, P.S.; Blann, A.D. Circulating endothelial cells, endothelial progenitor cells, and endothelial microparticles in cancer. *Neoplasia N. Y.* **2006**, *8*, 79–88. [CrossRef]
69. Gao, D.; Nolan, D.J.; Mellick, A.S.; Bambino, K.; McDonnell, K.; Mittal, V. Endothelial progenitor cells control the angiogenic switch in mouse lung metastasis. *Science* **2008**, *319*, 195–198. [CrossRef]
70. Moschetta, M.; Mishima, Y.; Kawano, Y.; Manier, S.; Paiva, B.; Palomera, L.; Aljawai, Y.; Calcinotto, A.; Unitt, C.; Sahin, I.; et al. Targeting vasculogenesis to prevent progression in multiple myeloma. *Leukemia* **2016**, *30*, 1103–1115. [CrossRef]
71. Moschetta, M.; Mishima, Y.; Sahin, I.; Manier, S.; Glavey, S.; Vacca, A.; Roccaro, A.M.; Ghobrial, I.M. Role of endothelial progenitor cells in cancer progression. *Biochim. Biophys. Acta* **2014**, *1846*, 26–39. [CrossRef]
72. Rajkumar, S.V.; Dimopoulos, M.A.; Palumbo, A.; Blade, J.; Merlini, G.; Mateos, M.-V.; Kumar, S.; Hillengass, J.; Kastritis, E.; Richardson, P.; et al. International Myeloma Working Group updated criteria for the diagnosis of multiple myeloma. *Lancet Oncol.* **2014**, *15*, e538–e548. [CrossRef]
73. Vacca, A.; Melaccio, A.; Sportelli, A.; Solimando, A.G.; Dammacco, F.; Ria, R. Subcutaneous immunoglobulins in patients with multiple myeloma and secondary hypogammaglobulinemia: A randomized trial. *Clin. Immunol.* **2018**, *191*, 110–115. [CrossRef] [PubMed]
74. Nucci, M.; Anaissie, E. Infections in Patients with Multiple Myeloma in the Era of High-Dose Therapy and Novel Agents. *Clin. Infect. Dis.* **2009**, *49*, 1211–1225. [CrossRef] [PubMed]

75. Ria, R.; Reale, A.; Solimando, A.G.; Mangialardi, G.; Moschetta, M.; Gelao, L.; Iodice, G.; Vacca, A. Induction therapy and stem cell mobilization in patients with newly diagnosed multiple myeloma. *Stem Cells Int.* **2012**, *2012*, 607260. [CrossRef] [PubMed]
76. Cicco, S.; Solimando, A.G.; Leone, P.; Battaglia, S.; Ria, R.; Vacca, A.; Racanelli, V. Suspected Pericardial Tuberculosis Revealed as an Amyloid Pericardial Mass. *Case Rep. Hematol.* **2018**, *2018*, 8606430. [CrossRef] [PubMed]
77. Solimando, A.G.; Sportelli, A.; Troiano, T.; Demarinis, L.; Di Serio, F.; Ostuni, A.; Dammacco, F.; Vacca, A.; Ria, R. A multiple myeloma that progressed as type I cryoglobulinemia with skin ulcers and foot necrosis: A case report. *Medicine* **2018**, *97*, e12355. [CrossRef]
78. Cicco, S.; Solimando, A.G.; Buono, R.; Susca, N.; Inglese, G.; Melaccio, A.; Prete, M.; Ria, R.; Racanelli, V.; Vacca, A. Right Heart Changes Impact on Clinical Phenotype of Amyloid Cardiac Involvement: A Single Centre Study. *Life* **2020**, *10*, 247. [CrossRef]
79. Di Marzo, L.; Desantis, V.; Solimando, A.G.; Ruggieri, S.; Annese, T.; Nico, B.; Fumarulo, R.; Vacca, A.; Frassanito, M.A. Microenvironment drug resistance in multiple myeloma: Emerging new players. *Oncotarget* **2016**, *7*, 60698–60711. [CrossRef]
80. Manier, S.; Sacco, A.; Leleu, X.; Ghobrial, I.M.; Roccaro, A.M. Bone marrow microenvironment in multiple myeloma progression. *J. Biomed. Biotechnol.* **2012**, *2012*, 157496. [CrossRef]
81. Solimando, A.G.; Vacca, A.; Ribatti, D. A Comprehensive Biological and Clinical Perspective Can Drive a Patient-Tailored Approach to Multiple Myeloma: Bridging the Gaps between the Plasma Cell and the Neoplastic Niche. *J. Oncol.* **2020**, *2020*, 1–16. [CrossRef]
82. Noonan, K.; Borrello, I. The immune microenvironment of myeloma. *Cancer Microenviron. Off. J. Int. Cancer Microenviron. Soc.* **2011**, *4*, 313–323. [CrossRef] [PubMed]
83. Saltarella, I.; Desantis, V.; Melaccio, A.; Solimando, A.G.; Lamanuzzi, A.; Ria, R.; Storlazzi, C.T.; Mariggiò, M.A.; Vacca, A.; Frassanito, M.A. Mechanisms of Resistance to Anti-CD38 Daratumumab in Multiple Myeloma. *Cells* **2020**, *9*, 167. [CrossRef] [PubMed]
84. Moschetta, M.; Basile, A.; Ferrucci, A.; Frassanito, M.A.; Rao, L.; Ria, R.; Solimando, A.G.; Giuliani, N.; Boccarelli, A.; Fumarola, F.; et al. Novel targeting of phospho-cMET overcomes drug resistance and induces antitumor activity in multiple myeloma. *Clin. Cancer Res. Off. J. Am. Assoc. Cancer Res.* **2013**, *19*, 4371–4382. [CrossRef] [PubMed]
85. Ferrucci, A.; Moschetta, M.; Frassanito, M.A.; Berardi, S.; Catacchio, I.; Ria, R.; Racanelli, V.; Caivano, A.; Solimando, A.G.; Vergara, D.; et al. A HGF/cMET autocrine loop is operative in multiple myeloma bone marrow endothelial cells and may represent a novel therapeutic target. *Clin. Cancer Res. Off. J. Am. Assoc. Cancer Res.* **2014**, *20*, 5796–5807. [CrossRef]
86. Frassanito, M.A.; Desantis, V.; Di Marzo, L.; Craparotta, I.; Beltrame, L.; Marchini, S.; Annese, T.; Visino, F.; Arciuli, M.; Saltarella, I.; et al. Bone marrow fibroblasts overexpress miR-27b and miR-214 in step with multiple myeloma progression, dependent on tumour cell-derived exosomes: Myeloma cell-derived exosomes and fibroblast miRNA expression. *J. Pathol.* **2019**, *247*, 241–253. [CrossRef]
87. Desantis, V.; Saltarella, I.; Lamanuzzi, A.; Melaccio, A.; Solimando, A.G.; Mariggiò, M.A.; Racanelli, V.; Paradiso, A.; Vacca, A.; Frassanito, M.A. MicroRNAs-Based Nano-Strategies as New Therapeutic Approach in Multiple Myeloma to Overcome Disease Progression and Drug Resistance. *Int. J. Mol. Sci.* **2020**, *21*, 3084. [CrossRef]
88. Solimando, A.G.; Da Via', M.C.; Leone, P.; Croci, G.; Borrelli, P.; Tabares Gaviria, P.; Brandl, A.; Di Lernia, G.; Bianchi, F.P.; Tafuri, S.; et al. Adhesion-Mediated Multiple Myeloma (MM) Disease Progression: Junctional Adhesion Molecule a Enhances Angiogenesis and Multiple Myeloma Dissemination and Predicts Poor Survival. *Blood* **2019**, *134*, 855. [CrossRef]
89. Saadatmand, S.; de Kruijf, E.M.; Sajet, A.; Dekker-Ensink, N.G.; van Nes, J.G.H.; Putter, H.; Smit, V.T.H.B.M.; van de Velde, C.J.H.; Liefers, G.J.; Kuppen, P.J.K. Expression of cell adhesion molecules and prognosis in breast cancer. *Br. J. Surg.* **2013**, *100*, 252–260. [CrossRef]
90. Moh, M.C.; Shen, S. The roles of cell adhesion molecules in tumor suppression and cell migration: A new paradox. *Cell Adhes. Migr.* **2009**, *3*, 334–336. [CrossRef]
91. Kelly, K.R.; Espitia, C.M.; Zhao, W.; Wendlandt, E.; Tricot, G.; Zhan, F.; Carew, J.S.; Nawrocki, S.T. Junctional adhesion molecule-A is overexpressed in advanced multiple myeloma and determines response to oncolytic reovirus. *Oncotarget* **2015**, *6*, 41275–41289. [CrossRef]

92. Teoh, G.; Anderson, K.C. Interaction of Tumor and Host Cells With Adhesion and Extracellular Matrix Molecules in the Development of Multiple Myeloma. *Hematol. Oncol. Clin. N. Am.* **1997**, *11*, 27–42. [CrossRef]
93. Braunstein, M.J.; Campagne, F.; Mukherjee, P.; Carrasco, D.R.; Sukhdeo, K.; Protopopov, A.; Anderson, K.C.; Batuman, O. Genome-Wide Profiling of Endothelial Progenitor Cells in Multiple Myeloma: Disease-Relevant Pathways and Overlaps with Common Cancer Biomarkers. *Blood* **2008**, *112*, 626. [CrossRef]
94. Gautier, L.; Cope, L.; Bolstad, B.M.; Irizarry, R.A. affy—analysis of Affymetrix GeneChip data at the probe level. *Bioinformatics* **2004**, *20*, 307–315. [CrossRef] [PubMed]
95. Ritchie, M.E.; Phipson, B.; Wu, D.; Hu, Y.; Law, C.W.; Shi, W.; Smyth, G.K. limma powers differential expression analyses for RNA-sequencing and microarray studies. *Nucleic Acids Res.* **2015**, *43*, e47. [CrossRef] [PubMed]
96. Gil Marques, F.; Poli, E.; Malaquias, J.; Carvalho, T.; Portêlo, A.; Ramires, A.; Aldeia, F.; Ribeiro, R.M.; Vitorino, E.; Diegues, I.; et al. Low doses of ionizing radiation activate endothelial cells and induce angiogenesis in peritumoral tissues. *Radiother. Oncol. J. Eur. Soc. Ther. Radiol. Oncol.* **2019**, *141*, 256–261. [CrossRef]
97. Galili, T. dendextend: An R package for visualizing, adjusting and comparing trees of hierarchical clustering. *Bioinformatics (Oxf. Engl.)* **2015**, *31*, 3718–3720. [CrossRef]
98. Zeileis, A.; Fisher, J.C.; Hornik, K.; Ihaka, R.; McWhite, C.D.; Murrell, P.; Stauffer, R.; Wilke, C.O. Colorspace: A Toolbox for Manipulating and Assessing Colors and Palettes. *arXiv* **2019**, arXiv:190306490.
99. Wickham, H. *Ggplot2: Elegant Graphics for Data Analysis*; Use R! Springer: Dordrecht, The Netherlands, 2009; ISBN 978-0-387-98141-3.
100. Abe, M. Targeting the interplay between myeloma cells and the bone marrow microenvironment in myeloma. *Int. J. Hematol.* **2011**, *94*, 334–343. [CrossRef]
101. Leich, E.; Weißbach, S.; Klein, H.-U.; Grieb, T.; Pischimarov, J.; Stühmer, T.; Chatterjee, M.; Steinbrunn, T.; Langer, C.; Eilers, M.; et al. Multiple myeloma is affected by multiple and heterogeneous somatic mutations in adhesion- and receptor tyrosine kinase signaling molecules. *Blood Cancer J.* **2013**, *3*, e102. [CrossRef]
102. Jridi, I.; Catacchio, I.; Majdoub, H.; Shahbazzadeh, D.; El Ayeb, M.; Frassanito, M.A.; Solimando, A.G.; Ribatti, D.; Vacca, A.; Borchani, L. The small subunit of Hemilipin2, a new heterodimeric phospholipase A2 from Hemiscorpius lepturus scorpion venom, mediates the antiangiogenic effect of the whole protein. *Toxicon Off. J. Int. Soc. Toxinology* **2017**, *126*, 38–46. [CrossRef]
103. Flati, V.; Pastore, L.I.; Griffioen, A.W.; Satijn, S.; Toniato, E.; D'Alimonte, I.; Laglia, E.; Marchetti, P.; Gulino, A.; Martinotti, S. Endothelial cell anergy is mediated by bFGF through the sustained activation of p38-MAPK and NF-kappaB inhibition. *Int. J. Immunopathol. Pharmacol.* **2006**, *19*, 761–773. [CrossRef] [PubMed]
104. Peddibhotla, S.S.D.; Brinkmann, B.F.; Kummer, D.; Tuncay, H.; Nakayama, M.; Adams, R.H.; Gerke, V.; Ebnet, K. Tetraspanin CD9 links junctional adhesion molecule-A to αvβ3 integrin to mediate basic fibroblast growth factor-specific angiogenic signaling. *Mol. Biol. Cell* **2013**, *24*, 933–944. [CrossRef] [PubMed]
105. Griffioen, A.W.; Damen, C.A.; Mayo, K.H.; Barendsz-Janson, A.F.; Martinotti, S.; Blijham, G.H.; Groenewegen, G. Angiogenesis inhibitors overcome tumor induced endothelial cell anergy. *Int. J. Cancer* **1999**, *80*, 315–319. [CrossRef]
106. Dirkx, A.E.M.; Oude Egbrink, M.G.A.; Kuijpers, M.J.E.; van der Niet, S.T.; Heijnen, V.V.T.; Bouma-ter Steege, J.C.A.; Wagstaff, J.; Griffioen, A.W. Tumor angiogenesis modulates leukocyte-vessel wall interactions in vivo by reducing endothelial adhesion molecule expression. *Cancer Res.* **2003**, *63*, 2322–2329. [PubMed]
107. Cook-Mills, J.M.; Deem, T.L. Active participation of endothelial cells in inflammation. *J. Leukoc. Biol.* **2005**, *77*, 487–495. [CrossRef] [PubMed]
108. Lauko, A.; Mu, Z.; Gutmann, D.H.; Naik, U.P.; Lathia, J.D. Junctional Adhesion Molecules in Cancer: A Paradigm for the Diverse Functions of Cell–Cell Interactions in Tumor Progression. *Cancer Res.* **2020**. [CrossRef]
109. Huang, H.; Langenkamp, E.; Georganaki, M.; Loskog, A.; Fuchs, P.F.; Dieterich, L.C.; Kreuger, J.; Dimberg, A. VEGF suppresses T-lymphocyte infiltration in the tumor microenvironment through inhibition of NF-κB-induced endothelial activation. *FASEB J. Off. Publ. Fed. Am. Soc. Exp. Biol.* **2015**, *29*, 227–238. [CrossRef] [PubMed]
110. Lazennec, G.; Richmond, A. Chemokines and chemokine receptors: New insights into cancer-related inflammation. *Trends Mol. Med.* **2010**, *16*, 133–144. [CrossRef]
111. Lanitis, E.; Irving, M.; Coukos, G. Targeting the tumor vasculature to enhance T cell activity. *Curr. Opin. Immunol.* **2015**, *33*, 55–63. [CrossRef]

112. Georganaki, M.; van Hooren, L.; Dimberg, A. Vascular Targeting to Increase the Efficiency of Immune Checkpoint Blockade in Cancer. *Front. Immunol.* **2018**, *9*, 3081. [CrossRef]
113. Salik, B.; Smyth, M.J.; Nakamura, K. Targeting immune checkpoints in hematological malignancies. *J. Hematol. Oncol. J Hematol. Oncol.* **2020**, *13*, 111. [CrossRef] [PubMed]
114. Georganaki, M.; Ramachandran, M.; Tuit, S.; Núñez, N.G.; Karampatzakis, A.; Fotaki, G.; van Hooren, L.; Huang, H.; Lugano, R.; Ulas, T.; et al. Tumor endothelial cell up-regulation of IDO1 is an immunosuppressive feed-back mechanism that reduces the response to CD40-stimulating immunotherapy. *Oncoimmunology* **2020**, *9*, 1730538. [CrossRef] [PubMed]
115. Shiao, S.L.; Kirkiles-Smith, N.C.; Shepherd, B.R.; McNiff, J.M.; Carr, E.J.; Pober, J.S. Human effector memory CD4+ T cells directly recognize allogeneic endothelial cells in vitro and in vivo. *J. Immunol. Baltim. Md 1950* **2007**, *179*, 4397–4404. [CrossRef] [PubMed]
116. Kochan, G.; Escors, D.; Breckpot, K.; Guerrero-Setas, D. Role of non-classical MHC class I molecules in cancer immunosuppression. *Oncoimmunology* **2013**, *2*, e26491. [CrossRef]
117. Goveia, J.; Rohlenova, K.; Taverna, F.; Treps, L.; Conradi, L.-C.; Pircher, A.; Geldhof, V.; de Rooij, L.P.M.H.; Kalucka, J.; Sokol, L.; et al. An Integrated Gene Expression Landscape Profiling Approach to Identify Lung Tumor Endothelial Cell Heterogeneity and Angiogenic Candidates. *Cancer Cell* **2020**, *37*, 21–36.e13. [CrossRef]
118. Barthel, S.R.; Gavino, J.D.; Descheny, L.; Dimitroff, C.J. Targeting selectins and selectin ligands in inflammation and cancer. *Expert Opin. Ther. Targets* **2007**, *11*, 1473–1491. [CrossRef]
119. Kong, D.-H.; Kim, Y.K.; Kim, M.R.; Jang, J.H.; Lee, S. Emerging Roles of Vascular Cell Adhesion Molecule-1 (VCAM-1) in Immunological Disorders and Cancer. *Int. J. Mol. Sci.* **2018**, *19*, 1057. [CrossRef]
120. Delfortrie, S.; Pinte, S.; Mattot, V.; Samson, C.; Villain, G.; Caetano, B.; Lauridant-Philippin, G.; Baranzelli, M.-C.; Bonneterre, J.; Trottein, F.; et al. Egfl7 promotes tumor escape from immunity by repressing endothelial cell activation. *Cancer Res.* **2011**, *71*, 7176–7186. [CrossRef]
121. Pannier, D.; Philippin-Lauridant, G.; Baranzelli, M.-C.; Bertin, D.; Bogart, E.; Delprat, V.; Villain, G.; Mattot, V.; Bonneterre, J.; Soncin, F. High expression levels of egfl7 correlate with low endothelial cell activation in peritumoral vessels of human breast cancer. *Oncol. Lett.* **2016**, *12*, 1422–1428. [CrossRef]
122. Salama, Y.; Heida, A.H.; Yokoyama, K.; Takahashi, S.; Hattori, K.; Heissig, B. The EGFL7-ITGB3-KLF2 axis enhances survival of multiple myeloma in preclinical models. *Blood Adv.* **2020**, *4*, 1021–1037. [CrossRef]
123. De Sanctis, F.; Ugel, S.; Facciponte, J.; Facciabene, A. The dark side of tumor-associated endothelial cells. *Semin. Immunol.* **2018**, *35*, 35–47. [CrossRef] [PubMed]
124. Zhang, B.; Karrison, T.; Rowley, D.A.; Schreiber, H. IFN-γ– and TNF-dependent bystander eradication of antigen-loss variants in established mouse cancers. *J. Clin. Investig.* **2008**, *118*, 1398–1404. [CrossRef] [PubMed]
125. Enzler, T.; Gillessen, S.; Manis, J.P.; Ferguson, D.; Fleming, J.; Alt, F.W.; Mihm, M.; Dranoff, G. Deficiencies of GM-CSF and interferon gamma link inflammation and cancer. *J. Exp. Med.* **2003**, *197*, 1213–1219. [CrossRef] [PubMed]
126. List, A.F. Vascular endothelial growth factor signaling pathway as an emerging target in hematologic malignancies. *Oncologist* **2001**, *6* (Suppl. 5), 24–31. [CrossRef]
127. Josephs, S.F.; Ichim, T.E.; Prince, S.M.; Kesari, S.; Marincola, F.M.; Escobedo, A.R.; Jafri, A. Unleashing endogenous TNF-alpha as a cancer immunotherapeutic. *J. Transl. Med.* **2018**, *16*, 242. [CrossRef]
128. Ray, A.; Song, Y.; Du, T.; Chauhan, D.; Anderson, K.C. Preclinical validation of Alpha-Enolase (ENO1) as a novel immunometabolic target in multiple myeloma. *Oncogene* **2020**, *39*, 2786–2796. [CrossRef]
129. Maishi, N.; Ohba, Y.; Akiyama, K.; Ohga, N.; Hamada, J.-I.; Nagao-Kitamoto, H.; Alam, M.T.; Yamamoto, K.; Kawamoto, T.; Inoue, N.; et al. Tumour endothelial cells in high metastatic tumours promote metastasis via epigenetic dysregulation of biglycan. *Sci. Rep.* **2016**, *6*, 28039. [CrossRef]
130. Lund, A.W.; Wagner, M.; Fankhauser, M.; Steinskog, E.S.; Broggi, M.A.; Spranger, S.; Gajewski, T.F.; Alitalo, K.; Eikesdal, H.P.; Wiig, H.; et al. Lymphatic vessels regulate immune microenvironments in human and murine melanoma. *J. Clin. Investig.* **2016**, *126*, 3389–3402. [CrossRef]
131. De Caterina, R.; Libby, P.; Peng, H.B.; Thannickal, V.J.; Rajavashisth, T.B.; Gimbrone, M.A.; Shin, W.S.; Liao, J.K. Nitric oxide decreases cytokine-induced endothelial activation. Nitric oxide selectively reduces endothelial expression of adhesion molecules and proinflammatory cytokines. *J. Clin. Investig.* **1995**, *96*, 60–68. [CrossRef]

132. Bouzin, C.; Brouet, A.; De Vriese, J.; Dewever, J.; Feron, O. Effects of vascular endothelial growth factor on the lymphocyte-endothelium interactions: Identification of caveolin-1 and nitric oxide as control points of endothelial cell anergy. *J. Immunol.* **2007**, *178*, 1505–1511. [CrossRef]
133. Secchiero, P.; Gonelli, A.; Celeghini, C.; Mirandola, P.; Guidotti, L.; Visani, G.; Capitani, S.; Zauli, G. Activation of the nitric oxide synthase pathway represents a key component of tumor necrosis factor–related apoptosis-inducing ligand–mediated cytotoxicity on hematologic malignancies. *Blood* **2001**, *98*, 2220–2228. [CrossRef] [PubMed]
134. Xu, W.; Liu, L.Z.; Loizidou, M.; Ahmed, M.; Charles, I.G. The role of nitric oxide in cancer. *Cell Res.* **2002**, *12*, 311–320. [CrossRef]
135. Gooden, M.; Lampen, M.; Jordanova, E.S.; Leffers, N.; Trimbos, J.B.; van der Burg, S.H.; Nijman, H.; van Hall, T. HLA-E expression by gynecological cancers restrains tumor-infiltrating CD8+ T lymphocytes. *Proc. Natl. Acad. Sci. USA* **2011**, *108*, 10656–10661. [CrossRef] [PubMed]
136. Rossi, E.; Sanz-Rodriguez, F.; Eleno, N.; Düwell, A.; Blanco, F.J.; Langa, C.; Botella, L.M.; Cabañas, C.; Lopez-Novoa, J.M.; Bernabeu, C. Endothelial endoglin is involved in inflammation: Role in leukocyte adhesion and transmigration. *Blood* **2013**, *121*, 403–415. [CrossRef] [PubMed]
137. Stalin, J.; Nollet, M.; Garigue, P.; Fernandez, S.; Vivancos, L.; Essaadi, A.; Muller, A.; Bachelier, R.; Foucault-Bertaud, A.; Fugazza, L.; et al. Targeting soluble CD146 with a neutralizing antibody inhibits vascularization, growth and survival of CD146-positive tumors. *Oncogene* **2016**, *35*, 5489–5500. [CrossRef]
138. Leone, P.; Solimando, A.G.; Malerba, E.; Fasano, R.; Buonavoglia, A.; Pappagallo, F.; De Re, V.; Argentiero, A.; Silvestris, N.; Vacca, A.; et al. Actors on the Scene: Immune Cells in the Myeloma Niche. *Front. Oncol.* **2020**, *10*, 599098. [CrossRef]
139. Scavelli, C.; Weber, E.; Aglianò, M.; Cirulli, T.; Nico, B.; Vacca, A.; Ribatti, D. Lymphatics at the crossroads of angiogenesis and lymphangiogenesis. *J. Anat.* **2004**, *204*, 433–449. [CrossRef]
140. Ribatti, D.; Nico, B.; Crivellato, E.; Vacca, A. The structure of the vascular network of tumors. *Cancer Lett.* **2007**, *248*, 18–23. [CrossRef]
141. Hunter, M.C.; Teijeira, A.; Halin, C. T Cell Trafficking through Lymphatic Vessels. *Front. Immunol.* **2016**, *7*. [CrossRef]
142. Eklund, L.; Bry, M.; Alitalo, K. Mouse models for studying angiogenesis and lymphangiogenesis in cancer. *Mol. Oncol.* **2013**, *7*, 259–282. [CrossRef]
143. Tomura, M.; Yoshida, N.; Tanaka, J.; Karasawa, S.; Miwa, Y.; Miyawaki, A.; Kanagawa, O. Monitoring cellular movement in vivo with photoconvertible fluorescence protein "Kaede" transgenic mice. *Proc. Natl. Acad. Sci. USA* **2008**, *105*, 10871–10876. [CrossRef] [PubMed]
144. Steele, M.M.; Churchill, M.J.; Breazeale, A.P.; Lane, R.S.; Nelson, N.A.; Lund, A.W. Quantifying Leukocyte Egress via Lymphatic Vessels from Murine Skin and Tumors. *J. Vis. Exp.* **2019**, 58704. [CrossRef] [PubMed]
145. Flores-Toro, J.A.; Luo, D.; Gopinath, A.; Sarkisian, M.R.; Campbell, J.J.; Charo, I.F.; Singh, R.; Schall, T.J.; Datta, M.; Jain, R.K.; et al. CCR2 inhibition reduces tumor myeloid cells and unmasks a checkpoint inhibitor effect to slow progression of resistant murine gliomas. *Proc. Natl. Acad. Sci. USA* **2020**, *117*, 1129–1138. [CrossRef] [PubMed]
146. Gibellini, L.; De Biasi, S.; Porta, C.; Lo Tartaro, D.; Depenni, R.; Pellacani, G.; Sabbatini, R.; Cossarizza, A. Single-Cell Approaches to Profile the Response to Immune Checkpoint Inhibitors. *Front. Immunol.* **2020**, *11*, 490. [CrossRef]
147. Progatzky, F.; Dallman, M.J.; Lo Celso, C. From seeing to believing: Labelling strategies for in vivo cell-tracking experiments. *Interface Focus* **2013**, *3*, 20130001. [CrossRef]
148. Ito, K.; Morimoto, J.; Kihara, A.; Matsui, Y.; Kurotaki, D.; Kanayama, M.; Simmons, S.; Ishii, M.; Sheppard, D.; Takaoka, A.; et al. Integrin $\alpha 9$ on lymphatic endothelial cells regulates lymphocyte egress. *Proc. Natl. Acad. Sci. USA* **2014**, *111*, 3080–3085. [CrossRef]
149. Schwab, S.R.; Cyster, J.G. Finding a way out: Lymphocyte egress from lymphoid organs. *Nat. Immunol.* **2007**, *8*, 1295–1301. [CrossRef]
150. Torcellan, T.; Hampton, H.R.; Bailey, J.; Tomura, M.; Brink, R.; Chtanova, T. In vivo photolabeling of tumor-infiltrating cells reveals highly regulated egress of T-cell subsets from tumors. *Proc. Natl. Acad. Sci. USA* **2017**, *114*, 5677–5682. [CrossRef]

151. Duhen, T.; Duhen, R.; Montler, R.; Moses, J.; Moudgil, T.; de Miranda, N.F.; Goodall, C.P.; Blair, T.C.; Fox, B.A.; McDermott, J.E.; et al. Co-expression of CD39 and CD103 identifies tumor-reactive CD8 T cells in human solid tumors. *Nat. Commun.* **2018**, *9*, 2724. [CrossRef]
152. Condeelis, J.; Weissleder, R. In vivo imaging in cancer. *Cold Spring Harb. Perspect. Biol.* **2010**, *2*, a003848. [CrossRef]
153. Vakoc, B.J.; Lanning, R.M.; Tyrrell, J.A.; Padera, T.P.; Bartlett, L.A.; Stylianopoulos, T.; Munn, L.L.; Tearney, G.J.; Fukumura, D.; Jain, R.K.; et al. Three-dimensional microscopy of the tumor microenvironment in vivo using optical frequency domain imaging. *Nat. Med.* **2009**, *15*, 1219–1223. [CrossRef] [PubMed]
154. Antonio, G.; Oronzo, B.; Vito, L.; Angela, C.; Antonel-la, A.; Roberto, C.; Giovanni, S.A.; Antonella, L. Immune system and bone microenvironment: Rationale for targeted cancer therapies. *Oncotarget* **2020**, *11*. [CrossRef] [PubMed]
155. Jung, K.; Heishi, T.; Khan, O.F.; Kowalski, P.S.; Incio, J.; Rahbari, N.N.; Chung, E.; Clark, J.W.; Willett, C.G.; Luster, A.D.; et al. Ly6Clo monocytes drive immunosuppression and confer resistance to anti-VEGFR2 cancer therapy. *J. Clin. Investig.* **2017**, *127*, 3039–3051. [CrossRef] [PubMed]
156. Bruns, O.T.; Bischof, T.S.; Harris, D.K.; Franke, D.; Shi, Y.; Riedemann, L.; Bartelt, A.; Jaworski, F.B.; Carr, J.A.; Rowlands, C.J.; et al. Next-generation in vivo optical imaging with short-wave infrared quantum dots. *Nat. Biomed. Eng.* **2017**, *1*, 56. [CrossRef]
157. Brown, E.B.; Campbell, R.B.; Tsuzuki, Y.; Xu, L.; Carmeliet, P.; Fukumura, D.; Jain, R.K. In vivo measurement of gene expression, angiogenesis and physiological function in tumors using multiphoton laser scanning microscopy. *Nat. Med.* **2001**, *7*, 864–868. [CrossRef]
158. Kirkpatrick, N.D.; Chung, E.; Cook, D.C.; Han, X.; Gruionu, G.; Liao, S.; Munn, L.L.; Padera, T.P.; Fukumura, D.; Jain, R.K. Video-rate resonant scanning multiphoton microscopy: An emerging technique for intravital imaging of the tumor microenvironment. *Intravital* **2012**, *1*. [CrossRef]
159. Yuan, F.; Salehi, H.A.; Boucher, Y.; Vasthare, U.S.; Tuma, R.F.; Jain, R.K. Vascular permeability and microcirculation of gliomas and mammary carcinomas transplanted in rat and mouse cranial windows. *Cancer Res.* **1994**, *54*, 4564–4568.
160. Fukumura, D.; Kloepper, J.; Amoozgar, Z.; Duda, D.G.; Jain, R.K. Enhancing cancer immunotherapy using antiangiogenics: Opportunities and challenges. *Nat. Rev. Clin. Oncol.* **2018**, *15*, 325–340. [CrossRef]
161. Martin, J.D.; Fukumura, D.; Duda, D.G.; Boucher, Y.; Jain, R.K. Reengineering the Tumor Microenvironment to Alleviate Hypoxia and Overcome Cancer Heterogeneity. *Cold Spring Harb. Perspect. Med.* **2016**, *6*. [CrossRef]
162. Li, J.; Chekkoury, A.; Prakash, J.; Glasl, S.; Vetschera, P.; Koberstein-Schwarz, B.; Olefir, I.; Gujrati, V.; Omar, M.; Ntziachristos, V. Spatial heterogeneity of oxygenation and haemodynamics in breast cancer resolved in vivo by conical multispectral optoacoustic mesoscopy. *Light Sci. Appl.* **2020**, *9*, 57. [CrossRef]
163. Pires, I.M.; Bencokova, Z.; Milani, M.; Folkes, L.K.; Li, J.-L.; Stratford, M.R.; Harris, A.L.; Hammond, E.M. Effects of acute versus chronic hypoxia on DNA damage responses and genomic instability. *Cancer Res.* **2010**, *70*, 925–935. [CrossRef] [PubMed]
164. Eales, K.L.; Hollinshead, K.E.R.; Tennant, D.A. Hypoxia and metabolic adaptation of cancer cells. *Oncogenesis* **2016**, *5*, e190. [CrossRef] [PubMed]
165. Zhang, L.; Huang, G.; Li, X.; Zhang, Y.; Jiang, Y.; Shen, J.; Liu, J.; Wang, Q.; Zhu, J.; Feng, X.; et al. Hypoxia induces epithelial-mesenchymal transition via activation of SNAI1 by hypoxia-inducible factor -1α in hepatocellular carcinoma. *BMC Cancer* **2013**, *13*, 108. [CrossRef] [PubMed]
166. Heddleston, J.M.; Li, Z.; McLendon, R.E.; Hjelmeland, A.B.; Rich, J.N. The hypoxic microenvironment maintains glioblastoma stem cells and promotes reprogramming towards a cancer stem cell phenotype. *Cell Cycle* **2009**, *8*, 3274–3284. [CrossRef]
167. Azad, M.B.; Chen, Y.; Henson, E.S.; Cizeau, J.; McMillan-Ward, E.; Israels, S.J.; Gibson, S.B. Hypoxia induces autophagic cell death in apoptosis-competent cells through a mechanism involving BNIP3. *Autophagy* **2008**, *4*, 195–204. [CrossRef]
168. Di Lernia, G.; Leone, P.; Solimando, A.G.; Buonavoglia, A.; Saltarella, I.; Ria, R.; Ditonno, P.; Silvestris, N.; Crudele, L.; Vacca, A.; et al. Bortezomib Treatment Modulates Autophagy in Multiple Myeloma. *J. Clin. Med.* **2020**, *9*, 552. [CrossRef]
169. Graham, K.; Unger, E. Overcoming tumor hypoxia as a barrier to radiotherapy, chemotherapy and immunotherapy in cancer treatment. *Int. J. Nanomedicine* **2018**, *13*, 6049–6058. [CrossRef]
170. Shannon, A.M.; Bouchier-Hayes, D.J.; Condron, C.M.; Toomey, D. Tumour hypoxia, chemotherapeutic resistance and hypoxia-related therapies. *Cancer Treat. Rev.* **2003**, *29*, 297–307. [CrossRef]

171. Brunetti, O.; Gnoni, A.; Licchetta, A.; Longo, V.; Calabrese, A.; Argentiero, A.; Delcuratolo, S.; Solimando, A.G.; Casadei-Gardini, A.; Silvestris, N. Predictive and Prognostic Factors in HCC Patients Treated with Sorafenib. *Med. Kaunas Lith.* **2019**, *55*, 707. [CrossRef]
172. Schaaf, M.B.; Garg, A.D.; Agostinis, P. Defining the role of the tumor vasculature in antitumor immunity and immunotherapy. *Cell Death Dis.* **2018**, *9*, 115. [CrossRef]
173. Longo, V.; Brunetti, O.; Gnoni, A.; Licchetta, A.; Delcuratolo, S.; Memeo, R.; Solimando, A.G.; Argentiero, A. Emerging role of Immune Checkpoint Inhibitors in Hepatocellular Carcinoma. *Med. Kaunas Lith.* **2019**, *55*, 698. [CrossRef] [PubMed]
174. Chouaib, S.; Messai, Y.; Couve, S.; Escudier, B.; Hasmim, M.; Noman, M.Z. Hypoxia promotes tumor growth in linking angiogenesis to immune escape. *Front. Immunol.* **2012**, *3*, 21. [CrossRef] [PubMed]
175. Solimando, A.G.; Da Vià, M.C.; Cicco, S.; Leone, P.; Di Lernia, G.; Giannico, D.; Desantis, V.; Frassanito, M.A.; Morizio, A.; Delgado Tascon, J.; et al. High-Risk Multiple Myeloma: Integrated Clinical and Omics Approach Dissects the Neoplastic Clone and the Tumor Microenvironment. *J. Clin. Med.* **2019**, *8*, 997. [CrossRef] [PubMed]
176. Michiels, C. Physiological and pathological responses to hypoxia. *Am. J. Pathol.* **2004**, *164*, 1875–1882. [CrossRef]
177. Brahimi-Horn, M.C.; Chiche, J.; Pouysségur, J. Hypoxia and cancer. *J. Mol. Med.* **2007**, *85*, 1301–1307. [CrossRef]
178. Luoto, K.R.; Kumareswaran, R.; Bristow, R.G. Tumor hypoxia as a driving force in genetic instability. *Genome Integr.* **2013**, *4*, 5. [CrossRef]
179. Gilkes, D.M.; Semenza, G.L.; Wirtz, D. Hypoxia and the extracellular matrix: Drivers of tumour metastasis. *Nat. Rev. Cancer* **2014**, *14*, 430–439. [CrossRef]
180. Rockwell, S.; Dobrucki, I.T.; Kim, E.Y.; Marrison, S.T.; Vu, V.T. Hypoxia and radiation therapy: Past history, ongoing research, and future promise. *Curr. Mol. Med.* **2009**, *9*, 442–458. [CrossRef]
181. Feng, J.; Byrne, N.M.; Al Jamal, W.; Coulter, J.A. Exploiting Current Understanding of Hypoxia Mediated Tumour Progression for Nanotherapeutic Development. *Cancers* **2019**, *11*, 1989. [CrossRef]
182. Koumenis, C.; Wouters, B.G. "Translating" tumor hypoxia: Unfolded protein response (UPR)-dependent and UPR-independent pathways. *Mol. Cancer Res. MCR* **2006**, *4*, 423–436. [CrossRef]
183. Wilson, W.R.; Hay, M.P. Targeting hypoxia in cancer therapy. *Nat. Rev. Cancer* **2011**, *11*, 393–410. [CrossRef] [PubMed]
184. Samples, J.; Willis, M.; Klauber-Demore, N. Targeting angiogenesis and the tumor microenvironment. *Surg. Oncol. Clin. N. Am.* **2013**, *22*, 629–639. [CrossRef] [PubMed]
185. Desantis, V.; Frassanito, M.A.; Tamma, R.; Saltarella, I.; Di Marzo, L.; Lamanuzzi, A.; Solimando, A.G.; Ruggieri, S.; Annese, T.; Nico, B.; et al. Rhu-Epo down-regulates pro-tumorigenic activity of cancer-associated fibroblasts in multiple myeloma. *Ann. Hematol.* **2018**, *97*, 1251–1258. [CrossRef] [PubMed]
186. Lugano, R.; Ramachandran, M.; Dimberg, A. Tumor angiogenesis: Causes, consequences, challenges and opportunities. *Cell. Mol. Life Sci.* **2020**, *77*, 1745–1770. [CrossRef]
187. Rao, L.; Giannico, D.; Leone, P.; Solimando, A.G.; Maiorano, E.; Caporusso, C.; Duda, L.; Tamma, R.; Mallamaci, R.; Susca, N.; et al. HB-EGF-EGFR Signaling in Bone Marrow Endothelial Cells Mediates Angiogenesis Associated with Multiple Myeloma. *Cancers* **2020**, *12*, 173. [CrossRef]
188. Winkler, F.; Kozin, S.V.; Tong, R.T.; Chae, S.-S.; Booth, M.F.; Garkavtsev, I.; Xu, L.; Hicklin, D.J.; Fukumura, D.; di Tomaso, E.; et al. Kinetics of vascular normalization by VEGFR2 blockade governs brain tumor response to radiation: Role of oxygenation, angiopoietin-1, and matrix metalloproteinases. *Cancer Cell* **2004**, *6*, 553–563. [CrossRef]
189. Goel, S.; Duda, D.G.; Xu, L.; Munn, L.L.; Boucher, Y.; Fukumura, D.; Jain, R.K. Normalization of the vasculature for treatment of cancer and other diseases. *Physiol. Rev.* **2011**, *91*, 1071–1121. [CrossRef]
190. Gerald, D.; Chintharlapalli, S.; Augustin, H.G.; Benjamin, L.E. Angiopoietin-2: An Attractive Target for Improved Antiangiogenic Tumor Therapy. *Cancer Res.* **2013**, *73*, 1649–1657. [CrossRef]
191. Rao, L.; De Veirman, K.; Giannico, D.; Saltarella, I.; Desantis, V.; Frassanito, M.A.; Solimando, A.G.; Ribatti, D.; Prete, M.; Harstrick, A.; et al. Targeting angiogenesis in multiple myeloma by the VEGF and HGF blocking DARPin®protein MP0250: A preclinical study. *Oncotarget* **2018**, *9*, 13366–13381. [CrossRef]
192. Chae, S.-S.; Kamoun, W.S.; Farrar, C.T.; Kirkpatrick, N.D.; Niemeyer, E.; de Graaf, A.M.A.; Sorensen, A.G.; Munn, L.L.; Jain, R.K.; Fukumura, D. Angiopoietin-2 Interferes with Anti-VEGFR2-Induced Vessel Normalization and Survival Benefit in Mice Bearing Gliomas. *Clin. Cancer Res.* **2010**, *16*, 3618–3627. [CrossRef]

193. Peterson, T.E.; Kirkpatrick, N.D.; Huang, Y.; Farrar, C.T.; Marijt, K.A.; Kloepper, J.; Datta, M.; Amoozgar, Z.; Seano, G.; Jung, K.; et al. Dual inhibition of Ang-2 and VEGF receptors normalizes tumor vasculature and prolongs survival in glioblastoma by altering macrophages. *Proc. Natl. Acad. Sci. USA* **2016**, *113*, 4470–4475. [CrossRef] [PubMed]
194. Facciabene, A.; Motz, G.T.; Coukos, G. T-regulatory cells: Key players in tumor immune escape and angiogenesis. *Cancer Res.* **2012**, *72*, 2162–2171. [CrossRef] [PubMed]
195. Liu, C.; Workman, C.J.; Vignali, D.A.A. Targeting regulatory T cells in tumors. *FEBS J.* **2016**, *283*, 2731–2748. [CrossRef] [PubMed]
196. Maenhout, S.K.; Thielemans, K.; Aerts, J.L. Location, location, location: Functional and phenotypic heterogeneity between tumor-infiltrating and non-infiltrating myeloid-derived suppressor cells. *Oncoimmunology* **2014**, *3*, e956579. [CrossRef] [PubMed]
197. Palazon, A.; Tyrakis, P.A.; Macias, D.; Veliça, P.; Rundqvist, H.; Fitzpatrick, S.; Vojnovic, N.; Phan, A.T.; Loman, N.; Hedenfalk, I.; et al. An HIF-1α/VEGF-A Axis in Cytotoxic T Cells Regulates Tumor Progression. *Cancer Cell* **2017**, *32*, 669–683.e5. [CrossRef] [PubMed]
198. Jain, R.K. Antiangiogenesis strategies revisited: From starving tumors to alleviating hypoxia. *Cancer Cell* **2014**, *26*, 605–622. [CrossRef] [PubMed]
199. Rolny, C.; Mazzone, M.; Tugues, S.; Laoui, D.; Johansson, I.; Coulon, C.; Squadrito, M.L.; Segura, I.; Li, X.; Knevels, E.; et al. HRG inhibits tumor growth and metastasis by inducing macrophage polarization and vessel normalization through downregulation of PlGF. *Cancer Cell* **2011**, *19*, 31–44. [CrossRef]
200. Chang, A.L.; Miska, J.; Wainwright, D.A.; Dey, M.; Rivetta, C.V.; Yu, D.; Kanojia, D.; Pituch, K.C.; Qiao, J.; Pytel, P.; et al. CCL2 Produced by the Glioma Microenvironment Is Essential for the Recruitment of Regulatory T Cells and Myeloid-Derived Suppressor Cells. *Cancer Res.* **2016**, *76*, 5671–5682. [CrossRef]
201. Hendry, S.A.; Farnsworth, R.H.; Solomon, B.; Achen, M.G.; Stacker, S.A.; Fox, S.B. The Role of the Tumor Vasculature in the Host Immune Response: Implications for Therapeutic Strategies Targeting the Tumor Microenvironment. *Front. Immunol.* **2016**, *7*, 621. [CrossRef]
202. Zhang, Y.; Zhang, Z. The history and advances in cancer immunotherapy: Understanding the characteristics of tumor-infiltrating immune cells and their therapeutic implications. *Cell. Mol. Immunol.* **2020**, *17*, 807–821. [CrossRef]
203. Conley, S.J.; Gheordunescu, E.; Kakarala, P.; Newman, B.; Korkaya, H.; Heath, A.N.; Clouthier, S.G.; Wicha, M.S. Antiangiogenic agents increase breast cancer stem cells via the generation of tumor hypoxia. *Proc. Natl. Acad. Sci. USA* **2012**, *109*, 2784–2789. [CrossRef] [PubMed]
204. Jain, R.K. Normalizing tumor microenvironment to treat cancer: Bench to bedside to biomarkers. *J. Clin. Oncol. Off. J. Am. Soc. Clin. Oncol.* **2013**, *31*, 2205–2218. [CrossRef] [PubMed]
205. Chung, A.S.; Kowanetz, M.; Wu, X.; Zhuang, G.; Ngu, H.; Finkle, D.; Komuves, L.; Peale, F.; Ferrara, N. Differential drug class-specific metastatic effects following treatment with a panel of angiogenesis inhibitors. *J. Pathol.* **2012**, *227*, 404–416. [CrossRef] [PubMed]
206. Colegio, O.R.; Chu, N.-Q.; Szabo, A.L.; Chu, T.; Rhebergen, A.M.; Jairam, V.; Cyrus, N.; Brokowski, C.E.; Eisenbarth, S.C.; Phillips, G.M.; et al. Functional polarization of tumour-associated macrophages by tumour-derived lactic acid. *Nature* **2014**, *513*, 559–563. [CrossRef] [PubMed]
207. Lorgis, V.; Maura, G.; Coppa, G.; Hassani, K.; Taillandier, L.; Chauffert, B.; Apetoh, L.; Ladoire, S.; Ghiringhelli, F. Relation between bevacizumab dose intensity and high-grade glioma survival: A retrospective study in two large cohorts. *J. Neurooncol.* **2012**, *107*, 351–358. [CrossRef] [PubMed]
208. Tolaney, S.M.; Boucher, Y.; Duda, D.G.; Martin, J.D.; Seano, G.; Ancukiewicz, M.; Barry, W.T.; Goel, S.; Lahdenrata, J.; Isakoff, S.J.; et al. Role of vascular density and normalization in response to neoadjuvant bevacizumab and chemotherapy in breast cancer patients. *Proc. Natl. Acad. Sci. USA* **2015**, *112*, 14325–14330. [CrossRef]
209. Gnoni, A.; Licchetta, A.; Memeo, R.; Argentiero, A.; Solimando, A.G.; Longo, V.; Delcuratolo, S.; Brunetti, O. Role of BRAF in Hepatocellular Carcinoma: A Rationale for Future Targeted Cancer Therapies. *Medicina* **2019**, *55*, 754. [CrossRef]
210. Krebs, M.; Solimando, A.G.; Kalogirou, C.; Marquardt, A.; Frank, T.; Sokolakis, I.; Hatzichristodoulou, G.; Kneitz, S.; Bargou, R.; Kübler, H.; et al. miR-221-3p Regulates VEGFR2 Expression in High-Risk Prostate Cancer and Represents an Escape Mechanism from Sunitinib In Vitro. *J. Clin. Med.* **2020**, *9*, 670. [CrossRef]

211. Plebanek, M.P.; Angeloni, N.L.; Vinokour, E.; Li, J.; Henkin, A.; Martinez-Marin, D.; Filleur, S.; Bhowmick, R.; Henkin, J.; Miller, S.D.; et al. Pre-metastatic cancer exosomes induce immune surveillance by patrolling monocytes at the metastatic niche. *Nat. Commun.* **2017**, *8*, 1319. [CrossRef]
212. Lee, W.S.; Yang, H.; Chon, H.J.; Kim, C. Combination of anti-angiogenic therapy and immune checkpoint blockade normalizes vascular-immune crosstalk to potentiate cancer immunity. *Exp. Mol. Med.* **2020**, *52*, 1475–1485. [CrossRef]
213. Makker, V.; Rasco, D.; Vogelzang, N.J.; Brose, M.S.; Cohn, A.L.; Mier, J.; Di Simone, C.; Hyman, D.M.; Stepan, D.E.; Dutcus, C.E.; et al. Lenvatinib plus pembrolizumab in patients with advanced endometrial cancer: An interim analysis of a multicentre, open-label, single-arm, phase 2 trial. *Lancet Oncol.* **2019**, *20*, 711–718. [CrossRef]

Publisher's Note: MDPI stays neutral with regard to jurisdictional claims in published maps and institutional affiliations.

© 2020 by the authors. Licensee MDPI, Basel, Switzerland. This article is an open access article distributed under the terms and conditions of the Creative Commons Attribution (CC BY) license (http://creativecommons.org/licenses/by/4.0/).

Review

Old Player-New Tricks: Non Angiogenic Effects of the VEGF/VEGFR Pathway in Cancer

Panagiotis Ntellas [1,2], Leonidas Mavroeidis [1,2], Stefania Gkoura [1,2], Ioanna Gazouli [1,2], Anna-Lea Amylidi [1,2], Alexandra Papadaki [1,2], George Zarkavelis [1,2], Davide Mauri [1,2], Georgia Karpathiou [3], Evangelos Kolettas [4,5], Anna Batistatou [6] and George Pentheroudakis [1,2,*]

[1] Department of Medical Oncology, University Hospital of Ioannina, 45500 Ioannina, Greece; ntellasp@gmail.com (P.N.); leo.mavroidis@gmail.com (L.M.); s.gkoura@uoi.gr (S.G.); ioannagazouli@gmail.com (I.G.); annalea.ami@gmail.com (A.-L.A.); alexpapadaki@yahoo.gr (A.P.); g.zarkavelis@uoi.gr (G.Z.); dmauri@uoi.gr (D.M.)
[2] Society for Study of Clonal Heterogeneity of Neoplasia (EMEKEN), 45445 Ioannina, Greece
[3] Department of Pathology, University Hospital of St-Etienne, 42055 Saint Etienne, France; georgia.karpathiou@chu-st-etienne.fr
[4] Laboratory of Biology, School of Medicine, Faculty of Health Sciences, University of Ioannina, 45110 Ioannina, Greece; ekoletas@uoi.gr
[5] Biomedical Research Division, Institute of Molecular Biology & Biotechnology, Foundation for Research & Technology, 45115 Ioannina, Greece
[6] Department of Pathology, University Hospital of Ioannina, 45500 Ioannina, Greece; abatista@uoi.gr
* Correspondence: gpenther@uoi.gr; Tel.: +30-26510-99394

Received: 28 September 2020; Accepted: 23 October 2020; Published: 27 October 2020

Simple Summary: Although VEGF-A is well characterized as the principal player of cancer angiogenesis, new data on the interplay with other components of the tumor microenvironment emerge. Here we review the effect of VEGF-A on cancer cells and immune cells as well as investigative and established combinational therapies of anti-angiogenic agents with immune checkpoint inhibitors. We thus elaborate the scientific rationale behind the development of these novel combinational approaches.

Abstract: Angiogenesis has long been considered to facilitate and sustain cancer growth, making the introduction of anti-angiogenic agents that disrupt the vascular endothelial growth factor/receptor (VEGF/VEGFR) pathway an important milestone at the beginning of the 21st century. Originally research on VEGF signaling focused on its survival and mitogenic effects towards endothelial cells, with moderate so far success of anti-angiogenic therapy. However, VEGF can have multiple effects on additional cell types including immune and tumor cells, by directly influencing and promoting tumor cell survival, proliferation and invasion and contributing to an immunosuppressive microenvironment. In this review, we summarize the effects of the VEGF/VEGFR pathway on non-endothelial cells and the resulting implications of anti-angiogenic agents that include direct inhibition of tumor cell growth and immunostimulatory functions. Finally, we present how previously unappreciated studies on VEGF biology, that have demonstrated immunomodulatory properties and tumor regression by disrupting the VEGF/VEGFR pathway, now provide the scientific basis for new combinational treatments of immunotherapy with anti-angiogenic agents.

Keywords: angiogenesis; VEGF; VEGFR; anti-angiogenesis; anti-angiogenic agents; tumor progression; immunosuppression; immunotherapy; immune-checkpoint inhibitors; combination therapy

1. Introduction

Over the last two decades, since the Cancer Genome Atlas Project (TCGA, https://www.cancer.gov/about-nci/organization/ccg/research/structural-genomics/tcga) started, our understanding of cancer biology has grown exponentially and has paved the way for new exciting treatment modalities. Indeed, the advancements in cancer research have brought the development of new anticancer drugs, radiation therapy devices, surgical techniques, diagnostic methods, prognostic and predictive biomarkers and prepared the ground for precision oncology, all of which have contributed to the survival and quality of life of cancer patients [1]. Despite these advances, the burden of cancer in developed societies remains high; malignancies are still the 2nd leading cause of death with 599,108 cancer-related deaths in the US alone, in 2017 [2]. Amongst the most important therapeutics that have driven cancer treatment the last two decades would be the introduction of anti-angiogenic therapy in 2004 and the emergence of immune-checkpoint inhibitors in 2011.

Already from the 1940s [3,4], the release of "blood-vessel growth-stimulating factors" that would have that the ability to induce new vessel growth, was hypothesized to confer a growth advantage to tumors [5]. Following an observation that rapidly growing tumors were heavily vascularized, Folkman et al. [6–8] were the first to isolate a factor from animal tumors that could stimulate angiogenesis and suggested, almost 5 decades ago, that "anti-angiogenesis" could be a strategy to treat cancer [5]. It was not until 1993 that the use of antibodies against the vascular endothelial growth factor (VEGF) in immune-deficient mice successfully suppressed tumor growth [5,9]. The murine anti-VEGF antibody used in the preclinical tumor models was humanized [10] and the recombinant antibody later known as bevacizumab was granted Federal Drug Administration (FDA) approval in 2004 for metastatic colorectal cancer [5]. Bevacizumab is an antibody against vascular endothelial growth factor A (VEGF-A) and following its success several other strategies to inhibit the vascular endothelial growth factor/receptor (VEGF/VEGFR) pathway were devised, namely: receptor tyrosine kinase inhibitors, anti-VEGFR2 antibodies and a VEGF trap (i.e., a soluble VEGF receptor). Bevacizumab also gained expanded approval for several different malignancies including non-small cell lung carcinoma (NSCLC) ovarian and renal cell carcinoma (RCC) [5,11,12]. However, anti-angiogenic therapy in general has managed to offer only modest survival benefits before resistance develops [8], and bevacizumab's benefit was only evident when combined with cytotoxic chemotherapy [13].

No doubt, VEGF is established as an indispensable regulating factor of angiogenesis, contributing to vascular homeostasis, and when dysregulated to disease, with proof of principle anti-VEGF therapy studies demonstrating anti-tumor efficacy by inducing regression of blood vessels [5,13,14]. Despite the moderate so far success of anti-angiogenic therapy, and while VEGF mainly targets endothelial cells, it has been demonstrated that this factor has multiple effects on additional cell types, including immune and tumor cells [13,15,16]; thus, implicating VEGF in diverse molecular pathogenic processes that drive tumor progression, unrelated to the stimulation of angiogenesis [5]. This review focusses on the effects of the VEGF/VEGFR pathway on non-endothelial cells and the resulting unconventional implications of anti- angiogenic agents, other than pruning of new blood vessels.

2. The VEGF/VEGFR Pathway

The process of vessel formation, either through vasculogenesis or angiogenesis, is regulated by numerous receptors that are predominantly expressed on endothelial cells [17,18]. VEGFRs are the most known and well-studied family of endothelial specific receptors, but others include the Tie and Ephrin (Eph) receptor families. While ephrin receptors are mainly involved in arterial-venous specification, VEGF receptors regulate endothelial differentiation and initiation of angiogenesis or vasculogenesis, and Tie receptors control later stages of vessel formation such as stabilization of the endothelial sprout [17,19,20]. In addition, the Notch signaling pathway is critical for the coordination of the multistep process of angiogenesis through specification of the tip or stalk cell phenotype [21].

The family of the VEGF receptors is comprised of the VEGFR1, VEGFR2 and VEGFR3 receptors, which although show similar overall structural organization, still they display differences in their mode of activation, signaling and biological effects [22]. VEGFRs contain multiple tyrosine residues in their cytoplasmic domain and possess intrinsic tyrosine kinase activity [17]. Binding of VEGFs to VEGFRs induces receptor homo- or hetero-dimerization, leading to autophosphorylation of the tyrosine residues. Phosphorylated tyrosines of the VEGFRs' intracellular domains act as binding sites for adaptor molecules, activating downstream signaling pathways [22]. Apart from VEGF-triggered signaling, VGFRs can also undergo non-VEGF-dependent activation, and this VEGFR non-canonical signaling can be induced by binding of non-VEGF ligands or shear-stress-activated cytoplasmic SRC tyrosine kinases [22]. All three VEGFR receptors can trigger cell survival and proliferation, similar to other growth factor receptors; however, they can also provide specific signals that mediate endothelial cell-specific functions for vessel formation [17]. VEGFR1 and 2 are primarily expressed on endothelial cells, while expression of VGFR3 is mainly restricted on lymphatic cells, although VGFR3 is also involved at the first stages of vessel formation in the embryo [17]. VEGFR1 is a high affinity tyrosine kinase receptor for VEGF-A, however, it displays weak ligand-dependent autophosphorylation [5,17,23] and has been suggested to act as a decoy receptor for VEGF-A, preventing it from binding with VEGFR2 [5,17]. VEGFR2 displays weaker VEGF-A binding affinity; however, VEGFR2 has been established as the main signaling receptor for VEGF-A promoting vascular endothelial cell mitogenesis, permeability and cell migration [5,17,24,25].

VEGFR signaling is also modulated by different co-receptors. Specifically, VEGFs as well as VEGFRs bind to co-receptors such as heparan sulphate proteoglycans (HSPGs) and neuropilins (NRPs), such as NRP1 and NRP2 [26]. Oddly, NRP receptors lack intrinsic catalytic activity, but can enhance endothelial cell activity, in response to VEGF signaling [17]. These interactions can influence VEGFR mediated responses, for example, by affecting the half-life of the receptor complex or VEGFR phosphorylation [26,27].

The family of vascular endothelial growth factors (VEGFs) includes VEGF-A, which is the first member described and is usually simply referred to as VEGF, as well as VEGF-B, VEGF-C, VEGF-D and placenta growth factor (PLGF) [15,22]. These structurally-related dimeric proteins are broadly expressed and play a central role in vascular homeostasis by binding to specific receptor tyrosine kinases, most notably VEGFRs [22]. VEGFs also display high affinity to VEGF co-receptors, namely NRP receptors and HSPGs [22]. VEGF-A binds both to VEGFR1 and 2, while VEGF-B and PLGF are selective for VEGFR1 [5,17,28]. VEGF-C and D are primarily ligands of VEGFR3 implicating them in the regulation of lymphangiogenesis, but they can also bind to VEGFR2 after being proteolytically processed [5,17,28,29]. Binding of VEGF-A to VEGFR2 is considered the main signaling event triggering angiogenesis [15], as highlighted by the embryonic lethality of mice lacking expression of either VEGF-A or VEGFR2 [30,31]; justifiably, most attention has been focused on VEGF-A [5]. The VEGF-A gene contains eight exons, that by alternative splicing give rise to different isoforms [32–35]. The VEGFxxx variants, where xxx denotes the number of aminoacids in VEGF-A protein, come up by alternative splicing of the exons 5–7 while alternative splicing of exon 8 give rise to either the VEGFxxxa or VEGFxxxb isoforms with pro-angiogenic and anti-angiogenic properties respectively [36,37] and potential predictive role on anti-VEGF treatment [35]. Indicative isoforms are the $VEGF_{121\,a}$, $VEGF_{121\,b}$, $VEGF_{165\,a}$, $VEGF_{165\,b}$, $VEGF_{189\,a}$, $VEGF_{189\,b}$, $VEGF_{206\,a}$ and $VEGF_{206\,b}$ with different profiles on activity and bioavailability with $VEGF\text{-}A_{165\,a}$ being the most extensively investigated [32–35]. Neo-angiogenesis and vascular permeability constitute the main pathogenic effects mediated by VEGF-A [5].

Although VEGF-A is the principal player that initiates sprouting angiogenesis, new vessel formation would not be possible without the joint action of the Notch signaling pathway [38]. VEGF-A promotes migration of endothelial cells towards a gradient of angiogenic factors in the tumor microenvironment. The leading role is taken by the tip cell that senses the external signals through extension of filopodia and increased expression of VEGFR2 [39]. However VEGF-A also induces the expression of the Notch ligand Delta-Like 4 (DLL4) in the tip cell which sequentially activates Notch signaling in the adjacent endothelial cells [40,41]. The latter results in a decrease of VEGFR2 [42] and increase of VEGFR1 expression in the neighbor cells and acquisition of a stalk cell phenotype [43]. Therefore, it is established an angiogenic front by tip cells that guide the new sprout and a thread of stalk cells that constitute the scaffold of the new vessel [21].

3. Autocrine Effects on Cancer Cells

The release of pro-angiogenic mediators, and of VEGF in particular, has been described for various solid tumors and hematologic malignancies [32,44–51]. Initially research on VEGF signaling focused on its survival and mitogenic effects towards endothelial cells [52,53], with stimulation of angiogenesis being considered the primary mechanism of VEGF mediated cancer progression and metastasis; not surprisingly, since VEGFRs were traditionally regarded to be restricted on the vascular endothelium [54]. However, over the years, expression of VEGFRs has been described on several types of non-endothelial cells, including cancer cells [55–57]. It is now conceivable that tumor-derived VEGF not only provides paracrine signaling for endothelial cells, but may also directly stimulate tumor growth in an autocrine manner [54,58]. Therefore, VEGF-blockade may act on multiple levels: antiangiogenic effects on the tumor vasculature and antineoplastic effects on the tumor cell population [16]. While these antineoplastic effects can be easier assessed in tumor cell lines, further investigation is warranted for their clinical relevance (Figure 1 and Table 1).

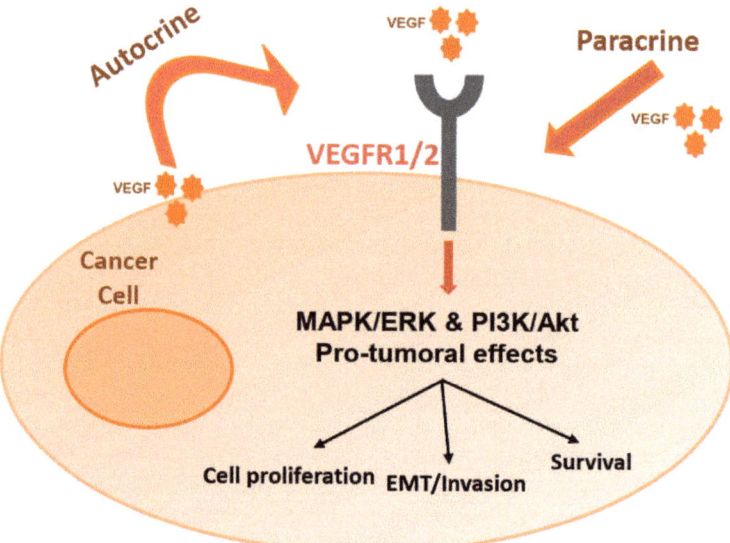

Figure 1. Non-angiogenic effects of the VEGF–VEGFR interaction in cancer cells. VEGF-mediated autocrine-paracrine loops directly influence and promote tumor cell survival, proliferation and invasion. VEGF(R): vascular endothelial growth factor (receptor); EMT: epithelial-mesenchymal transition; MAPK/ERK: Mitogen-activated protein kinase/extracellular signal-regulated kinase; PI3K/Akt: Phosphoinositide 3-kinase/Akt

Table 1. Effects of VEGF/VEGFR autocrine-paracrine signaling in cancer cells.

Cancer Type	Effects
Melanoma (VEGFR1/2; NRP1/2)	Enhances the proliferation of melanoma cells [59]. Mitigates melanoma cells migration (through a NRP1/VEGFR2-mediated response) [60].
Pancreatic (VEGFR1/2; NRP1)	Was shown to activate the MAPK/ERK pathway [44,61]. Stimulates cancer cell growth [44]. Promotes cancer cell migration and invasion, without affecting proliferation (VEGFR1-mediated effect) [61]. Promotes pancreatic cancer aggressiveness by TGFβ1-induced fibrosis and endothelial-to-mesenchymal transition (NRP1-mediated effect) [62].
NSCLC (VEGFR1/2; NRP1)	Induces PI3K/Akt and MAPK/ERK activation [63]. Stimulates tumor growth and proliferation of NRP1-expressing cells (VEGFR2/NRP1-mediated effect) [63].
SCLC (VEGFR2/3)	Promotes VEGFR2/3 activation resulting MAPK/ERK phosphorylation [64]. Induces cancer cell proliferation [64].
Colorectal (VEGFR1)	Promotes Akt and ERK phosphorylation [65]. Enhances survival and resistance to chemotherapy of cancer cells [65]. Was shown to enhance cellular migration and promote tumor progression and metastasis [66,67]. Was found to support the survival of cancer cells undergoing EMT [68,69].
Gastric (VEGFR1/2)	Stimulates tumor growth (VEGFR2-mediated response) [57,70].
Prostate (VEGFR1/2)	Was shown to enhance prostate cancer cells proliferation (VEGFR2-mediated effect) [71,72].
Glioblastoma (VEGFR1/2; NRP1)	Promotes MAPK/ERK, PI3K/Akt and PLC/PKC pathways activation [73,74]. Stimulates proliferation of glioma cells (VEGFR2-mediated response) [75]. Supports tumor growth (VEGFR1/2-mediated effect) [73].
Breast cancer (VEGFR1/2; NRP1)	Induces activation of the MAPK/ERK and PI3K/Akt pathways [76]. Supports tumor cells survival, stimulates their proliferation and contributes to mammary tumor growth [77–81]. Induces invasion and chemotaxis of breast cancer cells and enhances EMT [79–82]. Inhibits apoptosis and protects from chemotherapy [79,83]. Confers cancer stem cells traits in breast cancer cells and was found to drive cancer stem cells self-renewal [84,85].
Head & Neck (VEGFR2)	Regulates proliferation and invasion of head & neck cancer cells [86].
Bladder (VEGFR1/2)	Enhances survival and proliferation of bladder cancer cells (VEGFR2-mediated effect) [45,87].
Rhabdomyosarcoma (VEGFR1/2)	Increases cancer cell proliferation (VEGFR1-mediated effect) [52].
Ovarian (VEGFR2)	VEGFR2-phosphorylation has been corelated with ovarian cancer cell survival and proliferation [58].
Multiple Myeloma (VEGFR1)	Mediates activation of the MAPK/ERK, PI3 k/PKC and McL1/survivin pathways resulting in increased proliferation, migration and survival [88–92].

Abbreviations: EMT: epithelial–mesenchymal transition; VEGF(R): vascular endothelial growth factor (receptor); NRP: neuropilin; MAPK: Mitogen-activated protein kinase; ERK: extracellular signal-regulated kinase; TGFβ1: Transforming growth factor beta 1; PI3K: Phosphoinositide 3-kinase; Akt: Protein kinase B; PLC: phospholipase C; PKC: Protein kinase C; McL1: myeloid cell leukemia 1.

3.1. Melanoma

The hypothesis that the VEGF/VEGFR pathway would play an autocrine role in tumor progression began from the observation that many malignant cells co-express VEGF and its receptors [59,93]. In 1995, VGEFR2 was detected on three melanoma cell lines (i.e., MeWo, A375-metastatic, A375-wt), that were also known to co-express VEGF; intriguingly exogenous administration of VEGF increased the proliferation of A375 M melanoma cells in vitro [59]. Another study confirmed the expression of VEGFR2, VEGFR1, NRP1 (Neuropilin1), NRP2 (Neuropilin2) and production of $VEGF_{121}$, $VEGF_{165}$, $VEGF_{189}$ and PLGF in melanoma cell lines derived from primary or metastatic tumors (i.e., GR-Mel, ST-Mel, SN-Mel, PR-Mel, CN-Mel, TVMBO, SK-Mel-28, WM115, WM266-4, 13443-Mel, PDMel, PNP-Mel, PNM-Mel, LCP-Mel, LCM-Mel, GL-Mel, M14, LB-24, 397-Mel). Exposure of the VEGFR-expressing melanoma cells to $VEGF_{165}$ and PLGF-1 resulted on a proliferative response, while M14 cells lacking VEGFR1 and 2 were unresponsive [93]. In addition, stimulation of melanoma cells was inhibited by neutralizing anti-VEGF antibodies and was completely abolished with anti-PLGF antibodies, confirming the specificity of the response [93]. Likewise, NRP1 was reported to mitigate migration of melanoma cells (i.e., M14, GR-Mel) though VEGF-A-induced activation of VEGFR2 [60], or independently in response to PLGF, even in the absence of its high affinity receptor, VEFGR1 [94].

3.2. Pancreatic Cancer

Expression of VEGFRs has also been demonstrated in pancreatic cancer [44,61]. Analysis of pancreatic cancer tissues revealed concomitant over-expression of VEGF and of its high affinity receptors in 33% of pancreatic cancer patients [44]. VEGFR2 expression was observed in three pancreatic cell lines (AsPC-1, Capan-1 and MIAPaCa-2) and VEGFR1 mRNA was detected in four pancreatic cancer cell lines (AsPC-1, Capan-1, T3 M4 and PANC-1). Furthermore, radiolabeled VEGF was detected to bound to Capan-1 pancreatic cancer cells, which also exhibited enhanced MAPK activation and growth upon VEGF stimulation, demonstrating evidence of a VEGF/VEGFR2 autocrine signaling [44]. In another study, VEGFR1, but not VEGFR2, appeared to be ubiquitously expressed in pancreatic carcinoma cell lines (i.e., AsPC-1, BxPC3, CFPAC, HPAF2, MiaPaCa2, Panc-1, HS7665, Panc-48, L3.6 pl, FG) with concomitant expression of its ligands (i.e., VEGF-A and VEGF-B) [61]. Further analysis of the L3.6 p1 and Panc-1 cells revealed that both VEGF-A and VEGF-B induced ERK1/2 phosphorylation mediated through VEGFR1, as none of the cell lines examined were found to express VEGFR2. Use of a neutralizing antibody to VEGFR1 confirmed that the signaling was VEGFR1-dependent [61]. Further analysis demonstrated that migration and invasion of pancreatic cancer cells was promoted upon VEGFR1 stimulation, on the contrary, no effect was observed on cell proliferation [61]. NRP1 was also reported to contribute to pancreatic cancer aggressiveness by promoting transforming growth factor beta 1 (TGFβ1)-induced fibrosis and endothelial-to-mesenchymal cell transition, a process that serves as an important source of fibroblasts [62].

3.3. Lung Cancer

VEGFR-2, VEGFR-3, VEGF-A and VEGF-C have also been detected in small cell lung cancer (SCLC) cell lines (i.e., NCI-H82, H209, H510, H526 and H660) [64]. Stimulation by VEGF-A and VEGF-D induced phosphorylation of VEGFR2 and VEGFR3 respectively, as well as increased ERK1/2 phosphorylation and proliferation of these cells [64]. $VEGF_{165}$, as well as VEGFR1, VEGFR2, NRP1 and NRP2 were detected in several NSCLC cell lines examined (i.e., H460, H647, A549, SKMES1). VEGF-A was found to stimulate the proliferation of NRP1-expressing cells in the presence of VEGFR2 [63]. In addition, phosphorylation of the PI3K-mediator Akt and, to a lesser extent, of the MAPK's signaling proteins ERK1/2, was demonstrated in A549 and SKMES1 NSCLC cell lines treated with VEGF-A [63].

3.4. Gastrointestinal Cancer

Among the tumors of epithelial origin, expression of VEGFRs has been observed on those arising from the colon [66]. VEGFR1 has been detected in a series of colon cancer cell lines (i.e., HT29, SW480, SW620, ATCC, KM12 L4, KM12 SMLM2, GEO, RKO) [67], with evidence suggesting that the receptor is involved in processes that promote tumor progression and metastasis [66,67]. Likewise, upregulation of both VEGF and VEGFR1, but not of VEGFR2, has been detected in LIM1863 colon cancer cells undergoing an epithelial-to-mesenchymal cell transition (EMT). Importantly, VEGF/VEGFR1 autocrine interaction appeared to be necessary for the survival of the LIM1863 colon carcinoma cells after the induction of EMT [68,69]. Furthermore, RNAi-mediated depletion of VEGF decreased cell survival and enhanced sensitivity to chemotherapy of colorectal cancer cells (i.e., HCT116, SW480, HT29, HCP-1) by disrupting AKT and ERK1/2 signaling; notably, ribonucleic acid interference (RNAi)-mediated depletion of VEGFR1 replicated the effects of VEGF depletion on phospho-AKT and phospho-ERK1/2 levels [65]. Moreover, VEGF-A and VEGFR1/2 are widely expressed in gastric carcinoma cells (i.e., RF-1, RF-48, AGS-1, NCI-N87, NCI-SNU-1, NCI-SNU-5, NCI-SNU-16, KATO-III). Tumor growth was found to be enhanced in VEGFR2-positive cells after VEGF-A stimulation, but not in gastric adenocarcinoma cells expressing only VEGFR1 [57,70].

3.5. Prostate Cancer

VEGFR1 and 2 expression, with concomitant VEGF production has been observed in prostate cancer cells (i.e., LNCaP, PC3, DU145) [71,95], as well as in prostate cancer tissue specimens [71,72]. Malignant cells particularly displayed greater receptor expression compared to normal basal prostate cells [71]. Furthermore, the LNCaP prostate cancer cell line demonstrated 50% enhanced proliferation in the presence of VEGF$_{165}$, an effect that was abolished by a neutralizing antibody to VEGFR2, suggesting that the survival signals from VEGF are mediated specifically via VEGFR2 [71].

3.6. Gliomas

While several studies report that World Health Organization (WHO) grade IV gliomas (i.e., glioblastomas) secrete high levels of VEGFs, expression of VEGFRs on grade IV glioma cells (i.e., U118, U138, U343, U87) and primary glioblastoma cell lines has been mostly found to be weak [73,96,97]. Despite a low VEGFR2 and no VEGFR1 expression, drugs targeting the VEGF pathway demonstrated biological effects on cell proliferation, morphology and metabolism in the U87 glioma grade IV cell line [97]. Furthermore, higher VEGFR1 or VEGFR2 mRNA expression levels, in grade II, III and IV glioma patients, have been corelated with higher tumor grade and worse prognosis [73]. Additionally, activation of MAPK/ERK, PI3K/Akt and PLC/PKC pathways was found to be induced by VEGF through VEGFR2 and VEGFR1 signaling in a panel of grade III/IV glioma cell lines [73,74]. Likewise, in vivo studies indicate that VEGFR1 and VEGFR2 signaling support survival of orthotopic glioma bearing mice [73]. Furthermore, proliferation of glioblastoma stem-like cells was shown to be stimulated via VEGFR2 by exogenous VEGF in a dose-dependent matter, but not via VEGFR1. On the contrary VEGFR1 seemed to have a negative feedback effect on VEGFR2 when cells were exposed to higher concentrations of VEGF [75].

3.7. Breast Cancer

Production of VEGF and expression of VEGFR1 and 2 has been described in breast cancer tissues [98] and in several primary breast cancer cell lines [32], with in vitro studies demonstrating that ^{125}I-labeled VEGF can bind to T-47 D cells and by doing so to induce activation of the MAPK/ERK and PI3K/Akt pathways [76]. In addition, data from a transgenic mouse model with human VEGF$_{165}$ targeted to mammary epithelial cells, indicated that VEGF-A contributes to mammary tumor growth, not only through increased neovascularization, but also by stimulating the proliferation of tumor cells in an autocrine manner, and by inhibiting their apoptosis [77]. Specifically, expression of VEGFR1 has been reported in a panel of breast cancer cell lines (i.e., DU4475, MCF-7, T-47 D, SK-BR-3, MDA-MB-157, MDA-MB-175, MDA-MB-231, MDA-MB-435, MDA-MB-468, AU565, BT-474, BT-483, HCC38, UACC-812, ZR-75–1), followed by the observation that tumor cell growth is supported by selective VEGFR1 signaling and it is mediated by downstream activation of MAPK/ERK and PI3K/Akt pathways [99]. VEGFR2 expression has also been established in breast cancer specimens [98,100], along with concomitant VEGF expression [98,101]. Moreover, in a series of 142 invasive breast carcinomas, 64.5% of them tested positive for VEGFR2 expression and were also associated with the expression of Ki67 and topoisomerase-IIa proliferation indexes suggesting that VEGF may act as a growth factor via VEGFR2 in these cancer cells [102]. VEGFR2 phosphorylation in several breast cancer cell lines (i.e., MDA-MB-468, T47 d, MCF-7, HBL-100 and in a primary breast cancer culture) was enhanced by VEGF-A stimulation leading to activation of ERK1/2 and Akt pathways, indicating that the VEGFR/VEGF-A pathway might play crucial role in the regulation of survival and proliferation of breast cancer cells [78]. VEGF-A was also reported to drive self-renewal of breast and lung cancer stem cells by stimulating the VEGFR2/Stat3 signaling and inducing *Myc* and *Sox2* expression [84]. Likewise, the VEGF-A/NRP1 axis was suggested to confer cancer stem cell traits in breast cancer cells (i.e., MCF-7, MDA-MB-231) by activating the Wnt/β-catenin pathway [85]. In addition, the VEGF-A/NRP1 axis was associated with breast cancer progression by enhancing the

EMT process and NF-κB (nuclear factor kappa-light-chain-enhancer of activated B cells) and β-catenin signaling [82], with further evidence to support that neuropilin might also protect MDA-MB-231 breast cancer cells from apoptosis by autocrine stimulation of the PI3K-pathway in response to VEGF$_{165}$ [79]. Likewise, NRP1 gene silencing was reported to suppress the proliferation, promote apoptosis and increase the sensitivity of breast cancer cells (i.e., MCF-7, SK-BR-3) to chemotherapy [83].

3.8. Hematologic Malignancies

VEGF expression has been observed in hematologic malignancies [47], with evidence to suggest that VEGF triggers growth, survival and migration of leukemia and multiple myeloma (MM) cells [88,103,104]. The VEGF/VEGFR-induced activation of intracellular tyrosine kinase cascades in MM has been described since 2001 [89]. Specifically, the VEGF/VEGFR-triggered MAPK/ERK pathway was found to mediate MM cell proliferation, while the PI3 k/PKC–dependent cascade was associated with migration and the myeloid cell leukemia 1 (McL1)/survivin with survival [88,89]. VEGFR1 was found to be more widely expressed in MM cells compared to VEGFR2 [88,90,91]. Likewise, stromal derived VEGF-A was shown to induce VEGFR1-dependent proliferation of primary MM cells, while in vitro inhibition of MM cell lines (i.e., RPMI 8226, U266, ARP1, ARK) by bevacizumab resulted in a reduction of proliferation [92]. Recently, the junctional adhesion molecule-A (JAM-A) has emerged as a crucial mediator between MM plasma and medullary endothelial cells, and has been associated with poor prognosis of MM patients due to its role in invasion and metastasis [105,106]; while limited so far, evidence suggests that JAM-A could also interfere with the VEGF/VEGFR pathway [107]. Similarly, VEGF induced phosphorylation of VEGFR2 expressing leukemia cells (i.e., HL-60, HEL and primary leukemia cell lines), resulting in increased proliferation [51]. VEGF may also facilitate survival of leukemia cells by up-regulation of heat shock protein 90 (Hsp90), which deactivates significant pro-apoptotic molecules [89], and was also corelated with increased expression of the anti-apoptotic MCL-1 gene in B- chronic lymphocytic leukemia patients [108].

3.9. Other

Several other reports on a variety of additional malignancies suggest that VEGF may act in an autocrine loop fashion in cancer cells. For example, in head and neck (H&N) cancer, where VEGFR2 was detected in 109 H&N squamous cell tumors, with evidence to suggest that the receptor might regulate proliferation and invasion of H&N cancer cells (i.e., Hep2) [86]. VEGF-A, VEGFR1 and 2 expression is also present in bladder cancer, with VEGFR2 found particularly prominent in muscle invasive bladder cancer specimens [87]. Additionally, several bladder cancer cell lines exhibit VEGFR expression, with T24 cells displaying enhanced survival and proliferation, mediated by VEGFR2 in response to VEGF signaling [45]. Furthermore, expression of both VEGFR1 and VEGFR2 has been detected on multiple rhabdomyosarcoma cell lines (i.e., RH4, RH6, RH18, RH28, RD), with the VEGFR1-positive cell lines demonstrating increased proliferation upon VEGF$_{165}$ stimulation, while proliferation was halted after applying a blocking antibody against VEGFR1 [52]. VEGF-A has also been detected in the ovaries, both in normal and cancer tissues, and found to be secreted in malignant ascites, with epithelial cancer cells being identified as the source of VEGF-A [46,48,49]. VEGFR2 displayed a more prominent expression in ovarian cancer specimens and cell lines (A2774, SKOV3 ip1, HeyA8) as compared to normal ovarian samples where little to none VEGFR2 is detected [58]. VEGFR2 was also found phosphorylated in ovarian cancer cells and has been correlated with their proliferation and survival [58]. VEGFR1 on the contrary was largely absent [58].

3.10. VEGF Signaling on Cancer Cells: Stimulation of Survival and Migration

The signaling pathways activated by VEGF have been well characterized in endothelial cells [26,44]. VEGF-induced phosphorylation of VEGFRs is followed by downstream activation of MAPK/ERK, PI3K/Akt, PLC/PKC and other signaling pathways [26,44,74,79,109]. The activation of these pathways, brought by autocrine VEGF signaling and subsequent VEGFR dimerization, has also been observed in a variety of malignancies, promoting survival, proliferation and invasion of cancer cells [44,63,64,66,74,79,86,99]. Hypoxia can further provide these cancer cells with a survival and growth advantage by inducing the expression of VEGFs and VEGFRs [26,45,64,79]. Apart from the classical VEGF receptors, studies on neuropilin have highlighted its role as a critical co-receptor that facilitates VEGF signaling [79,110]. Indeed, NRP1 and NRP2 expression has been observed in cancer cells demonstrating a functional role [63,79,80,110]. Autocrine VEGF/VEGFR signaling was found to be enhanced by interaction with NRP1 in glioblastoma multiforme [54], while in NSCLC NRP1 overexpressing tumor cells exhibited significantly increased tumor growth [63]. In breast cancer, binding of VEGF to neuropilin enhanced cancer cell survival with additional evidence showing that NRP1 supports VEGF autocrine invasive function and chemotaxis of breast cancer cells [79–81].

EMT, the process by which epithelial cells can acquire mesenchymal features, has emerged as an integral process of cancer progression [111,112]. In addition, endothelial cells undergo a phenotypic switching, known as endothelial-to-mesenchymal cell transition (EndMT), which is essential during angiogenesis [113], with several EMT markers being associated with a pro-angiogenic phenotype [111]. In cancer, the process of EndMT produces cells with fibroblast-like properties which serve as cancer-associated fibroblasts facilitating tumor progression [114,115]. Furthermore, the crosstalk between VEGF and Notch pathways has been established to promote EndMT in endothelial cells of tumors [116], while the addition of VEGF was shown to induce EMT in A549 lung cancer cells [117] and elicit the appearance of EMT markers in pre-invasive prostate cancer cells [118]. These findings show the interdependent nature of angiogenesis, EndMT and EMT in promoting carcinogenesis [111,113]. Likewise, along with inducing tumor growth via an autocrine mechanism, evidence suggests that VEGFR expression in tumor cells also promotes their migration and induces EMT [51,100]. In breast cancer, the expression of EMT markers, including Twist1 and vimentin, was higher in tumors with greater VEGFR2 expression, while E-cadherin expression was lower in the same tumors [100]. Furthermore, VEGF signaling in breast cancer cells was found to promote changes stimulating their invasion [76]. Indeed, VEGF singling induced the expression of the CXCR4 chemokine receptor in breast cancer cells by employing the NRP1 receptor. This demonstrated that the VEGF pathway can direct the migration of cancer cells towards specific chemokines and promote breast carcinoma invasion, while no evidence was shown to suggest that this particular pathway would enhance the survival of these cancer cells [80]. In colorectal carcinoma, VEGF stimulation resulted in enhanced cell migration linked to the activation of focal adhesion components that regulate this process. Cell migration was effectively blocked by pharmacologic inhibition of VEGFR1 or Src kinase, suggesting that VEGFR1 promotes migration of tumor cells through a Src-dependent pathway [66]. In addition, metastatic colon cancer cells were found to be dependent on VEGFR1 signaling for their survival [68,69]. VEGFR1 activation by VEGF-A or VEGF-B was also found to promote migration and invasion of pancreatic carcinoma cell lines without appearing to enhance cancer cell proliferation [61]. Likewise, invasion and metastasis of pancreatic neuroendocrine tumors was suppressed with simultaneous inhibition of c-MET and VEGF signaling [119].

Nonetheless, solid conclusions on the role of the VEGF/VEGFR pathway in promoting autocrine stimulation of tumor cell migration and invasion, are difficult to be drawn, and are perhaps cell- and context-dependent, since contrary to the above, VEGF was demonstrated to negatively regulate tumor cell invasion and mesenchymal cell transition through a MET/VEGFR2 complex in glioblastoma mouse models [120]. Moreover, despite evidence suggesting that NRP1 is implicated in breast carcinoma invasion, NRP1 expression on prostate cells was strongly and negatively correlated with the ability of these cell lines to invade and migrate [95].

4. Immunomodulatory Effects of the VEGF/VEGFR Pathway

In addition to its various roles in angiogenesis and direct stimulation of tumor cells' survival, proliferation and invasion, VEGF can also have immunosuppressive effects [15,121]. Over the last several years, cancer immunotherapy has emerged as a major therapeutic modality, revolutionizing medical oncology [5,13,15]. Its success relies on the recruitment, expansion and effective anticancer activity of immune effector cells within the tumor microenvironment (TME) [13]. Despite rapidly transforming anticancer treatment and providing with durable responses, many patients do not derive benefit from this approach [5,13]. Human cancer cells can employ multiple immune inhibitory mechanisms, resulting to immune escape and likely explaining the lack of response observed in several cases [5]. One such mechanism relates to VEGF, hence combination with anti-angiogenic agents, is one of the many strategies currently under investigation to improve the response rates and duration of immunotherapies [5,15,122,123] (Table 2).

Table 2. Immunomodulatory effects of selected anti-angiogenic factors.

Anti-Angiogenic Agent		Functions
VEGF-A antibody	Bevacizumab	Decreases MDSCs and Tregs accumulation [124,125]. Enhances CTLs responses: It was shown to (a) increases the peripheral B- and T-cell compartments [126], (b) correlate with an increase in activated (CD8+ CD62 L+) CTLs, long-term effector memory (CD8+ CD27+) and central-memory (CD8+ C45 RA-CCR7+) CTLs [127,128] and (c) enhance antigen-specific T-cell migration [129] Improves DCs maturation and activation: It was shown to increase the percentage of activated and mature myeloid derived DC [127,130], and to reverse the VEGF inhibitory effects on DCs [131]. Induces vessel normalization, increases tumor vascular expression of ICAM1 and VCAM1 and T-cell tumor infiltration [132–135].
VEGFR1–3, PDGFR, c-KIT, FLT-3, CSF-1 R and RET mtTKI	Sunitinib	Enhances the Th1 immune response and inhibits the immunosuppressive Th2 response [136,137]. Decreases MDSCs and tumor Tregs compartments [136–140]. Induces endothelial activation and T-cell recruitment, by enhancing the expression of chemokines and adhesion molecules on tumor endothelial cells, resulting in a higher number of CD3+ T-cells in the tumor [141,142]. Enhances the percentage and number of intratumoral CD4 and CD8 T-cells and decreases the expression of inhibitory molecules (i.e., CTLA-4 and PD-1) on TILs [141,143].
VEGFR1–3, PDGFR and c-KIT mtTKI	Axitinib	Enhances the CD8+ T cells compartment [144]. Increases the antigen-presenting function of intratumoral DCs [145]. Reduces MDSCs levels [144] and inhibits their suppressive capacity [145].
VEFGR2 TKI	Apatinib	Increases the infiltration of CD8+ T cells and reduces the recruitment of TAMs [146]. Reduces the expression levels of inhibitory checkpoint molecules, such as Lag-3, PD-1 and Tim3 in CD8+ T cells [147]. Enhances the production of IFN-γ and IL-2 and promote the cytotoxicity of T cells [147].
Raf, VEGFR2, PDGFR, FLT3, RET and c-KIT mtTKI	Sorafenib	Reverses immunosuppression: It decreases MDSCs levels [148], Tregs and Th2-cells [149], and inhibits Tregs functions [150]. Upregulates tumor-specific effector T-cells functions [150] and induces Th1 dominance [149]. Reverses the VEGF inhibitory effects on DCs [131], but was also shown to inhibit the function of DCs [151] and inhibit the induction of antigen-specific T cells [151].

Abbreviations: (mt)TKI: (multi-target) tyrosine kinase inhibitor; PDGFR: platelet derived growth factor receptor; VEGF(R): vascular endothelial growth factor (receptor); FLT3: Fms-like tyrosine kinase-3; CSF-1 R: colony stimulating factor receptor; RET: glial cell-line derived neurotrophic factor receptor; DCs: dendritic cells; ICAM1: intercellular adhesion molecule-1; VCAM1: vascular cell adhesion molecule-1; Th(1/2): T helper cell (1/2); Lag3: lymphocyte activation gene 3 protein; Tim3: T-cell immunoglobulin mucin receptor 3; PD-1: programmed cell death protein 1; CTLs: cytotoxic T-lymphocytes; Tregs: T-regulatory cells; MDSCs: myeloid derived suppressor cells; TAMs; tumor associated macrophages.

4.1. Immune Cell Infiltration

Infiltration of tumors by immune cells is a multistep process involving trafficking of immune cells to the tumor blood vessels, adhesion to the endothelium and ultimately crossing the endothelial barriers into the TME [13,15]. Extravasation into the tumor tissue is dependent upon interactions with adhesion molecules expressed on the immune cells themselves and the luminal surface of the tumors' endothelial lining such as E-cadherin, intercellular adhesion molecule-1 (ICAM-1) and vascular cell adhesion molecule-1 (VCAM-1) [13,15,121]. VEGF is suggested to impair interactions between leukocytes and endothelial cells by downregulating the expression of these adhesion molecules or inhibiting their clustering [13,121,152–154]. Indeed, sunitinib treatment resulted in upregulated expression of ICAM-1 and VCAM-1 adhesion molecules on endothelial cells of tumor bearing mice [141], with several studies reporting that infiltration of TILs is markedly increased in animal tumor models and in humans after VEGF inhibition [5,129,132,141,155]. Furthermore, infiltration of immune cells into the TME is further hindered by the structurally and functionally abnormal tumor vessels [13]. It is suggested that judicious doses of anti-angiogenic agents have the potential to improve the effectiveness of immunotherapy by transiently restoring the abnormal tumor vasculature and thus increasing the infiltration of immune effector cells into the TME [13,14,156–158]. Likewise anti-angiogenic agents were found to induce the formation of high endothelial venules (HEVs) that further promote lymphocyte infiltration [159].

4.2. Effector T-cells

VEGFR expression has been detected on T-cells, with several reports suggesting that VEGF signaling can directly affect T-cells' development, homing and cytotoxic functions [15,121]. The activation of the MAPK/ERK and PI3K/Akt pathways after VEGF stimulation on CD4+CD45RO+ memory T-cells that express VEGFR1 and VEGFR2 provides evidence of a functional VEGF/VEGFR interaction [160]. Contrary to its suppressive role, VEGF induced, via VEGFR2, production of pro-inflammatory molecules, such as INF-γ and IL-2 and stimulated migratory responses in these memory CD4+ T-cells [160]. Nonetheless, mounting evidence support the suppressive effects of VEGF on effector T-cells [121,161]. Specifically, Ohm et al. [162] reported that VEGF can impede with the differentiation of hematopoietic progenitor cells in the thymus into CD8+ and CD4+ T-cells [15,162]. Furthermore, CD3+ T-cells' proliferation and cytotoxic effects were directly suppressed by VEGF upon its binding to VEGFR2 expressed on the activated effector T-cells' surface [163,164]. VEGF-A also contributes to CD8+ T-cells exhaustion, in a VEGFR2 and NFAT (nuclear factor of activated T-cells) dependent manner, by promoting the expression of checkpoint molecules such as programmed cell death protein 1 (PD-1), cytotoxic T-lymphocyte-associated protein 4 (CTLA-4), T-cell immunoglobulin mucin receptor 3 (TIM3) and lymphocyte activation gene 3 protein (LAG3); thus, resulting in the development of an immunosuppressive microenvironment that could be reverted upon VEGF-A/VEGFR inhibition by anti-angiogenic agents [165]. VEGF can also indirectly suppress effector T-cells functions by inducing Fas-Ligand expression on endothelial cells, resulting in a selective barrier that causes apoptosis of infiltrating CD8+ T-cells, but not of Tregs [166]. Interfering with the VEGF/VEGFR interaction on T-cells has shown promising results in enhancing anti-tumor immunity [121]. Notably, patients with metastatic colorectal cancer displayed increased B- and T-cell compartments after treatment with bevacizumab [126]. Decreased levels of pro-angiogenic mediators and inflammatory cytokines were also observed after addition of bevacizumab to concomitant chemotherapy in patients with NSCLC, resulting to improved DC activation and T-cell cytotoxicity [127]. Likewise, sunitinib, a multi-tyrosine kinase inhibitor, displayed increased Th1 responses by reducing the expression of inhibitory molecules including TGFβ, IL-10, Foxp3, PD-1 and CTLA4 [143].

4.3. Regulatory T-cells (Tregs)

Contrary to its inhibitory effects on effector T-cells, VEGF signaling seems to play a role in inducing and/or maintaining Foxp3+ regulatory T-cell populations (Tregs) in patients with cancer [167]. Regulatory T-cells exert immunosuppressive effects on effector T-cells [121,168], with evidence suggesting that VEGF induces Tregs proliferation through VEGFR2 activation [124]. Similarly, interaction of VEGF with NRP1 expressed on Tregs was found critical for tumor homing, since by abolishing NRP1 expression Tregs populations were reduced, resulting in CD8+ T-cells raise in melanoma mouse models [169]. Treatment with anti-angiogenic agents is considered to reverse VEGF induced promotion of Tregs; as expected, bevacizumab demonstrated inhibition of Treg accumulation in peripheral blood of patients with metastatic colorectal cancer [124]. Likewise, a decrease in regulatory T-cell numbers was evident after sunitinib treatment in tumor bearing mice and in patients with metastatic renal cancer [124,136,138,139]. It is therefore reasonable that anti-angiogenic agents are expected to modulate anti-tumor immunity by interfering with inhibitory Tregs [121].

4.4. Dendritic Cells (DCs)

One of the first described immunosuppressive functions of VEGF would be hindering dendritic cell (DC) maturation [15,170]. This is evident by the defective or reduced numbers of mature DCs reported in several malignancies to be inversely corelated with VEGF plasma concentrations [171–174]. Inhibition of NF-κB signaling is suggested to be the underlying mechanism that impairs DCs' differentiation and maturation, with various studies indicating this to be a direct consequence of VEGF binding to either VEGFR2 or VEGFR1 on DCs [15,170,175–177]; although NRP1 is also implicated [178], as well as PLGF binding to VEGFR1 [177,179]. Along with directly affecting DCs' maturation, VEGF was also found to upregulate PDL1 on DCs, resulting in inhibition of T-cells' expansion and function [180]. DCs are antigen-presenting cells, integral for a successful immune response, and thus targeting factors, such as VEGF, that interfere with DCs' differentiation, maturation and activation is a reasonable therapeutic strategy. Bevacizumab has shown promising results in reversing the VEGF-induced inhibition of differentiation of monocytes into DCs in vitro [131], as well as in restoring peripheral blood DC numbers in cancer patients and promoting their activation [127,130]. Sorafenib and sunitinib, two multi-kinase inhibitors, have also shown effects on DCs, although discrepancies lie among different studies making their exact role debatable and perhaps context-dependent [121,131,151].

4.5. Myeloid Derived Suppressor Cells (MDSCs)

Myeloid derived suppressor cells (MDSCs) form a heterogenous group of myeloid origin immune cells, that are frequently present in pathologic conditions characterized by chronic inflammation. Increased intratumoral VEGF concentration has been corelated with the presence of MDSCs [181], with several studies indicating that VEGF can promote the accumulation of MDSCs in tumors and peripheral blood of cancer patients, via VEGFR2-STAT3 activation, but not VEGFR1 [139,181,182]. MDSCs are known for their immunosuppressive properties [183] that stem from their ability to inhibit T-cell proliferation and activation, and when activated by VEGF, MDSCs could also stimulate the development of other immunosuppressive cells including Tregs [121,184–186]. Angiogenetic agents like sunitinib, axitinib, sorafenib and bevacizumab have demonstrated ability to constrain the MDSC compartment and reduce their suppressive capacity resulting in a more favorable microenvironment [125,139,140,143,148,187].

4.6. Tumor Associated Macrophages (TAMs)

Macrophages are important cells of the innate immunity and play a central role in inflammation [188]; however, macrophages that are present in the TME in high numbers, also known as tumor associated macrophages (TAMs), are suggested to display a tumor promoting phenotype [189,190]. VEGF and most likely PLGF are reported to act as chemoattractants for monocytes via activation of VEGFR1 [15,191]. VEGF-A could thus recruit macrophages to tumors with high VEGF expression and contribute to tumor growth by establishing an immunosuppressive microenvironment [190,192]. In addition to their immunosuppressive functions TAMs are implicated in the development of resistance to anti-VEGF agents [193]. Reducing the recruitment of TAMs or reprogramming M2-like TAMs towards an anticancer M1 phenotype [13] seems a reasonable strategy to reverse immunosuppression, as well as to deal with anti-VEGF resistance, especially in glioblastoma were increased TAMs have been correlated with poor prognosis and disease progress on bevacizumab [194–196].

4.7. Combinations of VEGF/VEGFR Inhibition with Cancer Immunotherapy

Cumulative evidence provides the rationale that anti-angiogenic treatment might augment the efficacy of immunotherapy and several recent pre-clinical models and clinical studies have tested this hypothesis. In a pre-clinical study, sunitinib was reported to exert potent complementary anti-tumor effects when combined with CD40-stimulating immunotherapy, by mediating DCs activation, reducing MDSCs and increasing endothelial activation that resulted in enhanced recruitment of cytotoxic T-cells [141]. Dual VEGF-A and Ang2 inhibition displayed enhanced anti-tumor immunity with PD1 blockade in breast, melanoma and pancreatic neuroendocrine tumor models [156]. Likewise, simultaneous blockade of PD-1 and VEGFR2, in a Colon-26 adenocarcinoma mouse model, induced a synergistic in vivo anti-tumor effect [197]. In a mouse model of SCLC, combined treatment with anti-VEGF and anti-PDL1 targeted therapy provided improved treatment outcome compared with anti-PDL1 or anti-VEGF monotherapy [198]. Co-administration of low-dose apatinib, a VEGFR2-TKI, with PDL1 inhibition resulted in reduced tumor growth, fewer metastases and prolonged survival of lung cancer mouse models [146]. Alleviated hypoxia, increased infiltration of CD8+ T-cells, reduced recruitment of TAMs and decreased TGFβ was observed with low-dose apatinib [146]. Anticancer activity of combining apatinib with anti-PD1 was also evident in a small cohort of pretreated patients with advanced NSCLC [146]. The treatment effect of axitinib, a TKI against VEGFR1/2/3, combined with CTLA4 blockade was investigated in a mouse melanoma model. Combination of anti-angiogenesis and checkpoint inhibition resulted to an increased anti-tumor effect and survival, partially due to enhanced immune response generated by an increased antigen-presenting function of intratumoral DCs in combination with a reduced suppressive capacity of intratumoral MDSCs [145].

Following the observation that metastatic melanoma patients with high levels of VEGF presented worse survival when treated with ipilimumab, a CTLA4 inhibitor [199], a phase I trial was conducted to investigate the combination of ipilimumab with bevacizumab. The trial demonstrated that VEGF-A blockade influences inflammation, lymphocyte trafficking and immune regulation, and was associated with favorable clinical outcome in metastatic melanoma patients [128,200]. Further analysis showed that the combination therapy elicited humoral immune responses against galectin-1, which exhibited protumor, pro-angiogenesis and immunosuppressive activities in 37.2% of treated patients [200]. The first ever phase III trial to successfully investigate the synergistic effect of immune-checkpoint inhibition with VEGF blockade was the Impower150 in NSCLC [201]. This pivotal study demonstrated that the addition of atezolizumab to bevacizumab plus chemotherapy significantly improved PFS and OS among patients with metastatic non-squamous NSCLC, regardless of PD-L1 expression and EGFR or ALK genetic alteration status [201]. Of note, while the quadruplet combination (i.e., atezolizumab, carboplatin, paclitaxel, bevacizumab) was superior to chemotherapy plus bevacizumab, the atezolizumab plus chemotherapy combination was not, thus supporting the modulatory role of anti-angiogenesis to immunotherapy. In metastatic renal cell carcinoma (mRCC) phase I studies combining a VEGF-TKI and PDL1/PD1 blockade, suggest that anti-angiogenesis could potentiate PDL1/PD1 inhibition. Specifically, tissues from patients with mRCC exhibited increased intra-tumoral CD8+ T-cells after combination treatment with bevacizumab and atezolizumab, a PDL1 inhibitor [129], while co-administration of axitinib plus pembrolizumab or avelumab, showed promising anti-tumor activity in patients with treatment-naive advanced RCC in phase I trials [202,203]. A similar but somewhat distinct therapeutic approach in a phase II trial combining dendritic cell-based immunotherapy with sunitinib, also demonstrated benefit for patients with mRCC [204]. Another phase II trial in RCC that compared atezolizumab plus bevacizumab against sunitinib, displayed enhanced efficacy for the combination in PDL1-positive patients, while sunitinib monotherapy had better results in in patients with predominant angiogenesis markers [205]. The phase III trial that followed, IMmotion151, confirmed prolonged PFS for the atezolizumab plus bevacizumab in the PDL1 positive population, however, longer follow-up is warranted to establish whether a survival benefit will emerge [206]. Likewise, two other phase III clinical trials have investigated the combination of an immune-checkpoint inhibitor with axitinib, in untreated patients with mRCC [207,208]. The JAVELIN Renal 101 trial reported a significantly longer progression free survival (PFS) for the combination of axitinib with avelumab, a PDL1 inhibitor, against sunitinib [207], while the KEYNOTE-426 trial resulted in both overall survival and PFS benefit for the combination of pembrolizumab with axitinib compared to sunitinib, regardless of PDL1 expression [208]. Both trials reported increased objective response rates for the combination therapy. VEGF inhibition and PDL1 blockade has also led to promising results in hepatocellular carcinoma. Specifically, a phase 1 b trial in patients with unrespectable hepatocellular carcinoma the combination of atezolizumab plus bevacizumab reported PFS benefit [209], leading to a phase III trial against sorafenib, where once again the combination treatment was superior in terms of OS and PFS [210] (Table 3).

Table 3. Phase III studies of immune-checkpoint inhibitors with anti-angiogenic agents.

Cancer Type	Immunotherapy	Anti-Angiogenic Agent	Indication	Year	Current Status	Identifier
Gastrointestinal	Atezolizumab (Anti-PDL1)	Bevacizumab	dMMR, Metastatic CRC	2016	Suspended	NCT02997228
	Nivolumab (Anti-PD1)	Bevacizumab	Metastatic CRC, 1st line	2018	Active, not recruiting	NCT03414983
	Sintilimab (Anti-PD1)	Bevacizumab	RAS-Mutant, Metastatic CRC, 1st line	2019	Not yet recruiting	NCT04194359
	HLX10 (Anti-PD1)	HLX04 (Anti-VEGF)	Metastatic CRC, 1st line	2020	Not yet recruiting	NCT04547166
	Atezolizumab (Anti-PDL1)	Bevacizumab	Advanced HCC, 1st line	2018	Active, not recruiting	NCT03434379
	HLX10 (Anti-PD1)	HLX04 (Anti-VEGF)	Advanced or Metastatic HCC, 1st line	2020	Not yet recruiting	NCT04465734
Genitourinary	Pembrolizumab (Anti-PD1)	Axitinib	Untreated, advanced RCC	2016	Active, not recruiting	NCT02853331
	Pembrolizumab (Anti-PD1)	Lenvatinib	Untreated, advanced RCC	2016	Active, not recruiting	NCT02811861
	Atezolizumab (Anti-PDL1)	Bevacizumab	Untreated, advanced RCC	2015	Active, not recruiting	NCT02420821
	Avelumab (Anti-PDL1)	Axitinib	Untreated, advanced RCC	2016	Active, not recruiting	NCT02684006
	Nivolumab (Anti-PD1)	Cabozantinib	Untreated, metastatic RCC	2019	Recruiting	NCT03793166
	Anlotinib (anti-PDL1)	TQB2450 (mtTKI)	Advanced RCC	2020	Recruiting	NCT04523272
	Toripalimab (anti-PD1)	Axitinib	Unresectable or Metastatic RCC, 1st line	2020	Recruiting	NCT04394975
Lung	Atezolizumab (Anti-PDL1)	Bevacizumab	Stage IV Non-Squamous NSCLC, 1st line	2015	Active, not recruiting	NCT02366143
	Atezolizumab (Anti-PDL1)	Bevacizumab	Stage IV Non-Squamous NSCLC, 1st line	2019	Recruiting	NCT04194203
	Sintilimab (Anti-PD1)	IBI305 (Anti-VEGF)	EGFR-mutated, TKI-resistant, Locally Advanced or Metastatic, non-squamous NSCLC	2019	Recruiting	NCT03802240
	HLX10 (Anti-PD1)	HLX04 (Anti-VEGF)	Stage IIIB/IIIC or IV non-squamous NSCLC	2019	Recruiting	NCT03952403
Gynecological	Atezolizumab (Anti-PDL1)	Bevacizumab	Platinum-Resistant, Recurrent, Ovarian, Fallopian Tube, or Peritoneal Cancer	2016	Recruiting	NCT02839707
	Atezolizumab (Anti-PDL1)	Bevacizumab	Platinum-Sensitive Relapse, Ovarian, Fallopian Tube, or Peritoneal Cancer	2016	Active, not recruiting	NCT02891824
	Atezolizumab (Anti-PDL1)	Bevacizumab	Stage III/IV Ovarian, Fallopian Tube, or Peritoneal Cancer	2017	Active, not recruiting	NCT03038100
	Atezolizumab (Anti-PDL1)	Bevacizumab	Persistent, Recurrent or Metastatic (Stage IVB) Cervical Cancer	2018	Recruiting	NCT03556839
	Pembrolizumab (Anti-PD1)	Bevacizumab	Persistent, Recurrent or Metastatic Cervical Cancer	2018	Active, not recruiting	NCT03635567
	Dostarlimab (Anti-PD1)	Bevacizumab	Stage III/IV Nonmucinous Ovarian Cancer, 1st line	2018	Recruiting	NCT03602859
	BCD-100 (Anti-PD1)	Bevacizumab	Advanced Cervical Cancer, 1st line	2019	Recruiting	NCT03912415

Abbreviations: PD(L)1: programmed death (ligand) 1; VEGF: vascular endothelial growth factor; mtTKI: multi-target tyrosine kinase inhibitor; HCC: hepatocellular carcinoma; NSCLC: non-small cell lung cancer; RCC: renal cell carcinoma; CRC: colorectal cancer; dMMR: deficient mismatch repair; EGFR: epidermal growth factor receptor.

5. Conclusions and Future Directions

Until recently, anti-angiogenic factors were considered to exert their anti-tumor effects by inhibiting the formation of new blood vessels; however, growing evidence suggests that inhibition of the VEGF/VEGFR pathway may have multiple effects including direct inhibition of tumor cell growth and immunostimulatory functions [15,16,54,58].

VEGF-mediated autocrine-paracrine loops that directly influence and promote tumor cell survival, proliferation and invasion have been identified in several cancers including lung, breast, prostate, bladder, colorectal, pancreatic, sarcomas, ovarian, melanoma, gliomas and hematopoietic malignancies. Moreover, this autocrine or paracrine loop represents an attractive therapeutic target [13]. Indeed, it has been shown that a natural occurring, soluble form of NRP1 can act as a $VEGF_{165}$ antagonist exhibiting anti-tumor activity in vivo [211]. Furthermore, in patients with inflammatory, locally advanced breast cancer, bevacizumab was reported to induce apoptosis in tumor cells, along with its inhibitory effects on VEGFR2 activation and permeability [212]. In gliomas, high doses of bevacizumab were suggested to have anticancer properties in vivo, not related to angiogenesis, as regression of glioma cells was demonstrated to occur independently from vascular regression [213]. Likewise, anti-VEGF treatment was explored in multiple myeloma (MM) cells. VEGF-A blockade caused cytostasis in MM cells, demonstrating that bevacizumab has a direct influence on major pathways critically activated in MM that is independent from its established effect on angiogenesis [92]. In a preclinical study, chronic exposure of colorectal cells to bevacizumab upregulated VEGF-A, -B, -C, PLGF, VEGFR1 expression and VEGFR1 phosphorylation, resulting to increased tumor cell migration and invasion and enhanced metastatic potential [214], highlighting the rationale of successfully blocking the VEGF/VEGFR pathway directly on tumor cells.

The immunosuppressive properties of VEGF likely stem from its role in initiating the wound healing process, which benefits from down-modulating cellular immunity and stimulating angiogenesis and tissue growth and repair [53,215]. Accordingly, the immunostimulatory effects of anti-angiogenetic agents can be condensed to at least four different functions: (a) preventing the VEGF-mediated inhibition of effector T-cells trafficking, proliferation and cytotoxic functions, thus enhancing T-cell mediated immune response; (b) restoring DCs' differentiation and maturation, thus promoting antigen presentation and T-cell activation; (c) hindering the recruitment of inhibitory cells in the TME, such as Tregs, MDSCs and M2-like TAMs; and (d) activating the tumor endothelium and inducing normalization of the disorganized, leaky and abnormal tumor vasculature that results to hypoxia, hinders effector T-cell infiltration and fosters immunosuppression in the TME [13,53,158,165,170,176,189,216]. Bevacizumab, in particular, has been found to relieve immunosuppression by decreasing MDSCs and Tregs populations, and also to enhance cytotoxic T-lymphocytes responses, improve DCs maturation and increase T-cells infiltration [124,125,127,128,130,132–135]. An inhibitory effect on MDSCs and/or Tregs compartments has also been displayed for sorafenib [148,150], axitinib [144,145] and sunitinib [136–140]. Likewise, a positive reinforcement on T-cells recruitment and functions has been observed for sunitinib [141,142], axitinib [144], apatinib [147] and sorafenib [150]. Another important function of sunitinib and apatinib is their ability to decrease the expression of inhibitory checkpoint molecules [141,143,147]. Furthermore, DCs' functions were enhanced with axitinib [145], while sorafenib provided contradictory results regarding DCs and T-cells regulation [131,151] (see Table 2).

Taking advantage of their immunomodulatory functions, several clinical trials in melanoma, RCC, NSCLC and hepatocellular carcinoma have successfully evaluated the combination of checkpoint inhibition with VEGF/VEGFR blockade, providing evidence of the efficacy of such an approach. Ongoing phase I to III clinical trials continue to explore the efficacy of a combinational strategy in a variety of different malignancies, including gynecological, gastrointestinal, genitourinary, central nervous system (CNS), lung and several others advanced solid tumors. Phase III clinical trials of an immune-checkpoint inhibitor with an anti-angiogenic agent combination that are currently active can be reviewed in Table 3. Ongoing research has the dynamic to expand the therapeutic indications of this strategy. However, despite, the synergetic therapeutic effect displayed so far in many trials, not all them reported positive outcomes; notably, in a phase I trial, the combination of tremelimumab, an anti-CTLA4 antibody with sunitinib induced severe toxicities, including kidney failure, in patients with mRCC [137], highlighting the importance of carefully designed clinical trials that will allow us to minimize toxicities and safely evaluate these combinations [13].

It is well accepted by now that not all patients will achieve benefit from immune-checkpoint inhibition, hence the need for identifying novel approaches [5]; at the same time, previously unappreciated studies on VEGF biology have demonstrated immunomodulatory properties and tumor regression by disrupting the VEGF/VEGFR pathway and now provide the scientific basis for new combinational treatments. To date therapeutic indications of anti-PD1/PDL1 combinations with anti-angiogenetic agents include RCC (pembrolizumab or avelumab plus axitinib), NSCLC (atezolizumab plus bevacizumab plus chemotherapy), hepatocellular carcinoma (atezolizumab plus bevacizumab). Ongoing and future studies will likely expand these therapeutic indications, while investigators have also the obligation to identify those patients that would safely benefit from such an approach.

Author Contributions: Conceptualization, G.P.; methodology, D.M., P.N.; investigation, P.N., G.Z., G.K., A.P.; data curation, P.N., L.M., S.G., I.G., A.-L.A.; writing—original draft preparation, P.N.; writing—review and editing, G.P., E.K., A.B.; supervision, G.P. All authors have read and agreed to the published version of the manuscript.

Funding: This research received no external funding.

Conflicts of Interest: The authors declare no conflict of interest.

References

1. Saijo, N.; Tamura, T.; Yamamoto, N.; Nishio, K. New strategies for cancer therapy in the 21st century. *Cancer Chemother. Pharmacol.* **2001**, *48*, S102–S106. [CrossRef]
2. Heron, M. Deaths: Leading Causes for 2017. *Nat. Vital Stat. Rep.* **2019**, *68*. Available online: https://www.cdc.gov/nchs/data/nvsr/nvsr68/nvsr68_06-508.pdf?fbclid=IwAR0ShnhUypiDEhFAJvjgxEjArda8ujdLSJj97y3cORXzUHlD_cLPdmzdSdY (accessed on 16 May 2020).
3. Ide, A.G.; Baker, N.H.; Warren, S.L. Vascularization of the Brown Pearce rabbit epithelioma transplant as seen in the transparent ear chamber. *Am. J. Roentgenol.* **1939**, *42*, 891–899.
4. Algire, G.H.; Chalkley, H.W.; Legallais, F.Y.; Park, H.D. Vasculae Reactions of Normal and Malignant Tissues in Vivo. I. Vascular Reactions of Mice to Wounds and to Normal and Neoplastic Transplants. *J. Natl. Cancer Inst.* **1945**, *6*, 73–85. [CrossRef]
5. Apte, R.S.; Chen, D.S.; Ferrara, N. VEGF in Signaling and Disease: Beyond Discovery and Development. *Cell* **2019**, *176*, 1248–1264. [CrossRef] [PubMed]
6. Sherwood, L.M.; Parris, E.E.; Folkman, J. Tumor Angiogenesis: Therapeutic Implications. *N. Engl. J. Med.* **1971**, *285*, 1182–1186. [CrossRef] [PubMed]
7. Folkman, J.; Merler, E.; Abernathy, C.; Williams, G. Isolation of a tumor factor responsible for angiogenesis. *J. Exp. Med.* **1971**, *133*, 275. [CrossRef] [PubMed]
8. Lugano, R.; Ramachandran, M.; Dimberg, A. Tumor angiogenesis: Causes, consequences, challenges and opportunities. *Cell. Mol. Life Sci.* **2019**, *77*, 1745–1770. [CrossRef] [PubMed]
9. Kim, K.J.; Li, B.; Winer, J.; Armanini, M.; Gillett, N.; Phillips, H.S.; Ferrara, N. Inhibition of vascular endothelial growth factor-induced angiogenesis suppresses tumour growth in vivo. *Nat. Cell Biol.* **1993**, *362*, 841–844. [CrossRef] [PubMed]
10. Presta, L.G.; Chen, H.; O'Connor, S.J.; Chisholm, V.; Meng, Y.G.; Krummen, L.; Winkler, M.; Ferrara, N. Humanization of an anti-vascular endothelial growth factor monoclonal antibody for the therapy of solid tumors and other disorders. *Cancer Res.* **1997**, *57*, 4593–4599. [PubMed]
11. Ferrara, N.; Adamis, A.P. Ten years of anti-vascular endothelial growth factor therapy. *Nat. Rev. Drug Discov.* **2016**, *15*, 385–403. [CrossRef] [PubMed]
12. Jayson, G.C.; Kerbel, R.; Ellis, L.M.; Harris, A.L. Antiangiogenic therapy in oncology: Current status and future directions. *Lancet* **2016**, *388*, 518–529. [CrossRef]
13. Fukumura, D.; Kloepper, J.; Amoozgar, Z.; Duda, D.G.; Jain, R.K. Enhancing cancer immunotherapy using antiangiogenics: Opportunities and challenges. *Nat. Rev. Clin. Oncol.* **2018**, *15*, 325–340. [CrossRef] [PubMed]

14. Jain, R.K. Antiangiogenesis Strategies Revisited: From Starving Tumors to Alleviating Hypoxia. *Cancer Cell* **2014**, *26*, 605–622. [CrossRef]
15. Khan, K.A.; Kerbel, R.S. Improving immunotherapy outcomes with anti-angiogenic treatments and vice versa. *Nat. Rev. Clin. Oncol.* **2018**, *15*, 310–324. [CrossRef] [PubMed]
16. Masood, R.; Cai, J.; Zheng, T.; Smith, D.L.; Hinton, D.R.; Gill, P.S. Vascular endothelial growth factor (VEGF) is an autocrine growth factor for VEGF receptor–positive human tumors. *Blood* **2001**, *98*, 1904–1913. [CrossRef]
17. Schweighofer, B.; Hofer, E. Signal transduction induced in endothelial cells by growth factor receptors involved in angiogenesis. *Thromb. Haemost.* **2007**, *97*, 355–363. [CrossRef]
18. Yancopoulos, G.D.; Davis, S.; Gale, N.W.; Rudge, J.S.; Wiegand, S.J.; Holash, J. Vascular-specific growth factors and blood vessel formation. *Nat. Cell Biol.* **2000**, *407*, 242–248. [CrossRef]
19. Eklund, L.; Olsen, B.R. Tie receptors and their angiopoietin ligands are context-dependent regulators of vascular remodeling. *Exp. Cell Res.* **2006**, *312*, 630–641. [CrossRef]
20. Héroult, M.; Schaffner, F.; Augustin, H.G. Eph receptor and ephrin ligand-mediated interactions during angiogenesis and tumor progression. *Exp. Cell Res.* **2006**, *312*, 642–650. [CrossRef]
21. Liu, Z.; Fan, F.; Wang, A.; Zheng, S.; Lu, Y. Dll4-Notch signaling in regulation of tumor angiogenesis. *J. Cancer Res. Clin. Oncol.* **2013**, *140*, 525–536. [CrossRef] [PubMed]
22. Simons, M.; Gordon, E.; Claesson-Welsh, E.G.L. Mechanisms and regulation of endothelial VEGF receptor signalling. *Nat. Rev. Mol. Cell Biol.* **2016**, *17*, 611–625. [CrossRef] [PubMed]
23. De Vries, C.; Escobedo, J.A.; Ueno, H.; Houck, K.; Ferrara, N.; Williams, L.T. The fms-like tyrosine kinase, a receptor for vascular endothelial growth factor. *Science* **1992**, *255*, 989–991. [CrossRef] [PubMed]
24. Holmqvist, K.; Cross, M.J.; Rolny, C.; Hägerkvist, R.; Rahimi, N.; Matsumoto, T.; Claesson-Welsh, L.; Welsh, M.J. The Adaptor Protein Shb Binds to Tyrosine 1175 in Vascular Endothelial Growth Factor (VEGF) Receptor-2 and Regulates VEGF-dependent Cellular Migration. *J. Biol. Chem.* **2004**, *279*, 22267–22275. [CrossRef] [PubMed]
25. Terman, B.I.; Dougher-Vermazen, M.; Carrion, M.E.; Dimitrov, D.; Armellino, D.C.; Gospodarowicz, D.; Böhlen, P. Identification of the KDR tyrosine kinase as a receptor for vascular endothelial cell growth factor. *Biochem. Biophys. Res. Commun.* **1992**, *187*, 1579–1586. [CrossRef]
26. Olsson, A.-K.; Dimberg, A.; Kreuger, J.; Claesson-Welsh, L. VEGF receptor signalling? In control of vascular function. *Nat. Rev. Mol. Cell Biol.* **2006**, *7*, 359–371. [CrossRef]
27. Favier, B.; Alam, A.; Barron, P.; Bonnin, J.; Laboudie, P.; Fons, P.; Mandron, M.; Herault, J.P.; Neufeld, G.; Savi, P.; et al. Neuropilin-2 interacts with VEGFR-2 and VEGFR-3 and promotes human endothelial cell survival and migration. *Blood* **2006**, *108*, 1243–1250. [CrossRef]
28. Pajusola, K.; Aprelikova, O.; Korhonen, J.; Kaipainen, A.; Pertovaara, L.; Alitalo, R.; Alitalo, K. FLT4 receptor tyrosine kinase contains seven immunoglobulin-like loops and is expressed in multiple human tissues and cell lines. *Cancer Res.* **1992**, *52*, 5738–5743.
29. Alitalo, K.; Tammela, T.; Petrova, T.V. Lymphangiogenesis in development and human disease. *Nat. Cell Biol.* **2005**, *438*, 946–953. [CrossRef]
30. Carmeliet, P.; Ferreira, V.; Breier, G.; Pollefeyt, S.; Kieckens, M.; Gertsenstein, M.; Fahrig, M.; Vandenhoeck, A.; Harpal, K.; Eberhardt, C.; et al. Abnormal blood vessel development and lethality in embryos lacking a single VEGF allele. *Nat. Cell Biol.* **1996**, *380*, 435–439. [CrossRef]
31. Shalaby, F.; Rossant, J.; Yamaguchi, T.P.; Gertsenstein, M.; Wu, X.-F.; Breitman, M.L.; Schuh, A.C. Failure of blood-island formation and vasculogenesis in Flk-1-deficient mice. *Nat. Cell Biol.* **1995**, *376*, 62–66. [CrossRef]
32. Speirs, V.; Atkin, S.L. Production of VEGF and expression of the VEGF receptors Flt-1 and KDR in primary cultures of epithelial and stromal cells derived from breast tumours. *Br. J. Cancer* **1999**, *80*, 898–903. [CrossRef] [PubMed]
33. Tischer, E.; Mitchell, R.; Hartman, T.; Silva, M.; Gospodarowicz, D.; Fiddes, J.C.; Abraham, J. The human gene for vascular endothelial growth factor. Multiple protein forms are encoded through alternative exon splicing. *J. Biol. Chem.* **1991**, *266*, 11947–11954. [PubMed]
34. Ferrara, N.; Davis-Smyth, T. The Biology of Vascular Endothelial Growth Factor. *Endocr. Rev.* **1997**, *18*, 4–25. [CrossRef]

35. Pentheroudakis, G.; Mavroeidis, L.; Papadopoulou, K.; Koliou, G.-A.; Bamia, C.; Chatzopoulos, K.; Samantas, E.; Mauri, D.; Efstratiou, I.; Pectasides, D.; et al. Angiogenic and Antiangiogenic VEGFA Splice Variants in Colorectal Cancer: Prospective Retrospective Cohort Study in Patients Treated With Irinotecan-Based Chemotherapy and Bevacizumab. *Clin. Color. Cancer* **2019**, *18*, e370–e384. [CrossRef]
36. Woolard, J.; Bevan, H.S.; Harper, S.J.; Bates, D.O. Molecular Diversity of VEGF-A as a Regulator of Its Biological Activity. *Microcirculation* **2009**, *16*, 572–592. [CrossRef] [PubMed]
37. Hilmi, C.; Guyot, M.; Pagès, G. VEGF Spliced Variants: Possible Role of Anti-Angiogenesis Therapy. *J. Nucleic Acids* **2011**, *2012*, 1–7. [CrossRef]
38. Hellström, M.; Phng, L.-K.; Gerhardt, H. VEGF and Notch Signaling. *Cell Adhes. Migr.* **2007**, *1*, 133–136. [CrossRef]
39. Phng, L.-K.; Gerhardt, H. Angiogenesis: A Team Effort Coordinated by Notch. *Dev. Cell* **2009**, *16*, 196–208. [CrossRef]
40. Liu, Z.-J.; Shirakawa, T.; Li, Y.; Soma, A.; Oka, M.; Dotto, G.P.; Fairman, R.M.; Velazquez, O.C.; Herlyn, M. Regulation of Notch1 and Dll4 by Vascular Endothelial Growth Factor in Arterial Endothelial Cells: Implications for Modulating Arteriogenesis and Angiogenesis. *Mol. Cell. Biol.* **2003**, *23*, 14–25. [CrossRef]
41. Lobov, I.B.; Renard, R.A.; Papadopoulos, N.; Gale, N.W.; Thurston, G.; Yancopoulos, G.D.; Wiegand, S.J. Delta-like ligand 4 (Dll4) is induced by VEGF as a negative regulator of angiogenic sprouting. *Proc. Natl. Acad. Sci. USA* **2007**, *104*, 3219–3224. [CrossRef]
42. Williams, C.K.; Li, J.-L.; Murga, M.; Harris, A.L.; Tosato, G. Up-regulation of the Notch ligand Delta-like 4 inhibits VEGF-induced endothelial cell function. *Blood* **2006**, *107*, 931–939. [CrossRef]
43. Harrington, L.S.; Sainson, R.C.; Williams, C.K.; Taylor, J.M.; Shi, W.; Li, J.-L.; Harris, A.L. Regulation of multiple angiogenic pathways by Dll4 and Notch in human umbilical vein endothelial cells. *Microvasc. Res.* **2008**, *75*, 144–154. [CrossRef] [PubMed]
44. Itakura, J.; Ishiwata, T.; Shen, B.; Kornmann, M.; Korc, M. Concomitant over-expression of vascular endothelial growth factor and its receptors in pancreatic cancer. *Int. J. Cancer* **2000**, *85*, 27–34. [CrossRef]
45. Wu, W.; Shu, X.; Hovsepyan, H.; Mosteller, R.D.; Broek, D. VEGF receptor expression and signaling in human bladder tumors. *Oncogene* **2003**, *22*, 3361–3370. [CrossRef]
46. Yamamoto, S.; Konishi, I.; Mandai, M.; Kuroda, H.; Komatsu, T.; Nanbu, K.; Sakahara, H.; Mori, T. Expression of vascular endothelial growth factor (VEGF) in epithelial ovarian neoplasms: Correlation with clinicopathology and patient survival, and analysis of serum VEGF levels. *Br. J. Cancer* **1997**, *76*, 1221–1227. [CrossRef] [PubMed]
47. Bellamy, W.T. Expression of vascular endothelial growth factor and its receptors in multiple myeloma and other hematopoietic malignancies. *Semin. Oncol.* **2001**, *28*, 551–559. [CrossRef]
48. Olson, T.; Mohanraj, D.; Carson, L.F.; Ramakrishnan, S. Vascular permeability factor gene expression in normal and neoplastic human ovaries. *Cancer Res.* **1994**, *54*, 276–280.
49. Paley, P.J.; Staskus, K.A.; Gebhard, K.; Mohanraj, D.; Twiggs, L.B.; Carson, L.F.; Ramakrishnan, S. Vascular endothelial growth factor expression in early stage ovarian carcinoma. *Cancer* **1997**, *80*, 98–106. [CrossRef]
50. Costache, M.; Ioana, M.; Iordache, S.; Ene, D.; Costache, C.A.; Săftoiu, A. VEGF expression in pancreatic cancer and other malignancies: A review of the literature. *Rom. J. Intern. Med.* **2015**, *53*, 199–208. [CrossRef]
51. Dias, S.; Hattori, K.; Zhu, Z.; Heissig, B.; Choy, M.; Lane, W.; Wu, Y.; Chadburn, A.; Hyjek, E.; Gill, M.; et al. Autocrine stimulation of VEGFR-2 activates human leukemic cell growth and migration. *J. Clin. Investig.* **2000**, *106*, 511–521. [CrossRef]
52. Gee, M.F.W.; Tsuchida, R.; Eichler-Jonsson, C.; Das, B.; Baruchel, S.; Malkin, D. Vascular endothelial growth factor acts in an autocrine manner in rhabdomyosarcoma cell lines and can be inhibited with all-trans-retinoic acid. *Oncogene* **2005**, *24*, 8025–8037. [CrossRef] [PubMed]
53. Chen, D.S.; Hurwitz, H. Combinations of Bevacizumab with Cancer Immunotherapy. *Cancer J.* **2018**, *24*, 193–204. [CrossRef] [PubMed]
54. Hamerlik, P.; Lathia, J.D.; Rasmussen, R.; Wu, Q.; Bartkova, J.; Lee, M.; Moudry, P.; Bartek, J.; Fischer, W.; Lukas, J.; et al. Autocrine VEGF–VEGFR2–Neuropilin-1 signaling promotes glioma stem-like cell viability and tumor growth. *J. Exp. Med.* **2012**, *209*, 507–520. [CrossRef] [PubMed]

55. Aesoy, R.; Sanchez, B.C.; Norum, J.H.; Lewensohn, R.; Viktorsson, K.; Linderholm, B. An Autocrine VEGF/VEGFR2 and p38 Signaling Loop Confers Resistance to 4-Hydroxytamoxifen in MCF-7 Breast Cancer Cells. *Mol. Cancer Res.* **2008**, *6*, 1630–1638. [CrossRef]
56. Miettinen, M.; Rikala, M.-S.; Rysz, J.; Lasota, J.; Wang, Z.-F. Vascular Endothelial Growth Factor Receptor 2 as a Marker for Malignant Vascular Tumors and Mesothelioma. *Am. J. Surg. Pathol.* **2012**, *36*, 629–639. [CrossRef] [PubMed]
57. Tian, X.; Song, S.; Wu, J.; Meng, L.; Dong, Z.; Shou, C. Vascular Endothelial Growth Factor: Acting as an Autocrine Growth Factor for Human Gastric Adenocarcinoma Cell MGC803. *Biochem. Biophys. Res. Commun.* **2001**, *286*, 505–512. [CrossRef] [PubMed]
58. Spannuth, W.A.; Nick, A.M.; Jennings, N.B.; Armaiz-Pena, G.N.; Mangala, L.S.; Danes, C.G.; Lin, Y.G.; Merritt, W.M.; Thaker, P.H.; Kamat, A.A.; et al. Functional significance of VEGFR-2 on ovarian cancer cells. *Int. J. Cancer* **2009**, *124*, 1045–1053. [CrossRef]
59. Liu, B.; Earl, H.M.; Baban, D.; Shoaibi, M.; Fabra, A.; Kerr, D.; Seymour, L. Melanoma Cell Lines Express VEGF Receptor KDR and Respond to Exogenously Added VEGF. *Biochem. Biophys. Res. Commun.* **1995**, *217*, 721–727. [CrossRef]
60. Ruffini, F.; D'Atri, S.; Lacal, P.M. Neuropilin-1 expression promotes invasiveness of melanoma cells through vascular endothelial growth factor receptor-2-dependent and -independent mechanisms. *Int. J. Oncol.* **2013**, *43*, 297–306. [CrossRef]
61. Wey, J.S.; Fan, F.; Gray, M.J.; Bauer, T.W.; Mccarty, M.F.; Somcio, R.; Liu, W.; Evans, D.B.; Wu, Y.; Hicklin, D.J.; et al. Vascular endothelial growth factor receptor-1 promotes migration and invasion in pancreatic carcinoma cell lines. *Cancer* **2005**, *104*, 427–438. [CrossRef] [PubMed]
62. Matkar, P.N.; Singh, K.K.; Rudenko, D.; Kim, Y.J.; Kuliszewski, M.A.; Prud'Homme, G.J.; Hedley, D.W.; Leong-Poi, H. Novel regulatory role of neuropilin-1 in endothelial-to-mesenchymal transition and fibrosis in pancreatic ductal adenocarcinoma. *Oncotarget* **2016**, *7*, 69489–69506. [CrossRef]
63. Barr, M.P.; Gray, S.G.; Gately, K.; Hams, E.; Fallon, P.G.; Davies, A.M.; Richard, D.J.; Pidgeon, G.P.; O'Byrne, K.J. Vascular endothelial growth factor is an autocrine growth factor, signaling through neuropilin-1 in non-small cell lung cancer. *Mol. Cancer* **2015**, *14*, 1–16. [CrossRef] [PubMed]
64. Tanno, S.; Ohsaki, Y.; Nakanishi, K.; Toyoshima, E.; Kikuchi, K. Human small cell lung cancer cells express functional VEGF receptors, VEGFR-2 and VEGFR-3. *Lung Cancer* **2004**, *46*, 11–19. [CrossRef]
65. Bhattacharya, R.; Ye, X.-C.; Wang, R.; Ling, X.; McManus, M.; Fan, F.; Boulbès, D.; Ellis, L.M. Intracrine VEGF Signaling Mediates the Activity of Prosurvival Pathways in Human Colorectal Cancer Cells. *Cancer Res.* **2016**, *76*, 3014–3024. [CrossRef]
66. Lesslie, D.P.; Summy, J.M.; Parikh, N.U.; Fan, F.; Trevino, J.G.; Sawyer, T.K.; Metcalf, C.A., III; Shakespeare, W.C.; Hicklin, D.J.; Ellis, L.M.; et al. Vascular endothelial growth factor receptor-1 mediates migration of human colorectal carcinoma cells by activation of Src family kinases. *Br. J. Cancer* **2006**, *94*, 1710–1717. [CrossRef]
67. Fan, F.; Wey, J.S.; Mccarty, M.F.; Belcheva, A.; Liu, W.; Bauer, T.W.; Somcio, R.J.; Wu, Y.; Hooper, A.; Hicklin, D.J.; et al. Expression and function of vascular endothelial growth factor receptor-1 on human colorectal cancer cells. *Oncogene* **2005**, *24*, 2647–2653. [CrossRef] [PubMed]
68. Bates, R.C.; Goldsmith, J.D.; Bachelder, R.E.; Brown, C.; Shibuya, M.; Oettgen, P.; Mercurio, A.M. Flt-1-Dependent Survival Characterizes the Epithelial-Mesenchymal Transition of Colonic Organoids. *Curr. Biol.* **2003**, *13*, 1721–1727. [CrossRef]
69. Bates, R.C.; Pursell, B.M.; Mercurio, A.M. Epithelial-Mesenchymal Transition and Colorectal Cancer: Gaining Insights into Tumor Progression Using LIM 1863 Cells. *Cells Tissues Organs* **2007**, *185*, 29–39. [CrossRef]
70. Zhang, H. Expression of vascular endothelial growth factor and its receptors KDR and Flt-1 in gastric cancer cells. *World J. Gastroenterol.* **2002**, *8*, 994–998. [CrossRef]
71. Jackson, M.W.; Roberts, J.S.; Heckford, S.E.; Ricciardelli, C.; Stahl, J.; Choong, C.; Horsfall, D.J.; Tilley, W.D. A potential autocrine role for vascular endothelial growth factor in prostate cancer. *Cancer Res.* **2002**, *62*, 854–859.
72. Ferrer, F.A.; Miller, L.J.; Lindquist, R.; Kowalczyk, P.; Laudone, V.P.; Albertsen, P.C.; Kreutzer, D.L. Expression of vascular endothelial growth factor receptors in human prostate cancer. *Urology* **1999**, *54*, 567–572. [CrossRef]

73. Szabo, E.; Schneider, H.; Seystahl, K.; Rushing, E.J.; Herting, F.; Weidner, K.M.; Weller, M. Autocrine VEGFR1 and VEGFR2 signaling promotes survival in human glioblastoma models in vitro and in vivo. *Neuro-Oncol.* **2016**, *18*, 1242–1252. [CrossRef] [PubMed]
74. Knizetova, P.; Hlobilkova, A.; Vancova, I.; Kalita, O.; Ehrmann, J.; Kolar, Z.; Bartek, J. Autocrine regulation of glioblastoma cell-cycle progression, viability and radioresistance through the VEGF-VEGFR2 (KDR) interplay. *Cell Cycle* **2008**, *7*, 2553–2561. [CrossRef]
75. Xu, C.; Wu, X.; Zhu, J. VEGF Promotes Proliferation of Human Glioblastoma Multiforme Stem-Like Cells through VEGF Receptor 2. *Sci. World J.* **2013**, *2013*, 1–8. [CrossRef] [PubMed]
76. Price, D.J.; Miralem, T.; Jiang, S.; Steinberg, R.; Avraham, H. Role of vascular endothelial growth factor in the stimulation of cellular invasion and signaling of breast cancer cells. *Cell Growth Differ. Mol. Boil. J. Am. Assoc. Cancer Res.* **2001**, *12*, 129–136.
77. Schoeffner, D.J.; Matheny, S.L.; Akahane, T.; Factor, V.; Berry, A.; Merlino, G.; Thorgeirsson, U.P. VEGF contributes to mammary tumor growth in transgenic mice through paracrine and autocrine mechanisms. *Lab. Investig.* **2005**, *85*, 608–623. [CrossRef]
78. Schmidt, M.; Hantel, P.; Kreienberg, R.; Waltenberger, J. Autocrine vascular endothelial growth factor signalling in breast cancer. Evidence from cell lines and primary breast cancer cultures in vitro. *Angiogenesis* **2005**, *8*, 197–204. [CrossRef]
79. Bachelder, R.E.; Crago, A.; Chung, J.; Wendt, M.A.; Shaw, L.M.; Robinson, G.; Mercurio, A.M. Vascular endothelial growth factor is an autocrine survival factor for neuropilin-expressing breast carcinoma cells. *Cancer Res.* **2001**, *61*, 5736–5740.
80. Bachelder, R.; Wendt, M.; Mercurio, A.M. Vascular endothelial growth factor promotes breast carcinoma invasion in an autocrine manner by regulating the chemokine receptor CXCR4. *Cancer Res.* **2002**, *62*, 7203–7206.
81. Bachelder, R.; Lipscomb, E.; Lin, X.; Wendt, M.; Chadborn, N.H.; Eickholt, B.J.; Mercurio, A.M. Competing autocrine pathways involving alternative neuropilin-1 ligands regulate chemotaxis of carcinoma cells. *Cancer Res.* **2003**, *63*, 5230–5233. [PubMed]
82. Luo, M.; Hou, L.; Li, J.; Shao, S.; Huang, S.; Meng, D.; Liu, L.; Feng, L.; Xia, P.; Qin, T.; et al. VEGF/NRP-1 axis promotes progression of breast cancer via enhancement of epithelial-mesenchymal transition and activation of NF-κB and β-catenin. *Cancer Lett.* **2016**, *373*, 1–11. [CrossRef] [PubMed]
83. Han, Z.; Jiang, G.; Zhang, Y.; Xu, J.; Chen, C.; Zhang, L.; Xu, Z.; Du, X. Effects of RNA interference-mediated NRP-1 silencing on the proliferation and apoptosis of breast cancer cells. *Mol. Med. Rep.* **2012**, *12*, 513–519. [CrossRef] [PubMed]
84. Zhao, D.; Pan, C.; Sun, J.; Gilbert, C.; Drews-Elger, K.; Azzam, D.J.; Picon-Ruiz, M.; Kim, M.; Ullmer, W.; El-Ashry, D.; et al. VEGF drives cancer-initiating stem cells through VEGFR-2/Stat3 signaling to upregulate Myc and Sox2. *Oncogene* **2014**, *34*, 3107–3119. [CrossRef]
85. Zhang, L.; Wang, H.; Li, C.; Zhao, Y.; Wu, L.; Du, X.; Han, Z. VEGF-A/Neuropilin 1 Pathway Confers Cancer Stemness via Activating Wnt/β-Catenin Axis in Breast Cancer Cells. *Cell. Physiol. Biochem.* **2017**, *44*, 1251–1262. [CrossRef] [PubMed]
86. Xu, H.-M.; Zhu, J.-G.; Gu, L.; Hu, S.-Q.; Wu, H. VEGFR2 Expression in Head and Neck Squamous Cell Carcinoma Cancer Cells Mediates Proliferation and Invasion. *Asian Pac. J. Cancer Prev.* **2016**, *17*, 2217–2221. [CrossRef]
87. Kopparapu, P.K.; Boorjian, S.A.; Robinson, B.D.; Downes, M.; Gudas, L.J.; Mongan, N.P.; Persson, J.L. Expression of VEGF and its receptors VEGFR1/VEGFR2 is associated with invasiveness of bladder cancer. *Anticancer Res.* **2013**, *33*, 2381–2390. Available online: http://ar.iiarjournals.org/content/33/6/2381.full (accessed on 18 June 2020).
88. Podar, K.; Anderson, K.C. The pathophysiologic role of VEGF in hematologic malignancies: Therapeutic implications. *Blood* **2005**, *105*, 1383–1395. [CrossRef]
89. Podar, K.; Tai, Y.-T.; Davies, F.E.; Lentzsch, S.; Sattler, M.; Hideshima, T.; Lin, B.K.; Gupta, D.; Shima, Y.; Chauhan, D.; et al. Vascular endothelial growth factor triggers signaling cascades mediating multiple myeloma cell growth and migration. *Blood* **2001**, *98*, 428–435. [CrossRef]
90. Kumar, S.; Witzig, T.E.; Timm, M.; Haug, J.; Wellik, L.; Fonseca, R.; Greipp, P.R.; Rajkumar, S.V. Expression of VEGF and its receptors by myeloma cells. *Leukemia* **2003**, *17*, 2025–2031. [CrossRef]

91. Vincent, L.; Jin, D.K.; Karajannis, M.A.; Shido, K.; Hooper, A.T.; Rashbaum, W.K.; Pytowski, B.; Wu, Y.; Hicklin, D.J.; Zhu, Z.; et al. Fetal Stromal–Dependent Paracrine and Intracrine Vascular Endothelial Growth Factor-A/Vascular Endothelial Growth Factor Receptor-1 Signaling Promotes Proliferation and Motility of Human Primary Myeloma Cells. *Cancer Res.* **2005**, *65*, 3185–3192. [CrossRef] [PubMed]
92. Attar-Schneider, O.; Drucker, L.; Zismanov, V.; Tartakover-Matalon, S.; Rashid, G.; Lishner, M. Bevacizumab attenuates major signaling cascades and eIF4E translation initiation factor in multiple myeloma cells. *Lab. Investig.* **2011**, *92*, 178–190. [CrossRef] [PubMed]
93. Lacal, P.M.; Failla, C.M.; Pagani, E.; Odorisio, T.; Schietroma, C.; Falcinelli, S.; Zambruno, G.; D'Atri, S. Human Melanoma Cells Secrete and Respond to Placenta Growth Factor and Vascular Endothelial Growth Factor. *J. Investig. Dermatol.* **2000**, *115*, 1000–1007. [CrossRef] [PubMed]
94. Pagani, E.; Ruffini, F.; Cappellini, G.C.A.; Scoppola, A.; Fortes, C.; Marchetti, P.; Graziani, G.; D'Atri, S.; Lacal, P.M. Placenta growth factor and neuropilin-1 collaborate in promoting melanoma aggressiveness. *Int. J. Oncol.* **2016**, *48*, 1581–1589. [CrossRef] [PubMed]
95. Qi, L.; Robinson, W.A.; Brady, B.M.; Glode, L.M. Migration and invasion of human prostate cancer cells is related to expression of VEGF and its receptors. *Anticancer Res.* **2003**, *23*, 3917–3922.
96. Mentlein, R.; Forstreuter, F.; Mehdorn, H.M.; Held-Feindt, J. Functional significance of vascular endothelial growth factor receptor expression on human glioma cells. *J. Neuro-Oncol.* **2004**, *67*, 9–18. [CrossRef]
97. Mesti, T.; Savarin, P.; Triba, M.N.; Le Moyec, L.; Ocvirk, J.; Banissi, C.; Carpentier, A.F. Metabolic Impact of Anti-Angiogenic Agents on U87 Glioma Cells. *PLoS ONE* **2014**, *9*, e99198. [CrossRef]
98. Rydén, L.; Linderholm, B.; Nielsen, N.H.; Emdin, S.; Jönsson, P.-E.; Landberg, G. Tumor Specific VEGF-A and VEGFR2/KDR Protein are Co-expressed in Breast Cancer. *Breast Cancer Res. Treat.* **2003**, *82*, 147–154. [CrossRef]
99. Wu, Y.; Hooper, A.T.; Zhong, Z.; Witte, L.; Böhlen, P.; Rafii, S.; Hicklin, D.J. The vascular endothelial growth factor receptor (VEGFR-1) supports growth and survival of human breast carcinoma. *Int. J. Cancer* **2006**, *119*, 1519–1529. [CrossRef]
100. Yan, J.-D.; Liu, Y.; Zhang, Z.-Y.; Liu, G.-Y.; Xu, J.-H.; Liu, L.-Y.; Hu, Y.-M. Expression and prognostic significance of VEGFR-2 in breast cancer. *Pathol.-Res. Pr.* **2015**, *211*, 539–543. [CrossRef]
101. Stendahl, M.; Emdin, S.; Bengtsson, N.O. Tumor-specific VEGF-A and VEGFR2 in postmenopausal breast cancer patients with long-term follow-up. Implication of a link between VEGF pathway and tamoxifen response. *Breast Cancer Res. Treat.* **2005**, *89*, 135–143. [CrossRef]
102. Nakopoulou, L.; Stefanaki, K.; Panayotopoulou, E.; Giannopoulou, I.; Athanassiadou, P.; Gakiopoulou-Givalou, H.; Louvrou, A. Expression of the vascular endothelial growth factor receptor-2/Flk-1 in breast carcinomas: Correlation with proliferation. *Hum. Pathol.* **2002**, *33*, 863–870. [CrossRef]
103. Lentzsch, S.; Chatterjee, M.; Gries, M.; Bommert, K.; Gollasch, H.; Dörken, B.; Bargou, R.C. PI3-K/AKT/FKHR and MAPK signaling cascades are redundantly stimulated by a variety of cytokines and contribute independently to proliferation and survival of multiple myeloma cells. *Leukemia* **2004**, *18*, 1883–1890. [CrossRef] [PubMed]
104. Ria, R.; Melaccio, A.; Racanelli, V.; Vacca, A. Anti-VEGF Drugs in the Treatment of Multiple Myeloma Patients. *J. Clin. Med.* **2020**, *9*, 1765. [CrossRef] [PubMed]
105. Solimando, A.G.; Da Via', M.C.; Leone, P.; Croci, G.; Borrelli, P.; Gaviria, P.T.; Brandl, A.; Di Lernia, G.; Bianchi, F.P.; Tafuri, S.; et al. Adhesion-Mediated Multiple Myeloma (MM) Disease Progression: Junctional Adhesion Molecule a Enhances Angiogenesis and Multiple Myeloma Dissemination and Predicts Poor Survival. *Blood* **2019**, *134*, 855. [CrossRef]
106. Ullah, T.R. The role of CXCR4 in multiple myeloma: Cells' journey from bone marrow to beyond. *J. Bone Oncol.* **2019**, *17*, 100253. [CrossRef] [PubMed]
107. Chatterjee, S.; Wang, Y.; Duncan, M.K.; Naik, U.P. Junctional Adhesion Molecule-A Regulates Vascular Endothelial Growth Factor Receptor-2 Signaling-Dependent Mouse Corneal Wound Healing. *PLoS ONE* **2013**, *8*, e63674. [CrossRef]
108. Veronese, L.; Tournilhac, O.; Verrelle, P.; Davi, F.; Dighiero, G.; Chautard, E.; Veyrat-Masson, R.; Kwiatkowski, F.; Goumy, C.; Gouas, L.; et al. Strong correlation between VEGF and MCL-1 mRNA expression levels in B-cell chronic lymphocytic leukemia. *Leuk. Res.* **2009**, *33*, 1623–1626. [CrossRef]

109. Guo, D.; Jia, Q.; Song, H.-Y.; Warren, R.S.; Donner, D.B. Vascular Endothelial Cell Growth Factor Promotes Tyrosine Phosphorylation of Mediators of Signal Transduction That Contain SH2 Domains. *J. Biol. Chem.* **1995**, *270*, 6729–6733. [CrossRef]
110. Soker, S.; Takashima, S.; Miao, H.Q.; Neufeld, G.; Klagsbrun, M. Neuropilin-1 Is Expressed by Endothelial and Tumor Cells as an Isoform-Specific Receptor for Vascular Endothelial Growth Factor. *Cell* **1998**, *92*, 735–745. [CrossRef]
111. Ribatti, D. Epithelial-mesenchymal transition in morphogenesis, cancer progression and angiogenesis. *Exp. Cell Res.* **2017**, *353*, 1–5. [CrossRef] [PubMed]
112. Markopoulos, G.S.; Roupakia, E.; Marcu, K.B.; Kolettas, E. Epigenetic Regulation of Inflammatory Cytokine-Induced Epithelial-To-Mesenchymal Cell Transition and Cancer Stem Cell Generation. *Cells* **2019**, *8*, 1143. [CrossRef] [PubMed]
113. Ghersi, G. Roles of molecules involved in epithelial/mesenchymal transition during angiogenesis. *Front. Biosci.* **2008**, *13*, 2335–2355. [CrossRef] [PubMed]
114. Zeisberg, E.M.; Potenta, S.; Xie, L.; Zeisberg, M.; Kalluri, R. Discovery of Endothelial to Mesenchymal Transition as a Source for Carcinoma-Associated Fibroblasts. *Cancer Res.* **2007**, *67*, 10123–10128. [CrossRef] [PubMed]
115. Potenta, S.; Zeisberg, E.M.; Kalluri, R. The role of endothelial-to-mesenchymal transition in cancer progression. *Br. J. Cancer* **2008**, *99*, 1375–1379. [CrossRef]
116. Holderfield, M.T.; Hughes, C.C.W. Crosstalk Between Vascular Endothelial Growth Factor, Notch, and Transforming Growth Factor-β in Vascular Morphogenesis. *Circ. Res.* **2008**, *102*, 637–652. [CrossRef]
117. Desai, S.; Laskar, S.; Pandey, B. Autocrine IL-8 and VEGF mediate epithelial–mesenchymal transition and invasiveness via p38/JNK-ATF-2 signalling in A549 lung cancer cells. *Cell. Signal.* **2013**, *25*, 1780–1791. [CrossRef]
118. Gonzalez-Moreno, O.; Lecanda, J.; Green, J.E.; Segura, V.; Catena, R.; Serrano, D.; Calvo, A. VEGF elicits epithelial-mesenchymal transition (EMT) in prostate intraepithelial neoplasia (PIN)-like cells via an autocrine loop. *Exp. Cell Res.* **2010**, *316*, 554–567. [CrossRef]
119. Sennino, B.; Ishiguro-Oonuma, T.; Wei, Y.; Naylor, R.M.; Williamson, C.W.; Bhagwandin, V.; Tabruyn, S.P.; You, W.-K.; Chapman, H.A.; Christensen, J.G.; et al. Suppression of Tumor Invasion and Metastasis by Concurrent Inhibition of c-Met and VEGF Signaling in Pancreatic Neuroendocrine Tumors. *Cancer Discov.* **2012**, *2*, 270–287. [CrossRef]
120. Lu, K.V.; Chang, J.P.; Parachoniak, C.A.; Pandika, M.M.; Aghi, M.K.; Meyronet, D.; Isachenko, N.; Fouse, S.D.; Phillips, J.J.; Cheresh, D.A.; et al. VEGF Inhibits Tumor Cell Invasion and Mesenchymal Transition through a MET/VEGFR2 Complex. *Cancer Cell* **2012**, *22*, 21–35. [CrossRef]
121. Yang, J.; Yan, J.; Liu, B.-R. Targeting VEGF/VEGFR to Modulate Antitumor Immunity. *Front. Immunol.* **2018**, *9*, 978. [CrossRef] [PubMed]
122. Ott, P.A.; Hodi, F.S.; Buchbinder, E.I. Inhibition of Immune Checkpoints and Vascular Endothelial Growth Factor as Combination Therapy for Metastatic Melanoma: An Overview of Rationale, Preclinical Evidence, and Initial Clinical Data. *Front. Oncol.* **2015**, *5*, 202. [CrossRef] [PubMed]
123. Vang, K.B.; Vang, K.B.; Castermans, K.; Popescu, F.; Zhang, Y.; Egbrink, M.G.A.O.; Mescher, M.F.; Farrar, M.A.; Griffioen, A.W.; Mayo, K.H. Enhancement of T-cell-Mediated Antitumor Response: Angiostatic Adjuvant to Immunotherapy against Cancer. *Clin. Cancer Res.* **2011**, *17*, 3134–3145. [CrossRef]
124. Terme, M.; Pernot, S.; Marcheteau, E.; Sandoval, F.; Benhamouda, N.; Colussi, O.; Dubreuil, O.; Carpentier, A.F.; Tartour, E.; Taïeb, J. VEGFA-VEGFR Pathway Blockade Inhibits Tumor-Induced Regulatory T-cell Proliferation in Colorectal Cancer. *Cancer Res.* **2012**, *73*, 539–549. [CrossRef]
125. Koinis, F.; Vetsika, E.K.; Aggouraki, D.; Skalidaki, E.; Koutoulaki, A.; Gkioulmpasani, M.; Georgoulias, V.; Kotsakis, A.; Vetsika, E.K.; Georgoulias, V. Effect of First-Line Treatment on Myeloid-Derived Suppressor Cells' Subpopulations in the Peripheral Blood of Patients with Non–Small Cell Lung Cancer. *J. Thorac. Oncol.* **2016**, *11*, 1263–1272. [CrossRef]
126. Manzoni, M.; Rovati, B.; Ronzoni, M.; Loupakis, F.; Mariucci, S.; Ricci, V.; Gattoni, E.; Salvatore, L.; Tinelli, C.; Villa, E.; et al. Immunological Effects of Bevacizumab-Based Treatment in Metastatic Colorectal Cancer. *Oncology* **2010**, *79*, 187–196. [CrossRef]

127. Martino, E.; Misso, G.; Pastina, P.; Costantini, S.; Vanni, F.; Gandolfo, C.; Botta, C.; Capone, F.; Lombardi, A.; Pirtoli, L.; et al. Immune-modulating effects of bevacizumab in metastatic non-small-cell lung cancer patients. *Cell Death Discov.* **2016**, *2*, 16025. [CrossRef]
128. Hodi, F.S.; Lawrence, D.; Lezcano, C.; Wu, X.; Zhou, J.; Sasada, T.; Zeng, W.; Giobbie-Hurder, A.; Atkins, M.B.; Ibrahim, N.; et al. Bevacizumab plus Ipilimumab in Patients with Metastatic Melanoma. *Cancer Immunol. Res.* **2014**, *2*, 632–642. [CrossRef]
129. Wallin, J.J.; Bendell, J.C.; Funke, R.; Sznol, M.; Korski, K.; Jones, S.; Hernandez, G.; Mier, J.; He, X.; Hodi, F.S.; et al. Atezolizumab in combination with bevacizumab enhances antigen-specific T-cell migration in metastatic renal cell carcinoma. *Nat. Commun.* **2016**, *7*, 12624. [CrossRef]
130. Osada, T.; Chong, G.; Tansik, R.; Hong, T.; Spector, N.; Kumar, R.; Hurwitz, H.I.; Dev, I.; Nixon, A.B.; Lyerly, H.K.; et al. The effect of anti-VEGF therapy on immature myeloid cell and dendritic cells in cancer patients. *Cancer Immunol. Immunother.* **2008**, *57*, 1115–1124. [CrossRef]
131. Alfaro, C.; Suarez, N.; Gonzalez, A.D.L.G.; Solano, S.; Erro, L.; Dubrot, J.; Palazon, A.; Hervasstubbs, S.; Gurpide, A.; Lopez-Picazo, J.M.; et al. Influence of bevacizumab, sunitinib and sorafenib as single agents or in combination on the inhibitory effects of VEGF on human dendritic cell differentiation from monocytes. *Br. J. Cancer* **2009**, *100*, 1111–1119. [CrossRef] [PubMed]
132. Shrimali, R.K.; Yu, Z.; Theoret, M.R.; Chinnasamy, D.; Restifo, N.P.; Rosenberg, S.A. Antiangiogenic Agents Can Increase Lymphocyte Infiltration into Tumor and Enhance the Effectiveness of Adoptive Immunotherapy of Cancer. *Cancer Res.* **2010**, *70*, 6171–6180. [CrossRef]
133. Lanitis, E.; Irving, M.; Coukos, G. Targeting the tumor vasculature to enhance T cell activity. *Curr. Opin. Immunol.* **2015**, *33*, 55–63. [CrossRef]
134. Goel, S.; Duda, D.G.; Xu, L.; Munn, L.L.; Boucher, Y.; Fukumura, D.; Jain, R.K. Normalization of the Vasculature for Treatment of Cancer and Other Diseases. *Physiol. Rev.* **2011**, *91*, 1071–1121. [CrossRef]
135. Wu, X.; Giobbie-Hurder, A.; Liao, X.; Lawrence, D.; McDermott, D.; Zhou, J.; Rodig, S.; Hodi, F.S. VEGF Neutralization Plus CTLA-4 Blockade Alters Soluble and Cellular Factors Associated with Enhancing Lymphocyte Infiltration and Humoral Recognition in Melanoma. *Cancer Immunol. Res.* **2016**, *4*, 858–868. [CrossRef] [PubMed]
136. Finke, J.; Rini, B.; Ireland, J.; Rayman, P.; Richmond, A.; Golshayan, A.; Wood, L.; Elson, P.; Garcia, J.; Dreicer, R.; et al. Sunitinib Reverses Type-1 Immune Suppression and Decreases T-Regulatory Cells in Renal Cell Carcinoma Patients. *Clin. Cancer Res.* **2008**, *14*, 6674–6682. [CrossRef]
137. Rini, B.; Stein, M.; Shannon, P.; Eddy, S.; Tyler, A.; Stephenson, J.J.; Catlett, L.; Huang, B.; Healey, D.; Gordon, M. Phase 1 dose-escalation trial of tremelimumab plus sunitinib in patients with metastatic renal cell carcinoma. *Cancer* **2010**, *117*, 758–767. [CrossRef] [PubMed]
138. Adotevi, O.; Pere, H.; Ravel, P.; Haicheur, N.; Badoual, C.; Merillon, N.; Medioni, J.; Peyrard, S.; Roncelin, S.; Verkarre, V.; et al. A Decrease of Regulatory T Cells Correlates With Overall Survival After Sunitinib-based Antiangiogenic Therapy in Metastatic Renal Cancer Patients. *J. Immunother.* **2010**, *33*, 991–998. [CrossRef]
139. Xin, H.; Zhang, C.; Herrmann, A.; Du, Y.; Figlin, R.; Yu, H. Sunitinib Inhibition of Stat3 Induces Renal Cell Carcinoma Tumor Cell Apoptosis and Reduces Immunosuppressive Cells. *Cancer Res.* **2009**, *69*, 2506–2513. [CrossRef]
140. Ko, J.S.; Zea, A.H.; Rini, B.I.; Ireland, J.L.; Elson, P.; Cohen, P.; Golshayan, A.; Rayman, P.A.; Wood, L.; Garcia, J.; et al. Sunitinib Mediates Reversal of Myeloid-Derived Suppressor Cell Accumulation in Renal Cell Carcinoma Patients. *Clin. Cancer Res.* **2009**, *15*, 2148–2157. [CrossRef]
141. Van Hooren, L.; Georganaki, M.; Huang, H.; Mangsbo, S.M.; Dimberg, A. Sunitinib enhances the antitumor responses of agonistic CD40-antibody by reducing MDSCs and synergistically improving endothelial activation and T-cell recruitment. *Oncotarget* **2016**, *7*, 50277–50289. [CrossRef] [PubMed]
142. Huang, H.; Langenkamp, E.; Georganaki, M.; Loskog, A.; Fuchs, P.F.; Dieterich, L.C.; Kreuger, J.; Dimberg, A. VEGF suppresses T-lymphocyte infiltration in the tumor microenvironment through inhibition of NF-κB-induced endothelial activation. *FASEB J.* **2014**, *29*, 227–238. [CrossRef]
143. Ozao-Choy, J.; Ma, G.; Kao, J.; Wang, G.X.; Meseck, M.; Sung, M.; Schwartz, M.; Divino, C.M.; Pan, P.-Y.; Chen, S.-H. The Novel Role of Tyrosine Kinase Inhibitor in the Reversal of Immune Suppression and Modulation of Tumor Microenvironment for Immune-Based Cancer Therapies. *Cancer Res.* **2009**, *69*, 2514–2522. [CrossRef] [PubMed]

144. Zhang, X.; Fang, X.; Gao, Z.; Chen, W.; Tao, F.; Cai, P.; Yuan, H.; Shu, Y.; Xu, Q.; Sun, Y.; et al. Axitinib, a selective inhibitor of vascular endothelial growth factor receptor, exerts an anticancer effect in melanoma through promoting antitumor immunity. *Anti-Cancer Drugs* **2014**, *25*, 204–211. [CrossRef] [PubMed]
145. Du Four, S.; Maenhout, S.K.; Niclou, S.P.; Thielemans, K.; Neyns, B.; Aerts, J.L. Combined VEGFR and CTLA-4 blockade increases the antigen-presenting function of intratumoral DCs and reduces the suppressive capacity of intratumoral MDSCs. *Am. J. Cancer Res.* **2016**, *6*, 2514–2531. [PubMed]
146. Zhao, S.; Ren, S.; Jiang, T.; Zhu, B.; Li, X.; Zhao, C.; Jia, Y.; Shi, J.; Zhang, L.; Liu, X.; et al. Low-dose apatinib optimizes tumor microenvironment and potentiates antitumor effect of PD-1/PD-L1 blockade in lung cancer. *Cancer Immunol. Res.* **2019**, *7*, 630–643. [CrossRef] [PubMed]
147. Yang, J.; Yan, J.; Shao, J.; Xu, Q.; Meng, F.; Chen, F.; Ding, N.; Du, S.; Zhou, S.; Cai, J.; et al. Immune-Mediated Antitumor Effect By VEGFR2 Selective Inhibitor For Gastric Cancer. *OncoTargets Ther.* **2019**, *12*, 9757–9765. [CrossRef]
148. Cao, M.; Xu, Y.; Youn, J.-I.; Cabrera, R.; Zhang, X.; Gabrilovich, D.; Nelson, D.R.; Liu, C. Kinase inhibitor Sorafenib modulates immunosuppressive cell populations in a murine liver cancer model. *Lab. Investig.* **2011**, *91*, 598–608. [CrossRef]
149. Nagai, H.; Mukozu, T.; Matsui, D.; Kanekawa, T.; Kanayama, M.; Wakui, N.; Momiyama, K.; Shinohara, M.; Iida, K.; Ishii, K.; et al. Sorafenib Prevents Escape from Host Immunity in Liver Cirrhosis Patients with Advanced Hepatocellular Carcinoma. *Clin. Dev. Immunol.* **2012**, *2012*, 1–8. [CrossRef]
150. Chen, M.-L.; Yan, B.-S.; Lu, W.-C.; Yu, S.-L.; Yang, P.-C.; Cheng, A.-L. Sorafenib relieves cell-intrinsic and cell-extrinsic inhibitions of effector T cells in tumor microenvironment to augment antitumor immunity. *Int. J. Cancer* **2013**, *134*, 319–331. [CrossRef]
151. Hipp, M.M.; Hilf, N.; Walter, S.; Werth, D.; Brauer, K.M.; Radsak, M.P.; Weinschenk, T.; Singh-Jasuja, H.; Brossart, P. Sorafenib, but not sunitinib, affects function of dendritic cells and induction of primary immune responses. *Blood* **2008**, *111*, 5610–5620. [CrossRef] [PubMed]
152. Dirkx, A.E.M.; Egbrink, M.G.A.O.; Kuijpers, M.J.E.; Van Der Niet, S.T.; Heijnen, V.V.T.; Steege, J.C.A.B.-T.; Wagstaff, J.; Griffioen, A.W. Tumor angiogenesis modulates leukocyte-vessel wall interactions in vivo by reducing endothelial adhesion molecule expression. *Cancer Res.* **2003**, *63*, 2322–2329. [PubMed]
153. Bouzin, C.; Brouet, A.; De Vriese, J.; Dewever, J.; Feron, O. Effects of Vascular Endothelial Growth Factor on the Lymphocyte-Endothelium Interactions: Identification of Caveolin-1 and Nitric Oxide as Control Points of Endothelial Cell Anergy. *J. Immunol.* **2007**, *178*, 1505–1511. [CrossRef] [PubMed]
154. Tromp, S.C.; Egbrink, M.G.A.O.; Dings, R.P.M.; Van Velzen, S.; Slaaf, D.W.; Hillen, H.F.P.; Tangelder, G.J.; Reneman, R.S.; Griffioen, A.W. Tumor angiogenesis factors reduce leukocyte adhesion in vivo. *Int. Immunol.* **2000**, *12*, 671–676. [CrossRef]
155. Dirkx, A.E.M.; Egbrink, M.G.A.O.; Castermans, K.; Van Der Schaft, D.W.J.; Thijssen, V.L.J.L.; Dings, R.P.M.; Kwee, L.; Mayo, K.H.; Wagstaff, J.; Ter Steege, J.C.A.B.; et al. Anti-angiogenesis therapy can overcome endothelial cell anergy and promote leukocyte-endothelium interactions and infiltration in tumors. *FASEB J.* **2006**, *20*, 621–630. [CrossRef]
156. Schmittnaegel, M.; Rigamonti, N.; Kadioglu, E.; Cassará, A.; Rmili, C.W.; Kiialainen, A.; Kienast, Y.; Mueller, H.-J.; Ooi, C.-H.; Laoui, D.; et al. Dual angiopoietin-2 and VEGFA inhibition elicits antitumor immunity that is enhanced by PD-1 checkpoint blockade. *Sci. Transl. Med.* **2017**, *9*, eaak9670. [CrossRef]
157. Hamzah, J.; Jugold, M.; Kiessling, F.; Rigby, P.J.; Manzur, M.; Marti, H.H.; Rabie, T.; Kaden, S.; Gröne, H.-J.; Hämmerling, G.J.; et al. Vascular normalization in Rgs5-deficient tumours promotes immune destruction. *Nat. Cell Biol.* **2008**, *453*, 410–414. [CrossRef] [PubMed]
158. Huang, Y.; Yuan, J.; Righi, E.; Kamoun, W.S.; Ancukiewicz, M.; Nezivar, J.; Santosuosso, M.; Martin, J.D.; Martin, M.R.; Vianello, F.; et al. Vascular normalizing doses of antiangiogenic treatment reprogram the immunosuppressive tumor microenvironment and enhance immunotherapy. *Proc. Natl. Acad. Sci. USA* **2012**, *109*, 17561–17566. [CrossRef]
159. Allen, E.; Jabouille, A.; Rivera, L.B.; Lodewijckx, I.; Missiaen, R.; Steri, V.; Feyen, K.; Tawney, J.; Hanahan, D.; Michael, I.P.; et al. Combined antiangiogenic and anti–PD-L1 therapy stimulates tumor immunity through HEV formation. *Sci. Transl. Med.* **2017**, *9*, eaak9679. [CrossRef]
160. Basu, A.; Hoerning, A.; Datta, D.; Edelbauer, M.; Stack, M.P.; Calzadilla, K.; Pal, S.; Briscoe, D.M. Cutting Edge: Vascular Endothelial Growth Factor-Mediated Signaling in Human CD45RO + CD4 + T Cells Promotes Akt and ERK Activation and Costimulates IFN-γ Production. *J. Immunol.* **2009**, *184*, 545–549. [CrossRef]

161. Li, Y.-L.; Zhao, H.; Ren, X.-B. Relationship of VEGF/VEGFR with immune and cancer cells: Staggering or forward? *Cancer Biol. Med.* **2016**, *13*, 206–214. [CrossRef] [PubMed]
162. Ohm, J.E.; Gabrilovich, D.I.; Sempowski, G.D.; Kisseleva, E.; Parman, K.S.; Nadaf, S.; Carbone, D.P. VEGF inhibits T-cell development and may contribute to tumor-induced immune suppression. *Blood* **2003**, *101*, 4878–4886. [CrossRef] [PubMed]
163. Gavalas, N.G.; Tsiatas, M.; Tsitsilonis, O.; Politi, E.; Ioannou, K.; Ziogas, A.C.; Rodolakis, A.; Vlahos, G.; Thomakos, N.; Haidopoulos, D.; et al. VEGF directly suppresses activation of T cells from ascites secondary to ovarian cancer via VEGF receptor type 2. *Br. J. Cancer* **2012**, *107*, 1869–1875. [CrossRef] [PubMed]
164. Ziogas, A.C.; Gavalas, N.G.; Tsiatas, M.; Tsitsilonis, O.; Politi, E.; Terpos, E.; Rodolakis, A.; Vlahos, G.; Thomakos, N.; Haidopoulos, D.; et al. VEGF directly suppresses activation of T cells from ovarian cancer patients and healthy individuals via VEGF receptor Type 2. *Int. J. Cancer* **2011**, *130*, 857–864. [CrossRef]
165. Voron, T.; Colussi, O.; Marcheteau, E.; Pernot, S.; Nizard, M.; Pointet, A.-L.; Latreche, S.; Bergaya, S.; Benhamouda, N.; Tanchot, C.; et al. VEGF-A modulates expression of inhibitory checkpoints on CD8+ T cells in tumors. *J. Exp. Med.* **2015**, *212*, 139–148. [CrossRef] [PubMed]
166. Motz, G.T.; Santoro, S.P.; Wang, L.-P.; Garrabrant, T.; Lastra, R.R.; Hagemann, I.S.; Lal, P.; Feldman, M.D.; Benencia, F.; Coukos, G. Tumor endothelium FasL establishes a selective immune barrier promoting tolerance in tumors. *Nat. Med.* **2014**, *20*, 607–615. [CrossRef]
167. Wada, J.; Suzuki, H.; Fuchino, R.; Yamasaki, A.; Nagai, S.; Yanai, K.; Koga, K.; Nakamura, M.; Tanaka, M.; Morisaki, T.; et al. The contribution of vascular endothelial growth factor to the induction of regulatory T-cells in malignant effusions. *Anticancer Res.* **2009**, *29*, 881–888.
168. Bettelli, E.; Carrier, Y.; Gao, W.; Korn, T.; Strom, T.B.; Oukka, M.; Weiner, H.L.; Kuchroo, V.K. Reciprocal developmental pathways for the generation of pathogenic effector TH17 and regulatory T cells. *Nat. Cell Biol.* **2006**, *441*, 235–238. [CrossRef]
169. Hansen, W.; Hutzler, M.; Abel, S.; Alter, C.; Stockmann, C.; Kliche, S.; Albert, J.; Sparwasser, T.; Sakaguchi, S.; Westendorf, A.M.; et al. Neuropilin 1 deficiency on CD4+Foxp3+ regulatory T cells impairs mouse melanoma growth. *J. Exp. Med.* **2012**, *209*, 2001–2016. [CrossRef]
170. Gabrilovich, D.I.; Chen, H.L.; Girgis, K.R.; Cunningham, H.T.; Meny, G.M.; Nadaf, S.; Kavanaugh, D.; Carbone, D.P. Production of vascular endothelial growth factor by human tumors inhibits the functional maturation of dendritic cells. *Nat. Med.* **1996**, *2*, 1096–1103. [CrossRef]
171. Boissel, N.; Rousselot, P.; Raffoux, E.; Cayuela, J.-M.; Maarek, O.; Charron, D.; Degos, L.; Dombret, H.; Toubert, A.; Réa, D. Defective blood dendritic cells in chronic myeloid leukemia correlate with high plasmatic VEGF and are not normalized by imatinib mesylate. *Leukemia* **2004**, *18*, 1656–1661. [CrossRef] [PubMed]
172. Lissoni, P.; Malugani, F.; Bonfanti, A.; Bucovec, R.; Secondino, S.; Brivio, F.; Ferrari-Bravo, A.; Ferrante, R.; Vigoré, L.; Rovelli, F.; et al. Abnormally enhanced blood concentrations of vascular endothelial growth factor (VEGF) in metastatic cancer patients and their relation to circulating dendritic cells, IL-12 and endothelin-1. *J. Boil. Regul. Homeost. Agents* **2001**, *15*, 140–144.
173. Strauss, L.; Volland, D.; Kunkel, M.; Reichert, T.E. Dual role of VEGF family members in the pathogenesis of head and neck cancer (HNSCC): Possible link between angiogenesis and immune tolerance. *Med. Sci. Monit.* **2005**, *11*, BR280–R292. [PubMed]
174. Della Porta, M.; Danova, M.; Rigolin, G.M.; Brugnatelli, S.; Rovati, B.; Tronconi, C.; Fraulini, C.; Rossi, A.R.; Riccardi, A.; Castoldi, G. Dendritic Cells and Vascular Endothelial Growth Factor in Colorectal Cancer: Correlations with Clinicobiological Findings. *Oncology* **2005**, *68*, 276–284. [CrossRef] [PubMed]
175. Mimura, K.; Kono, K.; Takahashi, A.; Kawaguchi, Y.; Fujii, H. Vascular endothelial growth factor inhibits the function of human mature dendritic cells mediated by VEGF receptor-2. *Cancer Immunol. Immunother.* **2006**, *56*, 761–770. [CrossRef] [PubMed]
176. Oyama, T.; Ran, S.; Ishida, T.; Nadaf, S.; Kerr, L.; Carbone, D.P.; Gabrilovich, D.I. Vascular endothelial growth factor affects dendritic cell maturation through the inhibition of nuclear factor-kappa B activation in hemopoietic progenitor cells. *J. Immunol.* **1998**, *160*, 1224–1232. [PubMed]
177. Dikov, M.M.; Ohm, J.E.; Ray, N.; Tchekneva, E.E.; Burlison, J.; Moghanaki, D.; Nadaf, S.; Carbone, D.P. Differential Roles of Vascular Endothelial Growth Factor Receptors 1 and 2 in Dendritic Cell Differentiation. *J. Immunol.* **2004**, *174*, 215–222. [CrossRef] [PubMed]

178. Oussa, N.A.E.; Dahmani, A.; Gomis, M.; Richaud, M.; Andreev, E.; Navab-Daneshmand, A.-R.; Taillefer, J.; Carli, C.; Boulet, S.; Sabbagh, L.; et al. VEGF Requires the Receptor NRP-1 To Inhibit Lipopolysaccharide-Dependent Dendritic Cell Maturation. *J. Immunol.* **2016**, *197*, 3927–3935. [CrossRef]
179. Lin, Y.-L.; Liang, Y.-C.; Chiang, B.-L. Placental growth factor down-regulates type 1 T helper immune response by modulating the function of dendritic cells. *J. Leukoc. Biol.* **2007**, *82*, 1473–1480. [CrossRef]
180. Curiel, T.J.; Wei, S.; Dong, H.; Alvarez, X.; Cheng, P.; Mottram, P.; Krzysiek, R.; Knutson, K.L.; Daniel, B.; Zimmermann, M.C.; et al. Blockade of B7-H1 improves myeloid dendritic cell–mediated antitumor immunity. *Nat. Med.* **2003**, *9*, 562–567. [CrossRef]
181. Karakhanova, S.; Link, J.; Heinrich, M.; Shevchenko, I.; Yang, Y.; Hassenpflug, M.; Bunge, H.; Von Ahn, K.; Brecht, R.; Mathes, A.; et al. Characterization of myeloid leukocytes and soluble mediators in pancreatic cancer: Importance of myeloid-derived suppressor cells. *Oncoimmunology* **2015**, *4*, e998519. [CrossRef] [PubMed]
182. Huang, Y.; Chen, X.; Dikov, M.M.; Novitskiy, S.V.; Mosse, C.A.; Yang, L.; Carbone, D.P. Distinct roles of VEGFR-1 and VEGFR-2 in the aberrant hematopoiesis associated with elevated levels of VEGF. *Blood* **2007**, *110*, 624–631. [CrossRef] [PubMed]
183. Gabrilovich, D.I.; Ostrand-Rosenberg, S.; Bronte, V. Coordinated regulation of myeloid cells by tumours. *Nat. Rev. Immunol.* **2012**, *12*, 253–268. [CrossRef] [PubMed]
184. Serafini, P.; Mgebroff, S.; Noonan, K.; Borrello, I. Myeloid-derived suppressor cells promote cross-tolerance in B-cell lymphoma by expanding regulatory T cells. *Cancer Res.* **2008**, *68*, 5439–5449. [CrossRef]
185. Evoron, T.; Emarcheteau, E.; Epernot, S.; Ecolussi, O.; Tartour, E.; Etaieb, J.; Eterme, M. Control of the Immune Response by Pro-Angiogenic Factors. *Front. Oncol.* **2014**, *4*, 70. [CrossRef]
186. Huang, B.; Pan, P.-Y.; Li, Q.; Sato, A.I.; Levy, D.E.; Bromberg, J.; Divino, C.M.; Chen, S.-H. Gr-1+CD115+ Immature Myeloid Suppressor Cells Mediate the Development of Tumor-Induced T Regulatory Cells and T-Cell Anergy in Tumor-Bearing Host. *Cancer Res.* **2006**, *66*, 1123–1131. [CrossRef]
187. Du Four, S.; Maenhout, S.K.; De Pierre, K.; Renmans, D.; Niclou, S.P.; Thielemans, K.; Neyns, B.; Aerts, J.L. Axitinib increases the infiltration of immune cells and reduces the suppressive capacity of monocytic MDSCs in an intracranial mouse melanoma model. *Oncoimmunology* **2015**, *4*, e998107. [CrossRef]
188. Mantovani, A.; Biswas, S.K.; Galdiero, M.R.; Sica, A.; Locati, M. Macrophage plasticity and polarization in tissue repair and remodelling. *J. Pathol.* **2012**, *229*, 176–185. [CrossRef]
189. Komohara, Y.; Fujiwara, Y.; Ohnishi, K.; Takeya, M. Tumor-associated macrophages: Potential therapeutic targets for anti-cancer therapy. *Adv. Drug Deliv. Rev.* **2016**, *99*, 180–185. [CrossRef]
190. DiNapoli, M.R.; Calderon, C.L.; Lopez, D.M. The altered tumoricidal capacity of macrophages isolated from tumor-bearing mice is related to reduce expression of the inducible nitric oxide synthase gene. *J. Exp. Med.* **1996**, *183*, 1323–1329. [CrossRef]
191. Barleon, B.; Sozzani, S.; Zhou, D.; Weich, H.; Mantovani, A.; Marme, D. Migration of human monocytes in response to vascular endothelial growth factor (VEGF) is mediated via the VEGF receptor flt-1. *Blood* **1996**, *87*, 3336–3343. [CrossRef] [PubMed]
192. Linde, N.; Lederle, W.; Depner, S.; Van Rooijen, N.; Gutschalk, C.M.; Mueller, M.M. Vascular endothelial growth factor-induced skin carcinogenesis depends on recruitment and alternative activation of macrophages. *J. Pathol.* **2012**, *227*, 17–28. [CrossRef] [PubMed]
193. Dalton, H.J.; Pradeep, S.; McGuire, M.; Hailemichael, Y.; Ma, S.; Lyons, Y.; Armaiz-Pena, G.N.; Previs, R.A.; Hansen, J.M.; Rupaimoole, R.; et al. Macrophages Facilitate Resistance to Anti-VEGF Therapy by Altered VEGFR Expression. *Clin. Cancer Res.* **2017**, *23*, 7034–7046. [CrossRef]
194. Kloepper, J.; Riedemann, L.; Amoozgar, Z.; Seano, G.; Susek, K.; Yu, V.; Dalvie, N.; Amelung, R.L.; Datta, M.; Song, J.W.; et al. Ang-2/VEGF bispecific antibody reprograms macrophages and resident microglia to anti-tumor phenotype and prolongs glioblastoma survival. *Proc. Natl. Acad. Sci. USA* **2016**, *113*, 4476–4481. [CrossRef] [PubMed]
195. Gabrusiewicz, K.; Liu, D.; Cortes-Santiago, N.; Hossain, M.B.; Conrad, C.A.; Aldape, K.D.; Fuller, G.N.; Marini, F.C.; Alonso, M.M.; Idoate, M.A.; et al. Anti-vascular endothelial growth factor therapy-induced glioma invasion is associated with accumulation of Tie2-expressing monocytes. *Oncotarget* **2014**, *5*, 2208–2220. [CrossRef]

196. Lu-Emerson, C.; Snuderl, M.; Kirkpatrick, N.D.; Goveia, J.; Davidson, C.; Huang, Y.; Riedemann, L.; Taylor, J.; Ivy, P.; Duda, D.G.; et al. Increase in tumor-associated macrophages after antiangiogenic therapy is associated with poor survival among patients with recurrent glioblastoma. *Neuro-Oncol.* **2013**, *15*, 1079–1087. [CrossRef]
197. Yasuda, S.; Sho, M.; Yamato, I.; Yoshiji, H.; Wakatsuki, K.; Nishiwada, S.; Yagita, H.; Nakajima, Y. Simultaneous blockade of programmed death 1 and vascular endothelial growth factor receptor 2 (VEGFR2) induces synergistic anti-tumour effect in vivo. *Clin. Exp. Immunol.* **2013**, *172*, 500–506. [CrossRef]
198. Meder, L.; Schuldt, P.; Thelen, M.; Schmitt, A.; Dietlein, F.; Klein, S.; Borchmann, S.; Wennhold, K.; Vlasic, I.; Oberbeck, S.; et al. Combined VEGF and PD-L1 Blockade Displays Synergistic Treatment Effects in an Autochthonous Mouse Model of Small Cell Lung Cancer. *Cancer Res.* **2018**, *78*, 4270–4281. [CrossRef]
199. Yuan, J.; Zhou, J.; Dong, Z.; Tandon, S.; Kuk, D.; Panageas, K.S.; Wong, P.; Wu, X.; Naidoo, J.; Page, D.B.; et al. Pretreatment Serum VEGF Is Associated with Clinical Response and Overall Survival in Advanced Melanoma Patients Treated with Ipilimumab. *Cancer Immunol. Res.* **2014**, *2*, 127–132. [CrossRef]
200. Wu, X.; Li, J.; Connolly, E.M.; Liao, X.; Ouyang, J.; Giobbie-Hurder, A.; Lawrence, D.; McDermott, D.; Murphy, G.; Zhou, J.; et al. Combined Anti-VEGF and Anti–CTLA-4 Therapy Elicits Humoral Immunity to Galectin-1 Which Is Associated with Favorable Clinical Outcomes. *Cancer Immunol. Res.* **2017**, *5*, 446–454. [CrossRef]
201. Socinski, M.A.; Jotte, R.M.; Cappuzzo, F.; Orlandi, F.; Stroyakovskiy, D.; Nogami, N.; Rodríguez-Abreu, D.; Moro-Sibilot, D.; Thomas, C.A.; Barlesi, F.; et al. Atezolizumab for First-Line Treatment of Metastatic Nonsquamous NSCLC. *N. Engl. J. Med.* **2018**, *378*, 2288–2301. [CrossRef] [PubMed]
202. Atkins, M.B.; Plimack, E.R.; Puzanov, I.; Fishman, M.N.; McDermott, D.F.; Cho, D.C.; Vaishampayan, U.; George, S.; Olencki, T.E.; Tarazi, J.C.; et al. Axitinib in combination with pembrolizumab in patients with advanced renal cell cancer: A non-randomised, open-label, dose-finding, and dose-expansion phase 1b trial. *Lancet Oncol.* **2018**, *19*, 405–415. [CrossRef]
203. Choueiri, T.K.; Larkin, J.; Oya, M.; Thistlethwaite, F.; Martignoni, M.; Nathan, P.; Powles, T.; McDermott, D.; Robbins, P.B.; Chism, D.D.; et al. Preliminary results for avelumab plus axitinib as first-line therapy in patients with advanced clear-cell renal-cell carcinoma (JAVELIN Renal 100): An open-label, dose-finding and dose-expansion, phase 1b trial. *Lancet Oncol.* **2018**, *19*, 451–460. [CrossRef]
204. Amin, A.; Dudek, A.Z.; Logan, T.F.; Lance, R.S.; Holzbeierlein, J.M.; Knox, J.J.; Master, V.A.; Pal, S.K.; Miller, W.H.; Karsh, L.I.; et al. Survival with AGS-003, an autologous dendritic cell-based immunotherapy, in combination with sunitinib in unfavorable risk patients with advanced renal cell carcinoma (RCC): Phase 2 study results. *J. Immunother. Cancer* **2015**, *3*. [CrossRef] [PubMed]
205. McDermott, D.F.; Atkins, M.B.; Motzer, R.J.; Rini, B.I.; Escudier, B.J.; Fong, L.; Joseph, R.W.; Pal, S.K.; Sznol, M.; Hainsworth, J.D.; et al. A phase II study of atezolizumab (atezo) with or without bevacizumab (bev) versus sunitinib (sun) in untreated metastatic renal cell carcinoma (mRCC) patients (pts). *J. Clin. Oncol.* **2017**, *35*, 431. [CrossRef]
206. Rini, B.I.; Powles, T.; Atkins, M.B.; Escudier, B.; McDermott, D.F.; Suarez, C.; Bracarda, S.; Stadler, W.M.; Donskov, F.; Lee, J.L.; et al. Atezolizumab plus bevacizumab versus sunitinib in patients with previously untreated metastatic renal cell carcinoma (IMmotion151): A multicentre, open-label, phase 3, randomised controlled trial. *Lancet* **2019**, *393*, 2404–2415. [CrossRef]
207. Motzer, R.J.; Penkov, K.; Haanen, J.; Rini, B.; Albiges, L.; Campbell, M.T.; Venugopal, B.; Kollmannsberger, C.; Negrier, S.; Uemura, M.; et al. Avelumab plus axitinib versus sunitinib for advanced renal-cell carcinoma. *N. Engl. J. Med.* **2019**, *380*, 1103–1115. [CrossRef]
208. Rini, B.I.; Plimack, E.R.; Stus, V.; Gafanov, R.; Hawkins, R.; Nosov, D.; Pouliot, F.; Alekseev, B.; Soulières, D.; Melichar, B.; et al. Pembrolizumab plus axitinib versus sunitinib for advanced renal-cell carcinoma. *N. Engl. J. Med.* **2019**, *380*. [CrossRef] [PubMed]
209. Lee, M.S.; Ryoo, B.Y.; Hsu, C.H.; Numata, K.; Stein, S.; Verret, W.; Hack, S.P.; Spahn, J.; Liu, B.; Abdullah, H.; et al. Atezolizumab with or without bevacizumab in unresectable hepatocellular carcinoma (GO30140): An open-label, multicentre, phase 1b study. *Lancet Oncol.* **2020**, *21*, 808–820. [CrossRef]
210. Finn, R.S.; Qin, S.; Ikeda, M.; Galle, P.R.; Ducreux, M.; Kim, T.Y.; Kudo, M.; Breder, V.; Merle, P.; Kaseb, A.O.; et al. Atezolizumab plus bevacizumab in unresectable hepatocellular carcinoma. *N. Engl. J. Med.* **2020**, *382*, 1894–1905. [CrossRef]

211. Gagnon, M.L.; Bielenberg, D.R.; Gechtman, Z.; Miao, H.Q.; Takashima, S.; Soker, S.; Klagsbrun, M. Identification of a natural soluble neuropilin-1 that binds vascular endothelial growth factor: In vivo expression and antitumor activity. *Proc. Natl. Acad. Sci. USA* **2000**, *97*, 2573–2578. [CrossRef] [PubMed]
212. Wedam, S.B.; Low, J.A.; Yang, S.X.; Chow, C.K.; Choyke, P.; Danforth, D.; Hewitt, S.M.; Berman, A.; Steinberg, S.M.; Liewehr, D.J.; et al. Antiangiogenic and antitumor effects of bevacizumab in patients with inflammatory and locally advanced breast cancer. *J. Clin. Oncol.* **2006**, *24*, 769–777. [CrossRef] [PubMed]
213. von Baumgarten, L.; Brucker, D.; Tirniceru, A.; Kienast, Y.; Grau, S.; Burgold, S.; Herms, J.; Winkler, F. Bevacizumab has differential and dose-dependent effects on glioma blood vessels and tumor cells. *Clin. Cancer Res.* **2011**, *17*, 6192–6205. [CrossRef] [PubMed]
214. Fan, F.; Samuel, S.; Gaur, P.; Lu, J.; Dallas, N.A.; Xia, L.; Bose, D.; Ramachandran, V.; Ellis, L.M. Chronic exposure of colorectal cancer cells to bevacizumab promotes compensatory pathways that mediate tumour cell migration. *Br. J. Cancer* **2011**, *104*, 1270–1277. [CrossRef]
215. Motz, G.T.; Coukos, G. The parallel lives of angiogenesis and immunosuppression: Cancer and other tales. *Nat. Rev. Immunol.* **2011**, *11*, 702–711. [CrossRef]
216. Jain, R.K. Normalizing tumor vasculature with anti-angiogenic therapy: A new paradigm for combination therapy. *Nat. Med.* **2001**, *7*, 987–989. [CrossRef]

Publisher's Note: MDPI stays neutral with regard to jurisdictional claims in published maps and institutional affiliations.

© 2020 by the authors. Licensee MDPI, Basel, Switzerland. This article is an open access article distributed under the terms and conditions of the Creative Commons Attribution (CC BY) license (http://creativecommons.org/licenses/by/4.0/).

Review

New Insights into Diffuse Large B-Cell Lymphoma Pathobiology

Antonio Giovanni Solimando [1,2,*], Tiziana Annese [3], Roberto Tamma [3], Giuseppe Ingravallo [4], Eugenio Maiorano [4], Angelo Vacca [1], Giorgina Specchia [5] and Domenico Ribatti [3,*]

[1] Department of Biomedical Sciences and Human Oncology, Section of Internal Medicine 'G. Baccelli', University of Bari Medical School, 70124 Bari, Italy; angelo.vacca@uniba.it
[2] Istituto di Ricovero e Cura a Carattere Scientifico-IRCCS Istituto Tumori "Giovanni Paolo II" of Bari, 70124 Bari, Italy
[3] Department of Basic Medical Sciences, Neurosciences, and Sensory Organs, University of Bari Medical School, 70124 Bari, Italy; tiziana.annese@uniba.it (T.A.); roberto.tamma@uniba.it (R.T.)
[4] Department of Emergency and Transplantation, Pathology Section, University of Bari Medical School, 70100 Bari, Italy; giuseppe.ingravallo@uniba.it (G.I.); eugenio.maiorano@uniba.it (E.M.)
[5] Department of Emergency and Transplantation, Hematology Section, University of Bari Medical School, 70100 Bari, Italy; giorgina.specchia@uniba.it
* Correspondence: antonio.solimando@uniba.it (A.G.S.); domenico.ribatti@uniba.it (D.R.); Tel.: +39-3395626475 (A.G.S.); +39-080.5478326 (D.R.)

Received: 16 June 2020; Accepted: 8 July 2020; Published: 11 July 2020

Abstract: Diffuse large B-cell lymphoma (DLBCL) is the most common non-Hodgkin lymphoma (NHL), accounting for about 40% of all cases of NHL. Analysis of the tumor microenvironment is an important aspect of the assessment of the progression of DLBCL. In this review article, we analyzed the role of different cellular components of the tumor microenvironment, including mast cells, macrophages, and lymphocytes, in the tumor progression of DLBCL. We examined several approaches to confront the available pieces of evidence, whereby three key points emerged. DLBCL is a disease of malignant B cells spreading and accumulating both at nodal and at extranodal sites. In patients with both nodal and extranodal lesions, the subsequent induction of a cancer-friendly environment appears pivotal. The DLBCL cell interaction with mature stromal cells and vessels confers tumor protection and inhibition of immune response while delivering nutrients and oxygen supply. Single cells may also reside and survive in protected niches in the nodal and extranodal sites as a source for residual disease and relapse. This review aims to molecularly and functionally recapitulate the DLBCL–milieu crosstalk, to relate niche and pathological angiogenic constitution and interaction factors to DLBCL progression.

Keywords: DLBCL; tumor microenvironment; angiogenesis; cell adhesion mediated drug resistance; tumor progression

1. Introduction

Diffuse large B-cell lymphoma (DLBCL) classified by the 2008 World Health Organization (WHO) classification as one of the B-cell lymphomas types is the most common non-Hodgkin B-cell lymphoma (NHL), accounting for about 40% of all cases of NHL [1]. DLBCL characteristically presents with advanced stage, both in nodal and in extranodal symptomatic disease, with a median age of 60, representing an important disease holding a practical objective of treatment represented by a curative approach, while minimizing the toxicity profile [2]. Most DLBCLs arise from germinal B cells at different stages of differentiation where recurrent genetic alterations contribute to the molecular pathogenesis of the disease [3,4]. The gene expression profiling technique allowed identifying at least two molecular subtypes of DLBCL with different prognoses [5]. The first is the lymphoma derived from normal germinal center B cells (GCB) and the second one is the lymphoma derived from activated B cells (ABCs) that arise from post-germinal center B cells that are blocked during plasmocytic differentiation. The two subtypes have different oncogenic mechanisms [6].

Specific markers, including CD10, LM02, and BCL6 are expressed in GCB patients who have a better response to conventional chemotherapy, whereas ABC patients express lower levels of BCL6 and are refractory to chemotherapy [5,7]. The ABC type showed constitutive activation of NF-κB which may be related to the presence of mutations of multiple genes regulating this pathway [8,9]. Constitutively activated STAT3 is correlated with a more advanced clinical stage and overall poor survival in DLBCL [10,11]. In ABC DLBCL, the activation of the Janus kinases (JAKs)/STAT3 pathway correlates with autocrine production of intereukin-6 (IL-6) and IL-10, which promotes cancer progression [12,13]. The STAT3 gene is a transcriptional target of BCL6 and is highly expressed and activated in ABC DLBCL and BCL6-negative normal germinal center B cells [12]. Moreover, STAT3 is strongly linked to tumor angiogenesis and metastasis and is related to poor prognosis in different tumors [14,15]. Activation of STAT3 contributes to hypoxia-inducible factor 1 alpha (HIF-1α) and vascular endothelial growth factor (VEGF) expression in tumor cells, while VEGF in turn activates STAT3 in endothelial cells. Finally, STAT3 inhibits the expression of the anti-angiogenesis transcription factor p53 [16].

In DLBCL, the gene expression signatures "stromal 1" and "stromal 2", related to extracellular matrix and angiogenesis-related genes, respectively, were identified [17]. Fibrosis and myelo-histiocytic infiltration, representing the "stromal 1" signature, correlated with a positive clinical outcome [18], while the "stromal 2" signature, characterized by increased vasculogenic activity, correlated with dismal prognosis in subjects treated with the R-CHOP (rituximab, cyclophosphamide, doxorubicin, vincristine, and prednisone) protocol [17]. Monocytic myeloid-derived suppressor cells and tumor-associated macrophages (TAMs) play a crucial role in the "stromal 2" signature [19,20]. Overall, as in other hematological niche addicted malignancies [21–24], the current evidence pinpoints that DLBCL disease progression is a multistep transformation process characterized by a complex vicious cycle between lymphoma cells and the tumor milieu.

Here, we show the latest findings on the disease evolution of DLBCL, by providing a specific focus on the role of new players within the cancer immune microenvironment in order to envision novel theragnostic windows.

2. Bridging the Gaps between Disease Biology and Clinical Translation: New and Old Tricks in DLBCL Classification

A correct diagnosis of DLBCL requires, in addition to the availability of qualitatively and quantitatively adequate tissue, a correct application of the most recent classification principles provided by the use of any ancillary diagnostic techniques. In particular, modern histopathological diagnostics of lymphomas requires knowledge and combination of morphological, phenotypic molecular, cytogenetic, and clinical profiling. This methodological approach constitutes the founding principle of the World Health Organization (WHO) and was translated into the "blue book" "WHO Classification of Tumors of the Hematopoietic and Lymphoid Tissues" [1]. Recent progresses in understanding the immunogenetic mechanisms and genetic molecular alterations of hematopoietic and in particular lymphoid neoplasms allowed a pathogenetic approach to the DLBCL taxonomy. Many lymphomas are considered distinct entities, characterized by immunophenotypic profiles and known genetic alterations, identifiable with laboratory techniques now widely used, with good reproducibility. DLBCL parallels the complex NHL biological architecture, being differentiated into the GCB type and the ABC/non-GC type, by means of an immunohistochemical algorithm, which is a distinction that can influence the therapeutic choice [1,25,26]. Furthermore, the co-expression of MYC and BCL2 identifies a new prognostic "subset" ("double-expressor" lymphomas) [27]. Although the understanding of the mutation scenario was also widened and deepened, the translational relevance in the clinical subset still represents an unmet medical need.

Recently, NGS studies uncovered different profiles of genomic alterations to be relevant in both the GCB and the non-GCB/ABC subtypes [28]. Alteration in histone-lysine N-methyltransferase (EZH2), as well as the translocation of BCL2 and GNA13 mutation, is a fundamental molecular fingerprint described in GCB. Conversely, MYD88, CD79a, CARD11, and TNFAIPA3 mutations play a pivotal role in non-GCB/ACB by activating the BCR and NF-κB pathways [25]. The importance of a subdivision of diffuse large B-cell lymphomas, NOS in the two groups (GCB and non-GCB/ABC), is confirmed. This distinction, with possible therapeutic consequences, can be obtained in routine diagnostics by applying an immunohistochemical algorithm based on a relatively simple and reliable antibody panel (CD10, BCL6, and IRF4/MUM1) [28]. Moreover, among the DLBCL NOS, the immunohistochemical co-expression of MYC and BCL2, deemed biologically and clinically relevant, identifies the category of double-expressor disease, harboring an unfavorable prognostic impact [27].

2.1. Molecular Pathogenesis: Novel Insights

Double/triple-hit high-grade B-cell lymphoma (HGBL-DH/TH) constitutes approximately 8% of DLBCL, harboring MYC, BCL2, and/or BCL6 translocations. Most of them belong to the GCB molecular subgroup and, clinically, despite the generally superior prognosis of GCB DLBCLs, patients with HGBL-DH/TH have a poor outcome [29]. Double-hit lymphomas show a distinct gene expression profile when dissected by RNA-seq. For example, 157 de novo GCB DLBCLs, including 25 HGBL-DH/TH BCL2, were analyzed to define gene expression differences between HGBL-DH/TH BCL2 and other GCB DLBCLs [30]. When RNA-seq was applied to RNA extracted from fresh frozen biopsy samples, 104 genes that were most significantly differentially expressed between HGBL-DH/TH BCL2 and other GCB DLBCLs were identified [30]. Double-hit gene signature-positive (DHITsig-pos) DLBCLs are characterized by a peculiar cell of origin and a distinct mutational landscape, after genetic feature association with DHITsig status. DHITsig-pos tumors were universally positive for CD10 staining, and the majority were MUM1 (IRF4)-negative. CD10+/MUM1− cases were significantly more frequent in DHITsig-pos tumors. Genes associated with the GC intermediate zone had higher expression within the DHITsis-pos tumors. These findings demonstrate that DHITsig-pos tumors are B cells transitioning from the GC dark zone to the GC light zone. Along with the expected enrichment of mutation in MYC and BCL2, mutations of genes involved in chromatin modification (e.g., CREBBP, EZH2, DDX3X, TP53, and KMT2D) were more frequently harbored by DHITsig-pos tumors [30–32]. Specifically, missense mutations in EZH2, DEAD-box helicase 3 X-linked (DDX3X), and lysine methyltransferase 2D (KMT2D), as well as both missense and truncating mutations in CREB-binding protein (CREBBP) and TP53, point toward different clinical features of the corresponding DLBCL subjects [30]. Moreover, DHITsig identified a group of DLBCL with peculiar clinical features. The variable molecular signatures, which identify HGBL and are constituted by the karyotype, the immunohistochemistry, and the DHIT signature [30], uncovered novel clinical scenarios to be driven by a still evolving genomic landscape in DLBCL, and they enable a rational patient management, based on consolidated [26,33] and novel therapeutic approaches [25,34].

In the frame of this thinking, regulation of chromatin status plays a pivotal role in the correct development and differentiation of mature B cells, and it is extensively investigated with therapeutic purposes. In B-cell tumors, a plethora of mutations affect genes involved in chromatin regulation and in normal B-cell development [31–33]. Specifically, EP300 and CREBP are main acetylation regulators and, therefore, modulate gene expression, as well as histone methylators such as KMT2D, SUZ12, and EZH2 [35–37]. These genes are mutated in 25–30% of DLBCL cases. Notably, CREBBP and EP300 positively modulate multiple biological programs in the germinal center, through acetylation of histone and nonhistone proteins. Moreover, CREBBP and EP300 mutations contribute to lymphomagenesis by perturbing the expression of genes that are relevant to normal biology (i.e., BCL6 and p53). Inactivation of CREBBP and EP300 rarely coexists in human DLBCL, suggesting that cells require a certain amount of acetyltransferase activity [38]. Remarkably, GC B cells essentially require a minimum amount of acetyltransferase activity [39] and CREBBP-mutated B cells are addicted to the residual activity of EP300, envisioning potential therapeutic windows driven by CREBBP-mutated GC B cells on EP300 [39]. Thus, double KO of CREBBP and EP300 is required to abrogate GC formation detected by BCL6 immune staining. Furthermore, CREBBP-deficient cells are preferentially sensitive to inhibitors targeting HAT/BRD domains of CREBBP/EP300 [39]. In DLBCL with CREBBP genetic inactivation by mutation, pharmacologic inactivation of EP300 may lead to lymphoma cell death. Additionally, EP300 polymorphism was uncovered to decrease the balance between acetylation and deacetylation in the tumor niche, impacting disease progression [40]. Epigenetic dysregulation can, therefore, represent one of the driver lesions in high-risk DLBCL, and the restoration of physiological chromatin remodeling is an attractive target for novel therapy.

In tumor patients, based on evidence from other solid [41–43] and hematological malignancies [44–47], cell-free DNA (cfDNA) and extracellular vesicles are released by tumor apoptotic cells; DLBCL makes no exception [48,49]. Circulating tumor DNA (ctDNA) is distinguished from other cfDNA by the presence of somatic mutations representative of tumor biology absent in normal cells [50]. Liquid biopsy was employed as a new tool for genotyping and evaluating minimal residual disease in DLBCL [51,52]. Kurtz et al. uncovered ctDNA^{high} DLBCL to be characterized by a prognostically unfavorable outcome [52]. Remarkably, the non-tumor cfDNA might additionally originate from the neoplastic site, expanding the concept of liquid biopsy to the microenvironment compartment [53]. Liquid biopsy and molecular deconvolution [51], dissecting the genomic architecture of hematological malignancies, are becoming tools able to predict the prognosis [7,54,55].

2.2. Molecular Prognostic Models

Efficient clinical prognostic tools were uncovered to be relevant in driving patient management. The international literature highlighted some molecular characteristics of DLBCLs that condition their prognosis and, in perspective, the therapy [1]. Adequate histological diagnosis must include in the report an evaluation of the parameters useful to guide the therapeutic choice in order to confirm the cell of origin, its immunophenotype, the presence of double expressors, and the proliferation index, as well as sometimes specific FISH characteristics addressed by BCL2, BCL6, MYC, and IG-heavy/kappa/lambda (IGH/IGK/IGL) DNA probes [1,56]. The prognostic impact of the biological characteristics holds relevant translational consequences. To this end, a proper stratification included specific characteristics of investigation on the cancer cells that were uncovered to be CD20- and/or CD79a-expressing B lymphocytes [26,57]; additionally, anti-CD5 was deemed important when expressed, thus allowing the identification of a clinically more aggressive CD5+ DLBCL subset [57]. Moreover, while characterizing the cell of origin phenotype, CD10, BCL6, and MUM1 play a pivotal role, by driving the GC-type identification, differentiated by CD10 and/or BCL6 expression in >30% of DLBCL cells, while their low expression, along with >30% expression of MUM1 documentation, indicates a non-GC-type [58].

MYC/BCL2 evaluation in DLBCL using immunohistochemical staining was employed to exactly define double expression and to identify subgroups with dismal prognosis, often belonging to the non-GC-type subgroup [59]. A percentage of cells with intense MYC positivity >70% is often associated with translocation [60].

The percentage of Ki67-positive tumor cells (clone MIB1) should also be considered. In the event of uneven distribution in the tissue, it is advisable to report a percentage value representative of the average, while signaling the uneven distribution of the positivity signal [61].

Several alternative prognostic models already exist for DLBCL. A new one was uncovered to be significant, showing, in 199 cases, the relevance of the immunohistochemistry according to the Hans algorithm and MYC/BCL2 evaluation. The cell of origin evaluated by Nanostring, FISH analysis assessing BCL2, BCL6, and c-MYC, and the targeted sequencing from a custom platform based on univariate analysis identifying gene mutations significantly correlated to poor or favorable prognosis [62]. According to that stratification system, the authors elaborated an m3D-IPI uncovering sex, age, extranodal sites, LDH, advanced stage, double hit, and mutation in KMT2D, PIM1, and MEF2B as being significantly related to high-risk disease in R-CHOP-treated patients. Despite statistically powered validation studies being required, this novel approach performed better than traditional IPI in this patient cohort (C-index 0.87 vs. 0.77, respectively). The increasing number of biological acquisitions, combined with clinical characteristics of patients, will allow a better treatment tailoring.

Recently, since limited data are available on comprehensive genetic signatures, Chapuy et al. proposed a novel molecular gene signature deconvoluting the DLBCL heterogeneity. While dissecting the complex genomic architecture, these authors uncovered an integrated approach combining analyses of recurrent mutations, somatic copy number alterations (SCNAs), and structural variants (SVs) to efficiently reveal DLBCL taxonomy, and they highlighted five genetically distinctive clusters (C1–C5) [63]. Specifically, these genetically distinct DLBCL subsets predict different outcomes, provide novel insights into lymphomagenesis, and suggest certain combinations of targeted therapies [63,64]. In more detail, among ABC DLBCLs, the C1 subtype DLBCL was deemed to be associated with favorable prognosis and was characterized by $MYD^{non-L265P}$, NOTCH2, and SPEN mutations, as well as BCL6 SVs, and this phenotype might origin from marginal-zone lymphoma and from an ancestor of extrafollicular origin [63]. Conversely, C5 subtype DLBCLs correlated with unfavorable clinical outcome, harboring $BCL2^{gain}$, $MYD88^{L265}$, $CD79B^{mut}$, and $TBL1XR1^{mut}$, and they were associated with extranodal tropism and genes overexpressed in the BCL2-overexpressing group [65,66]. Contrariwise, within the GCB DLBCLs, the C4 subtype was associated with more favorable PFS, and it was characterized by mutations in NF-κB, JAK/STAT, and RAS pathway components and histone genes. The C3 subgroup paralleled the C5 dismal prognosis, being associated with BCL2 SV and mutations, PTEN CN loss and mutation, and chromatin-modifying enzyme alterations. Lastly, Chapuy et al. also identified a remarkable feature from a C2 subtype with a distinct clinical trajectory, being composed by bi-allelic TP53 inactivation, 9p21.23/CDNKN2A copy loss, and increased genomic instability reflected by recurrent SCNAs and frequent genome doublings [63]. Next, to validate the genetic substrate in an independent dataset and develop a robust molecular classifier allowing prediction in new samples, Chapuy et al. also genetically confirmed identity-associated marker genes and biology of the C1–C5 DLBCL clusters in a combined larger cohort [32,67]. This independent analysis sanctioned a parsimonious probabilistic classifier able to prospectively identify the C1–C5 DLBCL subtypes in newly diagnosed patients [67].

2.3. Tumor Microenvironment and Angiogenesis

Based on several pieces of compelling evidence highlighting the impact of the DLBCL niche in nursing cancer cells, by promoting a favorable stromal environment, several prospective clinical studies are needed to validate the clinical utility of the stromal gene expression profile in DLBCL and dissect subtypes which would profit the most from anti-angiogenic and milieu-targeting strategies [68,69]. Nonetheless, it is well known that the presence of immune and inflammatory cells contributes to modulate tumor growth and invasion in hematological malignancies and DLBCL [70–72]. Analysis of the tumor microenvironment is an important aspect in the assessment of progression of DLBCL. Different components of the microenvironment are considered in DLBCL including mast cells and TAMs to establish several correlations among prognostic significance, stage-related tumor progression, and differences in treatment outcome [73,74].

Lymphomas include more than 40 lymphoproliferative disorders, and angiogenesis plays a critical role in their progression and prognosis [75,76].

The state-of-the-art knowledge of the crucial mechanisms promoting angiogenesis and mediating immunosuppression during DLBCL development, progression [77,78], and sensitivity to drugs [26,79] needs further in-depth analysis. Solid and hematological neoplasms propagate and progress through several vicious cycles, feeding into the surrounding tumoral milieu [80–83], and emergent knowledge pinpoints angiogenesis and immunosuppression as simultaneous processes in response to this reciprocal loop [84,85] and to a plethora of paracrine and exogenous stimuli [86–88]. Lymphoproliferative disorders [89,90] and DLBCL [91] are no exception. Accordingly, strategies combining anti-angiogenic therapy and immunotherapy seem to have the potential to tip the balance of the tumor microenvironment and improve the treatment response of lymphoid malignancies [21,22,92,93]. These pieces of evidence prompted an intense translational investigation aimed at targeting angiogenesis and the immune system in a coordinated fashion, based on the preclinical insights available [94,95].

2.4. Increased Vascularization, VEGF Expression and MicroRNA (miRNA)

The presence of an increased number of immature vessels in DLBCL compared with follicular lymphoma (FL) was demonstrated [96]. ABC DLBCL CD5+ showed higher microvascular density (MVD) than GCB DLBCL [97]. MVD was higher in CD5+ DLBCL in comparison with the CD5− subgroup [98].

Transformation from indolent B-cell lymphoma to aggressive DLBCL and poor prognostic subgroups within DLBCL is associated with increased VEGF expression [99]. In aggressive subtypes of DLBCL, VEGF-A-producing CD68+ VEGFR1+ myelo-monocytic cells are closely associated with newly formed blood vessels [68]. In DLBCL, the average MVD correlates with the intensity of VEGF, VEGFR-1, and VEGFR-2 expression in tumor cells [100]. Other studies in DLBCL found no correlation between MVD and VEGF expression [101]. The transcript level of the soluble isoforms of VEGF, such as VEGF121, has a major impact on the prognosis of ABC-like DLBCL, whereas low VEGF121 expression was associated with a significantly better survival than high expression [91]. Moreover, 57 genes involved in immune response and T-cell activation were decreased in patients with high VEGF121 expression in both ABC-like and GBC-like subtypes of DLBCL [91].

In a meta-analysis of eight studies conducted on 670 patients, positive VEGF expression in blood-circulating lymphocytes and lymph nodes correlated with shorter survival in newly diagnosed DLBCL [102]. In another study performed on 149 newly diagnosed DLBCLs, high serum VEGF level was associated with poorer prognosis [103]. VEGF-A- and VEGFR-1-negative patients had an improved overall survival compared to VEGF-A- and VEGFR-1-positive ones [104]. Polymorphism in the VEGFR-2 gene may be associated with better survival in DLBCL patients [105].

Borges et al. [106] demonstrated an association between increased expression of pro-angio miRs miR-126 and miR130a, along with anti-angio miR-328, and the subtype non-GCB. Moreover, they found higher levels of the anti-angio miR-16, miR-221, and miR-328 in patients with low MVD and a stromal 1 signature.

More recently, Lupino et al. [107] demonstrated that the overexpression of SPHK1, one of the two isozymes responsible for the production of sphingosine-1 phosphate (SP1), a bioactive sphingolipid metabolite acting as a potent inducer of angiogenesis [108], correlates with an angiogenic transcriptional program in DLBCL.

2.5. Correlations among Angiogenesis, VEGF Expression, and Response to Therapy

Immunodeficient mice engrafted with human DLBCL treated with antibodies against human or murine VEGFR-1 or VEGFR-2 showed a significant 50% reduction in tumor mass after treatment with human anti-VEGFR-1. By contrast, inhibition of murine VEGFR-1 resulted in a similar tumor reduction, but inhibition of human VEGFR-2 had no antitumor effect [109].

In patients affected by DLBCL treated with anthracycline-based chemotherapy, no correlation between increased MVD and VEGF expression in tumor cells was demonstrated. Moreover, high VEGF and VEGFR-1 expression identified a subgroup of patients affected by DLBCL with improved overall survival and progression-free survival [100]. In patients with DLBCL treated with R-CHOP, a high serum level of VEGF was associated with adverse outcome, having lower values in survivors than in non-survivors [110]. Additionally, high MVD determines a poor outcome in DLBCL in patients treated with R-CHOP [97]. Bevacizumab inhibits tumor growth, either alone or in combination with chemotherapy, in untreated DLBCL [111].

2.6. Targeting Angiogenesis and the Immune System in DLBCL: A Single-Center Experience

Recently, we demonstrated that there is a significant increase in tryptase-positive mast cells and CD68-positive TAMs, as well as a significant increase in MVD and a positive correlation in chemo-resistant non-responder when compared with chemo-sensitive responder DLBCL patients (Figure 1) [112].

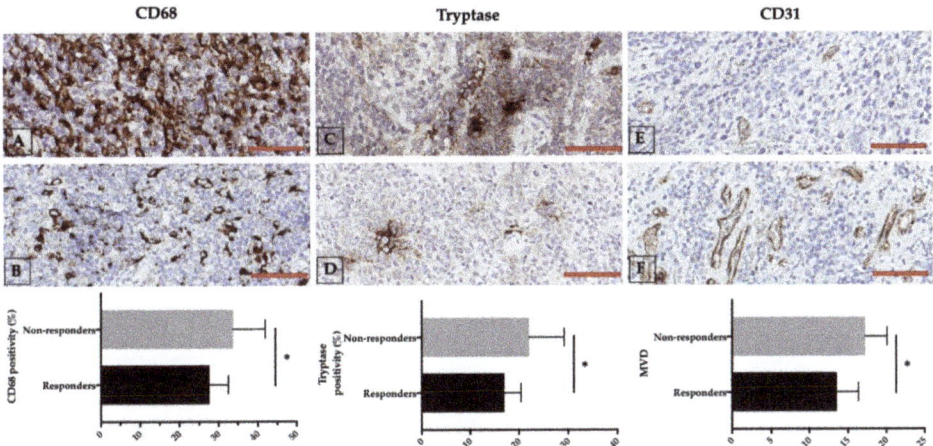

Figure 1. Non-responder (upper panels (**A,C,E**) and responder newly diagnosed diffuse large B-cell lymphoma (DLBCL) patients (middle panels (**B,D,F**) are characterized by different CD68, tryptase, and CD31 expression. Lower panels: respective comparison of non-responder and responder groups. Scale bar: 50 µm. * $p < 0.05$, assessed by Mann–Whitney test. Representative images from 29 untreated DLBCL patients are presented [112].

Moreover, we uncovered CD3-positive T cells to be decreased while comparing bulky (patients with bulky disease are defined by the presence of a large nodal tumor mass >10 cm or mediastinal disease) and non-bulky groups (Figure 2) [113], suggesting that a reduction in T cells in bulky disease patients contributes to loosen the immune control over the tumor, resulting in increased cell proliferation and large tumor masses [114].

Likewise, we demonstrated, comparing by means of RNA scope technology, STAT3 RNA expression in two selected groups of ABC DLBCL and GBC DLCBCL, that ABC tissue samples contained a significantly higher number of STAT3-positive cells than GBC tissue samples (Figure 3) [115].

Furthermore, through microscopic imaging, we uncovered tumor vessels in ABC samples but not GBC samples to be coated by FVIII- and STAT3-positive endothelial cells [115]. Evidence from our group revealed a positive correlation not only between STAT3 expression and CD3, CD8, and CD68, but also between D163-positive cells in the ABC and the GBC groups (Figure 4) [116].

Additionally, in the ABC group, we found also a positive correlation between CD8- and CD34- and between Ki67- and CD68/CD163-positive cells (Figure 5).

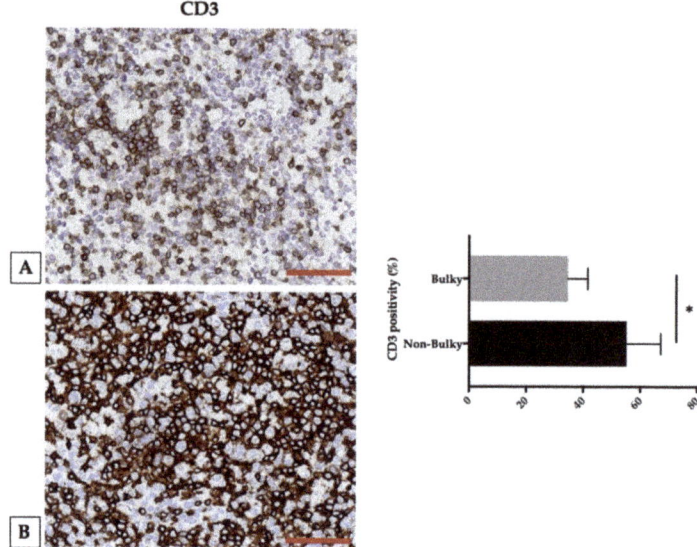

Figure 2. CD3 expression in bulky and non-bulky DLBCL. Left panel: (**A**) Representative image of CD3 expression in a case with bulky involvement. (**B**) Representative image of CD3 expression in a case with non-bulky DLBCL. Right panel: comparison between bulky and non-bulky disease groups with a significant difference between the groups in the CD3 infiltrate. Scale bar: 50 µm. * $p < 0.05$, assessed by Mann–Whitney test. Representative images from 29 untreated DLBCL patients are presented [113].

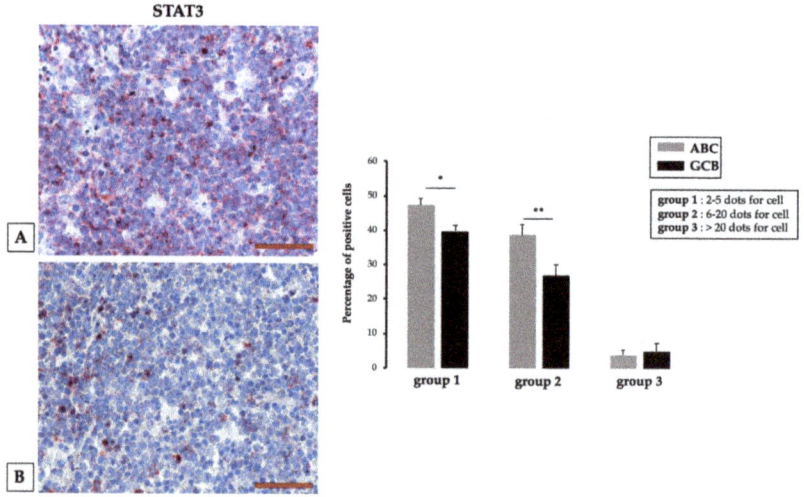

Figure 3. Left panel: different STAT3 expression in histological samples from activated B cell (ABC) (**A**) and germinal center B cell (GCB) (**B**) DLBCL assessed by RNAscope. Scale bar: 60 µm. Right panel: quantification of RNA ISH staining of STAT3 messenger RNA (mRNA) positivity in ABC and GCB DLBCL samples. The percentage of STAT3 mRNA expression significantly increases in the ABC group 1 and 2 tumor samples compared to GCB; * $p < 0.05$; ** $p < 0.01$, assessed by Mann–Whitney test. Representative images from 30 untreated DLBCL patients are presented [115].

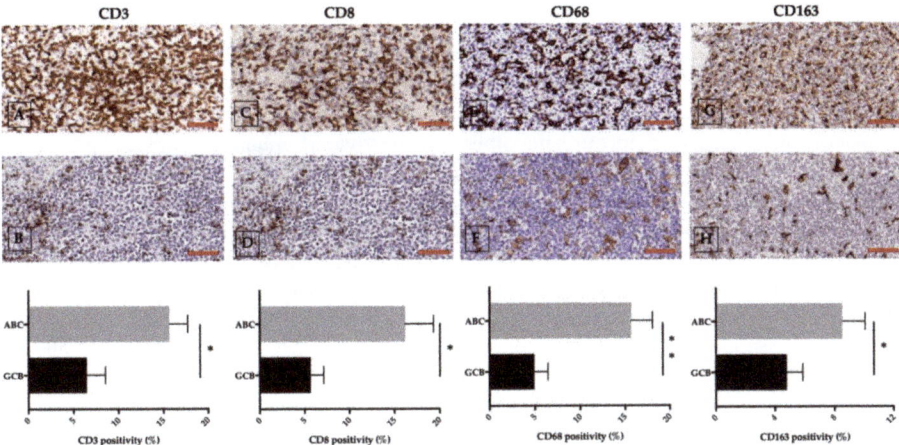

Figure 4. ABC (upper panel) and GCB (middle panel) DLBCL different expression of CD3 (**A,B**), CD8 (**C,D**) CD68 (**E,F**), and CD163 (**G,H**) assessed by immunohistochemical staining. The morphometric analysis is expressed as marker percentage positivity (lower panel). Scale bar: A–H 60 μm. Representative images from 60 untreated DLBCL patients are presented; * $p < 0.05$; ** $p < 0.01$, assessed by Mann–Whitney test [116].

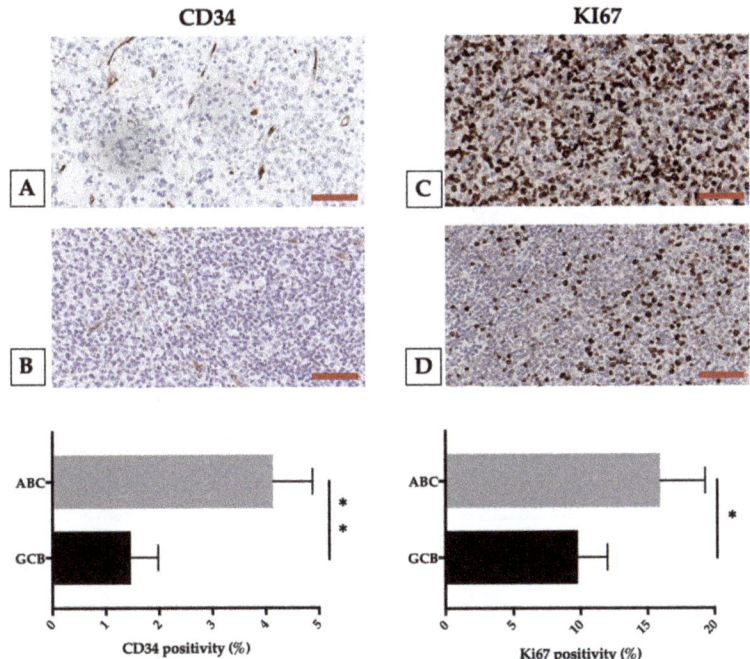

Figure 5. ABC (upper panel) and GCB (middle panel) DLBCL different expression of CD34 (**A,B**) and Ki67 (**C,D**) assessed by immunohistochemical staining. The morphometric analysis is expressed as marker percentage positivity (lower panel). Scale bar: A–D 60 μm. Representative images from 60 untreated DLBCL patients are presented; * $p < 0.05$; ** $p < 0.01$, assessed by Mann–Whitney test [116].

3. Discussion

Overall, data generated by our group corroborated previous findings, pointing toward a higher STAT3 expression being associated with higher CD163- and CD8-positive cell infiltration, which induces a strong angiogenic response in ABC DLBCL as compared with GCB DLBCL [116]. Preliminary results generated in our and other labs uncovered enhanced angiogenesis to be a strong regulator of lymphoproliferative disorder prognosis due to direct and indirect activation of cell survival [115–117]. The cell-adhesion-dependent DLBCL milieu interaction nurses DLBCL proliferation, by supporting immune-surveillance evasion [118]. Independent data provided compelling evidence that, in the intimate interaction between stromal cells, the malignant clone creates a permissive immune microenvironment within the lymphoma niche, which starts a vicious cycle hijacking anti-tumor activity [21,119,120]. Mechanistically, endothelial cells, by expressing TIM-3, HB-EGF [120–122], and a plethora of surface and soluble factors, prompt defective immunosurveillance and, in turn, allow for the persistence and proliferation of lymphoid neoplastic cells [123–125], envisioning novel therapeutic windows [126,127]. Moreover, the initial observation that the expression level of the adhesion molecules by the malignant lymphoma cells can predict disease outcome in extranodal DLBCL [128] prompted further investigation, especially in peculiar clinical disease phenotypes, such as DLBCLs involving the central nervous system (CNS) [128]. Remarkably, CNS spreading represents a paradigmatic extranodal localization with peculiar pathobiology involving adhesion molecule deregulated expression [129] and hyperactivation of the angiogenesis fueling pathway [130] along with a truncal genomic signature [131], which can contribute to drug sensitivity and resistance [132–134], as in other malignancies [135–137]. Therefore, given that the aberrant expression of adhesion molecules on bone marrow endothelial cells of patients with lymphoid and myeloid neoplasia was also discovered to predict poor clinical outcome [138–141], it is tempting to speculate a vicious cycle in DLBCL by paralleling the neoplastic cell behavior [128], whereby the described molecular signature [36,142] has more interactions among themselves than what would be expected for a random set of gene-encoding proteins drawn from the genome [143].

Based on these findings and on several pieces of compelling evidence investigating how the deregulated adhesion-mediated system would contribute to more aggressive disease, several attempts uncovered the junctional adhesion molecule role in mediating disease aggressiveness [141,144,145]. In line with previous results [128,146,147], preliminary data from our lab demonstrate that direct contact of environmental cells with DLBCL cells would enhance adhesion molecule levels, thus preventing both direct and indirect cell invasiveness and epithelial–mesenchymal transition and extra-nodal dissemination (unpublished data). Even more interesting, the cell adhesion molecule junctional adhesion molecule-A (JAM-A) presents remarkable features [148], whereby it can interact with itself if expressed on two opposing cell types. Furthermore, if JAM-A is shed by a cell, the soluble form of the JAM-A molecule can bind to cell-bound JAM-A, which in turn notably enhances its binding capacity [149–151]. Remarkably, consistently with Peng-Peng Xu et al. [128], JAM-A appears related to extra-nodal involvement in DLBCL, being selectively expressed in those cases. The therapeutic effects of blocking angiogenesis, the endothelial adhesion system, JAM-A, and its cognate shedding regulator ADAM17 were mainly observed in preclinical models but not in patients and, therefore, they must be interpreted with caution [151–155]. In a clinical setting, the adhesion system and neoangiogenesis, along with competent CD8 T cells and dendritic cells, had increased OS and time to progression [99]. Thus, it is likely that invasiveness potential, along with new blood vessel formation (i.e., angiogenesis) within the DLBCL environment, is a recognized hallmark of disease progression, mirroring cancer evasion from T-cell immune surveillance [156]. Endothelial progenitor cell trafficking was uncovered to be implicated in DLBCL progression [157,158], especially in early disease phases [100,159]. Several clinical trials in DLBCL tested the effects of angiogenesis-targeting agents, such as bevacizumab, which are used in combination with other agents, including B-cell-targeting agents [101,160,161]. Nonetheless, the lack of clinical effect in the randomized study gained by the addition of an anti-angiogenic approach to chemo-immunotherapy involving the tumor milieu might be predictive of the response to anti-angiogenesis in DLBCL, being beneficial in DLBCL with a high relative expression of a set of endothelial markers and angiogenic gatekeepers (the "stromal 2" subtype), correlating with enhanced vasculogenesis [17,69]. Furthermore, since compelling evidence pinpoints structural abnormalities in the endothelium as impairing antitumor immunity by forming barriers to immune surveillance [162], the tumor-associated endothelium is currently also described as a caretaker that synchronizes the entrance and egress of the immune cells within the neoplastic niche [163,164]. Therefore, while defining DLBCL also based on the quantity and quality of immune cell infiltrates might provide novel rationale to overcome the lack of clinical success gained by angiogenesis-targeting agents so far, identifying the abnormalities in the DLBCL endothelium impairing the crosstalk with adaptive immunity may also be valuable. Targeting these abnormalities can improve the success of immune-based therapies for different cancers, as well as DLBCL, by improving immune–vascular crosstalk for DLBCL, enhancing anti-lymphoma immunotherapy using anti-angiogenesis [165]. Thus, further studies of anti-angiogenic approaches in B-NHL and DLBCL should not be denied [161]. Indeed, while preventing secondary immunodeficiencies [166,167], this evidence provides the translational rationale to overcome the scanty effect of the anti-angiogenic approach in DLBCL obtained so far by novel angiogenesis targeting via RAS pathway inhibition, while combining immune-modulatory agents (IMiDs, i.e., lenalidomide) when appropriate [168–170]. Assuming the different angiogenic impacts on a given disease stage, it would be worth tailoring the vasculogenic manipulation in early DLBCL with the high-risk phenotype [78]. In this frame of thinking, one critical effect of corrupted angiogenesis is represented by disease dissemination, within and outside the original niche localization, driving intra- and extra-nodal adhesion-dependent manifestation in DLBCL. Finally, the judicious use of anti-angiogenics to normalize tumor vasculature might represent a strategy reprograming the tumor microenvironment to improve next-generation immunotherapy for DLBCL.

4. Conclusions

Lymphomas constitute a large group of more than 40 lymphoproliferative disorders, classified on the basis of morphologic, immunologic, genetic, and clinical criteria. The importance of the tumor milieu and angiogenesis in lymphoproliferative disorders was studied in relation to their impact on the prognosis of patients, suggesting high relevance in different types of lymphomas. Literature data concerning the angiogenesis of NHL are limited compared with HL, with most studies performed by retrospective immunohistochemical analysis, where evidence of correlation between cellular components of the microenvironment and increased vascularity was established. Within the different types of B-cell lymphomas, angiogenesis may be prominent in aggressive rather than indolent subtypes.

Current frontline DLBCL therapy although fairly successful (70–80% remission rates with the standard R-CHOP chemotherapy regimen) is frequently followed by relapse (40% of cases within 2–3 years), with an often refractory DLBCL. Anti-angiogenic therapy and microenvironment-directed therapy represent important tools for the treatment of human lymphomas. However, a significant number of patients are resistant, whereas those who respond have minimal benefits. Nevertheless, these new findings may point toward a potential Achilles heel of DLBCL which, in the future, might be exploited therapeutically in the relapsed/refractory setting and in extranodal dissemination.

Author Contributions: Conceptualization, D.R., A.G.S., G.S., and A.V.; methodology, T.A., R.T., G.I., and E.M.; software, R.T. and A.G.S.; validation, T.A., G.I., and A.V.; formal analysis, A.G.S., R.T., and G.I.; investigation, A.G.S.; resources, A.G.S., D.R., and A.V.; data curation, D.R.; writing—original draft preparation, A.G.S., T.A., and R.T., writing—review and editing, D.R., G.S., and A.V.; visualization, A.G.S., R.T., G.I., and E.M.; supervision, D.R., G.S., and A.V.; project administration, D.R. and A.G.S.; funding acquisition, D.R., A.G.S., G.S., and A.V. All authors read and agreed to the published version of the manuscript.

Funding: This research received no external funding.

Acknowledgments: This work was supported by the Association against Lymphomas "Il Sorriso di Antonio", Corato, Italy. This work was also supported in part by Associazione Italiana per la Ricerca sul Cancro (AIRC, Milan, Italy) through an Investigator Grant (no. 20441) and GLOBALDOC Project—CUP H96J17000160002 approved with A.D. n. 9 on 18 January 2017, from Puglia Region, financed under the Action Plan for Cohesion approved with Commission decision C (2016) 1417 of 3 March 2016 to A.G.S., as well as the Apulian Regional project: medicina di precisione to A.G.S.

Conflicts of Interest: The authors declare no conflict of interest.

Abbreviations

ADAM17	Disintegrin and metalloproteinase domain-containing protein 17
BCL2	B-Cell CLL/Lymphoma 2
BCL6	B-Cell Lymphoma 6 Protein
BCR	B-Cell Receptor
BRD	Bromodomain
CARD11	Caspase Recruitment Domain Family Member 11
CD68	cluster of differentiation 68
CN	Copy number
CREB	cAMP-response element-binding protein
CREBBP	CREB Binding Protein
DDX3X	DEAD-Box Helicase 3 X-Linked
DEAD	DEAD-box helicase family
EZH2	Enhancer Of Zeste Homolog 2
FISH	Fluorescence in situ hybridization
FVIII	Factor VIII
GNA13	G Protein Subunit Alpha 13
HAT	Histone Acetyltransferase
HB-EGF	Heparin-binding EGF-like growth factor
Ig	Immunoglobulin
IRF4	interferon regulatory factors 4
ISH	In situ hybridization
KMT2D	Lysine Methyltransferase 2D
KO	Knock out
LMO2	LIM domain only 2 (rhombotin-like 1)
m3D-IPI)	three-risk group model International Prognostic Index
MIB1	Mindbomb E3 Ubiquitin Protein Ligase 1
MUM1	melanoma associated antigen (mutated) 1
MYC	Myc-Related Translation/Localization Regulatory Factor
MYD	Myeloid Differentiation Primary Response Gene
MYD88	Myeloid Differentiation Primary Response Gene (88)
NF-κB	Nuclear Factor Kappa B Subunit
NGS	next generation sequensing
NOS	nitric oxide synthase
NOTCH2	Neurogenic Locus Notch Homolog Protein 2
OS	Overall Survival
PFS	Progressiob free Survival
PTEN	Phosphatase And Tensin Homolog
RAS	Rat Sarcoma Viral Oncogene Homolog
SPEN	SPEN family transcriptional repressor (
SPHK1	Sphingosine Kinase 1
STAT3	Signal Transducer And Activator Of Transcription 3
SUZ12	Suppressor Of Zeste 12 Protein Homolog
TIM-3	T-cell immunoglobulin mucin-3
TNFAIP3	Tumor necrosis factor, alpha-induced protein 3
TP53	Tumor Protein P53

References

1. WHO. *WHO Classification of Tumours of Haematopoietic and Lymphoid Tissues*, Revised 4th ed.; Swerdlow, S.H., Campo, E., Harris, N.L., Jaffe, E.S., Pileri, S.A., Stein, H., Thiele, J., Eds.; World Health Organization Classification of Tumours; International Agency for Research on Cancer: Lyon, France, 2017; ISBN 978-92-832-4494-3.
2. Liu, Y.; Barta, S.K. Diffuse large B-cell lymphoma: 2019 update on diagnosis, risk stratification, and treatment. *Am. J. Hematol.* **2019**, *94*, 604–616. [CrossRef]
3. Schneider, C.; Pasqualucci, L.; Dalla-Favera, R. Molecular pathogenesis of diffuse large B-cell lymphoma. *Semin. Diagn. Pathol.* **2011**, *28*, 167–177. [CrossRef] [PubMed]
4. Miao, Y.; Medeiros, L.J.; Li, Y.; Li, J.; Young, K.H. Genetic alterations and their clinical implications in DLBCL. *Nat. Rev. Clin. Oncol.* **2019**, *16*, 634–652. [CrossRef] [PubMed]
5. Alizadeh, A.A.; Eisen, M.B.; Davis, R.E.; Ma, C.; Lossos, I.S.; Rosenwald, A.; Boldrick, J.C.; Sabet, H.; Tran, T.; Yu, X.; et al. Distinct types of diffuse large B-cell lymphoma identified by gene expression profiling. *Nature* **2000**, *403*, 503–511. [CrossRef] [PubMed]
6. Dunleavy, K.; Pittaluga, S.; Czuczman, M.S.; Dave, S.S.; Wright, G.; Grant, N.; Shovlin, M.; Jaffe, E.S.; Janik, J.E.; Staudt, L.M.; et al. Differential efficacy of bortezomib plus chemotherapy within molecular subtypes of diffuse large B-cell lymphoma. *Blood* **2009**, *113*, 6069–6076. [CrossRef]
7. Rosenwald, A.; Wright, G.; Chan, W.C.; Connors, J.M.; Campo, E.; Fisher, R.I.; Gascoyne, R.D.; Muller-Hermelink, H.K.; Smeland, E.B.; Giltnane, J.M.; et al. The Use of Molecular Profiling to Predict Survival after Chemotherapy for Diffuse Large-B-Cell Lymphoma. *N. Engl. J. Med.* **2002**, *346*, 1937–1947. [CrossRef]
8. Davis, R.E.; Brown, K.D.; Siebenlist, U.; Staudt, L.M. Constitutive Nuclear Factor κB Activity Is Required for Survival of Activated B Cell–like Diffuse Large B Cell Lymphoma Cells. *J. Exp. Med.* **2001**, *194*, 1861–1874. [CrossRef]
9. Compagno, M.; Lim, W.K.; Grunn, A.; Nandula, S.V.; Brahmachary, M.; Shen, Q.; Bertoni, F.; Ponzoni, M.; Scandurra, M.; Califano, A.; et al. Mutations of multiple genes cause deregulation of NF-κB in diffuse large B-cell lymphoma. *Nature* **2009**, *459*, 717–721. [CrossRef]
10. Zl, W.; Yq, S.; Yf, S.; Zhu, J. High nuclear expression of STAT3 is associated with unfavorable prognosis in diffuse large B-cell lymphoma. *J. Hematol. Oncol. J. Hematol. Oncol.* **2011**, *4*, 31. [CrossRef] [PubMed]
11. Ok, C.Y.; Chen, J.; Xu-Monette, Z.Y.; Tzankov, A.; Manyam, G.C.; Li, L.; Visco, C.; Montes-Moreno, S.; Dybkaer, K.; Chiu, A.; et al. Clinical Implications of Phosphorylated STAT3 Expression in De Novo Diffuse Large B-cell Lymphoma. *Clin. Cancer Res.* **2014**, *20*, 5113–5123. [CrossRef]
12. Ding, B.B.; Yu, J.J.; Yu, R.Y.-L.; Mendez, L.M.; Shaknovich, R.; Zhang, Y.; Cattoretti, G.; Ye, B.H. Constitutively activated STAT3 promotes cell proliferation and survival in the activated B-cell subtype of diffuse large B-cell lymphomas. *Blood* **2008**, *111*, 1515–1523. [CrossRef]
13. Lam, L.T.; Wright, G.; Davis, R.E.; Lenz, G.; Farinha, P.; Dang, L.; Chan, J.W.; Rosenwald, A.; Gascoyne, R.D.; Staudt, L.M. Cooperative signaling through the signal transducer and activator of transcription 3 and nuclear factor-κB pathways in subtypes of diffuse large B-cell lymphoma. *Blood* **2008**, *111*, 3701–3713. [CrossRef] [PubMed]
14. Yu, H.; Kortylewski, M.; Pardoll, D. Crosstalk between cancer and immune cells: Role of STAT3 in the tumour microenvironment. *Nat. Rev. Immunol.* **2007**, *7*, 41–51. [CrossRef] [PubMed]
15. Doucette, T.A.; Kong, L.-Y.; Yang, Y.; Ferguson, S.D.; Yang, J.; Wei, J.; Qiao, W.; Fuller, G.N.; Bhat, K.P.; Aldape, K.; et al. Signal transducer and activator of transcription 3 promotes angiogenesis and drives malignant progression in glioma. *Neuro-Oncology* **2012**, *14*, 1136–1145. [CrossRef] [PubMed]
16. Gao, P.; Niu, N.; Wei, T.; Tozawa, H.; Chen, X.; Zhang, C.; Zhang, J.; Wada, Y.; Kapron, C.M.; Liu, J. The roles of signal transducer and activator of transcription factor 3 in tumor angiogenesis. *Oncotarget* **2017**, *8*. [CrossRef] [PubMed]
17. Lenz, G.; Wright, G.; Dave, S.S.; Xiao, W.; Powell, J.; Zhao, H.; Xu, W.; Tan, B.; Goldschmidt, N.; Iqbal, J.; et al. Stromal Gene Signatures in Large-B-Cell Lymphomas. *N. Engl. J. Med.* **2008**, *359*, 2313–2323. [CrossRef]
18. Sehn, L.H.; Gascoyne, R.D. Diffuse large B-cell lymphoma: Optimizing outcome in the context of clinical and biologic heterogeneity. *Blood* **2015**, *125*, 22–32. [CrossRef]

19. Azzaoui, I.; Uhel, F.; Rossille, D.; Pangault, C.; Dulong, J.; Le Priol, J.; Lamy, T.; Houot, R.; Le Gouill, S.; Cartron, G.; et al. T-cell defect in diffuse large B-cell lymphomas involves expansion of myeloid-derived suppressor cells. *Blood* **2016**, *128*, 1081–1092. [CrossRef]
20. Ji, H.; Niu, X.; Yin, L.; Wang, Y.; Huang, L.; Xuan, Q.; Li, L.; Zhang, H.; Li, J.; Yang, Y.; et al. Ratio of Immune Response to Tumor Burden Predicts Survival Via Regulating Functions of Lymphocytes and Monocytes in Diffuse Large B-Cell Lymphoma. *Cell. Physiol. Biochem.* **2018**, *45*, 951–961. [CrossRef]
21. Rudelius, M.; Rosenfeldt, M.T.; Leich, E.; Rauert-Wunderlich, H.; Solimando, A.G.; Beilhack, A.; Ott, G.; Rosenwald, A. Inhibition of focal adhesion kinase overcomes resistance of mantle cell lymphoma to ibrutinib in the bone marrow microenvironment. *Haematologica* **2018**, *103*, 116–125. [CrossRef]
22. Solimando, A.G.; Da Già, M.C.; Leone, P.; Borrelli, P.; Croci, G.A.; Tabares, P.; Brandl, A.; Di Lernia, G.; Bianchi, F.P.; Tafuri, S.; et al. Halting the vicious cycle within the multiple myeloma ecosystem: Blocking JAM-A on bone marrow endothelial cells restores the angiogenic homeostasis and suppresses tumor progression. *Haematologica* **2020**. [CrossRef] [PubMed]
23. Solimando, A.G.; Vacca, A.; Ribatti, D. A Comprehensive Biological and Clinical Perspective Can Drive a Patient-Tailored Approach to Multiple Myeloma: Bridging the Gaps between the Plasma Cell and the Neoplastic Niche. *J. Oncol.* **2020**, *2020*, 1–16. [CrossRef]
24. Plaks, V.; Kong, N.; Werb, Z. The cancer stem cell niche: How essential is the niche in regulating stemness of tumor cells? *Cell Stem Cell* **2015**, *16*, 225–238. [CrossRef]
25. Tilly, H.; Gomes da Silva, M.; Vitolo, U.; Jack, A.; Meignan, M.; Lopez-Guillermo, A.; Walewski, J.; André, M.; Johnson, P.W.; Pfreundschuh, M.; et al. Diffuse large B-cell lymphoma (DLBCL): ESMO Clinical Practice Guidelines for diagnosis, treatment and follow-up. *Ann. Oncol.* **2015**, *26*, v116–v125. [CrossRef] [PubMed]
26. Solimando, A.G.; Ribatti, D.; Vacca, A.; Einsele, H. Targeting B-cell non Hodgkin lymphoma: New and old tricks. *Leuk. Res.* **2016**, *42*, 93–104. [CrossRef] [PubMed]
27. Reagan, P.M.; Davies, A. Current treatment of double hit and double expressor lymphoma. *Hematol. Am. Soc. Hematol. Educ. Program* **2017**, *2017*, 295–297. [CrossRef] [PubMed]
28. Hans, C.P.; Weisenburger, D.D.; Greiner, T.C.; Gascoyne, R.D.; Delabie, J.; Ott, G.; Müller-Hermelink, H.K.; Campo, E.; Braziel, R.M.; Jaffe, E.S.; et al. Confirmation of the molecular classification of diffuse large B-cell lymphoma by immunohistochemistry using a tissue microarray. *Blood* **2004**, *103*, 275–282. [CrossRef]
29. Huang, W.; Medeiros, L.J.; Lin, P.; Wang, W.; Tang, G.; Khoury, J.; Konoplev, S.; Yin, C.C.; Xu, J.; Oki, Y.; et al. MYC/BCL2/BCL6 triple hit lymphoma: A study of 40 patients with a comparison to MYC/BCL2 and MYC/BCL6 double hit lymphomas. *Mod. Pathol. Off. J. U. S. Can. Acad. Pathol. Inc.* **2018**, *31*, 1470–1478. [CrossRef]
30. Ennishi, D.; Jiang, A.; Boyle, M.; Collinge, B.; Grande, B.M.; Ben-Neriah, S.; Rushton, C.; Tang, J.; Thomas, N.; Slack, G.W.; et al. Double-Hit Gene Expression Signature Defines a Distinct Subgroup of Germinal Center B-Cell-Like Diffuse Large B-Cell Lymphoma. *J. Clin. Oncol. Off. J. Am. Soc. Clin. Oncol.* **2019**, *37*, 190–201. [CrossRef]
31. Reddy, A.; Zhang, J.; Davis, N.S.; Moffitt, A.B.; Love, C.L.; Waldrop, A.; Leppa, S.; Pasanen, A.; Meriranta, L.; Karjalainen-Lindsberg, M.-L.; et al. Genetic and Functional Drivers of Diffuse Large B Cell Lymphoma. *Cell* **2017**, *171*, 481–494.e15. [CrossRef]
32. Schmitz, R.; Wright, G.W.; Huang, D.W.; Johnson, C.A.; Phelan, J.D.; Wang, J.Q.; Roulland, S.; Kasbekar, M.; Young, R.M.; Shaffer, A.L.; et al. Genetics and Pathogenesis of Diffuse Large B-Cell Lymphoma. *N. Engl. J. Med.* **2018**, *378*, 1396–1407. [CrossRef] [PubMed]
33. Cheah, C.Y.; Fowler, N.H.; Wang, M.L. Breakthrough therapies in B-cell non-Hodgkin lymphoma. *Ann. Oncol.* **2016**, *27*, 778–787. [CrossRef] [PubMed]
34. Ayyappan, S.; Maddocks, K. Novel and emerging therapies for B cell lymphoma. *J. Hematol. Oncol. J. Hematol. Oncol.* **2019**, *12*, 82. [CrossRef] [PubMed]
35. Pasqualucci, L.; Dominguez-Sola, D.; Chiarenza, A.; Fabbri, G.; Grunn, A.; Trifonov, V.; Kasper, L.H.; Lerach, S.; Tang, H.; Ma, J.; et al. Inactivating mutations of acetyltransferase genes in B-cell lymphoma. *Nature* **2011**, *471*, 189–195. [CrossRef] [PubMed]
36. Pasqualucci, L.; Trifonov, V.; Fabbri, G.; Ma, J.; Rossi, D.; Chiarenza, A.; Wells, V.A.; Grunn, A.; Messina, M.; Elliot, O.; et al. Analysis of the coding genome of diffuse large B-cell lymphoma. *Nat. Genet.* **2011**, *43*, 830–837. [CrossRef] [PubMed]

37. Morin, R.D.; Mendez-Lago, M.; Mungall, A.J.; Goya, R.; Mungall, K.L.; Corbett, R.D.; Johnson, N.A.; Severson, T.M.; Chiu, R.; Field, M.; et al. Frequent mutation of histone-modifying genes in non-Hodgkin lymphoma. *Nature* **2011**, *476*, 298–303. [CrossRef]
38. Pasqualucci, L.; Dalla-Favera, R. Genetics of diffuse large B-cell lymphoma. *Blood* **2018**, *131*, 2307–2319. [CrossRef]
39. Meyer, S.N.; Scuoppo, C.; Vlasevska, S.; Bal, E.; Holmes, A.B.; Holloman, M.; Garcia-Ibanez, L.; Nataraj, S.; Duval, R.; Vantrimpont, T.; et al. Unique and Shared Epigenetic Programs of the CREBBP and EP300 Acetyltransferases in Germinal Center B Cells Reveal Targetable Dependencies in Lymphoma. *Immunity* **2019**, *51*, 535–547.e9. [CrossRef]
40. Li, J.; Ding, N.; Wang, X.; Mi, L.; Ping, L.; Jin, X.; Liu, Y.; Ying, Z.; Xie, Y.; Liu, W.; et al. EP300 single nucleotide polymorphism rs20551 correlates with prolonged overall survival in diffuse large B cell lymphoma patients treated with R-CHOP. *Cancer Cell Int.* **2017**, *17*, 70. [CrossRef]
41. Crowley, E.; Di Nicolantonio, F.; Loupakis, F.; Bardelli, A. Liquid biopsy: Monitoring cancer-genetics in the blood. *Nat. Rev. Clin. Oncol.* **2013**, *10*, 472–484. [CrossRef]
42. Russano, M.; Napolitano, A.; Ribelli, G.; Iuliani, M.; Simonetti, S.; Citarella, F.; Pantano, F.; Dell'Aquila, E.; Anesi, C.; Silvestris, N.; et al. Liquid biopsy and tumor heterogeneity in metastatic solid tumors: The potentiality of blood samples. *J. Exp. Clin. Cancer Res. CR* **2020**, *39*, 95. [CrossRef] [PubMed]
43. Krebs, M.; Solimando, A.G.; Kalogirou, C.; Marquardt, A.; Frank, T.; Sokolakis, I.; Hatzichristodoulou, G.; Kneitz, S.; Bargou, R.; Kübler, H.; et al. miR-221-3p Regulates VEGFR2 Expression in High-Risk Prostate Cancer and Represents an Escape Mechanism from Sunitinib In Vitro. *J. Clin. Med.* **2020**, *9*, 670. [CrossRef]
44. Pantel, K.; Alix-Panabières, C. Liquid biopsy and minimal residual disease - latest advances and implications for cure. *Nat. Rev. Clin. Oncol.* **2019**, *16*, 409–424. [CrossRef] [PubMed]
45. Desantis, V.; Saltarella, I.; Lamanuzzi, A.; Melaccio, A.; Solimando, A.G.; Mariggiò, M.A.; Racanelli, V.; Paradiso, A.; Vacca, A.; Frassanito, M.A. MicroRNAs-Based Nano-Strategies as New Therapeutic Approach in Multiple Myeloma to Overcome Disease Progression and Drug Resistance. *Int. J. Mol. Sci.* **2020**, *21*, 3084. [CrossRef]
46. Di Lernia, G.; Leone, P.; Solimando, A.G.; Buonavoglia, A.; Saltarella, I.; Ria, R.; Ditonno, P.; Silvestris, N.; Crudele, L.; Vacca, A.; et al. Bortezomib Treatment Modulates Autophagy in Multiple Myeloma. *J. Clin. Med.* **2020**, *9*, 552. [CrossRef] [PubMed]
47. Leone, P.; Buonavoglia, A.; Fasano, R.; Solimando, A.G.; De Re, V.; Cicco, S.; Vacca, A.; Racanelli, V. Insights into the Regulation of Tumor Angiogenesis by Micro-RNAs. *J. Clin. Med.* **2019**, *8*, 2030. [CrossRef]
48. Larrabeiti-Etxebarria, A.; Lopez-Santillan, M.; Santos-Zorrozua, B.; Lopez-Lopez, E.; Garcia-Orad, A. Systematic Review of the Potential of MicroRNAs in Diffuse Large B Cell Lymphoma. *Cancers* **2019**, *11*, 144. [CrossRef]
49. Hutchinson, L. CtDNA—Identifying cancer before it is clinically detectable. *Nat. Rev. Clin. Oncol.* **2015**, *12*, 372. [CrossRef]
50. Snyder, M.W.; Kircher, M.; Hill, A.J.; Daza, R.M.; Shendure, J. Cell-free DNA Comprises an In Vivo Nucleosome Footprint that Informs Its Tissues-Of-Origin. *Cell* **2016**, *164*, 57–68. [CrossRef]
51. Rossi, D.; Diop, F.; Spaccarotella, E.; Monti, S.; Zanni, M.; Rasi, S.; Deambrogi, C.; Spina, V.; Bruscaggin, A.; Favini, C.; et al. Diffuse large B-cell lymphoma genotyping on the liquid biopsy. *Blood* **2017**, *129*, 1947–1957. [CrossRef]
52. Kurtz, D.M.; Scherer, F.; Jin, M.C.; Soo, J.; Craig, A.F.M.; Esfahani, M.S.; Chabon, J.J.; Stehr, H.; Liu, C.L.; Tibshirani, R.; et al. Circulating Tumor DNA Measurements As Early Outcome Predictors in Diffuse Large B-Cell Lymphoma. *J. Clin. Oncol. Off. J. Am. Soc. Clin. Oncol.* **2018**, *36*, 2845–2853. [CrossRef] [PubMed]
53. Diehl, F.; Schmidt, K.; Choti, M.A.; Romans, K.; Goodman, S.; Li, M.; Thornton, K.; Agrawal, N.; Sokoll, L.; Szabo, S.A.; et al. Circulating mutant DNA to assess tumor dynamics. *Nat. Med.* **2008**, *14*, 985–990. [CrossRef]
54. Pasqualucci, L.; Dalla-Favera, R. The genetic landscape of diffuse large B-cell lymphoma. *Semin. Hematol.* **2015**, *52*, 67–76. [CrossRef] [PubMed]
55. Solimando, A.G.; Da Vià, M.C.; Cicco, S.; Leone, P.; Di Lernia, G.; Giannico, D.; Desantis, V.; Frassanito, M.A.; Morizio, A.; Delgado Tascon, J.; et al. High-Risk Multiple Myeloma: Integrated Clinical and Omics Approach Dissects the Neoplastic Clone and the Tumor Microenvironment. *J. Clin. Med.* **2019**, *8*, 997. [CrossRef] [PubMed]

56. Ventura, R.A.; Martin-Subero, J.I.; Jones, M.; McParland, J.; Gesk, S.; Mason, D.Y.; Siebert, R. FISH analysis for the detection of lymphoma-associated chromosomal abnormalities in routine paraffin-embedded tissue. *J. Mol. Diagn. JMD* **2006**, *8*, 141–151. [CrossRef] [PubMed]
57. Dargent, J.-L.; Lespagnard, L.; Feoli, F.; Debusscher, L.; Greuse, M.; Bron, D. De novo CD5-positive diffuse large B-cell lymphoma of the skin arising in chronic limb lymphedema. *Leuk. Lymphoma* **2005**, *46*, 775–780. [CrossRef] [PubMed]
58. Lu, T.-X.; Miao, Y.; Wu, J.-Z.; Gong, Q.-X.; Liang, J.-H.; Wang, Z.; Wang, L.; Fan, L.; Hua, D.; Chen, Y.-Y.; et al. The distinct clinical features and prognosis of the $CD10^+MUM1^+$ and $CD10^-Bcl6^-MUM1^-$ diffuse large B-cell lymphoma. *Sci. Rep.* **2016**, *6*, 20465. [CrossRef]
59. Hu, S.; Xu-Monette, Z.Y.; Tzankov, A.; Green, T.; Wu, L.; Balasubramanyam, A.; Liu, W.; Visco, C.; Li, Y.; Miranda, R.N.; et al. MYC/BCL2 protein coexpression contributes to the inferior survival of activated B-cell subtype of diffuse large B-cell lymphoma and demonstrates high-risk gene expression signatures: A report from The International DLBCL Rituximab-CHOP Consortium Program. *Blood* **2013**, *121*, 4021–4031. [CrossRef]
60. Swerdlow, S.H. Diagnosis of "double hit" diffuse large B-cell lymphoma and B-cell lymphoma, unclassifiable, with features intermediate between DLBCL and Burkitt lymphoma: When and how, FISH versus IHC. *Hematol. Am. Soc. Hematol. Educ. Program* **2014**, *2014*, 90–99. [CrossRef]
61. Hasselblom, S.; Ridell, B.; Sigurdardottir, M.; Hansson, U.; Nilsson-Ehle, H.; Andersson, P.-O. Low rather than high Ki-67 protein expression is an adverse prognostic factor in diffuse large B-cell lymphoma. *Leuk. Lymphoma* **2008**, *49*, 1501–1509. [CrossRef]
62. Song, J.Y.; Perry, A.M.; Herrera, A.F.; Chen, L.; Skrabek, P.; Nasr, M.; Ottesen, R.; Nikowitz, J.; Bedell, V.; Murata-Collins, J.; et al. New Genomic Model Integrating Clinical Factors and Gene Mutations to Predict Overall Survival in Patients with Diffuse Large B-Cell Lymphoma Treated with R-CHOP. *Blood* **2018**, *132*, 346. [CrossRef]
63. Chapuy, B.; Stewart, C.; Dunford, A.J.; Kim, J.; Kamburov, A.; Redd, R.A.; Lawrence, M.S.; Roemer, M.G.M.; Li, A.J.; Ziepert, M.; et al. Molecular subtypes of diffuse large B cell lymphoma are associated with distinct pathogenic mechanisms and outcomes. *Nat. Med.* **2018**, *24*, 679–690. [CrossRef] [PubMed]
64. Bojarczuk, K.; Wienand, K.; Ryan, J.A.; Chen, L.; Villalobos-Ortiz, M.; Mandato, E.; Stachura, J.; Letai, A.; Lawton, L.N.; Chapuy, B.; et al. Targeted inhibition of $PI3K\alpha/\delta$ is synergistic with BCL-2 blockade in genetically defined subtypes of DLBCL. *Blood* **2019**, *133*, 70–80. [CrossRef] [PubMed]
65. Visco, C.; Tzankov, A.; Xu-Monette, Z.Y.; Miranda, R.N.; Tai, Y.C.; Li, Y.; Liu, W.; d'Amore, E.S.G.; Li, Y.; Montes-Moreno, S.; et al. Patients with diffuse large B-cell lymphoma of germinal center origin with BCL2 translocations have poor outcome, irrespective of MYC status: A report from an International DLBCL rituximab-CHOP Consortium Program Study. *Haematologica* **2013**, *98*, 255–263. [CrossRef]
66. Seibold, M.; Stühmer, T.; Kremer, N.; Mottok, A.; Scholz, C.-J.; Schlosser, A.; Leich, E.; Holzgrabe, U.; Brünnert, D.; Barrio, S.; et al. RAL GTPases mediate multiple myeloma cell survival and are activated independently of oncogenic RAS. *Haematologica* **2019**. [CrossRef]
67. Chapuy, B.; Stewart, C.; Wood, T.; Dunford, A.; Wienand, K.; Getz, G.; Shipp, M.A. Validation of the Genetically-Defined DLBCL Subtypes and Generation of a Parsimonious Probabilistic Classifier. *Blood* **2019**, *134*, 920. [CrossRef]
68. Ruan, J.; Leonard, J.P. Targeting angiogenesis: A novel, rational therapeutic approach for non-Hodgkin lymphoma. *Leuk. Lymphoma* **2009**, *50*, 679–681. [CrossRef]
69. Ciavarella, S.; Vegliante, M.C.; Fabbri, M.; De Summa, S.; Melle, F.; Motta, G.; De Iuliis, V.; Opinto, G.; Enjuanes, A.; Rega, S.; et al. Dissection of DLBCL microenvironment provides a gene expression-based predictor of survival applicable to formalin-fixed paraffin-embedded tissue. *Ann. Oncol. Off. J. Eur. Soc. Med. Oncol.* **2018**, *29*, 2363–2370. [CrossRef]
70. Kinugasa, Y.; Matsui, T.; Takakura, N. CD44 Expressed on Cancer-Associated Fibroblasts Is a Functional Molecule Supporting the Stemness and Drug Resistance of Malignant Cancer Cells in the Tumor Microenvironment: Tumor Stromal CD44. *Stem Cells* **2014**, *32*, 145–156. [CrossRef]
71. Frassanito, M.A.; Desantis, V.; Di Marzo, L.; Craparotta, I.; Beltrame, L.; Marchini, S.; Annese, T.; Visino, F.; Arciuli, M.; Saltarella, I.; et al. Bone marrow fibroblasts overexpress miR-27b and miR-214 in step with multiple myeloma progression, dependent on tumour cell-derived exosomes. *J. Pathol.* **2019**, *247*, 241–253. [CrossRef]

72. Nicholas, N.S.; Apollonio, B.; Ramsay, A.G. Tumor microenvironment (TME)-driven immune suppression in B cell malignancy. *Biochim. Biophys. Acta BBA Mol. Cell Res.* **2016**, *1863*, 471–482. [CrossRef] [PubMed]
73. Hedström, G.; Berglund, M.; Molin, D.; Fischer, M.; Nilsson, G.; Thunberg, U.; Book, M.; Sundström, C.; Rosenquist, R.; Roos, G.; et al. Mast cell infiltration is a favourable prognostic factor in diffuse large B-cell lymphoma. *Br. J. Haematol.* **2007**, *138*, 68–71. [CrossRef] [PubMed]
74. Cai, Q.; Liao, H.; Lin, S.; Xia, Y.; Wang, X.; Gao, Y.; Lin, Z.; Lu, J.; Huang, H. High expression of tumor-infiltrating macrophages correlates with poor prognosis in patients with diffuse large B-cell lymphoma. *Med. Oncol.* **2012**, *29*, 2317–2322. [CrossRef] [PubMed]
75. Kini, A.R. Angiogenesis in Leukemia and Lymphoma. In *Hematopathology in Oncology*; Finn, W.G., Peterson, L.C., Eds.; Cancer Treatment and Research; Kluwer Academic Publishers: Boston, MA, USA, 2004; Volume 121, pp. 221–238. ISBN 978-1-4020-7919-1.
76. Ribatti, D.; Nico, B.; Ranieri, G.; Specchia, G.; Vacca, A. The Role of Angiogenesis in Human Non-Hodgkin Lymphomas. *Neoplasia* **2013**, *15*, 231–238. [CrossRef]
77. Shain, K.H.; Dalton, W.S.; Tao, J. The tumor microenvironment shapes hallmarks of mature B-cell malignancies. *Oncogene* **2015**, *34*, 4673–4682. [CrossRef] [PubMed]
78. Buggy, J.J.; Elias, L. Bruton Tyrosine Kinase (BTK) and Its Role in B-cell Malignancy. *Int. Rev. Immunol.* **2012**, *31*, 119–132. [CrossRef]
79. Fornecker, L.-M.; Muller, L.; Bertrand, F.; Paul, N.; Pichot, A.; Herbrecht, R.; Chenard, M.-P.; Mauviex, L.; Vallat, L.; Bahram, S.; et al. Multi-omics dataset to decipher the complexity of drug resistance in diffuse large B-cell lymphoma. *Sci. Rep.* **2019**, *9*, 895. [CrossRef]
80. Di Marzo, L.; Desantis, V.; Solimando, A.G.; Ruggieri, S.; Annese, T.; Nico, B.; Fumarulo, R.; Vacca, A.; Frassanito, M.A. Microenvironment drug resistance in multiple myeloma: Emerging new players. *Oncotarget* **2016**, *7*, 60698–60711. [CrossRef]
81. Tomida, A.; Tsuruo, T. Drug resistance mediated by cellular stress response to the microenvironment of solid tumors. *Anticancer. Drug Des.* **1999**, *14*, 169–177.
82. Alizadeh, A.A.; Gentles, A.J.; Alencar, A.J.; Liu, C.L.; Kohrt, H.E.; Houot, R.; Goldstein, M.J.; Zhao, S.; Natkunam, Y.; Advani, R.H.; et al. Prediction of survival in diffuse large B-cell lymphoma based on the expression of 2 genes reflecting tumor and microenvironment. *Blood* **2011**, *118*, 1350–1358. [CrossRef]
83. Argentiero, A.; De Summa, S.; Di Fonte, R.; Iacobazzi, R.M.; Porcelli, L.; Da Vià, M.; Brunetti, O.; Azzariti, A.; Silvestris, N.; Solimando, A.G. Gene Expression Comparison between the Lymph Node-Positive and -Negative Reveals a Peculiar Immune Microenvironment Signature and a Theranostic Role for WNT Targeting in Pancreatic Ductal Adenocarcinoma: A Pilot Study. *Cancers* **2019**, *11*, 942. [CrossRef] [PubMed]
84. Facciabene, A.; Motz, G.T.; Coukos, G. T-regulatory cells: Key players in tumor immune escape and angiogenesis. *Cancer Res.* **2012**, *72*, 2162–2171. [CrossRef]
85. Leone, P.; Di Lernia, G.; Solimando, A.G.; Cicco, S.; Saltarella, I.; Lamanuzzi, A.; Ria, R.; Frassanito, M.A.; Ponzoni, M.; Ditonno, P.; et al. Bone marrow endothelial cells sustain a tumor-specific CD8+ T cell subset with suppressive function in myeloma patients. *Oncoimmunology* **2019**, *8*, e1486949. [CrossRef] [PubMed]
86. Ribatti, D.; Vacca, A.; Dammacco, F.; English, D. Angiogenesis and Anti-Angiogenesis in Hematological Malignancies. *J. Hematother. Stem Cell Res.* **2003**, *12*, 11–22. [CrossRef] [PubMed]
87. Desantis, V.; Frassanito, M.A.; Tamma, R.; Saltarella, I.; Di Marzo, L.; Lamanuzzi, A.; Solimando, A.G.; Ruggieri, S.; Annese, T.; Nico, B.; et al. Rhu-Epo down-regulates pro-tumorigenic activity of cancer-associated fibroblasts in multiple myeloma. *Ann. Hematol.* **2018**, *97*, 1251–1258. [CrossRef] [PubMed]
88. Tonia, T.; Mettler, A.; Robert, N.; Schwarzer, G.; Seidenfeld, J.; Weingart, O.; Hyde, C.; Engert, A.; Bohlius, J. Erythropoietin or darbepoetin for patients with cancer. *Cochrane Database Syst. Rev.* **2012**, *12*, CD003407. [CrossRef] [PubMed]
89. Suhasini, A.N.; Wang, L.; Holder, K.N.; Lin, A.-P.; Bhatnagar, H.; Kim, S.-W.; Moritz, A.W.; Aguiar, R.C.T. A phosphodiesterase 4B-dependent interplay between tumor cells and the microenvironment regulates angiogenesis in B-cell lymphoma. *Leukemia* **2016**, *30*, 617–626. [CrossRef]
90. Serafini, P.; Mgebroff, S.; Noonan, K.; Borrello, I. Myeloid-Derived Suppressor Cells Promote Cross-Tolerance in B-Cell Lymphoma by Expanding Regulatory T Cells. *Cancer Res.* **2008**, *68*, 5439–5449. [CrossRef]

91. Broséus, J.; Mourah, S.; Ramstein, G.; Bernard, S.; Mounier, N.; Cuccuini, W.; Gaulard, P.; Gisselbrecht, C.; Brière, J.; Houlgatte, R.; et al. VEGF121, is predictor for survival in activated B-cell-like diffuse large B-cell lymphoma and is related to an immune response gene signature conserved in cancers. *Oncotarget* **2017**, *8*, 90808–90824. [CrossRef]
92. Burger, J.A.; Ghia, P.; Rosenwald, A.; Caligaris-Cappio, F. The microenvironment in mature B-cell malignancies: A target for new treatment strategies. *Blood* **2009**, *114*, 3367–3375. [CrossRef]
93. Solimando, A.G.; Da Via', M.C.; Leone, P.; Croci, G.; Borrelli, P.; Tabares Gaviria, P.; Brandl, A.; Di Lernia, G.; Bianchi, F.P.; Tafuri, S.; et al. Adhesion-Mediated Multiple Myeloma (MM) Disease Progression: Junctional Adhesion Molecule a Enhances Angiogenesis and Multiple Myeloma Dissemination and Predicts Poor Survival. *Blood* **2019**, *134*, 855. [CrossRef]
94. Pizzi, M.; Boi, M.; Bertoni, F.; Inghirami, G. Emerging therapies provide new opportunities to reshape the multifaceted interactions between the immune system and lymphoma cells. *Leukemia* **2016**, *30*, 1805–1815. [CrossRef] [PubMed]
95. Scuto, A.; Kujawski, M.; Kowolik, C.; Krymskaya, L.; Wang, L.; Weiss, L.M.; Digiusto, D.; Yu, H.; Forman, S.; Jove, R. STAT3 inhibition is a therapeutic strategy for ABC-like diffuse large B-cell lymphoma. *Cancer Res.* **2011**, *71*, 3182–3188. [CrossRef] [PubMed]
96. Passalidou, E.; Stewart, M.; Trivella, M.; Steers, G.; Pillai, G.; Dogan, A.; Leigh, I.; Hatton, C.; Harris, A.; Gatter, K.; et al. Vascular patterns in reactive lymphoid tissue and in non-Hodgkin's lymphoma. *Br. J. Cancer* **2003**, *88*, 553–559. [CrossRef]
97. Cardesa-Salzmann, T.M.; Colomo, L.; Gutierrez, G.; Chan, W.C.; Weisenburger, D.; Climent, F.; Gonzalez-Barca, E.; Mercadal, S.; Arenillas, L.; Serrano, S.; et al. High microvessel density determines a poor outcome in patients with diffuse large B-cell lymphoma treated with rituximab plus chemotherapy. *Haematologica* **2011**, *96*, 996–1001. [CrossRef]
98. Woźnialis, N.; Gierej, B.; Popławska, L.; Ziarkiewicz, M.; Wilczek, E.; Kulczycka, E.; Ziarkiewicz-Wróblewska, B. Angiogenesis in CD5-positive Diffuse Large B Cell Lymphoma: A Morphometric Analysis. *Adv. Clin. Exp. Med.* **2016**, *25*, 1149–1155. [CrossRef]
99. Shipp, M.A.; Ross, K.N.; Tamayo, P.; Weng, A.P.; Kutok, J.L.; Aguiar, R.C.T.; Gaasenbeek, M.; Angelo, M.; Reich, M.; Pinkus, G.S.; et al. Diffuse large B-cell lymphoma outcome prediction by gene-expression profiling and supervised machine learning. *Nat. Med.* **2002**, *8*, 68–74. [CrossRef]
100. Gratzinger, D.; Zhao, S.; Tibshirani, R.J.; Hsi, E.D.; Hans, C.P.; Pohlman, B.; Bast, M.; Avigdor, A.; Schiby, G.; Nagler, A.; et al. Prognostic Significance of VEGF, VEGF Receptors, and Microvessel Density in Diffuse Large B Cell Lymphoma Treated with Anthracycline-Based Chemotherapy. *Blood* **2007**, *110*, 53. [CrossRef]
101. Jørgensen, J.M.; Sørensen, F.B.; Bendix, K.; Nielsen, J.L.; Olsen, M.L.; Funder, A.M.D.; D'amore, F. Angiogenesis in non-Hodgkin's lymphoma: Clinico-pathological correlations and prognostic significance in specific subtypes. *Leuk. Lymphoma* **2007**, *48*, 584–595. [CrossRef]
102. Jiang, L.; Sun, J.; Quan, L.-N.; Tian, Y.-Y.; Jia, C.-M.; Liu, Z.-Q.; Liu, A.-C. Abnormal vascular endothelial growth factor protein expression may be correlated with poor prognosis in diffuse large B-cell lymphoma: A meta-analysis. *J. Cancer Res. Ther.* **2016**, *12*, 605. [CrossRef]
103. Yoon, K.-A.; Kim, M.K.; Eom, H.-S.; Lee, H.; Park, W.S.; Sohn, J.Y.; Kim, M.J.; Kong, S.-Y. Adverse prognostic impact of vascular endothelial growth factor gene polymorphisms in patients with diffuse large B-cell lymphoma. *Leuk. Lymphoma* **2017**, *58*, 2677–2682. [CrossRef] [PubMed]
104. Ganjoo, K.N.; Moore, A.M.; Orazi, A.; Sen, J.A.; Johnson, C.S.; An, C.S. The importance of angiogenesis markers in the outcome of patients with diffuse large B cell lymphoma: A retrospective study of 97 patients. *J. Cancer Res. Clin. Oncol.* **2008**, *134*, 381–387. [CrossRef] [PubMed]
105. Kim, M.K.; Suh, C.; Chi, H.S.; Cho, H.S.; Bae, Y.K.; Lee, K.H.; Lee, G.-W.; Kim, I.-S.; Eom, H.-S.; Kong, S.-Y.; et al. VEGFA and VEGFR2 genetic polymorphisms and survival in patients with diffuse large B cell lymphoma. *Cancer Sci.* **2012**, *103*, 497–503. [CrossRef] [PubMed]
106. Borges, N.M.; do Elias, M.V.; Fook-Alves, V.L.; Andrade, T.A.; de Conti, M.L.; Macedo, M.P.; Begnami, M.D.; Campos, A.H.J.F.M.; Etto, L.Y.; Bortoluzzo, A.B.; et al. Angiomirs expression profiling in diffuse large B-Cell lymphoma. *Oncotarget* **2016**, *7*. [CrossRef]
107. Lupino, L.; Perry, T.; Margielewska, S.; Hollows, R.; Ibrahim, M.; Care, M.; Allegood, J.; Tooze, R.; Sabbadini, R.; Reynolds, G.; et al. Sphingosine-1-phosphate signalling drives an angiogenic transcriptional programme in diffuse large B cell lymphoma. *Leukemia* **2019**, *33*, 2884–2897. [CrossRef]

108. Pyne, S.; Pyne, N. Sphingosine 1-phosphate signalling via the endothelial differentiation gene family of G-protein-coupled receptors. *Pharmacol. Ther.* **2000**, *88*, 115–131. [CrossRef]
109. Wang, E.S.; Teruya-Feldstein, J.; Wu, Y.; Zhu, Z.; Hicklin, D.J.; Moore, M.A.S. Targeting autocrine and paracrine VEGF receptor pathways inhibits human lymphoma xenografts in vivo. *Blood* **2004**, *104*, 2893–2902. [CrossRef]
110. Aref, S.; Mabed, M.; Zalata, K.; Sakrana, M.; El Askalany, H. The Interplay Between C-Myc Oncogene Expression and Circulating Vascular Endothelial Growth Factor (sVEGF), Its Antagonist Receptor, Soluble Flt-1 in Diffuse Large B Cell Lymphoma (DLBCL): Relationship to Patient Outcome. *Leuk. Lymphoma* **2004**, *45*, 499–506. [CrossRef]
111. Stopeck, A.T.; Unger, J.M.; Rimsza, L.M.; Bellamy, W.T.; Iannone, M.; Persky, D.O.; Leblanc, M.; Fisher, R.I.; Miller, T.P. A phase II trial of single agent bevacizumab in patients with relapsed, aggressive non-Hodgkin lymphoma: Southwest oncology group study S0108. *Leuk. Lymphoma* **2009**, *50*, 728–735. [CrossRef]
112. Marinaccio, C.; Ingravallo, G.; Gaudio, F.; Perrone, T.; Nico, B.; Maoirano, E.; Specchia, G.; Ribatti, D. Microvascular density, CD68 and tryptase expression in human Diffuse Large B-Cell Lymphoma. *Leuk. Res.* **2014**, *38*, 1374–1377. [CrossRef]
113. Marinaccio, C.; Ingravallo, G.; Gaudio, F.; Perrone, T.; Ruggieri, S.; Opinto, G.; Nico, B.; Maiorano, E.; Specchia, G.; Ribatti, D. T cells, mast cells and microvascular density in diffuse large B cell lymphoma. *Clin. Exp. Med.* **2016**, *16*, 301–306. [CrossRef] [PubMed]
114. Song, M.-K.; Chung, J.-S.; Sung-Yong, O.; Lee, G.-W.; Kim, S.-G.; Seol, Y.-M.; Shin, H.-J.; Choi, Y.-J.; Cho, G.-J.; Shin, D.-H.; et al. Clinical impact of bulky mass in the patient with primary extranodal diffuse large b cell lymphoma treated with R-CHOP therapy. *Ann. Hematol.* **2010**, *89*, 985–991. [CrossRef] [PubMed]
115. Tamma, R.; Ingravallo, G.; Albano, F.; Gaudio, F.; Annese, T.; Ruggieri, S.; Lorusso, L.; Errede, M.; Maiorano, E.; Specchia, G.; et al. STAT-3 RNAscope Determination in Human Diffuse Large B-Cell Lymphoma. *Transl. Oncol.* **2019**, *12*, 545–549. [CrossRef] [PubMed]
116. Tamma, R.; Ingravallo, G.; Gaudio, F.; Annese, T.; Albano, F.; Ruggieri, S.; Dicataldo, M.; Maiorano, E.; Specchia, G.; Ribatti, D. STAT3, tumor microenvironment, and microvessel density in diffuse large B cell lymphomas. *Leuk. Lymphoma* **2020**, *61*, 567–574. [CrossRef] [PubMed]
117. Sircar, A.; Chowdhury, S.M.; Hart, A.; Bell, W.C.; Singh, S.; Sehgal, L.; Epperla, N. Impact and Intricacies of Bone Marrow Microenvironment in B-cell Lymphomas: From Biology to Therapy. *Int. J. Mol. Sci.* **2020**, *21*, 904. [CrossRef]
118. De Charette, M.; Houot, R. Hide or defend, the two strategies of lymphoma immune evasion: Potential implications for immunotherapy. *Haematologica* **2018**, *103*, 1256–1268. [CrossRef] [PubMed]
119. Miao, X.; Wu, Y.; Wang, Y.; Zhu, X.; Yin, H.; He, Y.; Li, C.; Liu, Y.; Lu, X.; Chen, Y.; et al. Y-box-binding protein-1 (YB-1) promotes cell proliferation, adhesion and drug resistance in diffuse large B-cell lymphoma. *Exp. Cell Res.* **2016**, *346*, 157–166. [CrossRef]
120. Huang, X.; Bai, X.; Cao, Y.; Wu, J.; Huang, M.; Tang, D.; Tao, S.; Zhu, T.; Liu, Y.; Yang, Y.; et al. Lymphoma endothelium preferentially expresses Tim-3 and facilitates the progression of lymphoma by mediating immune evasion. *J. Exp. Med.* **2010**, *207*, 505–520. [CrossRef]
121. Edwards, J.P.; Zhang, X.; Mosser, D.M. The expression of heparin-binding epidermal growth factor-like growth factor by regulatory macrophages. *J. Immunol. Baltim. Md 1950* **2009**, *182*, 1929–1939. [CrossRef]
122. Rao, L.; Giannico, D.; Leone, P.; Solimando, A.G.; Maiorano, E.; Caporusso, C.; Duda, L.; Tamma, R.; Mallamaci, R.; Susca, N.; et al. HB-EGF-EGFR Signaling in Bone Marrow Endothelial Cells Mediates Angiogenesis Associated with Multiple Myeloma. *Cancers* **2020**, *12*, 173. [CrossRef]
123. Tjin, E.P.M.; Groen, R.W.J.; Vogelzang, I.; Derksen, P.W.B.; Klok, M.D.; Meijer, H.P.; van Eeden, S.; Pals, S.T.; Spaargaren, M. Functional analysis of HGF/MET signaling and aberrant HGF-activator expression in diffuse large B-cell lymphoma. *Blood* **2006**, *107*, 760–768. [CrossRef]
124. Moschetta, M.; Basile, A.; Ferrucci, A.; Frassanito, M.A.; Rao, L.; Ria, R.; Solimando, A.G.; Giuliani, N.; Boccarelli, A.; Fumarola, F.; et al. Novel targeting of phospho-cMET overcomes drug resistance and induces antitumor activity in multiple myeloma. *Clin. Cancer Res. Off. J. Am. Assoc. Cancer Res.* **2013**, *19*, 4371–4382. [CrossRef] [PubMed]

125. Ferrucci, A.; Moschetta, M.; Frassanito, M.A.; Berardi, S.; Catacchio, I.; Ria, R.; Racanelli, V.; Caivano, A.; Solimando, A.G.; Vergara, D.; et al. A HGF/cMET autocrine loop is operative in multiple myeloma bone marrow endothelial cells and may represent a novel therapeutic target. *Clin. Cancer Res. Off. J. Am. Assoc. Cancer Res.* **2014**, *20*, 5796–5807. [CrossRef] [PubMed]
126. Rao, L.; De Veirman, K.; Giannico, D.; Saltarella, I.; Desantis, V.; Frassanito, M.A.; Solimando, A.G.; Ribatti, D.; Prete, M.; Harstrick, A.; et al. Targeting angiogenesis in multiple myeloma by the VEGF and HGF blocking DARPin® protein MP0250: A preclinical study. *Oncotarget* **2018**, *9*, 13366–13381. [CrossRef] [PubMed]
127. Siegler, E.; Li, S.; Kim, Y.J.; Wang, P. Designed Ankyrin Repeat Proteins as Her2 Targeting Domains in Chimeric Antigen Receptor-Engineered T Cells. *Hum. Gene Ther.* **2017**, *28*, 726–736. [CrossRef] [PubMed]
128. Xu, P.-P.; Sun, Y.-F.; Fang, Y.; Song, Q.; Yan, Z.-X.; Chen, Y.; Jiang, X.-F.; Fei, X.-C.; Zhao, Y.; Leboeuf, C.; et al. JAM-A overexpression is related to disease progression in diffuse large B-cell lymphoma and downregulated by lenalidomide. *Sci. Rep.* **2017**, *7*, 7433. [CrossRef] [PubMed]
129. Yu, W.; Si, M.; Li, L.; He, P.; Fan, Z.; Zhang, Q.; Jiao, X. Biomarkers Reflecting the Destruction of the Blood-Brain Barrier Are Valuable in Predicting the Risk of Lymphomas with Central Nervous System Involvement. *Onco Targets Ther.* **2019**, *12*, 9505–9512. [CrossRef]
130. Karar, J.; Maity, A. PI3K/AKT/mTOR Pathway in Angiogenesis. *Front. Mol. Neurosci.* **2011**, *4*, 51. [CrossRef]
131. Goodman, A.M.; Choi, M.; Wieduwilt, M.; Mulroney, C.; Costello, C.; Frampton, G.; Miller, V.; Kurzrock, R. Next-Generation Sequencing Reveals Potentially Actionable Alterations in the Majority of Patients with Lymphoid Malignancies. *JCO Precis. Oncol.* **2017**, 1–13. [CrossRef]
132. Todorovic Balint, M.; Jelicic, J.; Mihaljevic, B.; Kostic, J.; Stanic, B.; Balint, B.; Pejanovic, N.; Lucic, B.; Tosic, N.; Marjanovic, I.; et al. Gene Mutation Profiles in Primary Diffuse Large B Cell Lymphoma of Central Nervous System: Next Generation Sequencing Analyses. *Int. J. Mol. Sci.* **2016**, *17*, 683. [CrossRef]
133. Vaqué, J.P.; Martínez, N.; Batlle-López, A.; Pérez, C.; Montes-Moreno, S.; Sánchez-Beato, M.; Piris, M.A. B-cell lymphoma mutations: Improving diagnostics and enabling targeted therapies. *Haematologica* **2014**, *99*, 222–231. [CrossRef] [PubMed]
134. Lee, J.-H.S.; Vo, T.-T.; Fruman, D.A. Targeting mTOR for the treatment of B cell malignancies. *Br. J. Clin. Pharmacol.* **2016**, *82*, 1213–1228. [CrossRef] [PubMed]
135. Da Vià, M.C.; Solimando, A.G.; Garitano-Trojaola, A.; Barrio, S.; Munawar, U.; Strifler, S.; Haertle, L.; Rhodes, N.; Teufel, E.; Vogt, C.; et al. CIC Mutation as a Molecular Mechanism of Acquired Resistance to Combined BRAF-MEK Inhibition in Extramedullary Multiple Myeloma with Central Nervous System Involvement. *Oncologist* **2020**, *25*, 112–118. [CrossRef]
136. Holderfield, M.; Deuker, M.M.; McCormick, F.; McMahon, M. Targeting RAF kinases for cancer therapy: BRAF-mutated melanoma and beyond. *Nat. Rev. Cancer* **2014**, *14*, 455–467. [CrossRef] [PubMed]
137. Lamanuzzi, A.; Saltarella, I.; Desantis, V.; Frassanito, M.A.; Leone, P.; Racanelli, V.; Nico, B.; Ribatti, D.; Ditonno, P.; Prete, M.; et al. Inhibition of mTOR complex 2 restrains tumor angiogenesis in multiple myeloma. *Oncotarget* **2018**, *9*, 20563–20577. [CrossRef] [PubMed]
138. Solimando, A.G.; Brandl, A.; Mattenheimer, K.; Graf, C.; Ritz, M.; Ruckdeschel, A.; Stühmer, T.; Mokhtari, Z.; Rudelius, M.; Dotterweich, J.; et al. JAM-A as a prognostic factor and new therapeutic target in multiple myeloma. *Leukemia* **2018**, *32*, 736–743. [CrossRef]
139. Solimando, A.G.; Da Via', M.C.; Borrelli, P.; Leone, P.; Di Lernia, G.; Tabares Gaviria, P.; Brandl, A.; Pedone, G.L.; Rauert-Wunderlich, H.; Lapa, C.; et al. Central Function for JAM-a in Multiple Myeloma Patients with Extramedullary Disease. *Blood* **2018**, *132*, 4455. [CrossRef]
140. Mielcarek, M.; Sperling, C.; Shrappe, M.; Meyer, U.; Riehm, H.; Ludwig, W.-D. Expression of intercellular adhesion molecule 1 (ICAM-1) in childhood acute lymphoblastic leukaemia: Correlation with clinical features and outcome. *Br. J. Haematol.* **1997**, *96*, 301–307. [CrossRef]
141. Horstmann, W.G.; Timens, W. Lack of adhesion molecules in testicular diffuse centroblastic and immunoblastic B cell lymphomas as a contributory factor in malignant behaviour. *Virchows Arch.* **1996**, *429*, 83–90. [CrossRef]
142. Da Via', M.C.; Solimando, A.G.; Garitano-Trojaola, A.; Barrio, S.; Rodhes, N.; Strifler, S.; Teufel, E.; Lapa, C.; Einsele, H.; Beilhack, A.; et al. CIC-Mutation As a Potential Molecular Mechanism of Acquired Resistance to Combined BRAF/MEK Inhibition in CNS Multiple Myeloma. *Blood* **2018**, *132*, 3181. [CrossRef]
143. Pasqualucci, L. The genetic basis of diffuse large B-cell lymphoma. *Curr. Opin. Hematol.* **2013**, *20*, 336–344. [CrossRef] [PubMed]

144. Stopeck, A.T.; Gessner, A.; Miller, T.P.; Hersh, E.M.; Johnson, C.S.; Cui, H.; Frutiger, Y.; Grogan, T.M. Loss of B7.2 (CD86) and intracellular adhesion molecule 1 (CD54) expression is associated with decreased tumor-infiltrating T lymphocytes in diffuse B-cell large-cell lymphoma. *Clin. Cancer Res. Off. J. Am. Assoc. Cancer Res.* **2000**, *6*, 3904–3909.
145. Terol, M.J.; Tormo, M.; Martinez-Climent, J.A.; Marugan, I.; Benet, I.; Ferrandez, A.; Teruel, A.; Ferrer, R.; García-Conde, J. Soluble intercellular adhesion molecule-1 (s-ICAM-1/s-CD54) in diffuse large B-cell lymphoma: Association with clinical characteristics and outcome. *Ann. Oncol. Off. J. Eur. Soc. Med. Oncol.* **2003**, *14*, 467–474. [CrossRef] [PubMed]
146. Zhang, L.-H.; Kosek, J.; Wang, M.; Heise, C.; Schafer, P.H.; Chopra, R. Lenalidomide efficacy in activated B-cell-like subtype diffuse large B-cell lymphoma is dependent upon IRF4 and cereblon expression. *Br. J. Haematol.* **2013**, *160*, 487–502. [CrossRef]
147. Gascoyne, D.M.; Banham, A.H. The significance of FOXP1 in diffuse large B-cell lymphoma. *Leuk. Lymphoma* **2017**, *58*, 1037–1051. [CrossRef]
148. Solimando, A.; Brandl, A.; Katharina, M.; Graf, C.; Ritz, M.; Ruckdeschel, A.; Stühmer, T.; Rudelius, M.; Frassanito, M.A.; Andreas, R.; et al. JAM-A as a Prognostic Factor and New Therapeutic Target in Multiple Myeloma. *Blood* **2016**, *128*, 307. [CrossRef]
149. Ebnet, K. Junctional Adhesion Molecules (JAMs): Cell Adhesion Receptors With Pleiotropic Functions in Cell Physiology and Development. *Physiol. Rev.* **2017**, *97*, 1529–1554. [CrossRef]
150. Koenen, R.R.; Pruessmeyer, J.; Soehnlein, O.; Fraemohs, L.; Zernecke, A.; Schwarz, N.; Reiss, K.; Sarabi, A.; Lindbom, L.; Hackeng, T.M.; et al. Regulated release and functional modulation of junctional adhesion molecule A by disintegrin metalloproteinases. *Blood* **2009**, *113*, 4799–4809. [CrossRef]
151. Leech, A.O.; Cruz, R.G.B.; Hill, A.D.K.; Hopkins, A.M. Paradigms lost-an emerging role for over-expression of tight junction adhesion proteins in cancer pathogenesis. *Ann. Transl. Med.* **2015**, *3*, 184. [CrossRef]
152. Zhao, C.; Lu, F.; Chen, H.; Zhao, X.; Sun, J.; Chen, H. Dysregulation of JAM-A plays an important role in human tumor progression. *Int. J. Clin. Exp. Pathol.* **2014**, *7*, 7242–7248.
153. Scheller, J.; Chalaris, A.; Garbers, C.; Rose-John, S. ADAM17: A molecular switch to control inflammation and tissue regeneration. *Trends Immunol.* **2011**, *32*, 380–387. [CrossRef]
154. Katz, E.; Deehan, M.R.; Seatter, S.; Lord, C.; Sturrock, R.D.; Harnett, M.M. B cell receptor-stimulated mitochondrial phospholipase A2 activation and resultant disruption of mitochondrial membrane potential correlate with the induction of apoptosis in WEHI-231 B cells. *J. Immunol. Baltim. Md 1950* **2001**, *166*, 137–147. [CrossRef]
155. Jridi, I.; Catacchio, I.; Majdoub, H.; Shahbazzadeh, D.; El Ayeb, M.; Frassanito, M.A.; Solimando, A.G.; Ribatti, D.; Vacca, A.; Borchani, L. The small subunit of Hemilipin2, a new heterodimeric phospholipase A2 from Hemiscorpius lepturus scorpion venom, mediates the antiangiogenic effect of the whole protein. *Toxicon Off. J. Int. Soc. Toxinol.* **2017**, *126*, 38–46. [CrossRef]
156. Upadhyay, R.; Hammerich, L.; Peng, P.; Brown, B.; Merad, M.; Brody, J.D. Lymphoma: Immune evasion strategies. *Cancers* **2015**, *7*, 736–762. [CrossRef]
157. Tripodo, C.; Sangaletti, S.; Piccaluga, P.P.; Prakash, S.; Franco, G.; Borrello, I.; Orazi, A.; Colombo, M.P.; Pileri, S.A. The bone marrow stroma in hematological neoplasms–a guilty bystander. *Nat. Rev. Clin. Oncol.* **2011**, *8*, 456–466. [CrossRef]
158. Rafii, S.; Lyden, D.; Benezra, R.; Hattori, K.; Heissig, B. Vascular and haematopoietic stem cells: Novel targets for anti-angiogenesis therapy? *Nat. Rev. Cancer* **2002**, *2*, 826–835. [CrossRef]
159. Grunewald, M.; Avraham, I.; Dor, Y.; Bachar-Lustig, E.; Itin, A.; Jung, S.; Yung, S.; Chimenti, S.; Landsman, L.; Abramovitch, R.; et al. VEGF-induced adult neovascularization: Recruitment, retention, and role of accessory cells. *Cell* **2006**, *124*, 175–189. [CrossRef]
160. Seymour, J.F.; Pfreundschuh, M.; Trneny, M.; Sehn, L.H.; Catalano, J.; Csinady, E.; Moore, N.; Coiffier, B. R-CHOP with or without bevacizumab in patients with previously untreated diffuse large B-cell lymphoma: Final MAIN study outcomes. *Haematologica* **2014**, *99*, 1343–1349. [CrossRef]
161. Jiang, L.; Li, N. B-cell non-Hodgkin lymphoma: Importance of angiogenesis and antiangiogenic therapy. *Angiogenesis* **2020**. [CrossRef]
162. Joyce, J.A.; Fearon, D.T. T cell exclusion, immune privilege, and the tumor microenvironment. *Science* **2015**, *348*, 74–80. [CrossRef]

163. Galon, J.; Bruni, D. Approaches to treat immune hot, altered and cold tumours with combination immunotherapies. *Nat. Rev. Drug Discov.* **2019**, *18*, 197–218. [CrossRef]
164. Antonio, G.; Oronzo, B.; Vito, L.; Angela, C.; Antonel-la, A.; Roberto, C.; Giovanni, S.A.; Antonella, L. Immune system and bone microenvironment: Rationale for targeted cancer therapies. *Oncotarget* **2020**, *11*. [CrossRef]
165. Fukumura, D.; Kloepper, J.; Amoozgar, Z.; Duda, D.G.; Jain, R.K. Enhancing cancer immunotherapy using antiangiogenics: Opportunities and challenges. *Nat. Rev. Clin. Oncol.* **2018**, *15*, 325–340. [CrossRef] [PubMed]
166. Compagno, N.; Cinetto, F.; Semenzato, G.; Agostini, C. Subcutaneous immunoglobulin in lymphoproliferative disorders and rituximab-related secondary hypogammaglobulinemia: A single-center experience in 61 patients. *Haematologica* **2014**, *99*, 1101–1106. [CrossRef] [PubMed]
167. Vacca, A.; Melaccio, A.; Sportelli, A.; Solimando, A.G.; Dammacco, F.; Ria, R. Subcutaneous immunoglobulins in patients with multiple myeloma and secondary hypogammaglobulinemia: A randomized trial. *Clin. Immunol. Orlando Fl.* **2018**, *191*, 110–115. [CrossRef] [PubMed]
168. Ma, X.; Li, L.; Zhang, L.; Fu, X.; Li, X.; Wang, X.; Wu, J.; Sun, Z.; Zhang, X.; Feng, X.; et al. Apatinib in Patients with Relapsed or Refractory Diffuse Large B Cell Lymphoma: A Phase II, Open-Label, Single-Arm, Prospective Study. *Drug Des. Devel. Ther.* **2020**, *14*, 275–284. [CrossRef] [PubMed]
169. Wang, Y.; Deng, M.; Chen, Q.; Li, Y.; Guo, X.; Shi, P.; He, L.; Xie, S.; Yu, L.; Zhang, H.; et al. Apatinib exerts anti-tumor activity to non-Hodgkin lymphoma by inhibition of the Ras pathway. *Eur. J. Pharmacol.* **2019**, *843*, 145–153. [CrossRef] [PubMed]
170. Pfreundschuh, M.; Kloess, M.; Schmits, R.; Zeynalova, S.; Lengfelder, E.; Franke, A.; Steinhauer, H.; Reiser, M.; Clemens, M.; Nickenig, C.; et al. Six, Not Eight Cycles of Bi-Weekly CHOP with Rituximab (R-CHOP-14) Is the Preferred Treatment for Elderly Patients with Diffuse Large B-Cell Lymphoma (DLBCL): Results of the RICOVER-60 Trial of the German High-Grade Non-Hodgkin Lymphoma Study Group (DSHNHL). *Blood* **2005**, *106*, 13. [CrossRef]

© 2020 by the authors. Licensee MDPI, Basel, Switzerland. This article is an open access article distributed under the terms and conditions of the Creative Commons Attribution (CC BY) license (http://creativecommons.org/licenses/by/4.0/).

Review

Inducing Angiogenesis, a Key Step in Cancer Vascularization, and Treatment Approaches

Harman Saman [1,2,*], Syed Shadab Raza [3], Shahab Uddin [4] and Kakil Rasul [5]

1. Barts Cancer Institute, Queen Mary University of London, London E1 4NS, UK
2. Department of Medicine, Hazm Maubrairek Hospital, Ar-Rayyan PO Box 305, Qatar
3. Department of Stem Cell Biology and Regenerative Medicine, ERA University, Lucknow 226003, India; Drshadab@erauniversit.In
4. Translational Research Institute, Academic Health System, Hamad Medical Corporation, Doha 3050, Qatar; Skhan34@hamad.qa
5. National Cancer Care and Research, Hamad Medical Corporation, Doha 3050, Qatar; Krasul@hamad.qa
* Correspondence: HSaman@hamad.qa or h.saman@smd19.qmul.ac.uk; Tel.: +97-466506781

Received: 23 February 2020; Accepted: 17 April 2020; Published: 6 May 2020

Abstract: Angiogenesis is a term that describes the formation of new blood and lymphatic vessels from a pre-existing vasculature. This allows tumour cells to acquire sustenance in the form of nutrients and oxygen and the ability to evacuate metabolic waste. As one of the hallmarks of cancer, angiogenesis has been studied extensively in animal and human models to enable better understanding of cancer biology and the development of new anti-cancer treatments. Angiogenesis plays a crucial role in the process of tumour genesis, because solid tumour need a blood supply if they are to grow beyond a few millimeters in size. On the other hand, there is growing evidence that some solid tumour exploit existing normal blood supply and do not require a new vessel formation to grow and to undergo metastasis. This review of the literature will present the current understanding of this intricate process and the latest advances in the use of angiogenesis-targeting therapies in the fight against cancer.

Keywords: angiogenesis; cancer; VEGF; anticancer

---

1. Introduction

Under physiological conditions, angiogenesis is a highly regulated process. It plays crucial roles in embryogenesis, wound healing and the menstrual cycle [1]. Angiogenesis is also seen in non-malignant pathologies such as diabetic retinopathy, ischaemic diseases and autoimmune conditions such as connective tissue diseases and psoriasis [1].

In addition to providing nutrients and oxygen to the tumour and the removal of metabolic waste, new vessel formation also enables cancer cells to metastasize and proliferate to distant sites through entry into the newly formed blood and lymphatic system and subsequent extravasation [2]. A lack of adequate blood supply, on the other hand, could halt tumour growth, and might even lead to tumour shrinkage and sometimes cancer cell death [3]. Previous studies demonstrated that, in the absence of angiogenesis, tumours could grow to a maximum of 1–2 mm^3 in diameter before they stopped growing and died, whilst some tumour cells could grow beyond 2 mm^3 in size in angiogenesis-rich cell culture. The continued growth of cancer cells in angiogenesis-rich cell culture is explained by reproducing physiological properties in a three-dimensional cell culture model that provides controlled fluid perfusion that permits the regulation of oxygen intake, promoting a circulatory environment that is controlled by computer hardware [4].

2. Angiogenesis in Normal Tissue

The structure of the blood vessels depends on their size; small blood vessels are comprised of endothelial cells (EC), whereas in medium and large blood vessels, ECs are surrounded by pericytes (mural cells) [5]. In normal tissue, the process of neovascularization is tightly controlled. The process includes stepwise stages (Figure 1).

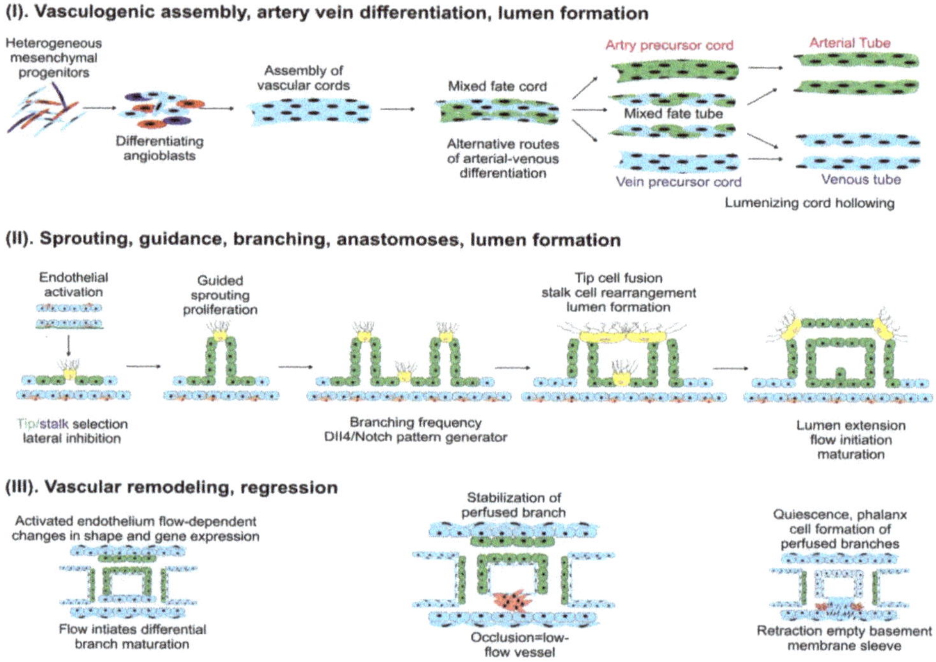

Figure 1. Steps of angiogenesis: (**I**)—Endothelial cell (EC) differentiated from angioblasts. (**II**)—sprouting, guidance, branching, anastomoses, lumen formation. (**III**)—vascular remodeling from a primitive (left box) towards a stabilized and mature vascular plexus (right box).

After this strictly controlled vessel formation, the normal vasculature becomes largely quiescent [5]. Angiogenesis is controlled by several growth factor stimulators and inhibitors. Angiogenic (stimulatory) growth factors include Fibroblast Growth Factor, Granulocyte Colony-Stimulating Factor, Interleukin-8, Transforming Growth Factors alpha and beta and Vascular Endothelial Growth Factor. Angiogenic inhibitors include Angiostatin, Interferons (alpha, beta and gamma), Endostatin, Interleukin-12 and retinoids [5]. Inhibitory factors are present within the extra-cellular matrix (ECM). At a molecular level, angiogenesis is normally controlled by a family of small none-coding RNA molecules that are collectively called angiomiRs. AngiomiRs are comprised of pro-angiogenic miRs and anti-angiogenic miRs (Table 1) [6]. A well-studied angiomiRs is miR-200b, which belongs to the miR-200 family [7]. miR-200b has antiangiogenic effects. Its expression is transiently turned down when new vessel formation is required, for example during wound-healing. Once the physiological demand subsides, miR-200b is expressed again to stop angiogenesis as a measure of tight control on new vessel formation. The downregulation of miR-200b in response to tissue hypoxia triggers epithelial to mesenchymal transition and modulates endothelial cell migration which result in new vessel formation [8]. There is evidence that the dysregulation of iR-200b contributes to oncogenesis and metastasis in some cancers, such as breast cancer [9].

Table 1. AngiomiR are none-coding RNAs that play an important role in angiogenesis in normal tissue, through their expression or silencing depending on physiological demand. The dysregulation of miR-200b is detected in some cancers. Different types of AngiomiR have specific effects on angiogenesis.

AngiomiR	Molecular Function	Reference
miR-15b, miR-16, miR-20a, miR-20b	Have no known functions. They might contribute in regulation of VEGF.	[10]
miR-21, miR-31	Triggers mobilisation of EC.	[11]
miR-17-92	Dysregulation of miR-17-92 in cancer cells promote growth.	[12]
miR-130a	Induces angiogenesis by supressing GAX and HOXA5	[13]
miR-296	Animal studies showed that by acting on HGS, miR-296 stimulate angiogenesis.	[14]
miR-320	Suppression of miR-320 in diabetic cells trigger angiogenesis by stimulating EC proliferation.	[15]
miR-210	In hypoxic cell culture, miR-210 promote EC proliferation and survival.	[16]
miR-378	Support tumour growth by improving vascularisation via angiogenesis.	[17]

3. Angiogenesis in Cancer, a Literature Review

In cancer, a switch to angiogenesis seems to be an imbalance between stimulatory and inhibitory factors that leads to a pro-angiogenic state [18]. This results from a state of a relatively poorly blood-supplied hyperplasia converted to an uncontrollable new vessel formation that ultimately causes malignant tumour progression. Researchers have investigated the molecular basis of pro- and inhibitory pathways with the view of better understanding oncogenesis and the development of anti-cancer treatment. The flip side of angiogenesis is poor tumour blood supply. Poor tumour blood supply is one of the postulated mechanisms of resistance to chemotherapy, due to the failure of an adequate delivery of cytotoxic drugs to the tumour site [19]. For example, for decades the five year overall survival of pancreatic cancer has not exceeded 5%, despite extensive research [20]. One explanation for this is that pancreatic cancer tissue is surrounded by dens stromal tissue that hinders the delivery of anticancer therapies. In contrast to antiangiogenetic treatment, vascular promotion therapy is investigated to promote tumour blood supply to facilitate the better delivery of cytotoxic drugs to the target tissue [20].

4. Pro- and Anti-Angiogenic Factors

Judah Folkman coined the phrase tumour angiogenesis and studied this process extensively [18]. He led the discovery of the first angiogenic factors. These factors trigger neovascularization through inducing angiogenesis switch [21]. As seen in Figure 2, tumour overgrowth is believed to be halted through maintaining an equilibrium between pro- and anti-angiogenesis factors, leading to a state of tumour dormancy [18,21].

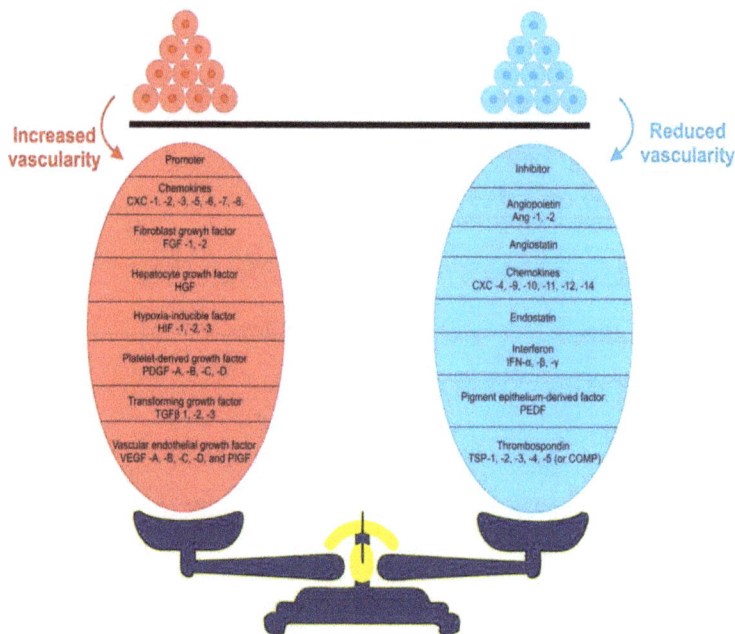

Figure 2. Maintaining homeostasis results from an equilibrium between promotors and inhibitors of angiogenesis.

Disturbance to this equilibrium results in increased angiogenesis, and thus uncontrollable tumour overgrowth [22]. Several angiogenic factors have been described. Of those, vascular endothelial growth factor A (VEGF-A) is a major regulator of angiogenesis both under normal conditions and in disease state [11]. VEGF-A belongs to family of gene factors that also encompasses VEGF-B, VEGF-C, VEGF-D, VEGF-E and placenta growth factor (PlGF). These growth factors have different levels of specificity and different affinities to tyrosine kinase receptors (VEGFR) 1,-2 and -3 [22]. The binding of VEGF-A to VEGFR 2 (predominantly found on EC of blood vessels) leads to angiogenesis, whereas VEGF-C and D preferentially bind to VEGFR-3, expressed predominantly on lymphatic EC, resulting in the proliferation of lymphatic vessels [23]. In cancer, the role of VEGF exceeds angiogenesis through a complex autocrine and paracrine signaling pathway; VEGF plays an important role in promoting the cancer stem cells' functionality and the initiation of tumour [24]. The upregulation of VEGF initiates tumourigenesis by contributing to the activation of epithelial–mesenchymal transition (EMT) [25]. EMT represents a key event in the process of new vessel formation. [26]. This because EMT leads to a loss of cell polarity and dramatic cytoskeletal changes, which lead to increased cell motility and loss of cells to cell adhesion by the loss of E-cadherin and ZO-1. The last two markers are associated with epithelial cells. EMT also results in the production of several proteolytic enzymes, including matrix metalloproteases and serine proteases that degrade the extracellular matrix (ECM). Several pathways involved in EMT support endothelial cell (EC) survival and proliferation [26]. These pathways invlove complex interactions between the cell membrane, ECM and intracellular regulatory signalling pathways. The resulting phenotypical changes caused by EMT promote cancer cell invasion of basement membrane, and eventually cancer cell metastasis [27].

Moreover, in hypoxic tumour, tumour-associated macrophages (TAMs), known for their protumour functions, secrete VEGF. VEGF interacts with key immune cells in the tumour micro-environment (TME), namely CD4+ forkhead box protein P3 (FOXP3) + regulatory T cells, which a strong suppressor of anticancer immunity. VEGF is able to attract these regulatory T cells to the TME using the chemoattractant neuropilin 1 (NRP1). Animal studies showed that removing NRP1 was associated with increased infiltration of TME with antitumoural CD8 T cells, with a reduction in tumour growth. Fibroblasts, that are present in abundance in TME and known to support tumour growth, also secrete VEGF. VEGF, in turn, stimulates fibronectin fibril assembly; the latter has a potent protumour effect within the TME [28].

Other angiogenesis promotors include platelet-derived growth factor (PDGF)-B and C and fibroblast growth factor (FGF)-1 and -2 [29]. Both groups of factors exert their effect, EC proliferation and migration, once they bind to their respective receptors on blood vessels' EC.

As seen in Figure 3, Tie1 and Tie2 are two signaling pathways that encompass the interaction between angiopoietins, tyrosine kinases (TK), VEGF and their receptors [29].

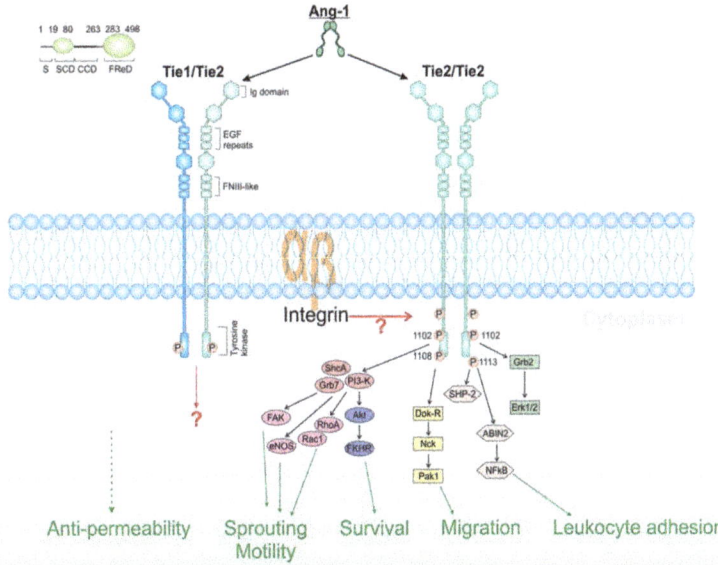

Figure 3. Ang1 and Ang2 bind to Tie2 with similar affinities; however, whereas Ang1 is an agonist, the ability of Ang2 to activate Tie2 appears to depend on the cell type and context. The activation of the Tie2 pathway results in the inhibition of apoptosis, cell survival and migration.

Other important regulators of angiogenesis are angiopoietins. Angiopoietins interact with Tie-2 TK receptor found on EC. Through cooperating with other angiogenesis factors, angiopoietins modulate the activity of the EC [29,30]. Angiopoietin-1 (Ang-1) and angiopoietin-2 (Ang-2) can form dimers, trimers and tetramers. Angiopoietin-1 can form higher-order multimers through its super clustering domain. It is believed that not all these structures bind with the TK receptor; the only activators of these receptors are at the tetramer level or higher [30].

Other promotors of angiogenesis include a wide range of polypeptides, metabolites and hormones that contribute to new blood vessels' formation in both physiological and disease state [30]. On the other hand, there is a wide range of antiangiogenetic factors that oppose the function of the promotors. Constituents and proteolytic fragments of the extra-cellular matrix (ECM) and the basement membrane represent potent angiogenesis inhibitors [31]. A well-studied angiogenesis inhibitor is thrombospondin-1 (TSP1), which is a large glycoprotein present in ECM [31]. Another matrix-derived angiogenesis inhibitor is a proteolytic product of collagen XVIII called endostatin [32]. Interferon-alpha and -beta and angiostatin, a cleavage product of plasmin, are other examples of angiogenesis inhibitors [33].

The activities of both angiogenesis promoters and inhibitors are regulated through a complex interaction of different pathways. The proangiogenic imbalance often occurs at the gene level due to the activation of oncogenes or inactivation of tumour suppressor genes, all the way to cell environmental factors such as hypoxia, hypoglycaemia, cellular nutrient deficiency and metabolic acidosis [34]. As part of multistage tumourigenesis, angiogenic switch arises from an imbalance between pro- and inhibitors of angiogenesis activity and level; this imbalance is driven from the tumour cells and the inflammatory cells that infiltrate the tumour [35]. The next section will focus on the mechanisms behind angiogenic switches in cancer and the different pathways involved.

5. Angiogenic Switch

In a seminal paper, Judah Folkman, Doug Hanahan and colleagues presented angiogenic switch in a transgenic Rip1Tag2 mouse module of pancreatic beta-cell carcinogenesis, during the progression from hyperplasia to heavily vascularised cancer [36]. Rip1Tag2 mice express the Simian Virus 40 large T antigen oncoprotein under the control of the rat insulin promoter. This leads to the overexpression of oncogene in pancreatic beta-cells of the islets of Langerhans, resulting in the development of beta-cell tumour. The study of Rip1Tag2 mice showed the phases of tumour genesis, from normal cells to hyperplasia, and adenoma to invasive carcinoma. VEGF-A was shown to be the main driver of EC proliferation, migration and tube formation, all essential components of angiogenesis [36]. Mice that overexpressed human VEGF-A165 in pancreatic beta-cells had angiogenesis at an early stage of tumourigenesis [37]. In contrast, inhibiting VEGF-A resulted in suppressing angiogenic switch and tumour growth [38,39]. Different techniques were used to inhibit VEGF-A, such as chemical inhibitors of VEGFR signaling or genetically depleting VEGF-A in beta-cells [39,40]. Figure 4 is a schematic representation of angiogenic switch in transgenic mice; it shows progression from dormant hyperplasia to growing hyper-vascularized tumours as the result of angiogenesis.

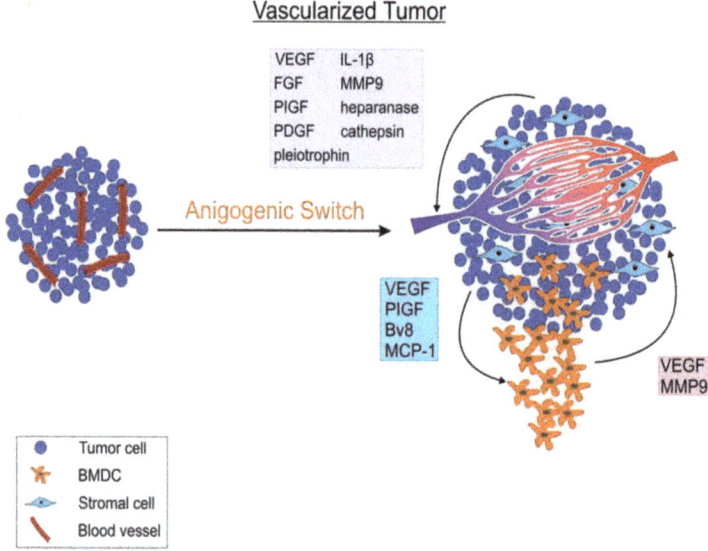

Figure 4. Angiogenic switch in transgenic mouse, showing progression from hyperplasia to hyper-vascularised tumour. The pro-angiogenic factors and proteases secreted by the tumour cells themselves (green box) and the cells of the immune system recruited to the tumour site (pink box), and the factors secreted by the tumour cells to recruit inflammatory cells (blue box).

Another important element of angiogenic switch is stromal cells of tumour microenvironment [41]. Through chemotaxis, cancer cells recruit innate immune cells. The immune cells contribute to angiogenesis via secreting pro-angiogenic factors. Using paracrine stimulation, tumour-associated macrophages (TAM) partake in the modulation of angiogenesis and tumour progression [42]. The cytokine/chemokine component of the tumour microenvironment determines the function of TAM. This function is either the M1 state of macrophages, which is an anticancer, or the M2 state, which suppresses immunity and promotes tumourigenesis via secreting pro-angiogenic cytokines and VEGF-A [43].

Endothelial progenitor cells (EPC), are also believed to play a role in angiogenesis [44]. Tumour-secreted factors recruit EPC from bone marrow to the tumour site to contribute to angiogenesis [45]. However, the exact role of EPC in angiogenesis remains to be fully understood [44,45]. Studies of mouse models of breast cancer have shown that myeloid progenitors differentiated to EC, leading to neovascularization [46]. This is further evidence of the role of the immune cells in promoting angiogenesis.

Importantly, not all tumours rely on new blood vessel formation to survive and grow [47], and therefore the angiogenic switch might never occur. Some tumours exploit the existing blood supply through a process named vessel co-option to support their growth and to enable metastasis. Vessel co-option has been observed in a number of tumours such as non-small cell lung cancer (NSCLC), glioblastoma and hepatocellular carcinoma [48,49]. Cancer cells seem to grow along existing vessels and/or invade the connective tissue that is present between the vessels, allowing the cancer cells to incorporate to the existing normal vasculature to begin hijacking the blood supply [50,51]. There is evidence that vessel co-option promotes cancer cell motility and metastasis and tumour dormancy [52]. Moreover, some tumours such as NSCLC, use both angiogenesis and vessel co-option simultaneously or sequentially (in no particular order) to acquire blood supply and venous and lymphatic drainage [53]. Moreover, there is growing evidence that increased vascularity, often measured through microvascular density, caused by vessel co-option, is associated with higher tumour grade and higher risk of metastasis [54]. Interestingly, bone marrow appears to be an important site for vessel co-option in both primary and secondary bone malignancies which, in turn, might explain the development of tumour dormancy in bones and the higher rate of chemoresistance [55,56].

In addition there is also evidence from preclinical studies that show that some tumors, such NSCLC and gliomas, never undergo angiogenic switch and rely only on vessel co-option [53,57]. In contrast, some tumors, for example hepatocellular carcinoma and liver metastases of the gastrointestinal tract, switch from using vessel co-option at early stages of tumourigenesis to angiogenesis at a later stage during tumour progression [58,59]. This progression from vessel co-option to angiogenesis is not an obligatory requirement of tumour progression and metastasis [60]. Moreover, preclinical and clinical studies showed that there are, at times, but not always, differences that exist between primary and secondary versions of of the same tumour in terms of their access to blood supply [61,62]. For example, when cells from angiogenic primary human breast tumors spread to the lung tissue, they switch to vessel co-option as a mode of accessing blood supply [62], which also functions as a resistant mechanism against antiangiogenic therapy [54].

6. Tumour Vasculature Modulation as a Therapeutic Option

Vascular Promotion Therapy

This approach is presumed to work though improving the delivery of cytotoxic agent(s) to the tumour (Figure 5).

An example of this is the use of Cilengitide and Verapamil in conjunction with Gemcitabine or Cisplatin to treat pancreatic ductal adenocarcinoma [63]. In high doses, Cilengitide, a selective inhibitor of integrins, leads to inhibition of the FAK/SRC/AKT pathway, causing apoptosis in EC. This drug was originally developed as an antiangiogenic agent. However, in clinical trials, it showed no efficacy in the treatment of glioblastoma. A low dosing of Cilengitide was, however, observed to be associated with the promotion of tumour angiogenesis [64]. Verapamil, a calcium channel blocker, causes vasodilatation, hence the increased blood flow to tumour. Cilengitide and Verapamil, in addition to Gemcitabine administered in a xenograft tumour model, through various schedules, mimicking human dosing regimens, was studied in trials by Wong et al. [63].

Vasculature Promoting Therapy

Producing less leaky blood vessels

Increased O_2 ↑↑

Figure 5. Promoting tumour blood supply to improve cytotoxic delivery to tumour. This approach might be particularly effective in tumours that are poorly supplied by blood, such as pancreatic cancer.

The impact of the therapy on the tumour blood flow was assessed by flow cytometry, imaging techniques and the concentration of the drugs in vital organs, tumour, and blood levels. The studies showed increased functional (less leaky) vessel formation, leading to an improved tumour blood supply to both highly and poorly vascularized tumors. This effect translated into tumour regression and improved survival in vivo models. The authors showed that vascular promotion increased the cell uptake of Gemcitabine with reduced side effects. The authors also argued that the promotion of vascularization improved the efficacy of Cisplatin due to better tumour blood perfusion, which improved cytotoxic delivery, leading to tumour regression in mice model. Future studies should deal with the impact of this approach in different tumour sites and their secondaries, address the wide variations in tumour behavior caused by intratumor heterogeneity and focus on the potential complications of promoting neovascularization, such as the risk of significant/life threatening bleeding, and its safety in vascular diseases.

7. Immune Modulation

As mentioned, the infiltration of tumour microenvironment with immune cells, importantly TAM, is associated with pro-angiogenic factor secretions by these cells. Several experiments studied the inhibition of TAM function or their complete removal from the tumour microenvironment. A study showed that treating K14-HPV/E2 mice with Zoledronic acid (ZA), a bisphosphonate used for skeletal metastasis with anti-inflammatory and anti-osteoclast properties, resulted in the suppressed mobilization of VEGF-A and, consequently, the inhibition of angiogenesis and tumourigenesis [65]. Other studies showed that treatment with ZA in advanced solid tumors was associated with a reduction in VEGF-A plasma levels [66]. The inhibition of neutrophils and macrophages to reverse angiogenic switch has been tested in preclinical trials but not applied in clinical settings [67].

Other immune modulated strategies that have been studied include: inhibition of Cyclooxygenase-2 (COX-2) expression by COX-2 inhibitors in pancreatic and cervical cancer [68] and Lenalidomide (an immunomodulatory drug) in advanced renal cancer [69], with benefits in phase two trials but no additional advantage in combination with standard cytotoxic protocols.

8. Anti-Angiogenic Therapy

The Food and Drug Administration (FDA) approved biological therapies in the form of tyrosine kinase inhibitors (TKIs), monoclonal antibodies and fusion peptides in non-small cell lung cancer, metastatic colorectal cancer, medullary thyroid cancer and renal cell cancer [70]. More specifically, targeting VEGF has become an important approach to stop tumour growth (Figure 6), and part of the treatment protocol of several tumour primaries, notably colon, non-small cell lung and renal cell cancers [71]. Several studies showed that arteriol formation and tortuosity, as well as venous dilation, are increased through VEGF expression [72]. Cell culture injected with adenovirus expressing VEGF undergo the induction of mother vessels (MV) and stabilized MV from normal capillaries and venules. In contrast, the inhibition of VEGF is shown to cause veins and arterioles to have fewer cleavage planes. For example, Aflibercept, a decoy receptor that binds VEGF-A, induces the rapid collapse of mother vessels (MV) to glomeruloid microvascular proliferations (GMP). VEGF inhibition, by anti-VEGF/VEGF receptor, is shown to restore vasculature within hours to normal microvessels by way of GMP [73]. GMP is believed to act as an intermediary step in MV reversion to normal microvessels after VEGF blockade [74].

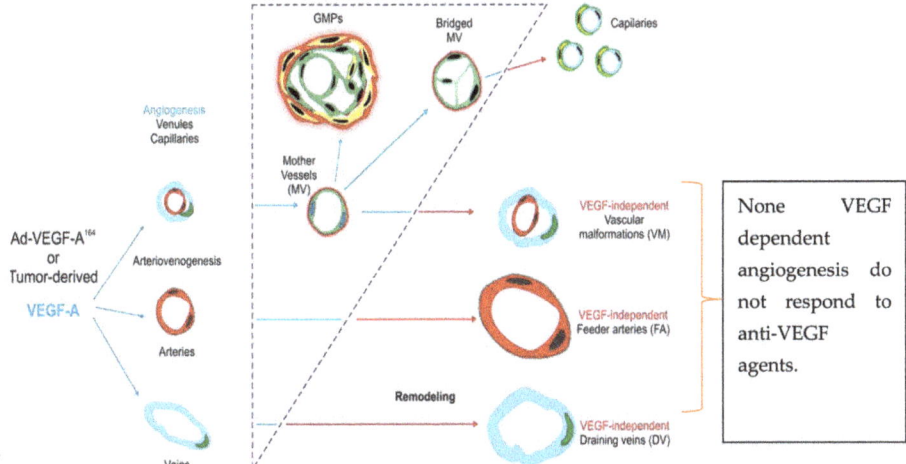

Figure 6. VEGF-A plays an important role in angiogenesis. The inhibition of VEGF-A prevents new vessel formation. VEGF-independent angiogenesis are not sensitive to the inhibition of VEGF-A.

Monoclonal antibodies such as Bevacizumab, which blocks the VEGF receptor, or small molecules such as Lapatinib, which inhibits TK downstream of VEGF, are examples of anti-VEGF treatment. Phase 1 trial of Bevacizumab showed that the drug was well tolerated and had good pharmacokinetic properties [75]. A phase 3 clinical trial of Bevacizumab in metastatic colorectal cancer (mCRC) showed a modest impact of 4 to 5 month improvement in overall survival (OS) in metastatic colon cancer [76]. In transgenic mouse models of non-squamous non-small cell lung cancer (nsNSCLC), Bevacizumab was shown to reduce the risk of brain metastasis, and therefore improve survival. This might translate into improved survival due to a reduction in the rate of brain metastases in patients with stage III nsNSCLC [77]. Despite prolonging the PFS of metastatic breast cancer, the FDA removed Bevacizumab from standard treatment protocol due to safety concerns [78].

Combining Bevacizumab with chemotherapy, in the first and second line settings of mCRC, improved OS [79]. The AVF2107g study showed an improvement in median survival from 15.6 to 20.3 months when combining Bevacizumab to irinotecan, bolus fluorouracil, and leucovorin, compared to placebo [76] in treatment-naïve mCRC patients. PFS, but not OS, was shown to improve in a randomized controlled trial of mCRC combining Bevacizumab with oxaliplatin-based chemotherapy as first-line treatment [80]. Another randomized controlled trial showed that adding Bevacizumab to fluorouracil and leucovorin improved PFS in patients with mCRC for whom first-line irinotecan was judged inappropriate due to their poor functional status [81]. The direct VEGFR2 antagonist, Ramucirumab, was approved in the treatment of advanced hepatocellular carcinoma (HCC) with high alpha-feto protein after progression to sorafenib [82]. Through binding to VEGF-B and placental growth factor, Ziv-aflibercept, a representative agent of the third type of angiogenesis inhibitor, composed of the extracellular domain of both VEGFR-1 and VEGFR-2 fused to the Fc region of IgG1, inhibits the pro-angiogenic effects of the VEGF/VEGFR signaling pathway [83]. Ziv-aflibercept, in combination with 5-fluorouracil, leucovorin and irinotecan (FOLFIRI) for mCRC, in patients resistant to or progressing after treatment with oxaliplatin, showed statistically significant improvements in PFS and OS [84].

Given the results of animal trials, this modest benefit of anti–VEGF-A/VEGFR therapy against human cancers has been relatively disappointing. One explanation for this modest effectiveness is that most cancer patients are elderly, frail and cannot tolerate high doses, in contrast to relatively healthy tumour-bearing mice that can be given higher doses [85]. Another possible reason is that tumour hypoxemia resulting from anti–VEGF-A/VEGFR therapy lead to the over-expression of matrix components that bind and sequester VEGF-A, rendering anti-VEGF drugs ineffective [86]. Hypoxia also might stimulate cancer cells to secrete other pro-angiogenic factors such as FGF, PDGF-B, PDGF-C, HGF, EGF, IL-8, IL-6, Ang-2, SDF1a, PDGF-C, CXCL6 and others, as well as their receptors [85,86]. Mobilisation from bone marrow to the tumour site of vascular progenitor cells and proangiogenic myelocytes are other mechanisms that might be responsible for the limited effectiveness of anti–VEGF-A/VEGFR therapy [87]. Another hindrance to anti-angiogenesis therapy is that the blood supply to the tumour is reduced, and this would lead to the impairment of the delivery of chemotherapy agents to the tumour, hence reducing their cytotoxic effects. Antiangiogenic treatment creates a hypoxic tumour microenvironment, which results in the tumour cells becoming more "aggressive" and promotes "escaping" of the tumour cells from the hypoxic environment to distant, normo-oxic, sites, i.e., metastasis [88]. Other mechanisms of therapy resistance involve the recruitment of pro-growth cells and molecules to the TME by the cancer cells as the result of tumour hypoxia, such as tumour-associated macrophages [89], tumour-associated fibroblasts (TAFs) [90], Tie2$^+$ monocytes [91], myeloid cells [92], pro-angiogenic bone-marrow-derived cells including CD11b$^+$ Gr1$^+$ and the overexpression of alternative angiogenic signaling molecules [93], including a fibroblast growth factor-2 [94], interleukin-8 (IL-8) [95], IL-17 [96], and angiopoietin 2 [97].

Vessels' co-option as a mechanism to attain blood supply by cancer cells is another resistant mechanism to anti-angiogenic treatment. Preclinical models demonstrated a switch from angiogenesis to vessel co-option during anti-angiogenic treatment [98,99]. The escaping anti-angiogenic agents' effect using vessel co-option is seen across a range of cancer types. For example, the modest response of glioma to bevacizumab is shown, in preclinical studies and clinical case reports, to be due to vessel co-option [100]. This could be intrinsic resistance or acquired during treatment with bevacizumab due to the switch from angiogenesis to the vessel co-option [101,102]. This switch from angiogenesis to the vessel co-option is also observed during the treatment of breast cancer with anti-angiogenic therapy. Pulmonary metastasis from breast cancer is shown to use the lung parenchymal blood supply for their survival and growth, which explains their resistance to anti-angiogenic therapy [61]. In addition, preclinical trials showed that after an initial response of xenograft model of hepatocellular carcinoma to sorafenib (a multi-kinase inhibitor with antiangiogenic properties), the tumour progressed within a month due to the large-scale co-option of sinusoidal and portal tract vessels [54]. Moreover, several studies showed that resistance to anti-angiogenic therapy in metastatic colorectal carcinoma (CRC) to the liver is likely secondary to the CRC cells co-option of pre-existing liver vessels; this can occur in the context of both intrinsic and acquired resistance [60].

9. Novel and Future Approaches to Modify Angiogenesis as Anti-Cancer Option

Targeting angiogenesis has shown limited effectiveness to date, but affirms Folkman's postulations. This limited success is likely caused by the heterogeneity of blood vessels, as some vessels are susceptible, whilst others are resistant, to the inhibition of VEGF/VEGFR. Furthermore, genomic instability would enable cancer cells to bypass the VEGF/VEGFR axis and stimulate new blood vessel growth using alternative signaling pathways. Future therapy should focus on targeting molecules, as well as VEGF, that are present on large blood vessels' EC lining. Targeting large vessels could stop the blood perfusion to the entire mass of the tumour, hence this would enhance the pruning of microvessels that are sensitive to the inhibition of VEGF/VEGFR. This concept was tested and supported by the findings of a study that utilised photodynamic energy to thrombose and subsequently blocked the main arteries and draining veins of a mouse ear tumour [103].

Another novel strategy is the use of nano-particles to deliver specific anti-angiogenic agents [104]. For example, endostatin, a protein that was extracted for the first time in 1996 from murine hemagioendothelioma (EOMA) cell culture medium [32]. Endostatin has a potent anti-angiogenic effect. The exact molecular anti-angiogenic mechanism(s) of endostatin are not fully understood and subject to investigation. In vitro and vivo studies showed that endostatins induce endothelial cell apoptosis, and suppress its proliferation and migration via a complex network of signaling [105]. However, there are important challenges in the clinical application of endostatin related to the chemical nature of this protein. These challenges include the short half-life and instability of the protein in vivo [106], the requirement of administering high volumes of endostatin to exert their anti-angiogenic effects, which in itself is associated with significant practical and cost implications [107], as well as technological challenges related to manufacturing a correctly folded and soluble protein to ensure adequate bioactivity within the tumour cells [108]. To overcome these challenges, nanotechnology has been utilized to manufacture nanoparticles as transporters of this protein [109]. Cancer cells are shown to readily uptake nano-particles, and therefore the anti-tumoural activity of endostatin is enhanced when delivered via nano-particles [110]. In addition, by adding nine amino acids to the N-terminal of recombinant human endostain, endostar is produced. Endostar is a more stable molecular bioengineered form of endostatin. This is because endostar is better at resisting degradation by proteolytic enzymes and more stable during temperature changes [111].

Two independent studies by Chen et al. [112] and Hu et al. [113] have confirmed that endostar carried by nanoparticles have a better anticancer activity than the conventional delivery system because of the improved release and longer half-life of endostar in target tumour. Chen et al. studied prepared particulate carriers (nanoparticles and microspheres) of poly (DL-lactide-co-glycolide) (PLGA) and poly (ethylene glycol) (PEG)-modified PLGA (PEG-PLGA) to promote a better delivery and release of endostar, as the nano-transporter enables high encapsulation, rapidly release and the higher cancer intracellular bioavailability of endostar.

As explained above, the vessel co-option acts as an important mechanism of resistance to anti-angiogenesis as well as an important source of blood supply that supports the growth of tumors. Therefore, the inhibition of the vessel co-option is the focus of many research groups, through targeting cell motility or adhesion pathways in tumour stroma. In in a mouse model of liver metastases, Frentzas et al. [60] showed that, by silencing the expression of actin-related protein 2/3 (ARP2/3), a protein complex involved in actin-mediated cell motility, and the vessel co-option, can be inhibited. Interestingly, preclinical trials showed improved tumour control when VEGF and vessel co-options are inhibited simultaneously compared to the blocking of VEGF signalling alone [114]. Another novel approach that has been tested in mouse models of brain-metastatic breast cancer and showed some promising results, is the inhibition of the adhesion of cancer cells to pre-existing blood vessels to block vessel co-option through inhibition of L1 Cell Adhesion Molecule (L1CAM) or the cell adhesion receptor β1 integrin [115]. Moreover, pre-clinical models of glioma, and metastases to the liver, lymph nodes or lungs that are vessel co-option-dependent, showed that blocking both the angiopoietin and VEGF pathways was more effective compared to the inhibition of VEGF alone [116,117]. The exact role of angiopoietin in the recruitment or maintenance of co-opted tumour vessels is not fully understood [118]. However, a phase 2 clinical trial of angiopoietin inhibition with and without bevacizumab in recurrent glioblastoma did not show any improvement in progression free survival (PFS) [118].

10. Conclusions

Excessive, insufficient or abnormal angiogenesis contributes to tumour survival, growth invasion and metastasis. Targeting single angiogenic (pro or inhibitory) molecules showed promising results in animal trials, but has been of limited success in human cancer. To date, despite their modest impact, anti VEGF continues to be one of the treatment lines of several solid malignancies. Nevertheless, it is believed that antiangiogenic monotherapy aiming at single molecule activity is insufficient to combat the myriad of angiogenic factors produced by cancer cells and its microenvironment and this would explain, at least partly, the modest effect of anti VEGF strategies. Future challenges include a detailed understanding of the many angio-modulating pathways in a more integrated manner to identify more holistic therapeutic approaches to improve survival rate in cancer patients.

Author Contributions: H.S. designing of manuscript, literature review, compiled the data, wrote the manuscript and supervised this manuscript. S.S.R. help in designing figures, reading and editing. S.U. assisted in preparing the designing the manuscript and help in writing and reading the manuscript. K.R. help design of the manuscript and contributed in writing and reading manuscript. All authors have read and agreed to the published version of the manuscript.

Funding: This research received no external funding.

Acknowledgments: The authors are thankful to Qatar National Library for its support for open access.

Conflicts of Interest: The authors declare no conflict of interest.

Abbreviations

Ang	Angiopoietin
CRC	Colorectal Cancer
EC	Endothelial cells
ECM	Extra-cellular matrix
EPC	Endothelial progenitor cells
FGF	Fibroblast growth factor
FOLFIRI	Fluorouracil, Leucovorin and Irinotecan
FOXP3	Forkhead box protein P3
HCC	Hepatocellular Carcinoma
NSCLC	Non-small Cell Lung Cancer
PDGF	Platelet-derived growth factor
PFS	Progression free survival
PlGF	Placenta growth factor
TAFs	Tumour associated fibroblasts
TK	Tyrosine kinases
TME	Tumour micro-environment
TSP1	Thrombospondin1
VEGF	Vascular endothelial growth factor
ZA	Zoledronic acid

References

1. Suh, D.Y. Understanding angiogenesis and its clinical applications. *Ann. Clin. Lab. Sci.* **2000**, *30*, 227–238. [PubMed]

2. Nishida, N.; Yano, H.; Nishida, T.; Kamura, T.; Kojiro, M. Angiogenesis in cancer. *Vasc. Health Risk Manag.* **2006**, *2*, 213–219. [CrossRef] [PubMed]
3. Ravi, R.; Mookerjee, B.; Bhujwalla, Z.M.; Sutter, C.H.; Artemov, D.; Zeng, Q.; Dillehay, L.E.; Madan, A.; Semenza, G.L.; Bedi, A. Regulation of tumor angiogenesis by p53-induced degradation of hypoxia-inducible factor 1alpha. *Genes. Dev.* **2000**, *14*, 34–44. [PubMed]
4. McCoy, R.J.; O'Brien, F.J. Influence of shear stress in perfusion bioreactor cultures for the development of three-dimensional bone tissue constructs: A review. *Tissue. Eng. Part B Rev.* **2010**, *16*, 587–601. [CrossRef]
5. Carmeliet, P. Angiogenesis in health and disease. *Nat. Med.* **2003**, *9*, 653–660. [CrossRef]
6. Kuehbacher, A.; Urbich, C.; Zeiher, A.M.; Dimmeler, S. Role of Dicer and Drosha for endothelial microRNA expression and angiogenesis. *Circ. Res.* **2007**, *101*, 59–68. [CrossRef]
7. Chan, Y.C.; Roy, S.; Khanna, S.; Sen, C.K. Downregulation of endothelial microRNA-200b supports cutaneous wound angiogenesis by desilencing GATA binding protein 2 and vascular endothelial growth factor receptor 2. *Arterioscler. Thromb. Vasc. Biol.* **2012**, *32*, 1372–1382. [CrossRef]
8. Chan, Y.C.; Khanna, S.; Roy, S.; Sen, C.K. miR-200b targets Ets-1 and is down-regulated by hypoxia to induce angiogenic response of endothelial cells. *J. Biol. Chem.* **2011**, *286*, 2047–2056. [CrossRef]
9. Zheng, Q.; Cui, X.; Zhang, D.; Yang, Y.; Yan, X.; Liu, M.; Niang, B.; Aziz, F.; Liu, S.; Yan, Q.; et al. miR-200b inhibits proliferation and metastasis of breast cancer by targeting fucosyltransferase IV and alpha1,3-fucosylated glycans. *Oncogenesis* **2017**, *6*, 358. [CrossRef]
10. Hua, Z.; Lv, Q.; Ye, W.; Wong, C.K.; Cai, G.; Gu, D.; Ji, Y.; Zhao, C.; Wang, J.; Yang, B.B.; et al. MiRNA-directed regulation of VEGF and other angiogenic factors under hypoxia. *PLoS ONE* **2006**, *1*, 116. [CrossRef]
11. Tsai, Y.H.; Wu, M.F.; Wu, Y.H.; Chang, S.J.; Lin, S.F.; Sharp, T.V.; Wang, H.W. The M type K15 protein of Kaposi's sarcoma-associated herpesvirus regulates microRNA expression via its SH2-binding motif to induce cell migration and invasion. *J. Virol.* **2009**, *83*, 622–632. [CrossRef] [PubMed]
12. Doganov, N.; Negentsov, N. Clinical trial of the laksafer preparation on patients following gynecological operations. *Akush. Ginekol.* **1989**, *28*, 47–51.
13. Chen, Y.; Gorski, D.H. Regulation of angiogenesis through a microRNA (miR-130a) that down-regulates antiangiogenic homeobox genes GAX and HOXA5. *Blood* **2008**, *111*, 1217–1226. [CrossRef] [PubMed]
14. Wurdinger, T.; Tannous, B.A.; Saydam, O.; Skog, J.; Grau, S.; Soutschek, J.; Weissleder, R.; Breakefield, X.O.; Krichevsky, A.M. miR-296 regulates growth factor receptor overexpression in angiogenic endothelial cells. *Cancer Cell* **2008**, *14*, 382–393. [CrossRef] [PubMed]
15. Byatt, G.; Dalrymple-Alford, J.C. Both anteromedial and anteroventral thalamic lesions impair radial-maze learning in rats. *Behav. Neurosci.* **1996**, *110*, 1335–1348. [CrossRef]
16. Pulkkinen, K.; Malm, T.; Turunen, M.; Koistinaho, J.; Yla-Herttuala, S. Hypoxia induces microRNA miR-210 in vitro and in vivo ephrin-A3 and neuronal pentraxin 1 are potentially regulated by miR-210. *FEBS Lett.* **2008**, *582*, 2397–2401. [CrossRef]
17. Ventura, A.; Young, A.G.; Winslow, M.M.; Lintault, L.; Meissner, A.; Erkeland, S.J.; Newman, J.; Bronson, R.T.; Crowley, D.; Stone, J.R.; et al. Targeted deletion reveals essential and overlapping functions of the miR-17 through 92 family of miRNA clusters. *Cell* **2008**, *132*, 875–886. [CrossRef]
18. Baeriswyl, V.; Christofori, G. The angiogenic switch in carcinogenesis. *Semin. Cancer Biol.* **2009**, *19*, 329–337. [CrossRef]
19. Saggar, J.K.; Yu, M.; Tan, Q.; Tannock, I.F. The tumor microenvironment and strategies to improve drug distribution. *Front. Oncol.* **2013**, *3*, 154. [CrossRef]
20. Bridges, E.; Harris, A.L. Vascular-promoting therapy reduced tumor growth and progression by improving chemotherapy efficacy. *Cancer Cell* **2015**, *27*, 7–9. [CrossRef]
21. Hanahan, D.; Weinberg, R.A. Retrospective: Judah Folkman (1933–2008). *Science* **2008**, *319*, 1055. [CrossRef] [PubMed]
22. Ferrara, N.; Gerber, H.P.; LeCouter, J. The biology of VEGF and its receptors. *Nat. Med.* **2003**, *9*, 669–676. [CrossRef] [PubMed]
23. Kubo, H.; Fujiwara, T.; Jussila, L.; Hashi, H.; Ogawa, M.; Shimizu, K.; Awane, M.; Sakai, Y.; Takabayashi, A.; Alitalo, K.; et al. Involvement of vascular endothelial growth factor receptor-3 in maintenance of integrity of endothelial cell lining during tumor angiogenesis. *Blood* **2000**, *96*, 546–553. [CrossRef] [PubMed]
24. Goel, H.L.; Mercurio, A.M. VEGF targets the tumour cell. *Nat. Rev. Cancer* **2013**, *13*, 871–882. [CrossRef] [PubMed]

25. Fantozzi, A.; Gruber, D.C.; Pisarsky, L.; Heck, C.; Kunita, A.; Yilmaz, M.; Meyer-Schaller, N.; Cornille, K.; Hopfer, U.; Bentires-Alj, M.; et al. VEGF-mediated angiogenesis links EMT-induced cancer stemness to tumor initiation. *Cancer Res.* **2014**, *74*, 1566–1575. [CrossRef]
26. Ghersi, G. Roles of molecules involved in epithelial/mesenchymal transition during angiogenesis. *Front. Biosci.* **2008**, *13*, 2335–2355. [CrossRef]
27. Polyak, K.; Weinberg, R.A. Transitions between epithelial and mesenchymal states: Acquisition of malignant and stem cell traits. *Nat. Rev. Cancer* **2009**, *9*, 265–273. [CrossRef]
28. Galdiero, M.R.; Bonavita, E.; Barajon, I.; Garlanda, C.; Mantovani, A.; Jaillon, S. Tumor associated macrophages and neutrophils in cancer. *Immunobiology* **2013**, *218*, 1402–1410. [CrossRef]
29. Friesel, R.E.; Maciag, T. Molecular mechanisms of angiogenesis: Fibroblast growth factor signal transduction. *FASEB J.* **1995**, *9*, 919–925. [CrossRef]
30. Augustin, H.G.; Koh, G.Y.; Thurston, G.; Alitalo, K. Control of vascular morphogenesis and homeostasis through the angiopoietin-Tie system. *Nat. Rev. Mol. Cell Biol.* **2009**, *10*, 165–177. [CrossRef]
31. Good, D.J.; Polverini, P.J.; Rastinejad, F.; Le Beau, M.M.; Lemons, R.S.; Frazier, W.A.; Bouck, N.P. A tumor suppressor-dependent inhibitor of angiogenesis is immunologically and functionally indistinguishable from a fragment of thrombospondin. *Proc. Natl. Acad. Sci. USA* **1990**, *87*, 6624–6628. [CrossRef] [PubMed]
32. O'Reilly, M.S.; Boehm, T.; Shing, Y.; Fukai, N.; Vasios, G.; Lane, W.S.; Flynn, E.; Birkhead, J.R.; Olsen, B.R.; Folkman, J. Endostatin: An endogenous inhibitor of angiogenesis and tumor growth. *Cell* **1997**, *88*, 277–285. [CrossRef]
33. O'Reilly, M.S.; Holmgren, L.; Shing, Y.; Chen, C.; Rosenthal, R.A.; Moses, M.; Lane, W.S.; Cao, Y.; Sage, E.H.; Folkman, J. Angiostatin: A novel angiogenesis inhibitor that mediates the suppression of metastases by a Lewis lung carcinoma. *Cell* **1994**, *79*, 315–328. [CrossRef]
34. Pugh, C.W.; Ratcliffe, P.J. Regulation of angiogenesis by hypoxia: Role of the HIF system. *Nat. Med.* **2003**, *9*, 677–684. [CrossRef]
35. North, S.; Moenner, M.; Bikfalvi, A. Recent developments in the regulation of the angiogenic switch by cellular stress factors in tumors. *Cancer Lett.* **2005**, *218*, 1–14. [CrossRef]
36. Folkman, J.; Watson, K.; Ingber, D.; Hanahan, D. Induction of angiogenesis during the transition from hyperplasia to neoplasia. *Nature* **1989**, *339*, 58–61. [CrossRef]
37. Gannon, G.; Mandriota, S.J.; Cui, L.; Baetens, D.; Pepper, M.S.; Christofori, G. Overexpression of vascular endothelial growth factor-A165 enhances tumor angiogenesis but not metastasis during beta-cell carcinogenesis. *Cancer Res.* **2002**, *62*, 603–608.
38. O'Reilly, T.; Lane, H.A.; Wood, J.M.; Schnell, C.; Littlewood-Evans, A.; Brueggen, J.; McSheehy, P.M. Everolimus and PTK/ZK show synergistic growth inhibition in the orthotopic BL16/BL6 murine melanoma model. *Cancer Chemother. Pharmacol.* **2011**, *67*, 193–200. [CrossRef]
39. Vajkoczy, P.; Menger, M.D.; Vollmar, B.; Schilling, L.; Schmiedek, P.; Hirth, K.P.; Ullrich, A.; Fong, T.A. Inhibition of tumor growth, angiogenesis, and microcirculation by the novel Flk-1 inhibitor SU5416 as assessed by intravital multi-fluorescence videomicroscopy. *Neoplasia* **1999**, *1*, 31–41. [CrossRef]
40. Bergers, G.; Javaherian, K.; Lo, K.M.; Folkman, J.; Hanahan, D. Effects of angiogenesis inhibitors on multistage carcinogenesis in mice. *Science* **1999**, *284*, 808–812. [CrossRef]
41. Bergers, G.; Song, S. The role of pericytes in blood-vessel formation and maintenance. *Neuro. Oncol.* **2005**, *7*, 452–464. [CrossRef] [PubMed]
42. Pollard, J.W. Tumour-educated macrophages promote tumour progression and metastasis. *Nat. Rev. Cancer* **2004**, *4*, 71–78. [CrossRef]
43. Zumsteg, A.; Christofori, G. Corrupt policemen: Inflammatory cells promote tumor angiogenesis. *Curr. Opin. Oncol.* **2009**, *21*, 60–70. [CrossRef] [PubMed]
44. Bertolini, F.; Shaked, Y.; Mancuso, P.; Kerbel, R.S. The multifaceted circulating endothelial cell in cancer: Towards marker and target identification. *Nat. Rev. Cancer* **2006**, *6*, 835–845. [CrossRef] [PubMed]
45. Bailey, A.S.; Willenbring, H.; Jiang, S.; Anderson, D.A.; Schroeder, D.A.; Wong, M.H.; Grompe, M.; Fleming, W.H. Myeloid lineage progenitors give rise to vascular endothelium. *Proc. Natl. Acad. Sci. USA* **2006**, *103*, 13156–13161. [CrossRef]
46. Nolan, D.J.; Ciarrocchi, A.; Mellick, A.S.; Jaggi, J.S.; Bambino, K.; Gupta, S.; Heikamp, E.; McDevitt, M.R.; Scheinberg, D.A.; Benezra, R.; et al. Bone marrow-derived endothelial progenitor cells are a major determinant of nascent tumor neovascularization. *Genes. Dev.* **2007**, *21*, 1546–1558. [CrossRef]

47. Kuczynski, E.A.; Vermeulen, P.B.; Pezzella, F.; Kerbel, R.S.; Reynolds, A.R. Vessel co-option in cancer. *Nat. Rev. Clin. Oncol.* **2019**, *16*, 469–493. [CrossRef]
48. Kurzrock, R.; Stewart, D.J. Exploring the Benefit/Risk Associated with Antiangiogenic Agents for the Treatment of Non-Small Cell Lung Cancer Patients. *Clin. Cancer Res.* **2017**, *23*, 1137–1148. [CrossRef]
49. Khasraw, M.; Ameratunga, M.; Grommes, C. Bevacizumab for the treatment of high-grade glioma: An update after phase III trials. *Expert Opin. Biol. Ther.* **2014**, *14*, 729–740. [CrossRef]
50. Vasudev, N.S.; Reynolds, A.R. Anti-angiogenic therapy for cancer: Current progress, unresolved questions and future directions. *Angiogenesis* **2014**, *17*, 471–494. [CrossRef]
51. Leenders, W.P.; Kusters, B.; de Waal, R.M. Vessel co-option: How tumors obtain blood supply in the absence of sprouting angiogenesis. *Endothelium* **2002**, *9*, 83–87. [CrossRef] [PubMed]
52. Winkler, F. Hostile takeover: How tumours hijack pre-existing vascular environments to thrive. *J. Pathol.* **2017**, *242*, 267–272. [CrossRef] [PubMed]
53. Szabo, V.; Bugyik, E.; Dezso, K.; Ecker, N.; Nagy, P.; Timar, J.; Tovari, J.; Laszlo, V.; Bridgeman, V.L.; Wan, E.; et al. Mechanism of tumour vascularization in experimental lung metastases. *J. Pathol.* **2015**, *235*, 384–396. [CrossRef] [PubMed]
54. Kuczynski, E.A.; Yin, M.; Bar-Zion, A.; Lee, C.R.; Butz, H.; Man, S.; Daley, F.; Vermeulen, P.B.; Yousef, G.M.; Foster, F.S.; et al. Co-option of Liver Vessels and Not Sprouting Angiogenesis Drives Acquired Sorafenib Resistance in Hepatocellular Carcinoma. *J. Natl. Cancer Inst.* **2016**, *108*. [CrossRef] [PubMed]
55. Raymaekers, K.; Stegen, S.; van Gastel, N.; Carmeliet, G. The vasculature: A vessel for bone metastasis. *Bonekey Rep.* **2015**, *4*, 742. [CrossRef]
56. Ghajar, C.M.; Peinado, H.; Mori, H.; Matei, I.R.; Evason, K.J.; Brazier, H.; Almeida, D.; Koller, A.; Hajjar, K.A.; Stainier, D.Y.; et al. The perivascular niche regulates breast tumour dormancy. *Nat. Cell Biol.* **2013**, *15*, 807–817. [CrossRef]
57. Baker, G.J.; Yadav, V.N.; Motsch, S.; Koschmann, C.; Calinescu, A.A.; Mineharu, Y.; Camelo-Piragua, S.I.; Orringer, D.; Bannykh, S.; Nichols, W.S.; et al. Mechanisms of glioma formation: Iterative perivascular glioma growth and invasion leads to tumor progression, VEGF-independent vascularization, and resistance to antiangiogenic therapy. *Neoplasia* **2014**, *16*, 543–561. [CrossRef]
58. Terayama, N.; Terada, T.; Nakanuma, Y. Histologic growth patterns of metastatic carcinomas of the liver. *Jpn. J. Clin. Oncol.* **1996**, *26*, 24–29. [CrossRef]
59. Kanai, T.; Hirohashi, S.; Upton, M.P.; Noguchi, M.; Kishi, K.; Makuuchi, M.; Yamasaki, S.; Hasegawa, H.; Takayasu, K.; Moriyama, N.; et al. Pathology of small hepatocellular carcinoma. A proposal for a new gross classification. *Cancer* **1987**, *60*, 810–819. [CrossRef]
60. Frentzas, S.; Simoneau, E.; Bridgeman, V.L.; Vermeulen, P.B.; Foo, S.; Kostaras, E.; Nathan, M.; Wotherspoon, A.; Gao, Z.H.; Shi, Y.; et al. Vessel co-option mediates resistance to anti-angiogenic therapy in liver metastases. *Nat. Med.* **2016**, *22*, 1294–1302. [CrossRef]
61. Bridgeman, V.L.; Vermeulen, P.B.; Foo, S.; Bilecz, A.; Daley, F.; Kostaras, E.; Nathan, M.R.; Wan, E.; Frentzas, S.; Schweiger, T.; et al. Vessel co-option is common in human lung metastases and mediates resistance to anti-angiogenic therapy in preclinical lung metastasis models. *J. Pathol.* **2017**, *241*, 362–374. [CrossRef] [PubMed]
62. Guerin, E.; Man, S.; Xu, P.; Kerbel, R.S. A model of postsurgical advanced metastatic breast cancer more accurately replicates the clinical efficacy of antiangiogenic drugs. *Cancer Res.* **2013**, *73*, 2743–2748. [CrossRef] [PubMed]
63. Wong, P.P.; Demircioglu, F.; Ghazaly, E.; Alrawashdeh, W.; Stratford, M.R.; Scudamore, C.L.; Cereser, B.; Crnogorac-Jurcevic, T.; McDonald, S.; Elia, G.; et al. Dual-action combination therapy enhances angiogenesis while reducing tumor growth and spread. *Cancer Cell* **2015**, *27*, 123–137. [CrossRef] [PubMed]
64. Reynolds, A.R.; Hart, I.R.; Watson, A.R.; Welti, J.C.; Silva, R.G.; Robinson, S.D.; Da Violante, G.; Gourlaouen, M.; Salih, M.; Jones, M.C.; et al. Stimulation of tumor growth and angiogenesis by low concentrations of RGD-mimetic integrin inhibitors. *Nat. Med.* **2009**, *15*, 392–400. [CrossRef] [PubMed]
65. Giraudo, E.; Inoue, M.; Hanahan, D. An amino-bisphosphonate targets MMP-9-expressing macrophages and angiogenesis to impair cervical carcinogenesis. *J. Clin. Invest.* **2004**, *114*, 623–633. [CrossRef]
66. Santini, D.; Vincenzi, B.; Dicuonzo, G.; Avvisati, G.; Massacesi, C.; Battistoni, F.; Gavasci, M.; Rocci, L.; Tirindelli, M.C.; Altomare, V.; et al. Zoledronic acid induces significant and long-lasting modifications of circulating angiogenic factors in cancer patients. *Clin. Cancer Res.* **2003**, *9*, 2893–2897.

67. Nozawa, H.; Chiu, C.; Hanahan, D. Infiltrating neutrophils mediate the initial angiogenic switch in a mouse model of multistage carcinogenesis. *Proc. Natl. Acad. Sci. USA* **2006**, *103*, 12493–12498. [CrossRef]
68. Bertagnolli, M.M.; Eagle, C.J.; Zauber, A.G.; Redston, M.; Solomon, S.D.; Kim, K.; Tang, J.; Rosenstein, R.B.; Wittes, J.; Corle, D.; et al. Celecoxib for the prevention of sporadic colorectal adenomas. *N. Engl. J. Med.* **2006**, *355*, 873–884. [CrossRef]
69. Aragon-Ching, J.B.; Li, H.; Gardner, E.R.; Figg, W.D. Thalidomide analogues as anticancer drugs. *Recent. Pat. Anticancer. Drug Discov.* **2007**, *2*, 167–174. [CrossRef]
70. Al-Abd, A.M.; Alamoudi, A.J.; Abdel-Naim, A.B.; Neamatallah, T.A.; Ashour, O.M. Anti-angiogenic agents for the treatment of solid tumors: Potential pathways, therapy and current strategies—A review. *J. Adv. Res.* **2017**, *8*, 591–605. [CrossRef]
71. Meadows, K.L.; Hurwitz, H.I. Anti-VEGF therapies in the clinic. *Cold Spring Harb. Perspect. Med.* **2012**, *2*. [CrossRef] [PubMed]
72. Hartnett, M.E.; Martiniuk, D.; Byfield, G.; Geisen, P.; Zeng, G.; Bautch, V.L. Neutralizing VEGF decreases tortuosity and alters endothelial cell division orientation in arterioles and veins in a rat model of ROP: Relevance to plus disease. *Invest. Ophthalmol. Vis. Sci.* **2008**, *49*, 3107–3114. [CrossRef] [PubMed]
73. Goffin, J.R.; Straume, O.; Chappuis, P.O.; Brunet, J.S.; Begin, L.R.; Hamel, N.; Wong, N.; Akslen, L.A.; Foulkes, W.D. Glomeruloid microvascular proliferation is associated with p53 expression, germline BRCA1 mutations and an adverse outcome following breast cancer. *Br. J. Cancer* **2003**, *89*, 1031–1034. [CrossRef] [PubMed]
74. Sitohy, B.; Chang, S.; Sciuto, T.E.; Masse, E.; Shen, M.; Kang, P.M.; Jaminet, S.C.; Benjamin, L.E.; Bhatt, R.S.; Dvorak, A.M.; et al. Early Actions of Anti-Vascular Endothelial Growth Factor/Vascular Endothelial Growth Factor Receptor Drugs on Angiogenic Blood Vessels. *Am. J. Pathol.* **2017**, *187*, 2337–2347. [CrossRef]
75. Gordon, M.S.; Margolin, K.; Talpaz, M.; Sledge, G.W., Jr.; Holmgren, E.; Benjamin, R.; Stalter, S.; Shak, S.; Adelman, D. Phase I safety and pharmacokinetic study of recombinant human anti-vascular endothelial growth factor in patients with advanced cancer. *J. Clin. Oncol.* **2001**, *19*, 843–850. [CrossRef]
76. Hurwitz, H.; Fehrenbacher, L.; Novotny, W.; Cartwright, T.; Hainsworth, J.; Heim, W.; Berlin, J.; Baron, A.; Griffing, S.; Holmgren, E.; et al. Bevacizumab plus irinotecan, fluorouracil, and leucovorin for metastatic colorectal cancer. *N. Engl. J. Med.* **2004**, *350*, 2335–2342. [CrossRef]
77. Ilhan-Mutlu, A.; Osswald, M.; Liao, Y.; Gommel, M.; Reck, M.; Miles, D.; Mariani, P.; Gianni, L.; Lutiger, B.; Nendel, V.; et al. Bevacizumab Prevents Brain Metastases Formation in Lung Adenocarcinoma. *Mol. Cancer Ther.* **2016**, *15*, 702–710. [CrossRef]
78. Montero, A.J.; Escobar, M.; Lopes, G.; Gluck, S.; Vogel, C. Bevacizumab in the treatment of metastatic breast cancer: Friend or foe? *Curr. Oncol. Rep.* **2012**, *14*, 1–11. [CrossRef]
79. Rosen, L.S.; Jacobs, I.A.; Burkes, R.L. Bevacizumab in Colorectal Cancer: Current Role in Treatment and the Potential of Biosimilars. *Target. Oncol.* **2017**, *12*, 599–610. [CrossRef]
80. Saltz, L.B.; Clarke, S.; Diaz-Rubio, E.; Scheithauer, W.; Figer, A.; Wong, R.; Koski, S.; Lichinitser, M.; Yang, T.S.; Rivera, F.; et al. Bevacizumab in combination with oxaliplatin-based chemotherapy as first-line therapy in metastatic colorectal cancer: A randomized phase III study. *J. Clin. Oncol.* **2008**, *26*, 2013–2019. [CrossRef]
81. Van Cutsem, E.; Cervantes, A.; Adam, R.; Sobrero, A.; Van Krieken, J.H.; Aderka, D.; Aranda Aguilar, E.; Bardelli, A.; Benson, A.; Bodoky, G.; et al. ESMO consensus guidelines for the management of patients with metastatic colorectal cancer. *Ann. Oncol.* **2016**, *27*, 1386–1422. [CrossRef] [PubMed]
82. Finn, R.S.; Ryoo, B.Y.; Merle, P.; Kudo, M.; Bouattour, M.; Lim, H.Y.; Breder, V.; Edeline, J.; Chao, Y.; Ogasawara, S.; et al. Pembrolizumab As Second-Line Therapy in Patients With Advanced Hepatocellular Carcinoma in KEYNOTE-240: A Randomized, Double-Blind, Phase III Trial. *J. Clin. Oncol.* **2020**, *38*, 193–202. [CrossRef] [PubMed]
83. Patel, A.; Sun, W. Ziv-aflibercept in metastatic colorectal cancer. *Biologics* **2014**, *8*, 13–25. [CrossRef] [PubMed]
84. Tang, P.A.; Moore, M.J. Aflibercept in the treatment of patients with metastatic colorectal cancer: Latest findings and interpretations. *Therap. Adv. Gastroenterol.* **2013**, *6*, 459–473. [CrossRef] [PubMed]
85. Sitohy, B.; Nagy, J.A.; Dvorak, H.F. Anti-VEGF/VEGFR therapy for cancer: Reassessing the target. *Cancer Res.* **2012**, *72*, 1909–1914. [CrossRef] [PubMed]
86. Kadenhe-Chiweshe, A.; Papa, J.; McCrudden, K.W.; Frischer, J.; Bae, J.O.; Huang, J.; Fisher, J.; Lefkowitch, J.H.; Feirt, N.; Rudge, J.; et al. Sustained VEGF blockade results in microenvironmental sequestration of VEGF by tumors and persistent VEGF receptor-2 activation. *Mol. Cancer Res.* **2008**, *6*, 1–9. [CrossRef] [PubMed]

87. Ferrara, N. Role of myeloid cells in vascular endothelial growth factor-independent tumor angiogenesis. *Curr. Opin. Hematol.* **2010**, *17*, 219–224. [CrossRef]
88. Abdalla, A.M.E.; Xiao, L.; Ullah, M.W.; Yu, M.; Ouyang, C.; Yang, G. Current Challenges of Cancer Anti-angiogenic Therapy and the Promise of Nanotherapeutics. *Theranostics* **2018**, *8*, 533–548. [CrossRef]
89. Ribatti, D. Mast cells and macrophages exert beneficial and detrimental effects on tumor progression and angiogenesis. *Immunol. Lett.* **2013**, *152*, 83–88. [CrossRef]
90. Raffaghello, L.; Vacca, A.; Pistoia, V.; Ribatti, D. Cancer associated fibroblasts in hematological malignancies. *Oncotarget* **2015**, *6*, 2589–2603. [CrossRef]
91. De Palma, M.; Venneri, M.A.; Galli, R.; Sergi Sergi, L.; Politi, L.S.; Sampaolesi, M.; Naldini, L. Tie2 identifies a hematopoietic lineage of proangiogenic monocytes required for tumor vessel formation and a mesenchymal population of pericyte progenitors. *Cancer Cell* **2005**, *8*, 211–226. [CrossRef] [PubMed]
92. Shojaei, F.; Wu, X.; Malik, A.K.; Zhong, C.; Baldwin, M.E.; Schanz, S.; Fuh, G.; Gerber, H.P.; Ferrara, N. Tumor refractoriness to anti-VEGF treatment is mediated by CD11b+Gr1+ myeloid cells. *Nat. Biotechnol.* **2007**, *25*, 911–920. [CrossRef] [PubMed]
93. Azam, F.; Mehta, S.; Harris, A.L. Mechanisms of resistance to antiangiogenesis therapy. *Eur. J. Cancer* **2010**, *46*, 1323–1332. [CrossRef]
94. Casanovas, O.; Hicklin, D.J.; Bergers, G.; Hanahan, D. Drug resistance by evasion of antiangiogenic targeting of VEGF signaling in late-stage pancreatic islet tumors. *Cancer Cell* **2005**, *8*, 299–309. [CrossRef] [PubMed]
95. Huang, D.; Ding, Y.; Zhou, M.; Rini, B.I.; Petillo, D.; Qian, C.N.; Kahnoski, R.; Futreal, P.A.; Furge, K.A.; Teh, B.T. Interleukin-8 mediates resistance to antiangiogenic agent sunitinib in renal cell carcinoma. *Cancer Res.* **2010**, *70*, 1063–1071. [CrossRef] [PubMed]
96. Chung, A.S.; Wu, X.; Zhuang, G.; Ngu, H.; Kasman, I.; Zhang, J.; Vernes, J.M.; Jiang, Z.; Meng, Y.G.; Peale, F.V.; et al. An interleukin-17-mediated paracrine network promotes tumor resistance to anti-angiogenic therapy. *Nat. Med.* **2013**, *19*, 1114–1123. [CrossRef] [PubMed]
97. Rigamonti, N.; Kadioglu, E.; Keklikoglou, I.; Wyser Rmili, C.; Leow, C.C.; De Palma, M. Role of angiopoietin-2 in adaptive tumor resistance to VEGF signaling blockade. *Cell Rep.* **2014**, *8*, 696–706. [CrossRef]
98. Keunen, O.; Johansson, M.; Oudin, A.; Sanzey, M.; Rahim, S.A.; Fack, F.; Thorsen, F.; Taxt, T.; Bartos, M.; Jirik, R.; et al. Anti-VEGF treatment reduces blood supply and increases tumor cell invasion in glioblastoma. *Proc. Natl. Acad. Sci. USA* **2011**, *108*, 3749–3754. [CrossRef]
99. Leenders, W.P.; Kusters, B.; Verrijp, K.; Maass, C.; Wesseling, P.; Heerschap, A.; Ruiter, D.; Ryan, A.; de Waal, R. Antiangiogenic therapy of cerebral melanoma metastases results in sustained tumor progression via vessel co-option. *Clin. Cancer Res.* **2004**, *10*, 6222–6230. [CrossRef]
100. Rossi, D.; Hermanowicz, M.; Serment, G.; Khouzami, A.; Bretheau, D.; Ducassou, J. Urethral diverticula in women. Apropos of 7 cases. *Ann. Urol.* **1989**, *23*, 352–353.
101. Navis, A.C.; Bourgonje, A.; Wesseling, P.; Wright, A.; Hendriks, W.; Verrijp, K.; van der Laak, J.A.; Heerschap, A.; Leenders, W.P. Effects of dual targeting of tumor cells and stroma in human glioblastoma xenografts with a tyrosine kinase inhibitor against c-MET and VEGFR2. *PLoS ONE* **2013**, *8*, 58262. [CrossRef] [PubMed]
102. Kunkel, P.; Ulbricht, U.; Bohlen, P.; Brockmann, M.A.; Fillbrandt, R.; Stavrou, D.; Westphal, M.; Lamszus, K. Inhibition of glioma angiogenesis and growth in vivo by systemic treatment with a monoclonal antibody against vascular endothelial growth factor receptor-2. *Cancer Res.* **2001**, *61*, 6624–6628.
103. Madar-Balakirski, N.; Tempel-Brami, C.; Kalchenko, V.; Brenner, O.; Varon, D.; Scherz, A.; Salomon, Y. Permanent occlusion of feeding arteries and draining veins in solid mouse tumors by vascular targeted photodynamic therapy (VTP) with Tookad. *PLoS ONE* **2010**, *5*, 10282. [CrossRef] [PubMed]
104. Mohajeri, A.; Sanaei, S.; Kiafar, F.; Fattahi, A.; Khalili, M.; Zarghami, N. The Challenges of Recombinant Endostatin in Clinical Application: Focus on the Different Expression Systems and Molecular Bioengineering. *Adv. Pharm. Bull.* **2017**, *7*, 21–34. [CrossRef]
105. Allen, R.T.; Cluck, M.W.; Agrawal, D.K. Mechanisms controlling cellular suicide: Role of Bcl-2 and caspases. *Cell Mol. Life Sci.* **1998**, *54*, 427–445. [CrossRef] [PubMed]
106. Qiu, B.; Ji, M.; Song, X.; Zhu, Y.; Wang, Z.; Zhang, X.; Wu, S.; Chen, H.; Mei, L.; Zheng, Y. Co-delivery of docetaxel and endostatin by a biodegradable nanoparticle for the synergistic treatment of cervical cancer. *Nanoscale Res. Lett.* **2012**, *7*, 666. [CrossRef]

107. Mohajeri, A.; Pilehvar-Soltanahmadi, Y.; Pourhassan-Moghaddam, M.; Abdolalizadeh, J.; Karimi, P.; Zarghami, N. Cloning and Expression of Recombinant Human Endostatin in Periplasm of Escherichia coli Expression System. *Adv. Pharm. Bull.* **2016**, *6*, 187–194. [CrossRef]
108. Boehm, T.; Folkman, J.; Browder, T.; O'Reilly, M.S. Antiangiogenic therapy of experimental cancer does not induce acquired drug resistance. *Nature* **1997**, *390*, 404–407. [CrossRef]
109. Luo, H.; Xu, M.; Zhu, X.; Zhao, J.; Man, S.; Zhang, H. Lung cancer cellular apoptosis induced by recombinant human endostatin gold nanoshell-mediated near-infrared thermal therapy. *Int. J. Clin. Exp. Med.* **2015**, *8*, 8758–8766.
110. Danafar, H.; Davaran, S.; Rostamizadeh, K.; Valizadeh, H.; Hamidi, M. Biodegradable m-PEG/PCL Core-Shell Micelles: Preparation and Characterization as a Sustained Release Formulation for Curcumin. *Adv. Pharm. Bull.* **2014**, *4*, 501–510. [CrossRef]
111. Jiang, L.P.; Zou, C.; Yuan, X.; Luo, W.; Wen, Y.; Chen, Y. N-terminal modification increases the stability of the recombinant human endostatin in vitro. *Biotechnol. Appl. Biochem.* **2009**, *54*, 113–120. [CrossRef]
112. Chen, W.; Hu, S. Suitable carriers for encapsulation and distribution of endostar: Comparison of endostar-loaded particulate carriers. *Int. J. Nanomed.* **2011**, *6*, 1535–1541. [CrossRef]
113. Hu, S.; Zhang, Y. Endostar-loaded PEG-PLGA nanoparticles: In vitro and in vivo evaluation. *Int. J. Nanomed.* **2010**, *5*, 1039–1048. [CrossRef]
114. Sennino, B.; Ishiguro-Oonuma, T.; Wei, Y.; Naylor, R.M.; Williamson, C.W.; Bhagwandin, V.; Tabruyn, S.P.; You, W.K.; Chapman, H.A.; Christensen, J.G.; et al. Suppression of tumor invasion and metastasis by concurrent inhibition of c-Met and VEGF signaling in pancreatic neuroendocrine tumors. *Cancer Discov.* **2012**, *2*, 270–287. [CrossRef] [PubMed]
115. Fredrickson, J.; Serkova, N.J.; Wyatt, S.K.; Carano, R.A.; Pirzkall, A.; Rhee, I.; Rosen, L.S.; Bessudo, A.; Weekes, C.; de Crespigny, A. Clinical translation of ferumoxytol-based vessel size imaging (VSI): Feasibility in a phase I oncology clinical trial population. *Magn. Reson. Med.* **2017**, *77*, 814–825. [CrossRef]
116. Cortes-Santiago, N.; Hossain, M.B.; Gabrusiewicz, K.; Fan, X.; Gumin, J.; Marini, F.C.; Alonso, M.M.; Lang, F.; Yung, W.K.; Fueyo, J.; et al. Soluble Tie2 overrides the heightened invasion induced by anti-angiogenesis therapies in gliomas. *Oncotarget* **2016**, *7*, 16146–16157. [CrossRef] [PubMed]
117. Koh, Y.J.; Kim, H.Z.; Hwang, S.I.; Lee, J.E.; Oh, N.; Jung, K.; Kim, M.; Kim, K.E.; Kim, H.; Lim, N.K.; et al. Double antiangiogenic protein, DAAP, targeting VEGF-A and angiopoietins in tumor angiogenesis, metastasis, and vascular leakage. *Cancer Cell* **2010**, *18*, 171–184. [CrossRef] [PubMed]
118. Reardon, D.A.; Lassman, A.B.; Schiff, D.; Yunus, S.A.; Gerstner, E.R.; Cloughesy, T.F.; Lee, E.Q.; Gaffey, S.C.; Barrs, J.; Bruno, J.; et al. Phase 2 and biomarker study of trebananib, an angiopoietin-blocking peptibody, with and without bevacizumab for patients with recurrent glioblastoma. *Cancer* **2018**, *124*, 1438–1448. [CrossRef] [PubMed]

© 2020 by the authors. Licensee MDPI, Basel, Switzerland. This article is an open access article distributed under the terms and conditions of the Creative Commons Attribution (CC BY) license (http://creativecommons.org/licenses/by/4.0/).

MDPI
St. Alban-Anlage 66
4052 Basel
Switzerland
Tel. +41 61 683 77 34
Fax +41 61 302 89 18
www.mdpi.com

Cancers Editorial Office
E-mail: cancers@mdpi.com
www.mdpi.com/journal/cancers

www.ingramcontent.com/pod-product-compliance
Lightning Source LLC
LaVergne TN
LVHW070506100526
838202LV00014B/1799